# PCR Cloning Protocols

# METHODS IN MOLECULAR BIOLOGY™

## John M. Walker, SERIES EDITOR

METHODS IN MOLECULAR BIOLOGY™

# PCR Cloning Protocols

## From Molecular Cloning to Genetic Engineering

Edited by

# Bruce A. White

*University of Connecticut Health Center, Farmington, CT*

Humana Press ✳ Totowa, New Jersey

© 1997 Humana Press Inc.
999 Riverview Drive, Suite 208
Totowa, New Jersey 07512

For additional copies, pricing for bulk purchases, and/or information about other Humana titles, contact Humana at the above address or at any of the following numbers: Tel.: 201-256-1699; Fax: 201-256-8341; E-mail: humana@interramp.com

This publication is printed on acid-free paper. ∞
ANSI Z39.48-1984 (American Standards Institute) Permanence of Paper for Printed Library Materials.

Cover illustration: Fig. 1 from Chapter 13 "Recombination and Site-Directed Mutagenesis Using Recombinant PCR," by Douglas H. Jones and Stanley C. Winistorfer

Printed in the United States of America. 10 9 8 7 6 5 4 3 2 1

Library of Congress Cataloging in Publication Data

Main entry under title:

Methods in molecular biology™.

PCR cloning protocols: from molecular cloning to genetic engineering/edited by Bruce A. White.
    p.   cm.—(Methods in molecular biology™; vol. 67)
    Includes index.
    ISBN 0-89603-343-0 (combbound) (alk. paper); ISBN 0-89603-443-6 (hardcover) (alk. paper)
    1. Molecular cloning—Laboratory manuals.  2. Polymerase chain reaction—Laboratory manuals.
I.   White, Bruce Alan.   II.   Series: Methods in molecular biology™ (Totowa, NJ); 67
QH442.2.P37   1997
574.87'3282—dc20                                                                96-30876
                                                                                        CIP

# Preface

The advent of PCR, with its power to amplify tiny amounts of DNA, quickly spawned the development of many analytical procedures that are widely used for detection, measurement, and characterization. However, creative investigators soon discovered the power of PCR for synthetic or preparative uses. This volume focuses on such preparative PCR protocols, which can be used in the cloning and modification of DNA.

*PCR Cloning Protocols: From Molecular Cloning to Genetic Engineering* is divided into seven parts, each containing a collection of chapters addressing a general approach or goal. Part I presents basic PCR protocols, emphasizing optimizing conditions for the amplification of DNA fragments of several kilobases in length. Part II offers several procedures for cloning PCR products, depending on whether a specific restriction site can be used in the cloning vector, the PCR product is to be gel purified before cloning, or the fragment needs to be inserted in one or both orientations. Part III includes several protocols involved in the mutagenesis of DNA, either site-directed or not, as well as several approaches to recombinant PCR, either for mutagenesis or building a custom gene, as well as one chapter describing a specific use of in vitro selection. Part IV addresses the frequent need to amplify and clone segments of DNA that are to the right, left, or scattered within a stretch of DNA (either vector, chromosomal, or cDNA) of known sequence. Part V describes the emerging use of PCR to create and screen libraries, and Part VI offers protocols for the creation of subtracted sublibraries and differential screening of libraries, as well as one chapter on the relatively new method of differential display. Finally, Part VII describes protocols that allow an investigator to discover or clone new genes based on their similarity to others within a gene family.

I hope that the 44 chapters of *PCR Cloning Protocols: From Molecular Cloning to Genetic Engineering* will encourage both new and established investigators to apply PCR-based methodology to the cloning and modification of DNA, either through the use of the protocols described herein, or by fostering individual creativity to modify and customize these protocols to best fit one's own needs.

Many thanks to Isolde Bates, Lori Capozzi, Judy Cheney, and Donna Zezzo for their help in preparing this volume.

*Bruce A. White*

# Contents

*vii*

Contents

# Contributors

JYRKI T. AATSINKI • *Biocenter Oulu and Department of Anatomy, University of Oulu, Finland*

SAILEN BARIK • *Department of Biochemistry, University of South Alabama, Mobile, AL*

SIMONA BARTL • *Department of Biological Sciences, University of North Carolina at Wilmington, NC*

JAMES BATTEY • *NCI-Naval Medical Oncology, Bethesda, MD*

DAVID BERTIOLI • *IACR-Rothamsted, Harpenden, UK*

SUSAN C. BOCK • *Microbiology Department, Temple University Medical School, Philadelphia, PA*

STEVE D. M. BROWN • *Department of Biochemistry, St. Mary's Hospital Medical School, London, UK*

GUSTAVO CAETANO-ANNOLLÈS • *Department of Plant Molecular Genetics, University of Tennessee, Knoxville, TN*

RINO CELLA • *Dipartmento de Genetica e Microbiologia, Università Pavia, Italy*

SUZANNE CHENG • *Roche Molecular Systems, Alameda, CA*

THIERRY G. COCHE • *Organogenesis Lab, University of Oxford, UK*

BIANCA M. CONTI-FINE • *Department of Biochemistry, University of Minnesota, St. Paul, MN*

COLLETE DUEZ • *Genetics Unit, Shriners Hospital, Montreal, Canada*

JEAN BAPTISTE DUMAS MILNE EDWARDS • *GENSET, Paris, France*

JEAN DUSART • *Genetics Unit, Shriners Hospital, Montreal, Canada*

FARZIN FARZANEH • *Department of Biological Anthropology, University of Cambridge, UK*

HELEN FORRESTER • *University of California, San Francisco, CA*

YVONNE J. FOSS • *Department of Biological Anthropology, University of Cambridge, UK*

FERNANDO GIBSON • *Department of Biochemistry, St. Mary's Hospital Medical School, London, UK*

KARL H. HECKER • *Department of Biological Sciences, Florida State University, Tallahassee, FL*

SARAH A. HESSE • *Department of Biological Anthropology, University of Cambridge, UK*

ROBERT M. HORTON • *Department of Biochemistry, University of Minnesota, St. Paul, MN*

MEIMEI HUI • *Department of Anatomy, University of Connecticut Health Center, Farmington, CT*

SHENG-HE HUANG • *Division of Infectious Disease, Children's Hospital, Los Angeles, CA*

RENE HUMMEL • *Division of Cancer Biology, Danish Cancer Society, Copenhagen, Denmark*

CHRISTINE ICARD-LIEPKALNS • *GENSET, Paris, France*

MITSUO ITAKURA • *Otsuka Department of Clinical and Molecular Nutrition, School of Medicine, University of Tokushima, Kuramoto, Tokushima, Japan*

HIROYUKI IWAHANA • *Otsuka Department of Clinical and Molecular Nutrition, School of Medicine, University of Tokushima, Kuramoto, Tokushima, Japan*

DOUGLAS H. JONES • *Department of Pediatrics, University of Iowa, Iowa City, IA*

AMBROSE Y. JONG • *Division of Infectious Disease, Children's Hospital, Los Angeles, CA*

MIGUEL JUNCOSA • *Instituto de Biología Fundemental, University Autónoma de Barcelona, Spain*

MICHAEL W. KING • *Department of Biochemistry and Molecular Biology, Indiana University School of Medicine, Terre Haute, IN*

LORI A. KOLMODIN • *Roche Molecular Systems, Alameda, CA*

W. MICHAEL KUEHL • *NCI-Navy Medical Oncology, Bethesda, MD*

MELISSA LAIL-TRECKER • *Department of Anatomy, University of Connecticut Health Center, Farmington, CT*

MICHAEL LARDELLI • *Department of Developmental Neuroscience, Uppsala University, Uppsala, Sweden*

HENRIK LEFFERS • *Division of Cancer Biology, Danish Cancer Society, Copenhagen, Denmark*

MEI LUO • *Dipartmento de Genetica e Microbiologia, Università Pavia, Italy*

JACQUES MALLET • *GENSET, Paris, France*

SCOTT F. MICHAEL • *Department of Microbiology, University of Alabama at Birmingham, AL*

ALAIN MOREAU • *Genetics Unit, Shriners Hospital, Montreal, Canada*

NIELS PALLISGAARD • *Division of Cancer Biology, Danish Cancer Society, Copenhagen, Denmark*

VÉRONIQUE PICARD • *Microbiology Department, Temple University Medical School, Philadelphia, PA*

JAMES C. PIERCE • *Department of Biological Science, Philadelphia College of Pharmacy and Science, Philadelphia, PA*

ANTONI PLANAS • *Instituto de Biología Fundemental, University Autónoma de Barcelona, Spain*

TINO PODSTUFKA • *Division of Cancer Biology, Danish Cancer Society, Copenhagen, Denmark*

JAUME PONS • *Instituto de Biología Fundemental, University Autónoma de Barcelona, Spain*

GENEVIVE PONT-KINGDON • *Biochemistry Department, University of Utah, Salt Lake City, UT*

GREGORY M. PRESTON • *Departments of Medicine and Biological Chemistry, Johns Hopkins University School of Medicine, Baltimore, MD. Present Address: Clinical Diagnostics, Johnson & Johnson, Rochester, NY*

ENRIQUE QUEROL • *Instituto de Biología Fundemental, University Autónoma de Barcelona, Spain*

IAN R. RADFORD • *University of California, San Francisco, CA*

RAGHAVANPILLAI RAJU • *Department of Biochemistry, University of Minnesota, St. Paul, MN*

PHILIPPE RAVASSARD • *GENSET, Paris, France*

HEIDI RIEDEL • *Division of Cancer Biology, Danish Cancer Society, Copenhagen, Denmark*

MIKKEL ROHDE • *Division of Cancer Biology, Danish Cancer Society, Copenhagen, Denmark*

KENNETH H. ROUX • *Department of Biological Science, Florida State University, Tallahassee, FL*

JUDITH T. SCHANKE • *Wisconsin Regional Primate Research Center, Madison, WI*

ALAN SHULDINER • *Johns Hopkins University School of Medicine, Baltimore, MD*

PAUL SOLLITTI • *Department of Biochemistry, Albert Einstein College of Medicine, Bronx, NY*

MICHAEL STRAUSS • *Division of Cancer Biology, Danish Cancer Society, Copenhagen, Denmark*

TORU TAKUMI • *Department of Pharmacology, Faculty of Medicine, Osaka University, Osaka, Japan*

KEITH TANNER • *Johns Hopkins University School of Medicine, Baltimore, MD*

ALESSANDRO TESTORI • *Department of Biochemistry, Albert Einstein College of Medicine, Bronx, NY*

MARK G. THOMAS • *Department of Biological Anthropology, University of Cambridge, UK*

ROBERT N. TRIGANO • *Plant Molecular Genetics, University of Tennessee, Knoxville, TN*

CRAIG TUERK • *Morehead State University, Morehead, KY*

OLIVIER VANDENAIRE • *GENSET, Paris, France*

KAI WANG • *Department of Molecular Biotechnology, University of Washington, Seattle, WA*

CEDRIC S. WESLEY • *The Rockefeller University, New York, NY*

UMADEVI V. WESLEY • *The Rockefeller University, New York, NY*

BRUCE A. WHITE • *Department of Anatomy, University of Connecticut Health Center, Farmington, CT*

J. FENTON WILLIAMS • *Roche Molecular Systems, Alameda, CA*

STANLEY C. WINISTORFER • *Department of Pediatrics, University of Iowa, Iowa City, IA*

# I

# PERFORMING AND OPTIMIZING PCR

# 1

## PCR

*Basic Principles and Routine Practice*

**Lori A. Kolmodin and J. Fenton Williams**

## 1. Introduction
### 1.1. PCR Definition

The polymerase chain reaction (PCR) is a primer-mediated enzymatic amplification of specifically cloned or genomic DNA sequences *(1)*. This PCR process, invented by Kary Mullis over 10 years ago, has been automated for routine use in laboratories worldwide. The template DNA contains the target sequence, which may be tens or tens of thousands of nucleotides in length. A thermostable DNA polymerase, *Taq* DNA polymerase, catalyzes the buffered reaction in which an excess of an oligonucleotide primer pair and four deoxynucleoside triphosphates (dNTPs) are used to make millions of copies of the target sequence. Although the purpose of the PCR process is to amplify template DNA, a reverse transcription step allows the starting point to be RNA *(2–5)*.

### 1.2. Scope of PCR Applications

PCR is widely used in molecular biology and genetic disease research to identify new genes. Viral targets, such as HIV-1 and HCV can be identified and quantitated by PCR. Active gene products can be accurately quantitated using RNA-PCR. In such fields as anthropology and evolution, sequences of degraded ancient DNAs can be tracked after PCR amplification. With its exquisite sensitivity and high selectivity, PCR has been used for wartime human identification and validated in crime labs for mixed-sample forensic casework. In the realm of plant and animal breeding, PCR techniques are used to screen for traits and to evaluate living four-cell embryos. Environmental and

From: *Methods in Molecular Biology, Vol. 67: PCR Cloning Protocols: From Molecular Cloning to Genetic Engineering* Edited by: B. A. White Humana Press Inc., Totowa, NJ

food pathogens can be quickly identified and quantitated at high sensitivity in complex matrices with simple sample preparation techniques.

## 1.3. PCR Process

The PCR process requires a repetitive series of the three fundamental steps that defines one PCR cycle: double-stranded DNA template denaturation, annealing of two oligonucleotide primers to the single-stranded template, and enzymatic extension of the primers to produce copies that can serve as templates in subsequent cycles. The target copies are double-stranded and bounded by annealing sites of the incorporated primers. The 3' end of the primer should complement the target exactly, but the 5' end can actually be a noncomplementary tail with restriction enzyme and promoter sites that will also be incorporated. As the cycles proceed, both the original template and the amplified targets serve as substrates for the denaturation, primer annealing, and primer extension processes. Since every cycle theoretically doubles the amount of target copies, a geometric amplification occurs. Given an efficiency factor for each cycle, the amount of amplified target, $Y$, produced from an input copy number, $X$, after $n$ cycles is

$$Y = X(1 + \text{efficiency})^n \qquad (1)$$

With this amplification power, 25 cycles could produce 33 million copies. Every extra 10 cycles produces 1024 more copies. Unfortunately, the process becomes self-limiting and amplification factors are generally between $10^5$- and $10^9$-fold. Excess primers and dNTPs help drive the reaction that commonly occurs in 10 m$M$ Tris-HCl buffer (pH 8.3 at room temperature). In addition, 50 m$M$ KCl is present to provide proper ionic strength and magnesium ion is required as an enzyme cofactor *(6)*.

The denaturation step occurs rapidly at 94–96°C. Primer annealing depends on the $T_m$, or melting temperature, of the primer:template hybrids. Generally, one uses a predictive software program to compute the $T_m$s based on the primer's sequence, their matched concentrations, and the overall salt concentration. The best annealing temperature is determined by optimization. Extension occurs at 72°C for most templates. PCR can also easily occur with a two temperature cycle consisting of denaturation and annealing/extension.

## 1.4. Carryover Prevention

PCR has the potential sensitivity to amplify single molecules, so PCR products that can serve as templates for subsequent reactions must be kept isolated after amplification. Even tiny aerosols can contain thousands of copies of carried-over target molecules that can convert a true negative into a false positive. In general, dedicated pipeters, plugged pipet tips, and separate work areas should be designated for pre and post-PCR work. As with any high sensitivity technique, the judicious and frequent use of positive and negative controls is required *(7–9)*. Through the use of dUTP instead of dTTP for all PCR samples, it is possible to design an internal biochemical mechanism to attack the PCR carry-over problem. PCR products will then be dU-containing and can be cloned, sequenced, and analyzed as usual. Pretreatment of each PCR reaction with uracil N-glycosylase will destroy any PCR product carried over from previous reactions, leaving the native T-containing sample ready for amplification *(10)*.

## 1.5. Hot Start

PCR is conceptualized as a process that begins when thermal cycling ensues. The annealing temperature sets the specificity of the reaction, assuring that the primary primer binding events are the ones specific for the target in question. In preparing a PCR reaction on ice or at room temperature, however, the reactants are all present for nonspecific primer annealing to any single-stranded DNA present. Since *Taq* DNA polymerase has some residual activity even at lower temperatures, it is possible to extend these misprimed hybrids and begin the PCR process at the wrong sites! By withholding a key reaction component, such as *Taq* DNA polymerase, until an elevated temperature can be reached, the possibility of mispriming is avoided. This can be accomplished by a manual addition of enzyme above 65–70°C during the first heating ramp to denaturation at 94°C. Alternatively, an inactive form of the enzyme AmpliTaq Gold can be added to all reactions to prevent misprimed extensions. Adding a pre-PCR heat step at 92–95°C for 9–12 min synchronously reactivates the enzyme and achieves an "invisible" hot start. In both cases, the lowest temperature experienced by the reaction components is the stringent primer annealing temperature, assuring best specificity *(11)*.

## 1.6. PCR Achievements

PCR has been used to speed the gene discovery process and for early detection of viral diseases. Single sperm cells to measure crossover frequencies can be analyzed and four-cell cow embryos can be typed. Trace forensic evidence of even mixed samples can be analyzed. Single copy amplification requires some care, but is feasible for both DNA and RNA targets. True needles in haystacks can be found simply by amplifying the needles. PCR facilitates cloning of DNA sequencing and forms a natural basis for cycle sequencing by the Sanger method *(12)*.

# 2. PCR Enzymes

## 2.1. AmpliTaq® DNA Polymerase

AmpliTaq DNA Polymerase (Perkin Elmer, Foster City, CA) is a highly characterized recombinant enzyme for PCR. It is produced in *E. coli* from the *Taq* DNA polymerase gene, thereby assuring high purity. It is commonly supplied and used as a 5 U/µL solution in buffered 50% glycerol.

## 2.2. Biophysical Properties

The enzyme is a 94-kDa protein with a 5'-3' polymerization activity that is most efficient in the 70–80°C range. This enzyme is very thermostable, with a half-life at 95°C of 35–40 min. In terms of thermal cycling, the half-life is approx 100 cycles. PCR products amplified using AmpliTaq DNA polymerase will often have single base overhangs on the 3' ends of each polymerized strand, and this artifact can be successfully exploited for use with T/A cloning vectors.

## 2.3. Biochemical Reactions

AmpliTaq DNA polymerase requires magnesium ion as a cofactor and catalyzes the extension reaction of a primed template at 72°C. The four dNTPs (consisting of dATP, dCTP, dGTP, and dTTP or dUTP) are used according to the base-pairing rule to extend the primer and thereby to copy the target sequence. Modified nucleotides (ddNTPs, biotin-11-dNTP, dUTP, deaza-dGTP, and fluorescently labeled dNTPs) can be incorporated into PCR products.

## 2.4. Associated Activities

AmpliTaq DNA polymerase has a fork-like structure-dependent, polymerization-enhanced, 5'-3' nuclease activity. This activity allows the polymerase to degrade downstream primers and indicates that circular targets should be linearized before amplification. In addition, this nuclease activity has been employed in a fluorescent signal-generating technique for PCR quantitation *(13)*. AmpliTaq DNA polymerase does not have an inherent 3'-5' exonuclease or proofreading activity, but produces amplicons of sufficiently high fidelity for most applications.

## 3. PCR Primers

### 3.1. Design Criteria

PCR primers are short oligodeoxyribonucleotides, or oligomers, that are designed to complement the end sequences of the PCR target amplicon. These synthetic DNAs are usually 15–25 nucleotides long and have ~50% G + C content. Since each of the two PCR primers is complementary to a different individual strand of the target sequence duplex, the primer sequences are not related to each other. In fact, special care must be taken to assure that the primer sequences do not form duplex structures with each other or hairpin loops within themselves. The 3' end of the primer must match the target in order for polymerization to be efficient, and allele-specific PCR strategies take advantage of this fact. To screen for mutants, a primer complementary to the mutant sequence is used and results in PCR positives, whereas the same primer will be a mismatch for the wild type and not amplify. The 5' end of the primer may have sequences that are not complementary to the target and that may contain restriction sites or promotor sites that are also incorporated into the PCR product. Primers with degenerate nucleotide positions every third base may be synthesized in order to allow for amplification of targets where only the amino acid sequence is known. In this case, early PCR cycles are performed with low, less stringent annealing temperatures, followed by later cycles with high, more stringent annealing temperatures.

A PCR primer can also be a homopolymer, such as oligo $(dT)_{16}$, which is often used to prime the RNA PCR process. In a technique called RAPDs (Randomly Amplified Polymorphic DNAs), single primers as short as decamers with random sequences are used to prime on both strands, producing a diverse array of PCR products that form a fingerprint of a genome *(14)*. Often, logically designed primers are less successful in PCR than expected, and it is usually advisable to try optimization techniques for a practical period of time before trying new primers frequently designed near the original sites.

### 3.2. $T_m$ *Predictions*

DNA duplexes, such as primer-template complexes, have a stability that depends on the sequence of the duplex, the concentrations of the two components, and the salt concentration of the buffer. Heat can be used to disrupt this duplex. The temperature at which half the molecules are single-stranded and half are double-stranded is called the $T_m$ of the complex. Because of the greater number of intermolecular hydrogen bonds, higher G + C content DNA has a higher $T_m$ than lower G + C content DNA. Often, G + C content alone is used to predict the $T_m$ of the DNA duplex; however, DNA duplexes with the same G + C content may have different $T_m$ values. Computer programs are available to perform more accurate $T_m$ predictions using sequence information (nearest neighbor analysis) and to assure optimal primer design (*see* Chapter 2, Section 3.1.4.).

Since the specificity of the PCR process depends on successful primer binding events at each amplicon end, the annealing temperature is selected based on the consensus of melting temperatures (within ~2–4°C) of the two primers. Usually the annealing temperature is chosen to be a few degrees below the consensus annealing temperatures of the primers. Different strategies are possible, but lower annealing temperatures should be tried first to assess the success of amplification of the target and then higher annealing temperatures can be investigated to find the stringency required for best product specificity.

## 4. PCR Samples

### 4.1. Types

The PCR sample type may be single- or double-stranded DNA of any origin—animal, bacterial, plant, or viral. RNA molecules, including total RNA, poly (A$^+$) RNA, viral RNA, tRNA, or rRNA, can also serve as templates for amplification after conversion to so-called complementary DNA (cDNA) by the enzyme reverse transcriptase (either MuLV or recombinant *Thermus thermophilus*, r*Tth* DNA polymerase).

### 4.2. Amount

The amount of starting material required for PCR can be as little as a single molecule, compared to the millions of molecules needed for standard cloning or molecular biological analysis. As a basis, up to nanogram amounts of DNA cloned template, up to microgram amounts of genomic DNA, or up to $10^5$ DNA target molecules are best for initial PCR testing.

### 4.3. Purity

Overall, the purity of the DNA sample to be subjected to PCR amplification need not be high. A single cell, a crude cell lysate, or even a small sample of degraded DNA template is usually adequate for successful amplification. The fundamental requirements of sample purity must be that the target contain at least one intact DNA strand encompassing the amplified region and that the impurities associated with the target be adequately dilute so as to not inhibit enzyme activity. However, for some applications, such as long PCR, it may be necessary to consider the quality and quantity of the DNA sample *(15,16)*. For example,

1. When more template molecules are available, there is less occurrence of false positives caused by either crosscontamination between samples or "carryover" contamination from previous PCR amplifications;
2. When the PCR amplification lacks specificity or efficiency, or when the target sequences are limited, there is a greater chance of inadequate product yield; and
3. When the fraction of starting DNA available to PCR is uncertain, it is increasingly difficult to determine the target DNA content *(17)*.

## 5. Thermal Cycling Considerations

### 5.1. PCR Vessels

PCR must be performed in vessels that are compatible with low amounts of enzyme and nucleic acids and that have good thermal transfer characteristics. Typically, polypropylene is used for PCR vessels and conventional, thick-walled microcentrifuge tubes are chosen for many thermal cycler systems. PCR is most often performed at a 10–100 µL reaction scale and requires the prevention of the evaporation/condensation processes in the closed reaction tube during thermal cycling. A mineral oil overlay or wax layer serves this purpose. More recently, 0.2 mL thin-walled vessels have been optimized for the PCR process and oil-free thermal cyclers have been designed that use a heated cover over the tubes held within the sample block.

## 5.2. Temperature and Time Optimization

It is essential that the reaction mixtures reach the denaturation, annealing, and extension temperatures in each thermal cycle. If insufficient hold time is specified at any temperature, the temperature of the sample will not be equilibrated with that of the sample block. Some thermal cycler designs time the hold interval based on block temperature, whereas others base the hold time on predicted sample temperature.

In a conventional thick-walled tube used in a cycler controlled by block temperature, a 60 s hold time is sufficient for equilibration. Extra time may be recommended at the 72°C extension step for longer PCR products. Using a thin-walled 0.2-mL tube in a cycler controlled by predicted sample temperature, only 15 s is required. To use existing protocols or to develop protocols for use at multiple labs, it is very important to choose hold times according to the cycler design and tube wall thickness.

## 6. Conditions for Successful PCR
## 6.1. Metal Ion Cofactors

Magnesium chloride is an essential cofactor for the DNA polymerase used in PCR, and its concentration must be optimized for every primer/template pair. Many components of the reaction bind magnesium ion, including primers, template, PCR products and dNTPs. The main 1:1 binding agent for magnesium ion is the high concentration of dNTPs in the reaction. Since it is necessary for free magnesium ion to serve as an enzyme cofactor in PCR, the total magnesium ion concentration must exceed the total dNTP concentration. Typically, to start the optimization process, 1.5 m$M$ magnesium chloride is added to PCR in the presence of 0.8 m$M$ total dNTPs. This leaves about 0.7 m$M$ free magnesium ion for the DNA polymerase. In general, magnesium ion should be varied in a concentration series from 1.5–4 m$M$ in 0.5-m$M$ steps *(1,17)*.

## 6.2. Substrates and Substrate Analogs

*Taq* DNA polymerase incorporates dNTPs very efficiently, but can also incorporate modified substrates when they are used as supplemental components in PCR. Digoxigenin-dUTP, biotin-11-dUTP, dUTP, c[7]deaza-dGTP, and fluorescently labeled dNTPs all serve as substrates for *Taq* DNA polymerase. For conventional PCR, the concentration of dNTPs remains balanced in equimolar ratios, but for mutagenesis, unequal concentrations should be used.

### 6.3. Buffers/Salts

The PCR buffer for *Taq* DNA polymerase consists of 50 m$M$ KCl and 10 m$M$ Tris-HCl, pH 8.3, at room temperature. This buffer provides the ionic strength and buffering capacity needed during the reaction. It is important to note that the salt concentration affects the $T_m$ of the primer/template duplex, and hence the annealing temperature. Cosolvents, such as DMSO and glycerol, have been successfully used in PCR buffers when the targets have very high denaturation temperatures *(18)*.

### 6.4. Cycles

The number of cycles of PCR should be optimized with respect to the number of input target copies. In a typical PCR, $10^{12}$ copies represents the plateau in the maximum amount of amplification possible. From a single copy, the most efficient PCR would reach plateau in 40 cycles ($10^{12} \approx 2^{40}$). PCR may be 80–95% efficient, so the amplification factors are nearer $(1.9)^n$, where $n$ is the number of cycles. It is usually advisable to run the minimum number of cycles needed to see the desired specific product, since unwanted nonspecific products will interfere if the number of cycles is excessive *(1,17)*.

### 6.5. Enzyme/Target

In a standard aliquot of *Taq* DNA polymerase used for a 100-µL reaction, there are about $10^{10}$ molecules. Each PCR sample should be evaluated for the number of target copies it contains or may contain. For example 1 ng of lambda DNA contains $1.8 \times 10^7$ copies. For low input copy number PCR, the enzyme is in great excess in early cycles. As the amplicon accumulates in later cycles, the enzyme becomes limiting and it may be necessary to give the extension process incrementally more time. Thermal cyclers can reliably perform this automatic segment extension procedure in order to maximize PCR yield *(1,17)*.

## 7. PCR Protocols

### 7.1. Reagents and Supplies

The protocol described below illustrates the basic principles and techniques of PCR and can be modified to suit other particular applications. The example chosen uses the Perkin-Elmer GeneAmplimer® HIV Primer pair, SK145 and SK431, in conjunction with the Perkin-Elmer GeneAmp® PCR Reagent Kit and PCR Carry-Over Prevention Kit (Perkin-Elmer), to amplify a 142-bp DNA fragment from the conserved *gag* region of HIV-1 using the AmpliWax® PCR Gem-facilitated hot start process *(11,19)*.

| Reagents | Stock concentrations |
|---|---|
| 10X PCR buffer II | 500 m$M$ KCl, 100 m$M$ Tris-HCl, pH 8.3 |
| MgCl$_2$ solution | 25 m$M$ |
| dNTPs | 10 m$M$ stocks of each of dATP, dCTP, dGTP; 20 m$M$ stock of dUTP; all neutralized to pH 7.0 with NaOH |
| Primer 1: SK145 | 25 μ$M$ in 10 m$M$ Tris-HCl, pH 8.3 |
| | 5'-AGTGGGGGGACATCAAGCAGCCATGCAAAT-3' |
| Primer 2: SK431 | 25 μ$M$ in 10 m$M$ Tris-HCl, pH 8.3 in 150 m$M$ NaCl, 30 m$M$ Tris-HCl, pH 7.5, 10 m$M$ EDTA, 1.0 m$M$ DTT, 0.05% Tween-20, 5% (v/v) glycerol. |
| | 5'-TGCTATGTCAGTTCCCCTTGGTTCTCT-3' |
| AmpErase™ UNG | Uracil N-glycosylase, 1.0 U/μL |
| AmpliTaq DNA polymerase | 5 U/μL in 100 m$M$ KCl, 20 m$M$ Tris-HCl, pH 8.0, 0.1 m$M$ EDTA, 1 m$M$ DTT, 0.5% Tween 20, 0.5% Nonidet P40, 50% (v/v) glycerol |
| HIV-1 positive control DNA | $10^5$ copies in 1 μg human placental DNA, 1 m$M$ EDTA, 10 m$M$ NaCl, 10 m$M$ Tris-HCl, pH 8.0 |
| AmpliWax PCR gems | Gem 100s for 50–100 μL reactions or Gem 50s for 20–50 μL reactions |

Use 0.5 mL Perkin-Elmer GeneAmp® PCR microcentrifuge tubes and the Perkin-Elmer GeneAmp® PCR instrument system

## 7.2. Methods

In the AmpliWax PCR gem-facilitated hot start process, a solid wax layer is formed over a subset of PCR reactants, called the lower reagent mix, that encompasses 30–50% of the total reaction mix volume. The remaining reactants, called the upper reagent mix, comprise the remaining 50–70% of the total reaction mix volume and are added above the wax layer. In the first thermal cycle, the wax layer melts during the temperature ramp to the denaturation temperature and is displaced by the more dense upper reagent mix. Thermal convection adequately mixes the combined lower and upper reagent mixes, whereas the melted wax layer acts as a vapor barrier during each PCR cycle.

1. Assemble the lower reagent mix shown in Table 1.
2. For 100 μL reactions, add 40 μL of the lower reagent mix (which can be made up as a batch mix) into the bottom of each GeneAmp PCR reaction tube. Avoid splashing liquid onto the tube. If any liquid is present on the tube walls, spin the tube briefly in a microcentrifuge.
3. Carefully add one AmpliWax PCR Gem 100 to each tube containing the lower reagent mix. Melt the wax gem by incubating each reaction tube at 75–80°C for 3–5 min. Solidify the wax at room temperature (25°C) for 3–5 min.
4. Assemble upper reagent mix as in Table 2.

**Table 1**
**Lower Reagent Mix**

| Reagent | Volume, 1X mix, μL | Final concentration, per 100 μL volume |
|---|---|---|
| Sterile water | 13.5 | N/A |
| 10X PCR buffer II | 4.0 | 1X |
| 25 m$M$ MgCl$_2$ | 10.0 | 2.5 m$M$ |
| 10 m$M$ dATP | 2.0 | 200 μ$M$ |
| 10 m$M$ dCTP | 2.0 | 200 μ$M$ |
| 10 m$M$ dGTP | 2.0 | 200 μ$M$ |
| 20 m$M$ dUTP | 2.0 | 400 μ$M$ |
| 25 μ$M$ primer 1 (SK145) | 2.0 | 0.5 μ$M$ |
| 25 μ$M$ primer 2 (SK431) | 2.0 | 0.5 μ$M$ |
| 1 U/μL AmpErase UNG | 0.5 | 1 U/reaction |
| Total volume | 40.0 | |

**Table 2**
**Upper Reagent Mix**

| Reagent | Volume, 1X mix, μL | Final concentration, per 100 μL volume |
|---|---|---|
| Sterile water | 43.0–52.9 | N/A |
| 10X PCR buffer II | 6.0 | 1X |
| 5 U/μL AmpliTaq DNA polymerase | 0.5 | 2.5 U/reaction |
| 1 U/μL AmpErase UNG | 0.5 | 1 U/reaction |
| 10$^3$ Copies/μL positive control DNA | 0.1–10.0 | 10$^2$–10$^4$ copies |
| Total volume | 60.0 | |

5. For 100 μL reactions, carefully aliquot 60 μL of the upper reagent mix to each GeneAmp PCR reaction tube above the wax layer. Avoid splashing liquid onto the tube wall. If any liquid is present on the tube wall, tap the tube gently to collect all droplets into the upper reagent layer. Do not spin the tube in a microcentrifuge, because this may dislodge the wax layer.

6. Amplify the PCR reactions within a programmable thermal cycler. For the Perkin Elmer DNA Thermal Cycler 480, program and run the following linked files:
   a. Step cycle file:  95°C for 1 min, 60°C for 2 min for 2 cycles; link to file (2).
   b. Step cycle file:  94°C for 1 min, 60°C for 1 min for 38 cycles; link to file (3).
   c. Time delay file:  60°C for 10 min for 1 cycle; link to file (4).
   d. Soak file:        10°C for "forever" (an infinite hold)

7. Very gently insert a pipet tip through the center of the solid wax layer to form a small hole. To withdraw the reaction sample, use a fresh tip.

## 8. Analysis of PCR Products

### 8.1. Ethidium Bromide/Agarose Gel Electrophoresis

PCR products can be easily and quickly analyzed using a 3% NuSieve GTG agarose (FMC Bioproducts, Rockland, ME) and 1% Seakem GTG agarose (FMC Bioproducts) gel run in either TBE (89 m$M$ Tris-borate, 2 m$M$ EDTA) or in TAE (40 m$M$ Tris-acetate, 2 m$M$ EDTA, pH ~8.5) stained with ~0.5 µg/mL ethidium bromide. Use a 123-base pair (bp) or 1-kilobase pair (kbp) ladder as a convenient marker for size estimates of the products *(20)*.

### 8.2. Other Analytical Methods

A variety of other detection methods are available for PCR product analysis, such as ethidium bromide-stained 8–10% polyacrylamide gels run in TBE buffer, Southern gels or dot/slot blots, subcloning and direct sequencing, HPLC analysis, and the use of 96-well microplates, to name a few. The reverse dot-blot method combines PCR amplification with nonradioactive detection *(21)*. A fluorescence assay has also been optimized to detect amplification using the 5'–3' nuclease activity of *Taq* DNA polymerase *(13)*.

## 9. PCR Troubleshooting

### 9.1. Most Obvious Pitfalls

Even though the PCR process has greatly enhanced scientific studies, a variety of problems with the process, easily revealed by ethidium bromide-stained agarose gel electrophoresis, can and may need to be considered when encountered. For example, unexpected molecular weight size bands (nonspecific banding) or smears can be produced. These unexpected products accumulate from enzymatic extension of primers that annealed to nonspecific target sites. Second, primer-dimer (approx 40–60 bp in length, the sum of the two primers) can be produced. Primer-dimer can arise during PCR amplification when the DNA template is left out of the reaction, too many amplification cycles are used, or the primers are designed with partial complementary at the 3' ends. An increase in primer-dimer formation will decrease the production of desired product. Third, *Taq* DNA polymerase, which lacks the 3' to 5' exonuclease "proofreading" activity, will occasionally incorporate the wrong base during PCR extension. The consequences of *Taq* misincorporations usually have little effect, but should be considered during each PCR amplification.

### 9.2. How to Investigate "Failures"

PCR amplification for user-selected templates and primers are considered "failures" when no product bands are observed, the PCR product band is multibanded, or the PCR product band is smeared. These "failures" can be investigated and turned into successful PCR by manipulation of a number of variables, such as enzyme and salt concentrations, denaturation and anneal/extend times and temperatures, primer design, and hot start procedures *(22)*.

When no desired PCR product band is observed, initially verify the enzyme addition and/or concentration by titrating the enzyme concentration. Second, the magnesium ion concentration is also critical, so care should be taken not to lower the magnesium ion molarity on addition of reagents (i.e., buffers containing EDTA will chelate out the magnesium ion). The denaturation and anneal/extend times and temperatures may be too high or too low, causing failures, and can be varied to increase reaction sensitivity. Finally, the chemical integrity of the primers should be considered. In cases where the PCR product band is multibanded, consider raising the anneal temperature in increments of 2° and/or review the primer design and composition.

If a smear of the PCR product band is seen on an ethidium-bromide-stained agarose gel, consider the following options initially, individually or in combinations: decreasing the enzyme concentration, lowering the magnesium ion concentration, lengthening and/or raising the denaturation time and temperature, shortening the extension time, reducing the overall cycle number, and decreasing the possiblity of carryover contamination. Finally, in PCR amplifications where the PCR product band is initially observed, and on later trials a partial or complete loss of the product band is observed, consider testing new aliquots of reagents and decreasing the possibility of carryover contamination.

## References

1. Innis, M. A., Gelfand, D. H., Sninsky, J. J., and White, T. J., eds. (1990) *PCR Protocols. A Guide to Methods and Applications.* Academic, San Diego, CA.
2. Mullis, K. B. and Faloona, F. A. (1987) Specific synthesis of DNA *in vitro* via a polymerase chain reaction. *Methods Enzymol.* **155,** 335–350.
3. Saiki, R. K., Gelfand, D. H., Stoffel, S., Scharf, S. J., Higuchi, R., Horn, G. T., Mullis, K. B., and Erlich, H. A. (1988) Primer-directed enzymatic amplification of DNA with a thermostable DNA polymerase. *Science* **239,** 487–491.
4. Saiki, R. K., Scharf, S. J., Faloona, F., Mullis, K. B., Horn, G. T., Erlich, H. A., and Arnheim, N. (1985) Enzymatic amplification of β-globin genomic sequences and restriction site analysis for diagnosis of sickle cell anemia. *Science* **230,** 1350–1354.
5. Scharf, S. J., Horn, G. T., and Erlich, H. A. (1986) Direct cloning and sequence analysis of enzymatically amplified genomic sequences. *Science* **233,** 1076–1087.
6. Wang, A. M., Doyle, M. V., and Mark, D. F. (1989) Quantitation of mRNA by the polymerase chain reaction. *Proc. Natl. Acad. Sci. USA* **86,** 9717–9721.

7. Kwok, S. and Higuchi, R. (1989) Avoiding false positives with PCR. *Nature* **339,** 237,238.

8. Orrego, C. (1990) Organizing a laboratory for PCR work. *PCR Protocols. A Guide to Methods and Applications* (Innis, M. A., Gelfand, D. H., Sninsky, J. J., and White, T. J., eds.), Academic, San Diego, CA, pp. 447–454.

9. Kitchin, P. A., Szotyori, Z., Fromholc, C., and Almond, N. (1990) Avoiding false positives. *Nature* **344,** 201.

10. Longo, N., Berninger, N. S., and Hartley, J. L. (1990) Use of uracil DNA glycosylase to control carry-over contamination in polymerase chain reactions. *Gene* **93,** 125–128.

11. Chou, Q., Russell, M., Birch, D. E., Raymond, J., and Bloch, W. (1992) Prevention of pre-PCR mis-priming and primer dimerization improves low-copy-number amplifications. *Nucleic Acids Res.* **20,** 1717–1723.

12. Innis, M. A., Myambo, K. B., Gelfand, D. H., and Brow, M. A. D. (1988) DNA sequencing with *Thermus aquaticus* DNA polymerases and direct sequencing of polymerase chain reaction-amplified DNA. *Proc. Natl. Acad. Sci. USA* **85,** 9436–9440.

13. Holland, P. M., Abramson, R. D., Watson, R., and Gelfand, D. H. (1991) Detection of specific polymerase chain reaction product by utilizing the 5'-3' exonuclease activity of *Thermus aquaticus* DNA polymerase. *Proc. Natl. Acad. Sci. USA* **88,** 7276–7280.

14. Sobral, B. W. S. and Honeycutt, R. J. (1993) High output genetic mapping of polyploids using Pcr generated markers. *Theor. Appl. Genetics* **86,** 105–112.

15. Cheng, S., Fockler, C., Barnes, W. M., and Higuchi, R. (1994) Effective amplification of long targets from cloned inserts and human genomic DNA. *Proc. Natl. Acad. Sci. USA* **91,** 5695–5699.

16. Cheng, S., Chen, Y., Monforte, J. A., Higuchi, R., and Van Houten, B. (1995) Template integrity is essential for PCR amplification of 20- to 30-kb sequences from genomic DNA. *PCR Meth. Appl.* **4,** 294–298.

17. Erlich, H. A., ed. (1989) *PCR Technology, Principles and Applications for DNA Amplification.* Stockton, New York.

18. Landre, P. A., Gelfand, D. H., and Watson, R. H. (1995) The use of cosolvents to enhance amplification by the polymerase chain reaction, in *PCR Strategies* (Innis, M. A., Gelfand, D. H., and Sninsky, J. J., eds.), Academic, San Diego, CA, pp. 3–16.

19. AmpliWax PCR Gem 100 and PCR Gem 50. Package Insert. **BIO-66,** 55631-10/93. Perkin Elmer Corporation, Foster City, CA.

20. Sambrook, J., Fritsch, E. F., and Maniatis, T. (1989) *Molecular Cloning: A Laboratory Manual,* 2nd ed. Cold Spring Harbor Laboratory, Cold Spring Harbor, NY, pp. 6.20,6.21, B.23,B.24.

21. Saiki, R. K., Walsh, P. S., Levenson, C. H., and Erlich, H. A. (1989) Genetic analysis of amplified DNA with immobilized sequence-specific oligonucleotide probes. *Proc. Natl. Acad. Sci. USA* **86,** 6230–6234.

22. Kolmodin, L., Cheng, S., and Akers, J. (1995) *GeneAmp XL PCR kit. Amplifications: A Forum for PCR Users* (The Perkin-Elmer Corporation) **13,** 1–5.

# 2

# XL PCR Amplification
# of Long Targets From Genomic DNA

## Suzanne Cheng and Lori A. Kolmodin

## 1. Introduction

Using EXtra Long Polymerase Chain Reaction (XL PCR) conditions, targets of up to 30 kb in size have been amplified from total human genomic DNA *(1)*. These conditions reflect our current understanding of key parameters for successful PCR amplification of long (>10 kb) targets *(1–4)*. Specifically, the single-stranded integrity of the template DNA must be protected during sample preparation and during thermal cycling. In each cycle, template strands must be completely denatured, extension times must be sufficiently long, and nucleotide misincorporations that could cause premature termination of strand synthesis must be removed. Primer design and reaction stringency must also be optimized for successful single-copy gene amplifications from complex genomic DNA. Optimal reaction conditions may be very system dependent, particularly as target length increases. With increased understanding of the key variables and subsequent improvements in long PCR technology, increasingly longer targets will be routinely amplifiable.

The ability to amplify longer targets by PCR has numerous applications. For example, larger steps may be taken in gene mapping and sequencing efforts. Amplification of increasingly larger segments of any gene can also enable the simultaneous characterization of several regions of interest, for both research and diagnostic purposes. XL PCR protocols may facilitate studies of apparently unclonable regions, and of certain viral genomes or recombinant phage λ clones that are not readily cultured *(3,5)*. XL PCR can also complement cloning approaches by providing larger quantities of longer inserts. As the technology develops further, the list of applications is expected to grow.

From: *Methods in Molecular Biology, Vol. 67: PCR Cloning Protocols: From Molecular Cloning to Genetic Engineering* Edited by: B. A. White  Humana Press Inc., Totowa, NJ

## 2. Materials

### 2.1. Primers

Use primers designed with high specificity (*see* Section 3.1. and Note 1). Primer sequences for a 17.7-kb region of the human b-globin gene cluster and 16.3-kb of the human mitochondrial genome are given in Note 2. Nontemplate sequences can be added at ends for subsequent restriction digestion and cloning (*see* Note 3).

### 2.2. Preparation of Total Genomic DNA from Cultured Cells

*See* Notes 4 and 5.

1. Source of template DNA, such as cultured cells available from the NIGMS Human Genetic Mutant Cell Repository (*see* Coriell Institute for Medical Research, Camden, NJ).
2. Phosphate-buffered saline or other balanced salt solution.
3. 10 m$M$ Tris-HCl, pH 8.2, 0.4$M$ NaCl, 2 m$M$ EDTA.
4. 10% SDS.
5. Proteinase K (2 mg/mL). Prepare fresh in 1% SDS, 2 m$M$ EDTA.
6. RNase A (10 mg/mL). Heat inactivate at 95–100°C for 20 min.
7. Saturated NaCl (~6$M$).
8. 95% (v/v) ethanol.
9. 70% (v/v) ethanol.
10. TE buffer: 10 m$M$ Tris-HCl, pH 7.5–8.0, 1 m$M$ EDTA.

### 2.3. Analysis of Template Integrity (Optional)

1. Agarose (e.g., as in Section 2.5.1.).
2. 50 m$M$ NaCl, 1 m$M$ EDTA.
3. 10X alkaline running buffer : 0.5$N$ NaOH, 10 m$M$ EDTA.
4. 6X Ficoll/bromocresol green solution: 0.3$N$ NaOH, 6 m$M$ EDTA, 18% Ficoll (type 400 from Pharmacia, Piscataway, NJ or Sigma, St. Louis, MO), 0.15% bromocresol green, 0.25% xylene cyanol FF.
5. 0.1$M$ Tris-HCl, pH 8.0, 1 m$M$ EDTA.
6. Ethidium bromide solution: 0.5 μg/mL in TAE buffer (40 m$M$ Tris-acetate, 2 m$M$ EDTA, pH ~8.5).

### 2.4. PCR Amplification

1. GeneAmp XL PCR Kit (Perkin-Elmer, Foster City, CA) consisting of: r$Tth$ DNA Polymerase, XL; XL Buffer II; dNTP blend; Mg(OAc)$_2$; and control template and primers for amplification of a 20.8-kb sequence from the phage λ genome (*see* Notes 6,7).
2. AmpliWax PCR Gems (Perkin-Elmer).
3. Perkin-Elmer GeneAmp PCR Instrument System and PCR tubes with which the XL PCR reagents have been optimized and quality control tested.

## 2.5. Product Analysis

1. SeaKem GTG agarose or SeaKem Gold agarose (FMC Bioproducts, Rockland, ME).
2. TAE buffer: 40 m$M$ Tris-acetate, 2 m$M$ EDTA pH 8.5.
3. TBE buffer: 89 m$M$ Tris-borate, 2 m$M$ EDTA.
4. Ethidium bromide (10 mg/mL). Store in dark. **Wear gloves when handling.**
5. Restriction endonucleases.
6. Micron-100 spin column (Amicon, Beverly, MA).

# 3. Methods
## 3.1. Primer Design

1. Choose primers with melting temperatures ($T_m$) of 60–70°C to allow the use of relatively high annealing temperatures (e.g., 62–70°C) for optimal reaction specificity. Sequences of 20–24 bases can work well if the GC-content is sufficiently high (50–60%), but longer primers (e.g., 25–30 bases) may be needed if the AT-content is higher.
2. As for standard PCR, primer sequences should not be self-complementary or complementary to each other. Regions of complementarity, particularly at the 3'-ends of the primers, may result in "primer-dimer" products (*see also* Note 3).
3. Primers used in standard PCR (with annealing temperatures of 5°C below the $T_m$) *(6)* may also work at higher annealing temperatures, particularly with longer incubation times. Primers that can be used for control XL PCR amplifications with human genomic DNA are given in Note 2.
4. Software programs to calculate melting temperatures include Oligo® (version 4.0 or 5.0 for PC or Macintosh, National Biosciences, Plymouth, MN) and Melt (in BASIC, from J. Wetmur, Mt. Sinai School of Medicine, New York, NY). Wu et al. *(7)* have developed an algorithm for an oligonucleotide's "effective priming temperature" ($T_p$) based on its "effective length" ($L_n$):

$$T_p \ (°C) = 22 + 1.46 \ L_n, \tag{1}$$

$$L_n = 2 \ (\text{number of G and C bases}) + (\text{number of A and T bases}) \tag{2}$$

5. Primers that have "matching" $T_m$s or $T_p$s (within 1–2°C) are more likely to have the same optimal annealing temperature. If the difference in $T_m$ is ≥3°C, the primer with the higher $T_m$ may anneal to secondary priming sites at the lower temperature optimal for annealing of the second primer.
6. In any PCR, nontarget sites within the genome that have sufficient complementarity to the 3' end of a primer sequence can be secondary priming sites. As target length increases, however, secondary priming sites within the target sequence are also likely. Shorter secondary products are likely to be more efficiently amplified than a long target, and thus may accumulate at the expense of the desired product. Consequently, specificity at the primer annealing step is critical for successful amplification of long targets.

   Whenever possible, candidate primers for XL PCR should be screened against available sequence databases, particularly against any known sequences within the target and related loci (e.g., for gene families). Avoid primers within interspersed repetitive elements, such as *Alu* sequences *(8)*. The program Oligo (*see* Section 3.1.4.) can be used to scan a template sequence for potential secondary priming sites. RightPrimer™ 1.01 (BioDisk Software, San Francisco, CA) can be used to screen sequences deposited in GenBank® (National Institutes of Health) for various target genomes to estimate the relative frequency of selected primer sequences.

## 3.2. Preparation of Total Genomic DNA from Cultured Cells

In general, take care to minimize single-stranded nicks and double-stranded breaks within long PCR template samples, particularly for targets longer than 15 kb. Successful amplification of long targets depends on the presence of intact copies of the single-stranded template, and the intactness of a DNA sample will depend on the method of preparation used. The method of choice may vary with the target and the source of template DNA (*see* Notes 4,5). The procedure of Miller et al. *(9)*, based on a high-salt extraction of proteins, has been adapted for isolation of DNA from cultured cells *(1)* as follows:

1. Use a hematocytometer to determine the cell density. Freshly grown cells should be kept on ice until used. Cells may also be frozen in growth medium with 5% (v/v) dimethylsulfoxide (DMSO) for future use.
2. Rinse cells once with phosphate-buffered saline or other balanced salt solution.
3. Thoroughly resuspend a pellet of $1–3 \times 10^6$ cells with 0.6 mL of 10 m$M$ Tris-HCl, pH 8.2, 0.4$M$ NaCl, 2 m$M$ EDTA in a 1.5-mL Eppendorf tube.
4. Add 40 µL of 10% SDS, 100 µL of 2 mg/mL proteinase K, and 7.5 µL of RNase A.
5. Mix thoroughly by inverting the tube several times or by vortexing briefly.
6. Incubate each sample for at least 1 h, but preferably overnight, in a 50°C shaking water bath for digestion of the proteins. Complete digestion is critical for efficient recovery of DNA in step 7.
7. Add 0.2 mL saturated NaCl, shake the tube vigorously for 15 s, then centrifuge at 10,000 rpm (7000$g$) for 15 min. The precipitated proteins should form a tight, white pellet. If the pellet is not tight, repeat the centrifugation step. Incomplete separation of the protein phase from the DNA will result from incomplete proteinase K/SDS treatment.
8. Carefully divide the clear supernatant containing the DNA into two new Eppendorf tubes, then add a 2X volume of 95% (v/v) ethanol at room temperature to each sample and gently invert each tube several times to mix well.
9. Centrifuge at 10,000$g$ for 10–15 min to pellet the precipitated DNA.
10. Remove the supernatant and rinse the pellet with 0.5–1.0 mL of 70% (v/v) ethanol. Drain excess fluid onto absorbent paper and allow the pellet to air-dry for 5–10 min before resuspending (*see* step 11). If the DNA becomes too dry, it will be difficult to resuspend in solution. If the amount of recovered DNA is sufficiently high, the precipitated DNA aggregate can be lifted out (using the tip of a Pasteur pipet that has been reshaped to form a hook with a Bunsen burner flame), rinsed in 70% ethanol, and air-dried for 5 min.

11. Based on expected yields (e.g., ~15–20 μg total human genomic DNA from ~3 × 10$^6$ cells), resuspend the genomic DNA in TE buffer to a final concentration of ≤300 ng/μL. Avoid highly concentrated DNA solutions that can be heterogeneous and too viscous to handle easily. The precipitated DNA may require overnight incubation at 4°C to fully solubilize. Do not vortex the sample to aid in resuspending the DNA, because the shearing forces may introduce double-stranded breaks.

12. Determine the DNA concentration from the absorbance reading (between 0.2 and 0.8, for accuracy) at 260 nm. If accurate DNA concentrations are critical, and residual RNA is present, a method based on ethidium bromide fluorescence is described in Note 8.

13. Store genomic DNA stocks in TE buffer at 4°C to minimize the introduction of nicks through repeated freeze-and-thaw cycles. Alternatively, divide each stock across several tubes and freeze. Thaw aliquots as needed and store at 4°C.

## 3.3. Analysis of Template Integrity (Optional)

The single-stranded integrity of a template DNA preparation can be qualitatively assessed using alkaline agarose gel electrophoresis, essentially as in ref. *10*. Use 0.3–0.5% gels to visualize the range of 2 to >30 kb (*see* Note 9).

1. Prepare a molten agarose solution in 50 m*M* NaCl, 1 m*M* EDTA.
2. When the solution has cooled (to ~50°C), add 0.1X-volume of 10X alkaline running buffer, then pour the gel.
3. Presoak the gel in 1X alkaline running buffer for 30 min to ensure pH equilibration.
4. Load samples with a 6X Ficoll/bromocresol green solution (*see* Section 2.3.4. or ref. *10*).
5. Run the gel at 0.5–1.8 V/cm (e.g., 3.5–5 h) using a peristaltic pump to circulate the buffer. The buffer may become quite warm, and both the buffer level and gel position should be checked periodically during the run.
6. Neutralize the gel by gently shaking in 0.1*M* Tris-HCl, pH 8.0, 1 m*M* EDTA for 30 min, then stain with ~0.5 μg/mL ethidium bromide in TAE buffer.

## 3.4. PCR Amplification

The GeneAmp XL PCR Kit (*see* Note 6) is designed to use the AmpliWax PCR Gem-facilitated hot start process *(11)*. In this process, a solid wax layer is formed over a subset of PCR reagents (lower reagent mix, 30–50% of the total reaction volume), with the remaining reagents (upper reagent mix, remaining 50–70% of the total reaction volume) added above the wax layer. During the temperature ramp to the first denaturation step, the wax layer melts and is displaced by the upper reagent mix which is more dense. Thermal convection suffices to completely mix the combined lower and upper reagent mixes, and the melted wax layer will serve as a vapor barrier during PCR amplification. Manual hot start processes (*see* Note 10) may also be used, although reproducibility may be lower. The wax-mediated process also helps to minimize contamination between samples.

**Table 1**
**Guidelines for Preparation of Lower Reagent Mix**

| Reagent | Volume, 1X mix, µL | Final concentration, per 100 µL volume |
|---|---|---|
| Sterile water | 14.0–15.2 | N/A |
| 3.3X XL Buffer II | 12.0 | 1X XL buffer II |
| 10 m$M$ dNTP blend | 8.0 | 800 µ$M$ dNTP blend (200 µ$M$ of each dNTP; *see* Note 11) |
| 25 m$M$ Mg(OAc)$_2$ | 4.4–5.2 | 1.1–1.3 m$M$ (*see* Note 12) |
| 25 µ$M$ primer RH1024 | 0.4–0.8 | 0.1–0.2 µ$M$, 10–20 pmol/reaction |
| 25 µ$M$ primer RH1053 | 0.4–0.8 | 0.1–0.2 µ$M$, 10–20 pmol/reaction |
| Total volume | 40.0 µL | |

The protocol below describes XL PCR amplification of a human genomic DNA target (e.g., using the primers RH1024 and RH1053 in Note 2) and a volume ratio of the lower:upper reagent mixes equal to 40:60. Each mix can be assembled as a master mix sufficient for multiple reactions and allowing for volume loss during aliquoting (e.g., a 10X master mix for nine reactions).

1. Assemble lower reagent mix as shown in Table 1. For a 100-µL reaction, place 40 µL of this lower reagent mix into the bottom of each MicroAmp teaction tube. Avoid splashing liquid onto the tube wall. If any liquid is present on the tube walls, spin the tube in a microcentrifuge briefly.
2. Carefully add a single AmpliWax PCR Gem 100 to each tube containing the lower reagent mix (*see* Note 13). Melt the wax by incubating the reaction tubes at 75–80°C for 3–5 min. Allow the wax layers to solidify completely as the tubes cool to room temperature.
3. Assemble upper reagent mix as shown in Table 2. For a 100-µL reaction, aliquot 60 µL of this upper reagent mix to each room temperature reaction tube, above the wax layer formed in step 2. Avoid splashing liquid onto the tube wall. If any liquid is present on the tube wall, tap the tube lightly to collect all droplets into the upper reagent layer. Do not spin the tube in a microcentrifuge, because this may dislodge the wax layer.
4. Amplify the PCR reactions in a programmable thermal cycler (*see* Note 17). For the Perkin-Elmer GeneAmp System 9600, program the following method:
   a. Hold: 94°C for 1 min (reagent mixing and initial template denaturation).
   b. Cycle: 94°C for 15 s (denaturation; *see* Note 18), 68°C for 12 min (annealing and extension; *see* Notes 19–21) for 20 cycles.
   c. Auto: 94°C for 15 s, 68°C for 12 min, adding 15 s/cycle (*see* Note 21) for 17 cycles (*see* Note 22).
   d. Hold: 70–72°C for 10 min (final completion of strand synthesis).
   e. Hold: 4°C until tubes are removed. (Use the "forever" software option.)

**Table 2**
**Guidelines for Preparation of Upper Reagent Mix**

| Reagent | Volume, 1X mix, μL | Final concentration, per 100 μL volume |
|---|---|---|
| Sterile water | 1.0–40.0 | N/A |
| 3.3X XL Buffer II | 18.0 | 1X XL Buffer II |
| 2 U/μL r*Tth* DNA Polymerase, XL | 1.0 | 2 U/reaction |
| Human genomic DNA | 1.0–40.0 | Up to 1 μg (*see* Notes 14–16) |
| Total volume | 60.0 μL | |

## 3.5. Product Analysis

1. Sample withdrawal: Gently insert a pipet tip through the center of the solid wax layer to form a small hole. If the tip becomes plugged during this procedure, use a fresh tip to withdraw the reaction sample.

2. The presence of PCR products can be quickly determined using a 0.6% SeaKem GTG agarose gel, run either in TAE or TBE buffer with ~0.5 μg/mL ethidium bromide. Ideally, each lane will contain a single band corresponding to each target (*see* Notes 23–25).

3. High-mol-wt products can be more accurately sized with a 0.3% SeaKem Gold agarose gel (*see* Note 9) run in TAE buffer with ~0.5 μg/mL ethidium bromide. Depending on the level of resolution needed, run the gel at 7 V/cm for 2 min and then either at 0.8 V/cm for up to 15 h, at 1.5 V/cm for up to 6 h, or at 5 V/cm for 1–2 h. Products may also be analyzed by pulsed field gel electrophoresis (*see* Note 26).

4. In general, XL PCR products may be analyzed directly by restriction digestion. If further manipulations, such as ligation and cloning, are planned, the reactions should be treated to remove the unincorporated dNTPs and the r*Tth* DNA polymerase, XL. If the polymerase and dNTPs are present during restriction digestions, recessed 3' termini may be filled in as they are created, eliminating such sites for ligation. One approach is to use a 100 spin column (Microcon), which will also remove buffer components and unused primers (*see also* Note 27).

## 4. Notes

1. Purification of primers by polyacrylamide gel electrophoresis does not appear to be generally necessary, although significant levels of truncated sequences could contribute to priming at nonspecific, secondary sites.

2. Two positive control primer sets for XL PCR with human genomic DNA are listed below (5'–3'). These primers were designed for use with a 67–68°C annealing temperature; at lower temperatures (e.g., 62°C), secondary products will also accumulate. The 16.3-kb human mitochondrial genome primers can be used to confirm the presence of low-copy DNA because the mitochondrial genome is present in many

copies per cell *(12)*. Longer products are likely to be lower in yield because of lower reaction efficiencies. Certain sequences may also be difficult to amplify regardless of the target length, perhaps because of their base composition and potential to form secondary structures. Consequently, the best controls for a troublesome system may be a series of primer pairs (e.g., one constant, the others variable in position) that define increasingly larger subsets of the target.

Primers for the human β-globin gene cluster *(1,13)* are:

RH1024 TTGAGACGCATGAGACGTGCAG, positions 44348–44369, and

RH1053 GCACTGGCTTAGGAGTTGGACT, which complements positions 61986–62007.

Primers for the human mitochondrial DNA genome *(14)* are

RH1065 TGAGGCCAAATATCATTCTGAGGGGC at positions 15149–15174, and

RH1066 TTTCATCATGCGGAGATGTTGGATGG, which complements positions 14816–14841.

3. For subsequent cloning of the PCR product, primers can be used to introduce recognition sites for restriction endonucleases. Such sites would be added to the 5'-end of the target-binding sequence, with an additional GC-rich "5'-clamp" of several bases for efficient binding of the restriction endonuclease and subsequent digestion *(15)*. These nontemplate bases at the 5'-end of the primer should not noticeably affect reaction specificity, but should be accounted for in determining the annealing temperature for the first few cycles (before the entire sequence has been incorporated into the template population).

4. A number of other methods are available to prepare highly intact DNA. As discussed in *(1)*, these include the Puregene® DNA Isolation Kit (Gentra Systems, Minneapolis, MN), QIAGEN® Genomic-tips (Chatsworth, CA), phenol-extraction (as in ref. *10,* with high quality phenol), and Megapore™ dialysis *(16)*. The Puregene DNA Isolation Kit is based on high-salt extraction of the proteins. It is important to fully resuspend the cell pellet with the lysis solution for the highest yields of DNA. The method does call for vigorous vortexing to mix the high-salt solution with the cell lysate. The benefits of this 15–20-s vortexing step for efficient precipitation of proteins and subsequent recovery of DNA outweigh the potential for damaging the DNA. In general, however, template stocks for long PCR should be handled gently. The QIAGEN Genomic-tips, based on an anion-exchange resin, are particularly useful for isolation of cosmid DNA without organic solvents. To avoid clogging the Genomic-tip, take care to not load too many cells and to thoroughly resuspend the dilute sample with buffer QBT, as recommended, before loading the sample onto the column. The Megapore method *(16)* utilizes dialysis through type HA 0.45-μm membranes (Millipore, Bedford, MA) membranes to remove denatured proteins and cellular debris. This approach is designed to generate very high-mol-wt DNA fragments for cloning.

5. A similar method that was not tested by the authors is the salt/chloroform extraction method of Mullenbach et al. *(17)*. The addition of chloroform facilitates the separation between the DNA and protein phases.

6. XL PCR Buffer II is composed of Tricine to maintain a protective pH during thermal cycling, and the cosolvents glycerol and dimethylsulfoxide (DMSO) to effectively lower melting and strand separation temperatures. The 3'-5'-exonuclease, or "proofreading," activity of r*Tth* DNA Polymerase, XL facilitates the completion of strand synthesis in each cycle by removing misincorporated nucleotides. The XL PCR amplification protocol uses relatively short denaturation times at moderately high temperatures to minimize template damage while ensuring complete denaturation. The XL PCR protocol also uses a wax-mediated hot start method, relatively high annealing temperatures, and reduced enzyme levels to enhance reaction specificity.

7. Different enzyme and buffer systems for long PCR are commercially available from TaKaRa Shuzo (Otsu, Shiga, Japan) and Boehringer Mannheim Biochemicals (Indianapolis, IN).

8. Ethidium bromide fluorescence is highly specific for double-stranded DNA. Prepare a standard curve using solutions of 10–400 ng of $\lambda$/*Hind*III DNA in 1.2-mL vol of 0.5 µg/mL ethidium bromide, 20 m$M$ $KH_2PO_4$, and 0.5 m$M$ EDTA (pH 11.8–12.0). The fluorescence of aliquots of template DNA can then be compared against these standards. The high pH is critical to minimize any RNA contribution to the fluorescence. An A-4 Filter Fluorometer (Optical Technology Devices, Elmsford, NY) can be used with a bandpass filter (360-nm max) for excitation and an interference glass filter at 610 nm for emission spectra (Bennett Van Houten, Sealy Center for Molecular Science, University of Texas Medical Branch, Galveston, TX, personal communication).

9. Higher molecular weight material will be better resolved on a 0.3% agarose gel than on a higher percentage gel. Since a 0.3% gel is fragile, use a high-strength agarose, such as SeaKem Gold. Chill the gel at 4°C before removing the comb; the walls of the wells may collapse, but the wells should be restored when filled with buffer. Alternatively, submerge the gel in chilled buffer before removing the comb.

10. For a manual hot start, assemble a single master mix comprised of all but one key reaction component (usually $Mg(OAc)_2$, dNTPs, or enzyme). Add a 75–80°C Hold (e.g., 5 min for up to 20–25 samples) before the initial denaturation step of the thermal cycling profile (*see* step 4 of Section 3.4.). Bring all samples to this Hold temperature for ~1 min in the thermal cycler, then add the remaining component to each reaction mix. Note that the volume of this addition should be large enough to minimize pipeting variations, yet small enough to minimize the cooling effect on the reaction mixture already within the tube. If the dNTPs or enzyme are the component that is withheld, dilute with 1X XL Buffer II to facilitate complete mixing.

11. XL PCR appears to be much more sensitive to the integrity of the dNTP solutions than is standard PCR. Stock dNTP solutions should be at pH 7.0–7.5. Reproduc-

ibility may be best if these solutions are aliquoted and subjected to a minimal number of freeze-thaw cycles. Changing dNTP stock solutions may require slight adjustment of the Mg(OAc)$_2$ concentration.

12. XL PCR amplifications can be quite sensitive to Mg(OAc)$_2$ levels, and each new target or primer pair may have a different optimal range for the Mg(OAc)$_2$ concentration. Titrations should be carried out in increments of 0.1 m$M$.

13. Tap or rotate the tube of AmpliWax PCR Gem 100s to empty a few beads onto either a clean sheet of weighing paper, into a clean weighing boat, or into the tube cap. Use a clean pipet tip to carefully direct a single bead into each tube containing the lower reagent mix.

14. In general, 50–100 ng of total human genomic DNA (with high average single-stranded molecular weight) will suffice for a 100-μL reaction. Excessive amounts of genomic DNA may contribute to the accumulation of nonspecific products.

15. If the template volume represents a large fraction of the final reaction volume, the DNA should be diluted in water or 10 m$M$ Tris-HCl, 0.1 m$M$ EDTA, to minimize chelation of the Mg(OAc)$_2$ in the final reaction mix (*see* Note 12).

16. Long PCR amplifications may be more sensitive to potential reaction inhibitors than shorter target amplifications. In such cases, the addition of 50–500 ng/μL nonacetylated BSA may enhance yields *(4)*, possibly by binding nonspecific inhibitors.

17. XL PCR amplifications are sensitive to the times and temperatures of denaturation and annealing, thus different types of thermal cyclers are likely to require adjustments to this thermal cycling profile. Profiles for all of the Perkin-Elmer GeneAmp PCR Instrument Systems are provided with the GeneAmp XL PCR Kit.

18. Complete denaturation of the template strands is critical for successful PCR amplifications. The presence of extended GC-rich regions may require use of 95–96°C denaturation temperatures, but the time should be kept short to minimize damage to the single-stranded template and loss of r*Tth* DNA polymerase, XL activity over the course of the PCR run.

19. Different primers may have different melting temperatures. The choice of annealing temperature should be based on the actual primer pair being used (*see* Section 3.1.). In general, the highest possible temperature should be used to minimize annealing to secondary priming sites. For reactions in which only the desired product is obtained, a lower annealing temperature may improve product yields.

20. This thermal cycling profile uses two temperatures, compared to the three temperatures typically used in standard PCR. Strand synthesis by r*Tth* DNA polymerase, XL is efficient between 60 and 70°C. Consequently, when an annealing temperature of ~62–70°C is used, the same temperature can be set for the extension phase of the cycle. If lower annealing temperatures are necessary, a third temperature step at 65–70°C should be added for efficient completion of strand synthesis, but sufficient time at the annealing temperature must be allowed for efficient priming before the reactions are raised to the extension temperature.

21. In general, use extension times sufficient for 30–60 s/kb of the target. In the two-temperature cycling profile, this applies to the total annealing-plus-extension time. As product accumulates, the ratio of template to polymerase molecules will increase, and overall reaction efficiency may decrease. The Auto feature of the Perkin-Elmer GeneAmp PCR System 9600 allows the extension time to be incrementally increased during later cycles of the run, which helps maintain reaction efficiency. The potential disadvantage of using very long extension times initially is that during early cycles, excessively long extension times may permit nonspecific products to accumulate.

22. The optimal total number of cycles will depend on the initial copy number of the template and the reaction efficiency. Reaction efficiency is generally higher for shorter (5–10 kb) vs longer (20–30 kb) targets. For example, from ~$10^4$ copies of human genomic DNA (37 ng, in a 50-μL reaction), the 16.3-kb multicopy mitochondrial genome target can be readily amplified with a total of 30–35 cycles, whereas the 17.7-kb single-copy β-globin target requires at least 35–37 cycles. The number of Auto cycles (GeneAmp PCR System 9600) will also depend on the initial copy number of the template and the target length.

23. Amplified target bands may be identified by size (gel mobility relative to standards), a Southern blot analysis (as in ref. *10*), and/or analytic restriction digestions. High-mol-wt smears tend to reflect high levels of nonspecific synthesis, as in the cases of excess r*Tth* DNA polymerase, XL or excess Mg(OAc)$_2$. Excessively long extension times or too many cycles of amplification can also result in the appearance of high-mol-wt bands or nonspecific smears. Low molecular weight secondary bands may reflect insufficient reaction specificity, and may be reduced by the use of a higher annealing temperature, and/or lower concentrations of template, primers, or r*Tth* DNA polymerase, XL. If accumulation of products other than the desired product is significant, the best solution may be to redesign one or both primers. Absence of any detectable product from a known template may indicate that the denaturation temperature was either too low for the template or too high for r*Tth* DNA polymerase, XL; that the annealing temperature was too high for the primer pair; or that either the polymerase or the Mg(OAc)$_2$ concentration was too low. If a Southern blot analysis or reamplification using primers located within the original target (nested PCR) reveals that the desired product is present at a very low level, the explanation may be that too few cycles of amplification were used given the starting copy number of the target (*see also* Notes 15,16).

24. Optimization of the denaturation or annealing temperatures should be made in increments of 1–2°C. Adjustments to enzyme concentrations can be made in increments of 0.5–1 U/100-μL reaction. Optimization of the Mg(OAc)$_2$ concentration should be carried out in increments of 0.1 m*M* (*see* Notes 21,22).

25. Carryover contamination (*see* Chapter 1, Section 1.4.) and dNTP stock solutions of poor quality can both significantly reduce the apparent optimal range for the Mg(OAc)$_2$ concentration. These problems may not be seen initially, but may become apparent during later amplifications with targets and primers previously observed to work well.

26. Resolution of high-mol-wt DNA (>50 kb) is best achieved using a pulsed field, for example, field inversion gel electrophoresis *(18)*. One such system is made by Hoefer (San Francisco, CA).

27. The 3'-ends of the PCR products may have an additional one or two nontemplated nucleotides *(19,20)*. Although the 3'-5'-exonuclease activity present in r*Tth* DNA polymerase, XL would be expected to remove these 3'-additions, there is evidence that a certain fraction of XL PCR product molecules have an additional 3'-A *(21)*. This fraction is likely to be less than that observed in standard PCR with *Taq* DNA polymerase, and using methods, such as the TA Cloning® Kit (Invitrogen, San Diego, CA), which take advantage of the 3'-A addition may therefore be inefficient. If necessary, the 3'-additions can be removed by incubation with *Pfu* DNA polymerase *(22)* or with the Klenow fragment of *Escherichia coli* DNA polymerase I *(15,20)*.

## Acknowledgments

We would like to thank Ben Van Houten (Sealy Center for Molecular Science) for the protocol to determine DNA concentrations by fluorescence and the high-salt extraction method for cultured cells, Robin Ma (University of British Columbia) for drawing our attention to salt/chloroform methods for DNA isolation, and Ann-Charlotte Stewart and Cosette Wheeler (University of New Mexico) for sharing their experiences with cloning XL PCR products. We also thank Julia Horak and J. Fenton Williams (Perkin-Elmer) for their enthusiastic support of this chapter.

## References

1. Cheng, S., Chen, Y., Monforte, J. A., Higuchi, R., and Van Houten, B. (1995) Template integrity is essential for PCR amplification of 20- to 30-kb sequences from genomic DNA. *PCR Meth. Appl.* **4,** 294–298.
2. Barnes, W. M. (1994) PCR amplification of up to 35 kb with high fidelity and high yield from λ bacteriophage templates. *Proc. Natl. Acad. Sci. USA* **91,** 2216–2220.
3. Cheng, S., Fockler, C., Barnes, W. M., and Higuchi, R. (1994) Effective amplification of long targets from cloned inserts and human genomic DNA. *Proc. Natl. Acad. Sci. USA* **91,** 5695–5699.
4. Cheng, S. (1995) Longer PCR amplifications, in *PCR Strategies* (Innis, M. A., Gelfand, D. H., and Sninsky, J. J., eds.), Academic, San Diego, CA, pp. 313–324.
5. Cheng, S., Chang, S.-Y., Gravitt, P., and Respess, R. (1994) Long PCR. *Nature* **369,** 684,685.
6. Innis, M. A. and Gelfand, D. H. (1990) Optimization of PCRs, in *PCR Protocols* (Innis, M. A., Gelfand, D. H., Sninsky, J. J., and White, T. J., eds.), Academic San Diego, CA, pp. 3–12.
7. Wu, D. Y., Ugozzoli, L., Pal, B. K., Qian, J., and Wallace, R. B. (1991) The effect of temperature and oligonucleotide primer length on the specificity and efficiency of amplification by the polymerase chain reaction. *DNA Cell Biol.* **10,** 233–238.

8. Schmid, C. W. and Jelinek, W. R. (1982) The Alu family of dispersed repetitive sequences. *Science* **216**, 1065–1070.

9. Miller, S. A., Dykes, D. D., and Polesky, H. F. (1988) A simple salting out procedure for extracting DNA from human nucleated cells. *Nucleic Acids Res.* **16**, 1215.

10. Sambrook, J., Fritsch, E. F., and Maniatis, T. (1989) *Molecular Cloning: A Laboratory Manual,* 2nd ed. Cold Spring Harbor Laboratory, Cold Spring Harbor, NY, pp. 6.20,6.21, 9.16–9.19, 9.34–9.57, and B.23,B.24.

11. Chou, Q., Russell, M., Birch, D. E., Raymond, J., and Bloch, W. (1992) Prevention of pre-PCR mis-priming and primer dimerization improves low-copy-number amplifications. *Nucleic Acids Res.* **20**, 1717–1723.

12. Robin, E. D. and Wong, R. (1988) Mitochondrial DNA molecules and virtual number of mitochondria per cell in mammalian cells. *J. Cell Physiol.* **136**, 507–513.

13. Kolmodin, L., Cheng, S., and Akers, J. (1995) GeneAmp® XL PCR Kit, in *Amplifications: A Forum for PCR Users* (The Perkin-Elmer Corporation), Issue **13**.

14. Cheng, S., Higuchi, R., and Stoneking, M. (1994) Complete mitochondrial genome amplification. *Nature Genet.* **7**, 350,351.

15. Scharf, S. J. (1990) Cloning with PCR, in *PCR Protocols* (Innis, M. A., Gelfand, D. H., Sninsky, J. J., and White, T. J., eds.), Academic, San Diego, CA, pp. 84–91.

16. Monforte, J. A., Winegar, R. A., and Rudd, C. J. (1994) Megabase genomic DNA isolation procedure for use in transgenic mutagenesis assays. *Environ. Mol. Mutagenesis* **23**, 46.

17. Mullenbach, R., Lagoda, P. J. L., and Welter, C. (1989) Technical tips: an efficient salt-chloroform extraction of DNA from blood and tissues. *Trends Genet.* **5**, 391.

18. Carle, G. F., Frank, M., and Olson, M. V. (1986) Electrophoretic separations of large DNA molecules by periodic inversion of the electric field. *Science* **232**, 65–68.

19. Clark, J. M. (1988) Novel nontemplated nucleotide addition reactions catalyzed by procaryotic and eucaryotic DNA polymerases. *Nucleic Acids Res.* **16**, 9677–9686.

20. Hu, G. (1993) DNA polymerase-catalyzed addition of nontemplated extra nucleotides to the 3' end of a DNA fragment. *DNA Cell Biol.* **12**, 763–770.

21. Stewart, A.-C. M., Gravitt, P. E., Cheng, S., and Wheeler, C. M. (1995) Generation of entire human papilloma virus genomes by long PCR: frequency of errors produced during amplification. *Genome Res.* **5**, 79–88.

22. Costa, G. L. and Weiner, M. P. (1994) Protocols for cloning and analysis of blunt-ended PCR-generated DNA fragments. *PCR Meth. Appl.* **3**, S95–S106.

# 3

## Amplification of DNA Sequences Up To 5 kb from Small Amounts of Genomic DNA Using Tub DNA Polymerase

### Helen B. Forrester and Ian R. Radford

## 1. Introduction

Standard PCR protocols can be used to amplify sequences of a few hundred base pairs in length from the DNA of a single cell *(1,2)*. However, the efficiency of standard PCR amplification using *Taq* DNA polymerase declines with sequence length, making it difficult to amplify long sequences from small amounts of genomic DNA.

The decreased efficiency of PCR amplification of long sequences is probably attributable to premature termination of synthesis following nucleotide misincorporation and/or depurination of nucleotides in the template DNA *(3,4)*. Naturally, the longer the sequence being amplified, the greater the chance of such events occurring. Consequently, when amplifying long DNA sequences it is important to use reaction conditions that discourage depurination. The rate of depurination of DNA has been shown to increase with rising temperature and with decreasing pH *(5)*. However, PCR reaction solutions are often buffered with Tris-HCl, which shows significant changes in pH with temperature. For example, a Tris solution of pH 8.5 at 25°C has a pH of <7.0 at 95°C. Accordingly, it has been suggested that PCR amplification of longer DNA sequences be performed in solutions containing Tris-HCl at an elevated pH (between 8.8 and 9.2 at 25°C) to allow for the decrease in pH with increasing temperature *(3,4)*.

Premature termination of synthesis caused by nucleotide misincorporation is a potential problem for amplification using either *Taq* or other thermostable enzymes, such as Tub DNA polymerase, which lack 3'-5' exonuclease activity and are thus unable to remove misincorporated nucleotides. However, Tub

From: *Methods in Molecular Biology, Vol. 67: PCR Cloning Protocols: From Molecular Cloning to Genetic Engineering* Edited by: B. A. White  Humana Press Inc., Totowa, NJ

DNA polymerase has been shown to amplify sequences from plasmid DNA >10 kb in length more efficiently than *Taq* DNA polymerase *(6)* and to amplify sequences of 4.8 kb in length from small amounts of genomic DNA *(7)*.

Incubation of *Taq* or Tub DNA polymerase-amplified products with the Klenow fragment of *Escherichia coli* DNA polymerase I was shown to markedly improve PCR yield *(7,8)*. This improvement may result from the removal of misincorporated bases by the 3'-5' exonuclease activity of Klenow fragment, thus allowing completion of synthesis by the associated DNA polymerase activity.

For both *Taq (9)* and Tub DNA polymerases *(7)* the efficiency of PCR amplification can be markedly influenced by the magnesium ion concentration. For example, a deviation of 0.5 m$M$ $MgCl_2$ from the optimal concentration can markedly affect the amount of product. This suggests a stringent dependency of successful amplification on $MgCl_2$ concentration. Accordingly, the optimal $MgCl_2$ concentration for each pair of primers should be determined by titration of the reaction. By amplifying target sequence with Tub DNA polymerase at the optimal magnesium chloride concentration and then treating the product with the Klenow fragment of *E. coli* DNA polymerase I, it is possible to amplify DNA sequences of at least 4.8 kbp with high efficiency from the equivalent of the DNA content of one human cell *(7)*. This chapter outlines the procedure to find the optimal $MgCl_2$ concentration and to amplify sequences using Tub DNA polymerase.

## 2. Materials
### 2.1. DNA Amplification

1. Purified genomic DNA (20 ng/µL).
2. Plasmid containing DNA sequence of interest, to act as a positive control for PCR and as a hybridization probe (optional).
3. Oligonucleotide (20–25-mers with approx 50% GC content) primers A (sense complementary primer) and B (antisense complementary primer).
4. Tub DNA polymerase (Amersham, Arlington Heights, IL).
5. 10X buffer (provided with Tub DNA polymerase): 500 m$M$ Tris-HCl, pH 9.0, 200 m$M$ $(NH_4)_2SO_4$, 7 m$M$ $MgCl_2$.
6. dNTP solution (1.25 m$M$ of each dNTP).
7. $MgCl_2$ solutions (10 and 25 m$M$).
8. EDTA solutions (10 and 25 m$M$).
9. *E. coli* DNA polymerase I large (Klenow) fragment.
10. DNA thermal cycler.

### 2.2. Reagents for Southern Blot Analysis

1. Positively charged nylon membrane (e.g., GeneScreen Plus, DuPont NEN, Boston, MA).
2. 0.4$M$ NaOH.

3. 2X SSC: 0.3$M$ sodium chloride, 0.03$M$ sodium citrate, pH 7.0.
4. $^{32}$P-labeled DNA probe, specific for amplified sequence.
5. Hybridization buffer (as recommended for nylon membrane).
6. 2X SSC, 0.1% SDS.
7. 0.5X SSC, 1% SDS.
8. 0.5X SSC, 0.1% SDS.
9. Hybridization oven and bottle, or 65°C water bath and heat sealable plastic.
10. Blotting paper.
11. X-ray film, intensifying screen, and film exposure cassette.
12. Agarose gel electrophoresis reagents and equipment.

## 3. Methods

### 3.1. Handling Procedures to Avoid PCR Crosscontamination

1. Reaction solutions should be prepared in a laminar flow hood.
2. A dedicated pipet should be used for all pipeting before amplification and a different pipet should be used for all pipeting after the amplification is completed.
3. Aerosol-resistant tips should be used for all pipeting.
4. Disposable gloves should be worn when preparing the reaction solutions and should be changed frequently.

### 3.2. Optimization of the Magnesium Ion Concentration for Each Set of Primers

DNA is amplified in 50 µL of solution containing 50 m$M$ Tris-HCl, pH 9.0, 20 m$M$ (NH$_4$)$_2$SO$_4$, 200 m$M$ of each dNTP, 1.0 m$M$ of each primer, 0.6 U of Tub DNA polymerase (Amersham), and different concentrations of MgCl$_2$. The magnesium ion concentration is decreased by addition of EDTA or increased by addition of MgCl$_2$ solution.

All amplification samples are prepared on ice and left on ice until they are transferred to the DNA thermal cycler, in order to minimize DNA extension from nonspecifically bound primers. Alternatively, hot-start PCR may be used.

1. Prepare PCR reaction mix stock for MgCl$_2$ titration by adding in order (*see* Note 1): 202.7 µL water; 52.5 µL 10X buffer; 84 µL dNTP solution; 26.3 µL 20 m$M$ solution of primer A; 26.3 µL 20 m$M$ solution of primer B; 2.1 µL Tub DNA polymerase (3 U/µL); and 52.5 µL DNA (genomic [20 ng/µL] or plasmid [0.2 ng/µL]) (*see* Note 2).
2. Prepare final reaction solutions testing ten different MgCl$_2$ concentrations by adding the materials shown in Table 1.
3. Add oil or wax to each tube if necessary (*see* Note 3).
4. Transfer reaction tubes to a thermal cycler and amplify using the following conditions (*see* Note 4): 95°C for 5 min, 64°C for 7 min (initial denaturation cycle); 95°C for 1 min, 64°C for 7 min for 34 main cycles.
5. If oil or wax was used, separate PCR solution (*see* Note 5).

**Table 1**
**Reaction Tube Preparation**
**to Determine Appropriate MgCl$_2$ Concentration**

| PCR tube no. | MgCl$_2$ concentration, m$M$ | Volume water, μL | Volume 10 m$M$ EDTA | Volume 25 m$M$ MgCl$_2$, μL | Reaction mix volume, μL | Total volume, μL |
|---|---|---|---|---|---|---|
| 1 | 0.5 | 6.6 | 1 μL | — | 42.5 | 50 |
| 2 | 0.7 | 7.6 | — | — | 42.5 | 50 |
| 3 | 1.0 | 6.1 | — | 1.5 (10 m$M$) | 42.5 | 50 |
| 4 | 1.5 | 6 | — | 1.6 | 42.5 | 50 |
| 5 | 2.0 | 5 | — | 2.6 | 42.5 | 50 |
| 6 | 2.5 | 4 | — | 3.6 | 42.5 | 50 |
| 7 | 3.0 | 3 | — | 4.6 | 42.5 | 50 |
| 8 | 3.5 | 2 | — | 5.6 | 42.5 | 50 |
| 9 | 4.0 | 1 | — | 6.6 | 42.5 | 50 |
| 10 | 4.5 | — | — | 7.6 | 42.5 | 50 |

6. Add Klenow fragment of *E. coli* DNA polymerase I (2 U/50 μL) to each reaction solution.
7. Incubate samples for 30 min at 37°C and then heat inactivate Klenow fragment at 65°C for 15 min.
8. Add 2 μL of 10X loading buffer to 15 μL of each PCR sample.
9. Separate DNA on a 1% agarose gel.
10. Visualize DNA by UV illumination of ethidium bromide-stained gel.
11. Determine MgCl$_2$ concentration that gave the highest ratio of product to background amplification (*see* Note 2).

### 3.3. Amplification Efficiency

To determine the sensitivity of product amplification and detection, amplify various dilutions of genomic DNA (*see* Note 6).

1. Dilute genomic DNA (starting at a concentration of 20 ng/μL) as shown in Table 2.
2. Prepare a stock reaction mix for 26 samples; and aliquot into 25 tubes by adding successively (*see* Note 1): (696.8–$y$) μL water (for the value of $y$, *see* Note 7); $y$ μL 25 m$M$ MgCl$_2$ (for MgCl$_2$ concentrations higher than 0.7 m$M$) or 25 m$M$ EDTA (for MgCl$_2$ concentrations lower than 0.7 m$M$); 130 μL 10X buffer; 208 μL 1.25 m$M$ dNTP solution; 65 μL primer A; 65 μL primer B; and 5.2 μL Tub DNA polymerase (3 U/μL). (Add 45 μL of reaction mix to each reaction tube.)
3. Prepare reaction tubes as shown in Table 3.
4. Amplify products as for MgCl$_2$ concentration titration (Section 3.2., steps 3–5).
5. Treat amplified products with Klenow fragment (Section 3.2., steps 6,7).

**Table 2**
**Dilution of Genomic DNA**

| Concentration of DNA, pg/μL | Volume of water, μL | Concentration of DNA added, pg/μL | Volume of DNA added, μL |
|---|---|---|---|
| 2000 | 90 | 20,000 | 10 |
| 200 | 90 | 2000 | 10 |
| 20 | 90 | 200 | 10 |
| 2 | 90 | 20 | 10 |
| 0.2 | 90 | 2 | 10 |

6. Separate samples on a 1% agarose gel (Section 3.2., steps 8–10).
7. Analyze PCR products by Southern blotting (*see* Section 3.4.).

### 3.4. Southern Blot Analysis of PCR Products

1. Transfer DNA from 1% agarose gel in $0.4M$ NaOH onto a nylon membrane by capillary action for 4–8 h.
2. Wash membrane with 2X SSC.
3. Dry bake or crosslink DNA to membrane according to manufacturer's instructions.
4. Hybridize membrane-bound DNA with random-primed, $^{32}$P-labeled DNA sequence of interest, using conditions recommended by the nylon membrane's manufacturer.
5. Wash membrane as recommended and blot off excess fluid.
6. Enclose membrane in polyethylene food wrap and expose for several days (time depends on the activity of the membrane) to X-ray film at −70°C, with an intensifying screen.
7. Remove the film and develop it.

## 4. Notes

1. The reagents should be added in the order listed with DNA added last in order to decrease the chance of contaminating the other reagents with DNA. When reactions containing different DNA concentrations are being set up, dispense the lowest concentration first.
2. The $MgCl_2$ concentration may be optimized using either genomic DNA or a plasmid containing the sequence of interest. It will be easier to initially amplify the sequence of interest from plasmid DNA. However, the optimal $MgCl_2$ concentration for amplification may differ between plasmid and genomic DNA. Adjustment to the annealing/extension time used for plasmid DNA may also be necessary when amplifying from genomic DNA. It is also possible to optimize the conditions for sequences that are difficult to amplify by using Southern blot analysis to detect the amplification products. If PCR product yield is poor or nonexistent for all $MgCl_2$ concentrations tested, check the reagents by amplify-

**Table 3**
**Reaction Tube Preparation**
**to Determine Amplification Efficiency**

| PCR tube no. | Amount of DNA, pg | Equivalent cell number (human) | Volume of water, μL | Reaction mix, μL | DNA dilution (pg/μL) added | Volume of DNA added, μL | Total volume, μL |
|---|---|---|---|---|---|---|---|
| 1 | no DNA | 0 | 5 | 45 | no DNA | — | 50 |
| 2 | 1 | <0.2 | — | 45 | 0.2 | 5 | 50 |
| 3 | 1 | <0.2 | — | 45 | 0.2 | 5 | 50 |
| 4 | 1 | <0.2 | — | 45 | 0.2 | 5 | 50 |
| 5 | 1 | <0.2 | — | 45 | 0.2 | 5 | 50 |
| 6 | 1 | <0.2 | — | 45 | 0.2 | 5 | 50 |
| 7 | 1 | <0.2 | — | 45 | 0.2 | 5 | 50 |
| 8 | 1 | <0.2 | — | 45 | 0.2 | 5 | 50 |
| 9 | 1 | <0.2 | — | 45 | 0.2 | 5 | 50 |
| 10 | 1 | <0.2 | — | 45 | 0.2 | 5 | 50 |
| 11 | 1 | <0.2 | — | 45 | 0.2 | 5 | 50 |
| 12 | 5 | 0.5–1 | 2.5 | 45 | 2 | 2.5 | 50 |
| 13 | 5 | 0.5–1 | 2.5 | 45 | 2 | 2.5 | 50 |
| 14 | 5 | 0.5–1 | 2.5 | 45 | 2 | 2.5 | 50 |
| 15 | 5 | 0.5–1 | 2.5 | 45 | 2 | 2.5 | 50 |
| 16 | 10 | 1–2 | — | 45 | 2 | 5 | 50 |
| 17 | 10 | 1–2 | — | 45 | 2 | 5 | 50 |
| 18 | 10 | 1–2 | — | 45 | 2 | 5 | 50 |
| 19 | 10 | 1–2 | — | 45 | 2 | 5 | 50 |
| 20 | 10 | 1–2 | — | 45 | 2 | 5 | 50 |
| 21 | 100 | 30–35 | — | 45 | 20 | 5 | 50 |
| 22 | 100 | 30–35 | — | 45 | 20 | 5 | 50 |
| 23 | 1000 | 300–350 | — | 45 | 200 | 5 | 50 |
| 24 | 100,000 | 3300–3350 | — | 45 | 20,000 (20 ng/μL) | 5 | 50 |
| 25 | 1 ng of plasmid DNA | Approx $10^8$ copies | — | 45 | 200 | 5 | 50 |

ing smaller stretches of the sequence of interest using primers A and B with different partners.

3. To avoid evaporation of the amplification reaction solution, it may be necessary to add a drop of light mineral oil (e.g., Sigma [St. Louis, MO] cat no. M 3516) to the tube. Wax beads (Perkin Elmer, Norwalk, CT) may also be used for this purpose. In some DNA thermal cycling machines, such as the GeneAmp PCR system 9600 (Perkin Elmer), condensation problems are avoided by additional heating of the tubes from above. There are also special PCR tubes (Sarstedt, Newton, NC) with inserts that prevent evaporation.

4. The cycling conditions outlined in this chapter were optimized for a Gene Machine (Innovonics, Melbourne, Australia) thermal cycler, which uses three waterbaths and a mechanical arm to move samples between baths. Different DNA thermal cyclers will probably require slight modification to the denaturation and annealing/extension times. In addition, the thickness of the tubes used for PCR could have an effect on the times used, because of differences in the insulating effects of the plastic. According to Perkin-Elmer (personal communication) the equivalent cycling conditions in the GeneAMP PCR system 9600 using Perkin-Elmer PCR tubes would be: 94°C, 4 min, 64°C, 7 min (initial denaturation cycle); 94°C, 15 s, 64° C, 7 min (approx 1 min/kb amplified) for 34 main cycles. DNA thermal cyclers that have a rapid change in temperature between steps may be better for the amplification of long sequences *(3)*.

5. The PCR sample can be separated from the mineral oil by adding a drop of chloroform to the tube. The aqueous and organic phases then invert and the former can be removed with a pipet. Alternatively, the PCR sample can be removed by placing a pipet tip under the mineral oil and then expelling the air through the tip. This action displaces oil from the end of the pipet tip. The PCR sample can then be placed in another tube after removal of any remaining oil on the tip by wiping it with a tissue.

6. Amplification sensitivity and accuracy of DNA concentration estimation can be determined in a single experiment if amplifying a single copy sequence (i.e., two copies/diploid cell). In 3 pg of human genomic DNA, on average, there would be one copy of the sequence of interest. Therefore, in nine samples containing 1 pg each of human genomic DNA on average, three would contain the sequence of interest and potentially give rise to PCR product. If PCR product is found in, on average, three samples out of nine, then the DNA estimate was probably correct and the procedure used is capable of amplification from a single template copy. If none of the samples shows amplification, the DNA concentration may be lower than estimated or the efficiency of amplification may be poor.

7. The volume $y$ μL used to adjust $MgCl_2$ concentration is determined by:

$$y = (x - 0.7)/\ 25 \times r \times 50 \qquad (1)$$

where:

   $x$ = optimal $MgCl_2$ concentration (m$M$);
   $r$ = the number of reaction mix samples (calculated to be approx 5% more than the number of samples being set up);
   0.7 = initial $MgCl_2$ concentration (m$M$) of the Tub DNA polymerase buffer;
   25 = the concentration of EDTA or $MgCl_2$ solution (m$M$) added to adjust the $MgCl_2$ concentration; and
   50 = volume of each reaction (μL).
For negative values of $y$, add 25 m$M$ EDTA.
For positive values of $y$, add 25 m$M$ $MgCl_2$.

## Acknowledgment

This project was funded by the Peter MacCallum Cancer Institute, Melbourne, Australia.

## References

1. Li, H., Gyllensten, U. B., Cui, X., Saiki, R. K., Erlich, H. A., and Arnheim, N. (1988) Amplification and analysis of DNA sequences in single human sperm and diploid cell. *Nature* **335,** 414–417.
2. Saiki, R. K., Gelfand, D. H., Stoffel, S., Scharf, S. J., Higuchi, R., Horn, G. T., Mullis, K. B., and Erlich, H. A. (1988) Primer-direct enzymatic amplification of DNA with a thermostable DNA polymerase. *Science* **239,** 487–491.
3. Barnes, W. M. (1994) PCR amplification of up to 35-kb DNA with high fidelity and high yield from λ bacteriophage templates. *Proc. Natl. Acad. Sci. USA* **91,** 2216–2220.
4. Cheng, S., Fockler, C., Barnes, W. M., and Higuchi, R. (1994) Effective amplification of long targets from cloned inserts and human genomic DNA. *Proc. Natl. Acad. Sci. USA* **91,** 5695–5699.
5. Lindahl, T., and Nyberg, B. (1972) Rate of depurination of native deoxyribonucleic acid. *Biochemistry* **11,** 3610–3618.
6. Kainz, P., Schmiedlechner, A., and Strack, H. B. (1992) In vitro amplification of DNA fragments >10 kb. *Anal. Biochem.* **202,** 46–49.
7. Forrester, H. B., Radford, I. R., and Deacon, N. J. (1994) Amplification of low copy number, large DNA sequences in human genomic DNA using Tub DNA polymerase. *BioTechniques* **17,** 20–22.
8. Ledbetter, S. A., Garcia-Heras, J., and Ledbetter, D. H. (1990) "PCR-Karyotype" of human chromosomes in somatic cell hybrids. *Genomics* **8,** 614–622.
9. Ohler, L. D., and Rose, E. A. (1992) Optimization of long-distance PCR using a transposon-based model system. *PCR Methods Applic.* **2,** 51–59.

# 4

## One-Step Optimization
## Using Touchdown and Stepdown PCR

### Kenneth H. Roux and Karl H. Hecker

### 1. Introduction

Polymerase chain reaction (PCR) optimization and troubleshooting can consume considerable energy and resources because of the finicky and often unpredictable nature of the reactions. Small variations in any of the many variables in a given reaction can have a pronounced effect on the resultant amplicon profile. Reactions that are too stringent yield negligible product and reactions that are not stringent enough yield artifactual amplicons. Variables include concentrations of $Mg^{2+}$, $H^+$, dNTPs, primers, and template, as well as cycling parameters. With regard to the latter, the value selected for the annealing temperatures is most critical. Unfortunately, even with the most sophisticated algorithms (i.e., OLIGO) it is often difficult to predict the amplification optima *a priori* leaving no other choice but to employ empirical determination.

Touchdown (TD) PCR *(1,2)* and its sister technique, stepdown (SD) PCR *(2)*, represent a markedly different approach that, in a single amplification regimen, inherently compensates for suboptimal reagent concentrations and less than perfect cycling parameters. Instead of guessing (or using imprecise calculations) to arrive at an appropriate temperature for the primer extension segment of the cycle, one can cast a wider net by using progressively lower annealing temperatures over consecutive cycles. The goal is to select a broad range of annealing temperatures that begins above the estimated $T_m$ and ends below it (*see* Note 1). Typically, one runs a TD PCR program at 2 cycles/°C declining over a 10–20°C range at 1°C intervals. In this way, the first primer-template hybridizations and primer extensions

From: *Methods in Molecular Biology, Vol. 67: PCR Cloning Protocols: From Molecular Cloning to Genetic Engineering* Edited by: B. A. White Humana Press Inc., Totowa, NJ

will be those with the highest specificity, i.e., presumably, the combination that gives the desired amplicon. Although the annealing temperature continues to drop in subsequent cycles to levels that normally would promote spurious amplification, the desired product, having already experienced several cycles of amplification, will be in a position to out-compete most lower $T_m$ (spurious) amplicons. If, for example, there is only a 3°C difference between the $T_m$ of the target amplicon and the $T_m$ of the first-primed spurious amplicon, the desired product will have undergone up to a 64–fold ($2^6$) amplification.

Our experience has been that TD PCR is applicable to a wide range of PCR situations *(2,3)*. At one extreme, TD PCR generally yields a single strong amplicon from genomic DNA even when the primer-template combinations are grossly mismatched (*see* Note 2). Mismatching might occur when attempting to amplify specific members from a complex multigene family, using nucleotide sequence information deduced from an amino acid sequence, or amplifying across species lines *(4–6)*. Primer-template base pair mismatches are permissible and can even be near (but probably not at) the 3' end of the primer *(3)*. TD PCR can also compensate for suboptimal buffer composition (e.g., $Mg^{2+}$ concentration) *(2)*. On the other hand, reactions that are already optimal, as assessed by conventional PCR, will usually yield equally strong amplicons even when using a broad temperature range TD PCR protocol in which the annealing temperature dips well below the $T_m$. Stated another way, TD PCR appears to greatly aid marginal reactions while not imposing significant penalties on already robust reactions. Hence, TD PCR can be used routinely in lieu of conventional PCR and need not be viewed solely as an optimization procedure *(7)*.

One potential drawback to TD PCR stems from the complexity of the programming (*see* Note 3). Because of the numerous *(10–20)* annealing temperatures used, a large segment of the programming capacity of conventional thermal cyclers can be encumbered. Also, attempts to adjust the annealing temperature range can involve considerable reprogramming (*see* Note 4). Some newer thermal cyclers circumvent these problems by permitting the programming of automatic incremental temperature changes in progressive cycles. We have recently tested modified versions of TD PCR, which we term SD PCR, that utilize simplified programming *(2)*. For SD PCR, one uses fewer but larger annealing temperature steps with proportionately more cycles per step. For example, a program might consist of three or four steps, at three to four cycles per step, with 3–5°C temperature differences between steps. SD PCR is not quite as universally applicable as TD PCR, but is adequate for many applications. The ease in programming may frequently be worth the trade-off.

## 2. Materials

### 2.1. Touchdown PCR With Mismatched Primer-Template Pairs

1. Template DNA (rabbit genomic liver 125 ng/μL).
2. Primers. Stock solutions are at 200 μg/mL in H₂O. The following primer pairs yield a 445-bp amplicon. The sites of mismatches are in capital letters. The sequence of the corresponding genomic homologous strand (shown in brackets) are for comparative purposes (*see* Note 5 for comments on degenerate primer design).
   a.  Primers: 5'cttgccaGtaatatAcgccctgcTaaCTtg3';
       5'ggatcttctgttgatgtctgactGttGgAg 3'
   b.  Homologous genomic sequences: [5'cttgccaAtaatatCcgccctgcCaaTCtg 3'];
       [5'ggatcttctgttgatgtctgactAttTgTg 3'];
3. 10X PCR buffer: 500 m*M* KCl, 100 m*M* Tris-HCl, pH 9.0, 1% Triton X-100, 2 m*M* of each dNTP (A, T, C, and G), and 15 m*M* MgCl₂ (*see* Note 6).
4. Ampli*Taq* DNA Polymerase (Perkin Elmer, Norwalk, CT).
5. Sterile mineral oil.
6. Standard wall 0.6-mL capped conical tubes.
7. Equipment and reagents for 1.5% agarose gel electrophoresis.

### 2.2. Stepdown PCR with a Mismatched Degenerate Primer

1. Template DNA (rabbit genomic liver 125 ng/μL).
2. Primers: Prepare stock solutions at 200 μg/mL. The following primer pair yields a 703-bp amplicon. The sites of mismatches are in capital letters. Degeneracies are separated by a slash and are in parenthesis. The sequence of the genomic homologous strand corresponding to the primer with degeneracies is presented (shown in brackets) for comparative purposes.
   a.  Primers: 5'agggatcgggtgaaaggggtctcagc3'; 5'ttAtgagcattcat(a/G)aacttctggagg 3'.
   b.  Homologous genomic sequence: 5'agggatcgggtgaaaggggtctcagc3'; [5'ttGtgagc-attcatAaacttctggagg 3'].
3. 10X PCR buffer. 500 m*M* KCl, 100 m*M* Tris-HCl, pH 9.0, 1% Triton X-100, 2 m*M* of each dNTP (A, T, C, and G), and 15 m*M* MgCl₂.
4. Ampli*Taq* DNA Polymerase.
5. Equipment and reagents for 1.5% agarose gel electrophoresis.

## 3. Methods

Programming of the thermal cycler for TD and SD PCR are described first, followed by specific PCR conditions used for both reactions.

### 3.1. TD PCR Programming

1. Set thermal cycler to denature for 1 min at 94°C, anneal for 2 min, and primer extend for 3 min at 74°C.
2. Follow the cycling program with a 7-min primer extension step and a 4°C soak step (*see* Note 7).

**Table 1**
**PCR Master Mix**

| Components | Stock concentration | Amount per 50 μL reaction, μL | Final concentration |
|---|---|---|---|
| dNTP mix | 2.0 m*M*, ea. | 5 | 0.2 m*M* |
| PCR buffer | 10X | 5 | 1X |
| Primer 1 | 200 ng/μL | 1 | 4.0 ng/μL |
| Primer 2 | 200 ng/μL | 1 | 4.0 ng/μL |
| Template | 125 ng/μL | 3 | 7.5 ng/μL |
| MgCl$_2$ | 25 m*M* | 3 | 1.5 m*M* |
| H$_2$O | — | 29.5 | — |
| | | Subtotal: 47.5 | |

3. Set the annealing stage for 2 cycles/°C beginning at 55°C and decreasing at 1°C increments to 41°C (i.e., 30 total cycles in 15 steps) to be followed by ten additional cycles at 40°C.

## 3.2. SD PCR Programming

1. Set thermal cycler to denature for 1 min at 94°C, anneal for 2 min, and primer extend for 3 min at 74°C as detailed below. Follow the cycling program by a 7-min primer extension step and a 4°C soak step (*see* Note 7). Program the annealing stage for six cycles per temperature step beginning at 70°C and decreasing at 3°C increments to 58°C (i.e., 30 total cycles in five steps) to be followed by ten additional cycles at 55°C (*see* Note 8 for programming considerations).
2. Analyze and reamplify as described below.

## 3.3. PCR Setup

1. Set up master mix for 50 μL reactions as indicated in Table 1.
2. Dispense master mix to 0.6-mL standard wall PCR tubes. Add 50 μL mineral oil to each tube and place in thermal cycler (Perkin Elmer DNA Thermal Cycler). Begin initial cycle and add 2.5 μL (1.25 U) of a 1:10 dilution of polymerase to each tube only after the temperature exceeds 80°C in the thermal cycler (i.e., the hot start protocol, *see* Note 9). Cap tubes and continue cycling.
3. Following amplification, monitor results by running 3 μL on a 1.5% agarose/ethidium bromide gel and view by UV illumination. If a product is not evident or the desired amplicon is of insufficient amount, amplify for an additional 5–10 cycles (with TD PCR, you are not sure exactly how many nonproductive and suboptimal cycles preceded the start of efficient amplification) at lowest annealing temperature or consider repeating the amplification with the TD PCR annealing temperature range shifted downward by 5°C.
4. If the desired product is still not evident, consider conventional or TD nested PCR on a 1:100 to 1:1000 dilution of the initial TD PCR reaction.

## 4. Notes

1. The $T_m$ for the primer-template combination can be roughly estimated using the formula:

$$T_m = 2(A + T) + 4(G + C) \qquad (1)$$

   For primer-template combinations with known or suspected mismatches, 5–20°C should be subtracted from the normal annealing temperature. Of course, more sophisticated programs, such as OLIGO Primer Analysis Software (National Biosciences, Inc., Plymouth, MN) *(10,11)*, may also be used to calculate the $T_m$.

2. An estimation of the $T_m$ is particularly difficult when using primers and templates containing mismatched base pairs. We have successfully amplified, to a single intensely staining band, primer-template pairs containing 3–5 mismatches with the template *(3)*. In all cases, we used a TD PCR program in which the annealing temperatures dropped from 55–41°C at 2 cycles/°C.

3. If using a thermal cycler that has a programmable automatic temperature variation feature, set the annealing stages to decline by 0.5°C/cycle. For a standard thermal cycler, program 2 cycles/°C drop (for example, cycles 1 and 2, 65°C; cycles 3 and 4, 64°C, and so forth). In both instances, the TD portion of the program should be followed by ten cycles at a fixed annealing temperature about 10°C below the estimated $T_m$. One should bear in mind that for situations in which the template is not fully complementary to the primers, once amplification commences, the amplicon will be fully complementary to the primers and thus will have a greater $T_m$ than initially estimated. On the other hand, we have noted that final stage amplification at 10°C or more below the estimated $T_m$ (rather than the 4–5°C usually recommended for standard PCR) can significantly increase the yield of otherwise marginal reactions *(2)*. Avoid the temptation of adding too many cycles to this terminal fixed annealing temperature stage of the program. Excessive cycling can degrade the product and lead to spurious banding and high-mol-wt smearing *(14)*.

4. A convenient way to adjust the TD temperature range segment of a conventional thermal cycler (i.e., having linked or sequential file programming) is to program files covering a wide range of annealing temperatures (20–25°C). The specific subset of files to be used in any given amplification protocol can be bypassed by simply linking the initial 5 min denaturation file to the file having the highest annealing temperature to be used and linking the file containing the lowest annealing temperature in the TD range to the terminal primer extension file. Be sure to keep note of these changes because they must be undone before the next alteration in the range. If this approach will tie up too many files, you may wish to reprogram those files containing the segments to be deleted from the range (say 5°C from the bottom) and using the freed file capacity to add new files to the top (beginning) of the program. Again, the initial denaturation stage and terminal primer extension steps must be linked to the beginning and end of the new range, respectively. Thermal cyclers in which individual files cannot be linked but which rely on a single long multistep program are even less versatile and may require

complete reprogramming. Fortunately, TD and SD PCR are very forgiving and a single temperature range can be applied to a wide variety of situations.

5. Design the best primer set based on the information available. When designing primers to amplify genes of uncertain complementarity, try to cluster the sequence of greatest certainty near the 3' ends of the primers. Degeneracy derived from multiple nucleotides or inosine residues at positions of uncertainty are permissible *(4,5,8)* but not necessary. Note that some polymerases other than *Taq* cannot prime from inosine-containing primers *(9)*.

6. Because of the minimal effort involved, it is generally advantageous to vary one of the buffer components (usually $Mg^{2+}$) during the initial optimization.

7. Most time and temperature characteristics of the cycling program (denaturation, primer extension) will be the same as conventional PCR for your system if using primer-template combinations other than the examples described.

8. The temperature range of SD PCR may be divided into 3–5 more or less equal increments (steps) and be programmed accordingly. If you can afford to tie up the programming capacity of the thermal cycler, more steps are better than fewer. Proportion the total number of cycles to be used in the SD segment of the program equally among the steps. Add ten cycles at a fixed temperature well below the $T_m$ as described above. When background problems are expected to be minimal, a simple two step SD PCR protocol can still be advantageous. Here, the initial stage has perhaps six to ten cycles at a fixed temperature 5°C above that which would normally be used in standard PCR (i.e., slightly above the calculated $T_m$). Even though a full 30 cycles at this elevated temperature would not be expected to yield a detectable amplicon, we have found *(2)* that sufficient, highly specific amplification is occurring to allow the desired amplicon to dominate the amplification throughout the remainder of the standard temperature cycles, thus reducing the possibility that unwanted amplicons will be generated.

9. Because TD PCR is based on the use of high temperature to prevent spurious priming, it is imperative that hot start procedures be followed *(12,13)*. If multiple samples are to be run, add an extended denaturation step to the beginning of the program or use the hold option.

## References

1. Don, R. H., Cox, P. T., Wainwright, B. J., Baker, K., and Mattick, J. S. (1991) 'Touchdown' PCR to circumvent spurious priming during gene amplification. *Nucleic Acids Res.* **19**, 4008.

2. Hecker, K. H. and Roux, K. H. (1996) High and low annealing temperatures increase both specificity and yield in TD and SD PCR. *BioTechniques* **20**, 478–485.

3. Roux, K. H. (1994) Using mismatched primer-template pairs in TD PCR. *BioTechniques* **16**, 812–814.

4. Knoth, K., Roberds, S. Poteet, C., and Tamkun, M. (1988) Highly degenerate inosine-containing primers specifically amplify rare cDNA using the polymerase chain reaction. *Nucleic Acids Res.* **16**, 10932.

5. Patil, R. V. and Dekker, E. E. (1990) PCR amplification of an *Escherichia coli* gene using mixed primers containing deoxyinosine at ambiguous positions in degenerate amino acid codons. *Nucleic Acids Res.* **18,** 3080.
6. Batzer, M. A., Carlton J. E., and Deininger, P. L. (1991) Enhanced evolutionary PCR using oligonucleotides with inosine at the 3'-terminus. *Nucleic Acids Res.* **19,** 5081.
7. Roux, K. H. (1995) Optimization and troubleshooting in PCR. *PCR Meth. Applic.* **4,** S185–S194.
8. Peterson, M. G., Inostroza, J., Maxon, M. E., Flores, O., Adomon, A., Reinberg, D., and. Tjian, R. (1991) Structure and functional properties of human general transcription factor IIE. *Nature* **354,** 369–373.
9. Knittel, T. and Picard, D. (1993) PCR with degenerate primers containing deoxyinosine fails with *Pfu* DNA polymerase. *PCR Meth. Applic.* **2,** 346,347.
10. Rychlik, W. and Spencer, W. J. (1989) A computer program for choosing optimal oligonucleotides for filter hybridization, sequencing and in vitro amplification of DNA. *Nucleic Acids Res.* **17,** 8543–8551.
11. Rychlik, W. (1994) New algorithm for determining primer efficiency in PCR and sequencing. *J. NIH Res.* **6,** 78.
12. D'Aquila, R. T., Bechtel, L. J., Viteler, J. A., Eron, J. J., Gorczyca, P., and Kaplin, J. C. (1991) Maximizing sensitivity and specificity of PCR by preamplification heating. *Nucleic Acids Res.* **19,** 3749.
13. Erlich, H. A., Gelfand, D., and Sninsky, J. J. (1991) Recent advances in the polymerase chain reaction. *Science* **252,** 1643–1651.
14. Bell, D. A. and DeMarini, D. (1991) Excessive cycling converts PCR products to random-length higher molecular weight fragments. *Nucleic Acids Res.* **19,** 5079.

# 5

# GC-Rich Template Amplification by Inverse PCR

*DNA Polymerase and Solvent Effects*

**Alain Moreau, Colette Duez, and Jean Dusart**

## 1. Introduction

The amplification of GC-rich templates by any PCR method is usually a difficult task and despite the development of modified methods and conditions, this type of amplification still remains a specific case approach. Problems usually observed with GC-rich DNA are constraint of template amplification by stable secondary structures that stall or reduce the DNA polymerase progress, and the presence of secondary annealing sites giving rise to nonspecific amplified bands. This latter point is not exclusive to GC-rich templates but is frequently encountered in other types of templates. In order to design a more general method for GC-rich templates, different DNA polymerases were compared in combination with different organic solvents with the purpose of abolishing stable secondary structures *(1)*. Our attention focused on the inverse polymerase chain reaction (iPCR) used to perform site-directed mutagenesis *(1,2)*. This very attractive method requires a single pair of primers and involves the amplification of the whole recombinant plasmid, a difficult step with high GC-content DNA. Inverse PCR also proves useful in cloning missing parts of genes by using a self-ligated genomic DNA fragment as template.

A recent survey of the literature showed the absence of comparative studies regarding the use of different DNA polymerases in the amplification of GC-rich DNA with or without the addition of organic solvents, such as dimethylsulfoxide (DMSO) *(3,4)*, formamide *(3,5,6)*, and tetramethylammonium chloride (TEMAC) *(7)*. Furthermore, little is known of the exact role of these chemicals in PCR. It was suggested that these compounds primarily affect the annealing kinetics as well as the efficiency of the DNA polymerase used. In

From: *Methods in Molecular Biology, Vol. 67: PCR Cloning Protocols: From Molecular Cloning to Genetic Engineering* Edited by: B. A. White Humana Press Inc., Totowa, NJ

order to identify critical parameters involved in iPCR with GC-rich templates, we analyzed the influence of DNA polymerases in combination with the afore-mentioned solvents *(1)*. The results obtained allowed us to improve iPCR for difficult template amplifications by either iPCR or standard PCR. Our iPCR method can be divided into two steps: amplification of the whole recombinant plasmid and iPCR product purification and ligation. This method was used to perform site-directed mutagenesis by amplification of a 4.8-kb plasmid derived from pUC18 and containing a 1980-bp insert, the gene encoding the extracel-lular DD-carboxypeptidase from *Actinomadura* R39, a 74% GC-content DNA *(14)*. Different DNA polymerases were tested according to the manufacturer's specifications. However, correct amplification was not detected with any DNA polymerase tested *(1)*. The efficiency of amplification by addition of DMSO, formamide, or TEMAC in the reaction mixture was evaluated according to the conditions described above. Analysis of PCR products in presence of these organic solvents revealed that only Vent™ DNA polymerase (New England Biolabs, Beverly, MA) amplified the 4.8-kb plasmid *(1)*. We then focused our attention on the conditions to amplify GC-rich DNA templates with Vent DNA polymerase. In addition, the $Mg^{2+}$ concentration was increased to obtain a 10-m$M$ final concentration. This allowed us to amplify large DNA fragments in the PCR assay.

## 2. Materials

1. pBlueScript™ vector (Stratagene, La Jolla, CA) or any other pUC derivative plas-mid used to clone the gene to be mutated (*see* Note 1).
2. 25–50 ng of DNA template (*see* Note 2).
3. 2 μ$M$ of each oligonucleotide, primers A and B corresponding respectively to the sense (coding) and antisense (noncoding) message (*see* Note 3).
4. 10X DNA polymerase buffer (provided by the manufacturer of the Vent DNA polymerase (*see* Note 4) and Vent DNA polymerase (*see* Note 5).
5. dNTPs: Mix 10 m$M$ each (Pharmacia LKB, Piscataway, NJ).
6. 100 m$M$ $MgSO_4$.
7. Fresh, deionized formamide (Sigma, St. Louis, MO).
8. Sterile water.
9. Light mineral oil (Sigma).
10. Reagents for agarose gel electrophoresis (Life Technologies [Gibco-BRL], Gaithersburg, MD).
11. Sephadex G-50 fine (Pharmacia LKB), siliconized wool and 1-mL syringe.
12. T4 polynucleotide kinase (PNK) (New England Biolabs).
13. Reagents for ligation:10X ligation buffer (Boehringer Mannheim, Indianapolis, IN); T4 DNA ligase (Boehringer Mannheim, Mannheim, Germany).
14. 400-μL and 1.5-mL sterile Eppendorf tube.
15. Thermocycler.

# 3. Methods

## 3.1. Inverse PCR

1. Perform iPCR by adding in a 400-µL sterile Eppendorf tube the following reagents (*see* Note 6): 5 µL DNA template (25–50 ng); 10 µL 10X Vent DNA polymerase buffer; 8 µL MgSO$_4$ (100 m$M$); 2 µL dNTPs (10 m$M$ each); 4 µL primer A (50 pmol/µL); 4 µL primer B (50 pmol/µL); 10 µL Formamide; 56 µL sterile H$_2$O; and 1 µL Vent DNA polymerase (2 U/µL); to a total volume of 100 µL.
2. Overlay the sample with 50 µL of light mineral oil.
3. Submit the samples to a standard three-step cycling protocol according to the following parameters (*see* Note 7): 95°C for 1 min (initial denaturation) (1 cycle); 94°C for 30 s (denaturation); XX°C, 1 min (annealing); 72°C for Y min (extension) 30 cycles; 72°C for 10 min (final extension) (1 cycle).

## 3.2. iPCR Product Purification and Ligation

1. Separate the iPCR reaction from the light mineral oil by simply pipeting only the reaction volume from the bottom of the tube into a new 1.5-mL sterile Eppendorf tube.
2. Take 10–20 µL aliquot of the iPCR reaction to visualize the product by agarose gel electrophoresis (0.7% agarose, *see* Note 8).
3. Purify the iPCR product by passing the remaining iPCR reaction through a Sephadex G-50 spun column of 1-mL and elute with 100 µL of TE buffer or H$_2$O.
4. Perform the phosphorylation and ligation reaction by adding the following reagents in a 1.5-mL sterile Eppendorf tube: 10 µL aliquot of purified iPCR reaction; 2 µL 10X ligation buffer; 0.5 µL T4 polynucleotide kinase (10 U/µL); and 7.5 µL H$_2$O sterile; to a total volume of 20 µL.
5. Incubate the mixture 15 min at 37°C, then again add 0.5 µL of T4 polynucleotide kinase and incubate for another 15 min at 37°C.
6. Add 1 µL of T4 DNA ligase (1U/µL) to the reaction mixture and incubate overnight at 4°C, followed by 16°C for 3 h.
7. Transform competent *Escherichia coli* cells with 5-µL aliquot of the ligation mixture.

# 4. Notes

1. The targeted DNA is initially cloned in a vector, in general, a pUC-derived plasmid. Since the difficulty of amplification increases with plasmid length, it is better to avoid unnecessarily large plasmids and clone only a part of the gene flanking the targeted DNA. Once mutated by iPCR, this part will then be used to reconstruct the whole gene. The latter step reduces the subsequent sequencing necessary to check the integrity of the mutated DNA fragment.
2. The template concentration used to perform iPCR is about 25–50 ng. Higher concentrations will increase the background because of the wild-type plasmid that easily transforms *E. coli*. Lower template concentration reduces the final amount of amplified material, thus requiring more PCR cycles. These additional

cycles contribute to the introduction of more errors by DNA polymerases. Under our conditions, 30 cycles are sufficient to amplify GC-rich templates.

3. The primer design is a crucial step. In iPCR, the primers to be used are oriented in inverted tail-to-tail direction, i.e., one primer corresponding to the coding sense (5'-3') whereas the other is antisense (3'-5'). Usually, one primer harbors the mutation, which can be a substitution, deletion, or insertion of one or more nucleotides. The selection of this pair of primers is sometimes difficult, but can be simplified by using one of several new computer programs for oligonucleotide selection *(12,13)*. However, this initial step is frequently overlooked, resulting in great difficulties in template amplification (not just GC-rich ones). There are three basic rules to follow to avoid primer design problems:
   a. Elimination of duplex formation at the 3' ends with either or both primers, plus elimination of hairpin structure formation within primers.
   b. Design of primers with $T_m$ (°C) close to each other, i.e., less than a 10°C difference. The addition of any organic solvent to the PCR reaction will decrease the $T_m$ for a specific primer; and its partner primer bearing the mutation will also show reduced $T_m$ depending on the introduced mutation.
   c. Location of the chosen mutation in the middle of the primer in order to maintain the internal stability of the oligonucleotide.

4. The buffer supplied by the manufacturer is 1X: 20 m$M$ Tris-HCl, pH 8.8 (at 25°C), 10 m$M$ KCl, 10 m$M$ (NH$_4$)$_2$SO$_4$, 2 m$M$ MgSO$_4$, and 0.1% Triton X-100.

5. The choice of DNA polymerase is a key point for the amplification of the whole recombinant plasmid. Among several DNA polymerases tested, our choices were Vent DNA polymerase *(8–10)* and *Pfu* DNA polymerase (Stratagene) *(11)*. Indeed, these two enzymes produce almost exclusively blunt ends, whereas *Taq* DNA polymerase requires additional manipulations to obtain blunt ends. Furthermore, Vent and *Pfu* DNA polymerases are more accurate during iPCR than *Taq* DNA polymerase. We recommend testing the first iPCR reactions without the addition of organic solvents. It is difficult to choose the solvent and its optimal concentration without empirical assay. Furthermore, it has been noted that most DNA polymerases are sensitive to organic solvents, especially formamide. However, the use of 10% formamide is suggested in combination with Vent DNA polymerase *(1)*. This DNA polymerase proved to be the most robust enzyme tested in the presence of formamide, because other DNA polymerases (*Taq* and *Pfu*) could not amplify DNA under similar conditions (Moreau, A., unpublished observations). Another possibility is the use of *Pfu* DNA polymerase without any solvent addition. In some cases, *Pfu* DNA polymerase gives rise to good amplification with GC-rich templates but does not tolerate formamide concentrations greater than 2.5%.

6. We observed that the Mg$^{2+}$ concentration is very important in obtaining proper amplification with the Vent DNA polymerase, especially with large plasmids (5-kb). In the absence of amplification products, we recommend the modification of Mg$^{2+}$ concentration by supplemental addition of increasing amounts of Mg$^{2+}$ in the iPCR reaction. The optimum Mg$^{2+}$ concentration usually occurs in a narrow range with the Vent DNA polymerase. Among a choice of several organic

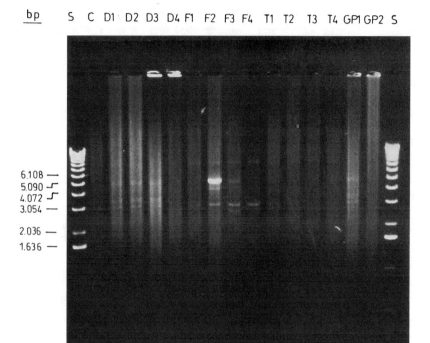

**Fig. 1.** Solvent effects on the Amplification by iPCR with Vent DNA polymerase of a 4.8-kb plasmid by iPCR, containing a GC-rich (74%) 1980-bp insert. Samples covered with mineral oil (50 µL) were submitted to 30 amplification cycles on a Biometra Trio-Thermoblock™: 1 min denaturation at 95°C, 1 min annealing at 55°C, and 5 min extension at 72°C, followed by a final 10 min extension at 72°C. A 10-µL aliquot of each sample was loaded onto a 0.7% agarose gel. Lanes: S, 1-kb ladder size standard; C, control iPCR under standard conditions without organic solvent; D1, D2, D3, and D4, iPCR in presence of 8, 10, 15, and 20% DMSO, respectively; F1, F2, F3, and F4, iPCR in presence of 5, 10, 15, and 20% formamide, respectively; T1, T2, T3 and T4, iPCR in presence of $10^{-2}$, $10^{-3}$, $10^{-4}$, and $10^{-5}M$ TEMAC, respectively; $G_1$, iPCR in presence of 8% DMSO with 3 µg of T4 gene 32 protein; $G_2$, iPCR under standard conditions with 3 µg of T4 gene 32 protein (reproduced with permission from ref. *1*).

solvents, 10% formamide was the most useful addition to correctly amplify the 4.8-kb plasmid (Fig. 1). Higher formamide concentrations appeared detrimental for the iPCR. It was also observed that addition of <8% DMSO failed to amplify the plasmid but higher concentrations did not significantly improve the iPCR, even with addition of T4 gene 32 protein, a single-stranded binding protein used to overcome secondary structures *(15,16)*. The addition of TEMAC at concentrations from $10^{-2}$–$10^{-5}M$ was not effective (Fig. 1). The iPCR assay was repeated with different pairs of primers, in different locations and each time, the 4.8-kb plasmid was successfully amplified according to the above described conditions with 10% formamide and Vent DNA polymerase.

7. We choose to perform iPCR with the *Actinomadura* R39 DD-carboxypeptidase (74% GC-rich) with an annealing temperature of 55°C/1 min, and 5 min extension time at 72°C. This iPCR has been performed with the following two primers: 5'-GCCCTCGGCGGCGGTACCGCAT-3' ($T_m$=70°C), which anneals perfectly with the target sequence and 5'-GTGGTCGAGGCCCA<u>C</u>A<u>C</u>CGGGACGATG-3'($T_m$=61.6°C) introducing two noncontiguous mismatches in the sequence (since these substitutions are not contiguous, we have considered them as three mismatches). The annealing temperature has also been reduced, based on the guideline of $T_m$ decrease of both primers by about 0.6°C/% of formamide added to the iPCR reaction. We used the extension time of 1 min per-kb of plasmid for the polymerization at 72°C, and a final 10 min polymerization at 72°C. The guideline of 1 min/kb for the extension at 72°C is suitable when the plasmid size is 5-kb or less, whereas with larger plasmid size, the extension time should be increased on an empirical basis. The initial denaturation is an important step, but prolonging it may be detrimental for longer templates.

8. An aliquot of the iPCR reaction is visualized by agarose gel electrophoresis. The presence of minor bands of smaller size than the wild-type plasmid is not a real problem. Most of them will generate partial plasmids missing key regions of replication or antibiotic resistance and will be easily eliminated after transformation. However, several methods are available to purify PCR products. We use a Sephadex G-50 spun-column to remove unincorporated dNTPs. A purified aliquot is then phosphorylated and ligated. It is not unusual to find frameshifts +1 or −1 at the junction of each primer after isolation of plasmid DNA from transformants obtained through iPCR product ligation. However, we observed that a majority of transformants are the result of correct ligation.

## References

1. Moreau, A., Duez, C., and Dusart, J. (1994) Improvement of GC-rich template amplification by inverse PCR. *BioTechniques* **17,** 233,234.
2. Ochman, H., Gerber, A. S., and Hartl, D. L. (1988) Genetic applications of an inverse polymerase chain reaction. *Genetics* **120,** 621–623.
3. Bookstein, R., Lai, C.-C., To, H., and Lee, W. H. (1990) PCR-based detection of a polymorphic BamHI site in intron I of the human retinoblastoma (RB) gene *Nucleic Acids Res.* **18,** 1666.
4. Winship, P. R. (1989) An improved method for directly sequencing PCR amplified material using dimethyl sulfoxide. *Nucleic Acids Res.* **17,** 1266.
5. Sarkar, G., Kapelner, S., and Sommer, S. S. (1990) Formamide can dramatically improve the specificity of PCR. *Nucleic Acids Res.* **18,** 7465.
6. Schuchard, M., Sarkar, G., Ruesink, T., and Spelsberg, T. C. (1993) Two-step "hot" PCR amplification of GC-rich avian c-myc sequences. *BioTechniques* **14**, 390–394.

7. Hung, T., Mak, K., and Fong, K. (1990) A specificity enhancer for polymerase chain reaction. *Nucleic Acids Res.* **18,** 4953.
8. Eckert, K. A. and Kunkel, T. A. (1991) DNA polymerase fidelity and the DNA polymerase chain reaction. *PCR Meth. Applic.* **1,** 17–24.
9. Kong, H. M., Kucera, R. B., and Jack W. E. (1993) Characterization of a DNA from the hyperthermophile *Archaea thermococcus litoralis. J. Biol. Chem.* **268,** 1965–1975.
10. Mattila, P., Korpela, J., Tenkanen, T., and Pitkänen, K. (1991) Fidelity of DNA synthesis by the *Thermococcus litoralis* DNA polymerase–an extremely heat stable enzyme with proofreading activity. *Nucleic Acids Res.* **19,** 4967–4993.
11. Lundberg, K. S., Shoemaker, D. D., Adams, M. W. W., Short, J. M., Sorge, J. A., and Mathur, E. J. (1991) High fidelity amplification using thermostable DNA polymerase isolated from *Pyrococcus furiosus. Gene* **108,** 1–6.
12. Breslauer, K. J., Frank, R., Blöcker, H., and Marky, L. A. (1986) Predicting DNA duplex stability from the base sequence. *Proc. Natl. Acad. Sci. USA* **83,** 3746–3750.
13. Freier, S. M., Klerzek, R., Jaeger, J. A., Sugimoto, N., Caruthers, M. H., Neilson, T., and Turner, D. H. (1986) Improved free-energy parameters for predictions of RNA duplex stability. *Proc. Natl. Acad. Sci. USA* **83,** 9373–9377.
14. Granier, B., Duez, C., Lepage, S., Englebert, S., Dusart, J., Dideberg, O., Van Beeumen, J., Frère, J.-M., and Ghuysen, J.-M. (1992) Primary and predicted secondary structures of the *Actinomadura* R39 extracellular DD-peptidase, a penicillin-binding protein (PBP) related to the *Escherichia coli* PBP4. *Biochem. J.* **282,** 781–788.
15. Alberts, B. and Sternglanz, R. (1977) Recent excitement in the DNA replication problem. *Nature* **269,** 655–661.
16. Bittner, M., Burke, R. L., and Alberts, B. M. (1979) Purification of the T4 gene 32 protein free from detectable deoxyribonuclease activities. *J. Biol. Chem.* **254,** 9565–9572.

# 6

## Coupled One-Step Reverse Transcription and Polymerase Chain Reaction Procedure for Cloning Large cDNA Fragments

**Jyrki T. Aatsinki**

## 1. Introduction

Although *Thermus aquaticus (Taq)* and *Thermus thermophilus (Tth)* DNA polymerases have the ability to reverse transcribe RNA to complementary DNA (cDNA) and, subsequently, amplify the target cDNA, they are not usually the first choices for reverse transcription-polymerase chain reactions (RT-PCR) *(1–4)*. Since they only synthesize short cDNA fragments, their use is not widespread. In general, avian myeloblastosis virus (AMV), or moloney murine leukemia virus (M-MLV) reverse transcriptases (RTs) are used to reverse transcribe RNA to cDNA templates for PCR. Previous coupled methods are also unable to amplify large cDNA fragments and, thus, they are suitable only for the detection of gene expression *(5–8)*. The one-step RT-PCR procedure presented here was developed to amplify large cDNA fragments suitable for cloning full-length open reading frames (ORFs) encoding rat LH/CG receptor isoforms *(9,10)*.

The one-step RT-PCR procedure was first optimized for its specificity. Low concentrations of dNTP (0.2 m$M$ of each), MgCl$_2$ (1.5 m$M$), primer (10 pmol), and a relatively high annealing temperature (55°C) were used, since these conditions have been found to enhance specific amplification. The commercially available PCR buffer (10 m$M$ Tris-HCl, pH 8.8, at 25°C, 50 m$M$ KCl, and 0.1% Triton X-100) was found to be suitable for primer extension by AMV-RT, although it differed in its constituents from the recommendations of the manufacturer. To assure that primer extension was completed, long extension times were used, both in reverse transcription and PCR (60 min and 10 min +

From: *Methods in Molecular Biology, Vol. 67: PCR Cloning Protocols: From Molecular Cloning to Genetic Engineering* Edited by: B. A. White  Humana Press Inc., Totowa, NJ

59 s/cycle, respectively). Possible aggregates and secondary structures were eliminated by denaturing both primers and RNA at 65°C, for 15 min, before starting the amplification. Subsequent to a 1 h incubation at 42°C, the temperature was raised to 95°C for 5 min to dissociate RNA-cDNA hybrids. Finally, RT-PCR products could be easily cloned for in vitro translation studies and for transfections in different cell lines, since suitable restriction enzyme sites were synthesized at both ends. The construction of clones, including library screening and restriction mapping, by conventional methods is very laborious and difficult. Examples of other applications of the present coupled RT-PCR procedure, in addition to cDNA cloning and the detection of gene expression, include the detection of viral RNA and the quantitation of mRNA.

## 2. Materials

*See* Note 1.

1. Total RNA isolated by the guanidinium isothiocyanate/cesium chloride method (*11; see* Note 2).
2. Oligonucleotide primers, typically 32–35 nucleotides long and with a 40–60% G + C content, are designed to have internal unique restriction sites and an additional 8–9 nucleotide complementary sequence at the 5' end of the restriction sites. These added nucleotides are important for helping restriction enzymes to cleave the RT-PCR product. When creating a new restriction site, select a sequence that requires as few changes as possible. Primer-dimer formation during PCR is best avoided by using primers having noncomplementary 3' ends.
3. Ribonuclease inhibitor Inhibit-ACE (5 prime-3 prime, Boulder, CO).
4. 10X PCR buffer: 500 m*M* KCl, 100 m*M* Tris-HCl, pH 8.8, at 25°C, 15 m*M* MgCl$_2$, 1.0% Triton X-100 (Finnzymes, Espoo, Finland).
5. Deoxynucleoside triphosphates (dNTPs) (Pharmacia, Uppsala, Sweden).
6. AMV-RT (Promega, Madison, MI).
7. Dynazyme DNA polymerase (Finnzymes).
8. Mineral oil (Sigma, St. Louis, MO).
9. DNA thermal cycler (Perkin-Elmer/Cetus, Norwalk, CT).
10. Microcentrifuge.
11. Wizard PCR product purification column (Promega).
12. Restriction enzymes (e.g., *Bam*HI and *Eco*RI) (Pharmacia).
13. Plasmid DNA, pUCBM20 (Boehringer Mannheim GmbH, Mannheim, Germany).
14. *Escherichia coli* JM109 strain.
15. T4 DNA ligase (high concentration; 5 U/µL) (Boehringer Mannheim GmbH).
16. Alkaline phosphatase (Promega).
17. Reagents and supplies for agarose gel electrophoresis: *See* preparation of mixtures and use of equipment in the laboratory manual (*11*).
18. Reagents for DNA precipitation: 4*M* NaCl, absolute ethanol, 70% ethanol.
19. Reagents and supplies for molecular cloning: 1*M* CaCl$_2$, isopropyl-β-D-thiogalactoside (IPTG), 5-bromo-4-chloro-3-indolyl-β-D-galactopyranoside (X-gal),

Luria-Bertani medium, Luria-Bertani medium plates, bacterial growth incubator. *See* preparation of mixtures in the laboratory manual *(11)*.

## 3. Methods

### 3.1. Coupled One-Step RT-PCR

1. Dilute primers to a concentration of 20 ng/μL in sterile distilled $H_2O$. Add 5 μL of each primer to a reaction tube containing a volume of sterile distilled $H_2O$ sufficient to bring the total reaction volume to 100 μL after the addition of the reactants from steps 3 and 4. Add 1 U of Inhibit-ACE and incubate for 20–30 min at room temperature.

2. Prepare stock mixtures of 10X PCR buffer and 10 m$M$ dNTPs (2.5 m$M$ of each dNTP) and incubate for 20–30 min at room temperature after adding 1.5 U of Inhibit-ACE/100 μL of stock mixture.

3. Add total RNA (0.1–50 μg) to the reaction tube (step 1), denature at 65°C for 15 min and cool to 4°C using a programmed DNA thermal cycler (*see* Note 3).

4. Add 10 μL of 10X PCR buffer and 8 μL of 10 m$M$ dNTPs from step 2 to the reaction tube. Add 8 U of AMV-RT and 2.5 U of DNA polymerase, mix carefully, and collect by brief centrifugation.

5. Overlay with mineral oil and incubate at 42°C for 1 h to allow reverse transcription.

6. Step 5 is linked to PCR cycles. Initial denaturation at 95°C for 5 min, followed by 30 cycles consisting of denaturation at 95°C for 1 min, primer annealing at 55°C for 2 min, and extension at 72°C for 10 min + 59 s/cycle.

7. Take 5–20 μL of the RT-PCR product, add gel loading buffer and size fractionate on a 1% ethidium bromide-stained agarose gel. Gel electrophoresis of the RT-PCR products are shown in Fig. 1 (*see* Notes 4–8).

### 3.2. Cloning of RT-PCR Product

1. Purify the RT-PCR product using a Wizard PCR product purification column following the manufacturer's instructions.

2. Digest the purified RT-PCR product with restriction enzymes using a several-fold excess of enzyme and long incubation times. Usually, 50 U of restriction enzyme can be added at the beginning of digestion and a further 20 U during the incubation, which can be done overnight (*see* Note 9).

3. Purify the sample as in step 1.

4. Precipitate the sample by adding 1/20 vol of 4$M$ NaCl and 2 vol of cold absolute ethanol, allow to stand overnight at −20°C, or for 30 min at −80°C. Centrifuge at 10,000–12,000$g$ for 20 min at +4°C and discard the supernatant. Add 1 mL of 70% ethanol to the pellet and centrifuge as before. Dissolve the dried pellet in 20 μL of sterile distilled $H_2O$.

5. Prepare plasmid DNA by digesting with suitable restriction enzymes for 2–3 h using 10 U/1 μg of DNA. Dephosphorylate the DNA according to the manufacturer's instructions if using a single restriction enzyme. Purify the sample as described in steps 1 and 4.

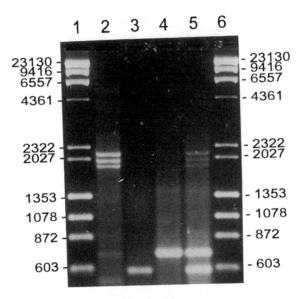

**Fig. 1.** Coupled one-step RT-PCR amplification of three different cDNA species from rat testis total RNA. Lane 2 shows LH/CG receptor isoforms amplified using CAATTTTG<u>G</u>AAT<u>T</u>CTAGTGAGTTAACGCTCTCG as the reverse primer, GGGAGCTC<u>GA</u>AT<u>T</u>CAGGCTGGCGGGCCATGGGGCGG as the forward primer (mismatched nucleotides, to create internal restriction sites, are underlined), and 10 μg of total RNA as the template. Lane 3 shows the carbonic anhydrase II RT-PCR product amplified using GAGCACTATCCAGGTCACACATTCCAG as the reverse primer, ACTGGCACAAGGAGTTCCCCATTGCCA as a forward primer, and 1 μg of total RNA as the template. Lane 4 shows the β-actin RT-PCR product amplified using GATGCCACAG<u>A</u>ATTCCATACCCAGGAAGGAAGGC as the reverse primer, GCCGCCCTAGG<u>AT</u>CCAGGGTGTGATGGTGGGTAT as the forward primer (mismatched nucleotides, to create internal restriction sites, are underlined), and 0.1 μg of total RNA as the template. Lane 5 shows the simultaneous amplification of three different cDNA species using the above mentioned primers and 10 μg of total RNA as the template. Other conditions for RT-PCR are as mentioned in Section 3., except for the simultaneous amplification of the three different cDNA species, where 20 ng of both β-actin primers and 100 ng of each of the LH/CG receptor and carbonic anhydrase II primers were used as amplimers. RT-PCR products (18 μL of mixture for lanes 2–4 and 36 μL of mixture for lane 5) were size-fractionated on a 1% ethidium bromide-stained agarose gel. Lanes 1 and 6 show the molecular weight standard (λ DNA/*Hin*dIII digest and φX174 DNA/*Hae*III digest).

6. Set up three ligation mixtures using 3:1, 1:1, and 1:3 molar ratios of insert and plasmid DNA. Add sterile distilled $H_2O$ to final volume of 8 μL, heat at 45°C for 5 min, and cool to 16°C. Add 1 μL of 10X ligation buffer, 1 μL of high concentration T4 DNA ligase (5 U/μL), and incubate overnight at 16°C.

7. Transform competent cells using standard laboratory protocols *(11)*. Pick up positive clones and analyze them by restriction digestion and agarose gel electrophoresis. Determine the nucleotide sequences of some clones by the double-stranded dideoxy sequencing method *(12)*.

## 4. Notes

1. In Section 2., the commercial sources of the different products are mentioned solely because they have been tested and used in this one-step RT-PCR procedure.
2. The coupled one-step RT-PCR procedure presented here uses total RNA isolated by the guanidinium isothiocyanate/cesium chloride method *(11)* as a template. Commercial total RNA purification kits can also be used for obtaining template RNA.
3. It is always necessary to test the amount of RNA for optimal amplification. If the total RNA contains high amounts of the target mRNA, efficient amplification is obtained with nanogram amounts of RNA. On the other hand, if the total RNA contains only very small amounts of the target mRNA, up to 50 μg of the total RNA can be used to obtain an efficient amplification.
4. If no specific bands are visible on the ethidium bromide-stained agarose gel, use a gradually increasing amount of reverse transcriptase. Do not use excess amounts of *Taq* DNA polymerase since this has been reported to lower the amount of specific RT-PCR product *(10)*.
5. If no specific bands are visible on the ethidium bromide stained agarose gel after optimization, prepare a Southern blot *(13)* of the gel. Hybridize with an appropriate probe that does not contain overlapping sequences with the primers used for the RT-PCR.
6. If, after Southern blotting, a specific hybridization signal is obtained, use nested PCR to produce visible bands on the ethidium bromide-stained agarose gel. Prepare a new pair of primers located further outside the region where the first primer set was designed. This new primer set does not need modification (e.g., restriction sites) and can be shorter (about 22–25 nucleotides). Use these new primers in RT-PCR; take 1–5 μL of the RT-PCR product and use it as a template and the original modified oligonucleotides as primers for the second round of PCR.
7. If no specific bands are seen after procedures described in Notes 3–6, check the total RNA used in the experiments. Use primers of abundant mRNA (e.g., β-actin), instead of your primers, in the coupled one-step RT-PCR procedure. A positive signal in the control reaction leaves only two possibilities for explaining negative results; the sample RNA does not contain the target template RNA, or the primers anneal inefficiently to the template RNA. Try a new pair of primers located in a different region of the cDNA, since primers are sometimes chosen in a region of secondary structure, causing difficulties in priming.
8. Negative controls in RT-PCR should be done to eliminate the possibility of potential DNA contamination. Prepare two control samples following the above procedure, but omit the template RNA in one control and omit RT in the second.

9. The procedure for directional cloning of RT-PCR products is also included in this chapter because it has been problematic for many laboratories. In the present procedure, the use of a several-fold excess of restriction enzymes and long incubation times are critical for optimal results. Commercial methods for cloning PCR products (e.g., T/A cloning and blunt end DNA ligation kits) are also recommended, although they are expensive to use.

## References

1. Jones, M. D. and Foulkes, N. S. (1989) Reverse transcription of mRNA by *Thermus aquaticus* DNA polymerase. *Nucleic Acids Res.* **17,** 8387,8388.
2. Shaffer, A. L., Wojnar, W., and Nelson, W. (1990) Amplification, detection, and automated sequencing of gibbon interleukin-2 mRNA by *Thermus aquaticus* DNA polymerase reverse transcription and polymerase chain reaction. *Anal. Biochem.* **190,** 292–296.
3. Tse, W. T. and Forget, B. G. (1990) Reverse transcription and direct amplification of cellular RNA transcripts by *Taq* polymerase. *Gene* **88,** 293–296.
4. Myers, T. W. and Gelfand, D. H. (1991) Reverse transcription and DNA amplification by a *Thermus thermophilus* DNA polymerase. *Biochemistry* **30,** 7661–7666.
5. Goblet, C., Prost, E., and Whalen, R. G. (1989) One-step amplification of transcripts in total RNA using the polymerase chain reaction. *Nucleic Acids Res.* **17,** 2144.
6. Singer-Sam, J., Robinson, M. O., Bellvé, A. R., Simon, M. I., and Riggs, A. D. (1990) Measurement by quantitative PCR of changes in HPRT, PGK-1, PGK-2, APRT, MTase, and Zfy gene transcripts during mouse spermatogenesis. *Nucleic Acids Res.* **18,** 1255–1259.
7. Zafra, F., Hengerer, B., Leibrock. J., Thoenen, H., and Lindholm, D. (1990) Activity dependent regulation of BDNF and NGF mRNAs in the rat hippocampus is mediated by non-NMDA glutamate receptors. *EMBO J.* **9,** 3545–3550.
8. Wang, R.-F., Cao, W.-W., and Johnson M. G. (1992) A simplified, single tube, single buffer system for RNA-PCR. *BioTechniques* **12,** 702–704.
9. Aatsinki, J. T., Pietilä, E. M., Lakkakorpi, J. T., and Rajaniemi, H. J. (1992) Expression of the LH/CG receptor gene in rat ovarian tissue is regulated by an extensive alternative splicing of the primary transcript. *Mol. Cell. Endocrinol.* **84,** 127–135.
10. Aatsinki, J. T., Lakkakorpi, J. T., Pietilä, E. M., and Rajaniemi, H. J. (1994) A coupled one-step reverse transcription PCR procedure for generation of full-length open reading frames. *BioTechniques* **16,** 282–288.
11. Sambrook, J., Fritsch, E. F., and Maniatis, T. (1989) *Molecular Cloning: A Laboratory Manual,* 2nd ed., Cold Spring Harbor Laboratory Press, Cold Spring Harbor, NY.
12. Sanger, F., Nicklen, S., and Coulson, A. R. (1977) DNA sequencing with chain-terminating inhibitors. *Proc. Natl. Acad. Sci. USA* **74,** 5463–5467.
13. Southern, E. M. (1975) Detection of specific sequences among DNA fragments separated by gel electrophoresis. *J. Mol. Biol.* **98,** 503–517.

# I

# CLONING PCR PRODUCTS

# 7

## Using T4 DNA Polymerase to Generate Clonable PCR Products

### Kai Wang

### 1. Introduction

Polymerase chain reaction (PCR) mediated through *Taq* DNA polymerase has become a simple and routine method for cloning, sequencing, and analyzing genetic information from very small amounts of materials *(1)*. *Taq* DNA polymerase, like some other DNA polymerases, lacks 3' to 5' exonuclease activity and will add nontemplate-directed nucleotides to the ends of double-stranded DNA fragments. Because of the strong preference of the *Taq* polymerase for dATP, the nucleotide added is almost exclusively an adenosine *(2)*. This results in generating "ragged" unclonable amplification products *(2,3)*. Restriction endonuclease sites are often incorporated into the amplification primers so that clonable PCR products can be generated by restriction enzyme cleavage *(4)*. However, the possible secondary sites located within amplified products often complicate the cloning and interpretation of PCR results. A cloning system exploiting the template-independent terminal transferase activity of *Taq* polymerase has been reported *(5–7)*. However, a special vector with thymidine (T) overhanging ends has to be used in the process.

T4 DNA polymerase has very strong exonuclease and polymerase activities in a broad range of reaction conditions *(8)*. By adapting its strong enzymatic activities, a simple and efficient method to generate clonable PCR fragments with T4 DNA polymerase has been developed *(9)*. The T4 DNA polymerase not only repairs the ends of the PCR products, but also removes the remaining primers in the reaction with its strong single-stranded exonuclease activity. Therefore, this method usually does not require multiple sample handling, buffer changes, or gel purification steps. Instead, a simple alcohol precipitation step is used to purify the PCR products.

From: *Methods in Molecular Biology, Vol. 67: PCR Cloning Protocols: From Molecular Cloning to Genetic Engineering* Edited by: B. A. White  Humana Press Inc., Totowa, NJ

The blunt-end cloning protocol can be modified for sticky-end cloning. Even though this may increase cloning efficiency to a certain extent, a purification step, to remove excess deoxynucleotides from PCR reactions, has to be added before adding T4 DNA polymerase.

## 2. Materials

### 2.1. PCR

1. DNA template containing the sequence of interest.
2. Oligonucleotide primers.
3. *Taq* DNA polymerase (Perkin-Elmer, Cetus, Norwalk, CT).
4. 10X PCR and enzymatic repair buffer: 500 m$M$ Tris-HCl, pH 9.0, 25 m$M$ MgCl$_2$, 500 m$M$ NaCl, and 5 m$M$ DTT. Commercial 10X PCR buffer also works well.
5. 1.5 m$M$ 10X deoxynucleotides (dNTP) solution. Concentrated stock solution (100 m$M$) can be obtained from Pharmacia (Piscataway, NJ) or Boehringer Mannheim (Indianapolis, IN).
6. Gel electrophoresis and PCR equipment.

### 2.2. End Repair and Blunt-End Cloning

1. Enzymes (T4 DNA polymerase, T4 polynucleotide kinase and T4 DNA ligase). Enzymes can be purchased from Boehringer Mannheim, Life Technologies (Gibco-BRL, Gaithersburg, MD) or any other provider.
2. 1 m$M$ of ATP solution. Concentrated stock solution (100 m$M$) can be obtained from Boehringer Mannheim.
3. Isopropyl alcohol.
4. Vector (blunt end and dephosphorylated).
5. 10X Ligase buffer: 660 m$M$ Tris-HCl, pH 7.6, 66 m$M$ MgCl$_2$, 10 m$M$ ATP, 1 m$M$ spermidine and 10 m$M$ DTT. Commercially available 10X ligase buffers also works well.
6. TE buffer: 10 m$M$ Tris-HCl, pH 7.5, 1 m$M$ EDTA.
7. 5$M$ NaCl.

### 2.3. End Repair and Sticky-End Cloning

1. Sephacryl S-400 spin column. A commercial spin column (MicroSpin S-400HR) can be obtained from Pharmacia.
2. Enzymes (T4 DNA polymerase, T4 polynucleotide kinase, and T4 DNA ligase). Enzymes can be purchased from Boehringer Mannheim, Life Technologies, or any other provider.
3. ATP and dNTPs.
4. Isopropyl alcohol.
5. Vector (digested and dephosphorylated).
6. 10X Ligase buffer: 660 m$M$ Tris-HCl, pH 7.6, 66 m$M$ MgCl$_2$, 10 m$M$ ATP, 1 m$M$ spermidine and 10 m$M$ DTT. Commercially available 10X ligase buffers also works well.

7. 0.5*M* EDTA, pH 8.0.
8. 5*M* NaCl.

## 3. Method

### 3.1. Primer Design

For blunt-end cloning, no special primer is needed. However, secondary structure and stretch of homopolymer should be avoided. For sticky-end cloning, depending on restriction site selected, specific sequences should be included so that compatible ends can be generated after T4 DNA polymerase treatment (*see* Section 3.4.).

### 3.2. PCR Reaction

1. Prepare the following in a PCR reaction tube: 5 μL of 10X PCR buffer; 0.25 μg of genomic DNA; 1 μ*M* of each primer; 0.15 m*M* of each deoxynucleotide (5 μL of 1.5 m*M* stock dNTP solution); 1 U *Taq* polymerase; and deionized H$_2$O to a final volume of 50 μL.
2. Amplification conditions largely depend on the specific applications. However, a general cycling profile can be used in most experiments: 94°C for 7 min (initial denaturation); 94°C for 30 s (amplification), 55°C for 45 s, 72°C for 90 s; 72°C for 10 min (extension).
3. Examine the PCR amplification results with agarose gel electrophoresis (*see* Note 1).

### 3.3. End Repair and Blunt-End Cloning

1. Add the following to PCR reaction tubes directly to repair the PCR products (*see* Note 2): 1 U of T4 DNA polymerase; 1 μL of 4 m*M* dNTP solution (optional) (*see* Note 3); 5 U of T4 polynucleotide kinase (*see* Note 4); and 1 μL of 1 m*M* ATP.
2. Incubate the reaction tubes at 25°C (room temperature) for 20 min (*see* Notes 5,6). Stop the reactions by adding 3 ml of 0.5*M* EDTA, pH 8.0.
3. Incubate the reaction tubes at 70°C for 10 min to inactive the enzymes.
4. Precipitate the PCR products by adding 5 μL of 5*M* NaCl and 60 μL of isopropyl alcohol *(8)* (*see* Note 7).
5. Resuspend the DNA fragments in 20 μL of TE or water.
6. Take 2 μL of DNA (containing approx 20–50 ng of PCR product) and mix with ligase and vector for ligation (*see* Notes 8,9): 1 μL of 10X ligase buffer; 1 μL of T4 ligase; 2 μL of repaired DNA; dephosphorylated vector (60–150 ng); and add deionized H$_2$O to a final volume of 10 μL.
7. Incubate at 16°C overnight.
8. Dilute the ligation reaction fivefold in TE buffer. Use 2 μL of the diluted ligation reaction for transformation (*see* Note 10).

### 3.4. End Repair and Sticky-End Cloning

An *Eco*RI site is used as an example in the following protocol (Fig. 1). However, depending on the desired cloning site, a different combination of dNTP should be added in the "repair" reaction (step 2).

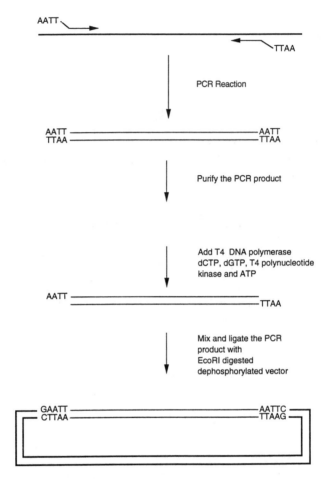

**Fig. 1.** A brief outline of the strategy used to generate sticky-end PCR products with T4 DNA polymerase.

1. Spin through the PCR reaction mixture (40 μL) in a pre-equilibrated Sephacryl S-400HR spin column (*see* Note 11) to remove excess dNTP.
2. Add the following to the column-purified PCR fragments to generate sticky ends: 5 μL of 10X PCR buffer; 1 U of T4 DNA polymerase; 5 U of T4 polynucleotide kinase; 1 μL of 4 m*M* dCTP and dGTP; 1 μL of 1 m*M* ATP; and add deionized H₂O to a final volume of 50 μL.
3. Incubate the reaction tubes at 25°C (room temperature) for 20 min. Stop the reaction by adding 3 μL of 0.5*M* EDTA, pH 8.0.
4. Heat inactivate the enzymes by placing the reaction tubes at 70°C for 10 min.
5. Precipitate the PCR products by adding 5 μL of 5*M* NaCl and 60 μL of isopropyl alcohol *(8)*.
6. Resuspend the DNA fragments in 20 μL of TE or water.

7. Take 2 μL of DNA (containing approx 20–50 ng of DNA) and mix with ligase and *Eco*RI digested, dephosphorylated vector for cloning: 1 μL of 10X ligase buffer; 1 μL of T4 ligase; 2 μL of repaired DNA; dephosphorylated vector (60–150 ng); and add deionized H$_2$O to a final volume of 10 μL

8. Incubate at 16°C overnight.

9. Dilute the ligation reaction fivefold in TE buffer. Use 2 μL of the diluted ligation reaction for transformation (*see* Note 10).

## 4. Notes

1. In case of multiple PCR products from a single reaction, the specific products should be purified by gel electrophoresis based on size after repair reaction. Several different methods can be used to purify DNA fragments from agarose gel, such as phenol extraction from low-melting gel *(8)*, "glassmilk" method, or simple low-speed centrifugation *(10)*. The phenol extraction method has been found to be less expensive and able to recover a sufficient amount of clean DNA for cloning.

2. This protocol utilizes one buffer for all the enzymes that include *Taq* polymerase in PCR reaction, T4 polymerase, and T4 polynucleotide kinase in end-repair reaction. Therefore, a slightly higher concentration of reagents and enzymes can be added in the reaction.

3. T4 DNA polymerase can be added directly into the PCR reaction tube without providing additional nucleotides. However, T4 DNA polymerase balances its exonuclease and polymerase activities based on the concentration of available deoxynucleotides. Depending on the length of amplification products, number of amplification cycles, and nucleotide sequence composition of amplified region, the remaining nucleotide concentration after PCR amplification may be different from experiment to experiment. In order to avoid unnecessary confusion, supplemental nucleotides are routinely added for end-repair reaction.

4. T4 polynucleotide kinase is not needed when vector used has not been treated with phosphatase previously. However, dephosphorylated vector should be used to increase the cloning efficiency.

5. Room temperature (25°C) was chosen for the reaction since T4 DNA polymerase has excessive exonuclease activity at 37°C.

6. The T4 polynucleotide kinase works well at room temperature as opposed to the higher reaction temperature (37°C) regularly used *(8)*.

7. Although the PCR products purified directly by alcohol precipitation after end repairing are sufficient for routine cloning, passing the repaired PCR product mixtures through a gel filtration column prior to the alcohol precipitation can greatly enhance the cloning efficiency.

8. Digested and dephosphorylated cloning vectors can be purchased directly from Pharmacia (Ready-To-Go pUC18 cloning kit).

9. In the ligation reaction, we routinely used 1:1 molar ratio between vector (dephosphorylated) and insert. Generally more than 200 recombinant clones can be obtained with 0.4 μL of the ligation reaction. Therefore, a single ligation reaction for each PCR product is sufficient for most applications.

10. Because of its reliability and high transformation efficiency, commercial CaCl$_2$-treated competent cells are used for the transformation step. The bacteria strain we routinely used is DH10B (BRL; MAX efficiency DH10B competent cells). However, various competent cells can be purchased from companies, such as Stratagene and Life Technologies (Gibco/BRL).

11. Prepacked Sephacryl S-400HR spin columns (MicroSpin S-400HR) can be purchased from Pharmacia. Alternatively, the spin columns can be prepared from bulk gel filtration matrix (Sephacryl S-400HR, Pharmacia) as described in other protocol books *(8)*. The filtration medium (prepacked or bulk filtration matrix) contains 20% alcohol; therefore, the spin columns should be washed and equilibrated with TE.

12. The blunt-end cloning protocol has been used routinely in the lab, even though the sticky-end method gives better cloning efficiency. The blunt-end cloning is simple and usually generates sufficient amounts of recombinant clones for further analyses, such as sequencing.

13. When large amounts of PCR products need to be cloned and analyzed, a low-cost 96-well format minispin column that can be used to purify 96 DNA samples simultaneously has been described *(11)*.

## References

1. Saiki, R. K., Scharf, S., Faloona, F., Mullis, K. B., Horn, G. T., Erlich, H. A. and Arnheim, N. (1985) Primer-directed enzymatic amplification of DNA with a thermostable DNA polymerase. *Science* **230,** 1350–1354.
2. Clark, J. M. (1988) Novel non-templated nucleotide addition reactions catalyzed by procaryotic and eucaryotic DNA polymerases. *Nucleic Acids Res.* **16,** 9677–9686.
3. Scharf, S. (1990) *PCR Protocols: A Guide to Methods and Applications,* Academic, San Diego, CA.
4. Jung, V., Pestka, S. B., and Pestka, S. (1990) Efficient cloning of PCR generated DNA containing terminal restriction endonuclease recognition sites. *Nucleic Acids Res.* **18,** 6156.
5. Mead, D. A., Pey, N. K., Herrnstadt, C., Marcil, R. A., and Smith, L. M. (1991) A universal method for the direct cloning of PCR amplified nucleic acid. *Bio/Technology* **9,** 657.
6. Kovalic, D., Kwak, J., and Weisblum, B. (1991) General method for direct cloning of DNA fragments generated by the polymerase chain reaction. *Nucleic Acids Res.* **19,** 4560.
7. Marchuk, D., Drumm, M., Saulino, A., and Collins, F. (1991) Construction of T-vectors, a rapid and general system for direct cloning of unmodified PCR products. *Nucleic Acids Res.* **19,** 1154.
8. Sambrook, J., Fritsch, E. F., and Maniatis, T. (1989) *Molecular Cloning: A Laboratory Manual,* 2nd ed., Cold Spring Harbor Laboratory, Cold Spring Harbor, NY.
9. Wang, K., Koop, B. F., and Hood, L. (1994) A simple method using T4 DNA polymerase to clone polymerase chain reaction products. *BioTechniques* **17,** 236–238.
10. Heery, D. M., Gannon, F., and Powell, R. (1990) A simple method for subcloning DNA fragments from gel slices. *Trends Genet.* **6,** 173.
11. Wang, K., Gan, L., Boysen, C., and Hood, L. (1995) A microtiter plate based high throughput DNA purification method. *Anal. Biochem.* **226,** 85–90.

# 8

# Rapid (Ligase-Free) Subcloning of PCR Products

## Alan R. Shuldiner and Keith Tanner

## 1. Introduction

The polymerase chain reaction (PCR) is a versatile, widely used method for the production of a very large number of copies of a specific DNA molecule *(1,2)*. For some applications, it is advantageous to subclone the PCR product into a plasmid vector for subsequent replication in bacteria *(3–6)*. Subcloning the PCR product into a plasmid vector has several advantages: The amplified fragment can be sequenced with greater reliability, only one allele is sequenced per clone, and the vector containing the PCR product may be used for other molecular biological experiments, e.g., in vitro transcription, transfection, and further amplification in bacteria.

Although conventional strategies, such as blunt-end or sticky-end ligation, can be very successful for subcloning most DNA fragments, the subcloning of DNA generated by PCR is often very difficult *(7–9)*. We describe a rapid and versatile method to subclone PCR products directionally into a specific site of virtually any plasmid vector *(8,9)*. Ligase-free subcloning of PCR products has several advantages over conventional strategies: It does not require DNA ligase, it requires only four primers, two of which are complementary to the plasmid vector, and therefore may be used repeatedly, and it may be accomplished in a single day. Typically, hundreds to thousands of colonies per transformation are obtained (approximate efficiency $5 \times 10^3$ to $5 \times 10^4$ colonies/microgram of PCR insert). With this method, PCR products of up to 1.7 kb in length have been subcloned successfully.

From: *Methods in Molecular Biology, Vol. 67: PCR Cloning Protocols: From Molecular Cloning to Genetic Engineering* Edited by: B. A. White  Humana Press Inc., Totowa, NJ

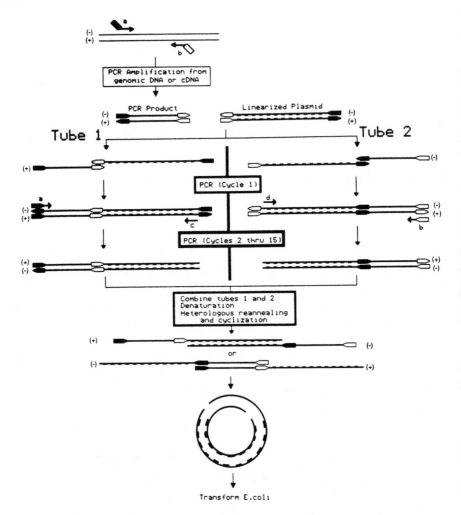

**Fig. 1.** Schematic of ligase-free subcloning. The 5' addition sequences of primers a and b are designated by closed and open boxes, respectively. DNA sequences corresponding to the PCR-amplified product are shown as straight lines, whereas DNA sequences corresponding to the plasmid vector are shown as hatched lines.

## 1.1. Theory of the Method

### 1.1.1. Plasmid Linearization and Primer Design

The first step in ligase-free subcloning is to linearize the plasmid vector at the desired site with the appropriate restriction endonuclease(s). The second step is to perform PCR on genomic DNA or the cDNA of interest. Like conventional PCR, the primers (primers a and b in Fig. 1) must contain sequences at their 3' ends (approx 20–25 nucleotides) that are complementary to opposite strands of the target sequence at a predetermined distance from each other. For ligase-free subcloning, primers a and b must also contain sequences at their 5' ends, approx 24 nucleotides in length (designated the 5' addition sequences), that are identical to each of the 3' ends of the linearized plasmid (opened and closed boxes in Fig. 1; also, *see* example in Fig. 2). Since the two 3' ends of the linearized plasmid are different from each other, the PCR fragment may be subcloned directionally simply by choosing the appropriate 5' addition sequence for each primer.

## 1.1.2. PCR Amplification and Ligation by Overlap Extension

PCR amplification is accomplished using primers a and b, which results in large amounts of the PCR product containing the 5' addition sequences at each end. Next, the PCR product is freed from excess primers by ultrafiltration and divided into two tubes each containing the linearized plasmid vector. A second PCR reaction is performed. Tube 1 also contains primer a, the same primer that was used during the initial amplification of the PCR product, and primer c, a primer that is complementary to the strand of the plasmid vector that contains the primer a addition sequence (the [+] strand in Fig. 1), in a region internal to the primer a addition sequence (Figs. 1 and 2). Similarly, tube 2 contains primer b, the same primer that was used for the initial amplification of the PCR product, and primer d, a primer that is complementary to the strand of the plasmid vector that contains the primer b addition sequence (the [−] strand in Fig. 1), in a region internal to the primer b addition sequence (Figs. 1 and 2).

During the first cycle of the second PCR, denaturation and annealing result in hybridization of the 3' ends of the PCR product (the complements of the 5' addition sequences) to the complementary 3' ends of the linearized plasmid (Fig. 1). During the extension step of the first cycle, these overlapping 3' ends act as templates for each other. Extension results in the "ligation" of the PCR fragment to the linearized plasmid (ligation by overlap extension) *(10,11)*.

Amplification during subsequent PCR cycles (cycles 2–20) with the respective primers in each tube results in large amounts of linear double-stranded DNA that contain the plasmid vector ligated to the PCR insert at one end (tube 1 in Fig. 1), or to the other end (tube 2 in Fig. 1). Although not shown in the figure, hybridization may also occur between the 5' ends of the PCR product and the complementary 5' ends of the linearized plasmid. However, these hybrid products cannot act as templates for *Taq* DNA polymerase.

## 1.1.3. Denaturation, Heterologous Annealing, and Cyclization

After the second PCR reaction, the reaction mixtures in tubes 1 and 2 are combined and denaturation of double-stranded DNA into single-stranded DNA

Nucleotide sequence of pGEM4Z linearized with SmaI:

                                                        primer d
                                                  primer a
5'-GGGGATCCTCTAGAGTCGACCTGCaggcatgcaagcttgtctccct**atagtgtgattagagc**...( 3 Kb)...ttggatttaggtgacactag**AATACGAATTCGAGCTCGGTACCC**-3'

3'-**CCCCTAGGAGATCTCAGCTGGAC**gtccgtacgttcgaacagaggatatcactcagcataatctcg...( 3 Kb)...**aacgtaatcgactgtgatatc**TTATGCTTAAGCTCGAGCCATGGG-5'
                       primer b                                                      primer c

5' addition sequences should be identical to the 3' ends (upper case bold and underlined nucleotides):

                    Primer a:    5'-AATACGAATTCGAGCTCGGTACCC.....3'

                    Primer b:    5'-GCAGGTCGACTCTAGAGGATCCCC.....3'

Primers c and d should be complementary to opposite strands of the plasmid internal to the 5' addition sequences
(lower case bold and underlined nucleotides):

                    Primer c:    5'-tatagtgtcacctaaatccaa-3'

                    Primer d:    5'-tatagtgagtgctattagagc-3'

(Note that primers c and d do not need to be immediately adjacent to primers a and b.)

**Fig. 2.** Example of primer design for ligase-free subcloning into the *SmaI* site of pGEM4Z (Promega Biotec, Madison, WI).

72

is accomplished by treating with alkali. After neutralization and dilution of the denatured DNA, the single-stranded DNA may anneal into several different products. Complementary DNA strands may reanneal to form the same linear double-stranded products that were present in tubes 1 and 2 before the denaturation step (not shown in Fig. 1). Alternatively, single-stranded DNA products from tube 1 may anneal with single-stranded DNA products from tube 2, resulting in DNAs that are partly double-stranded and contain long single-stranded 5' or 3' overhangs (heterologous annealing in Fig. 1). Since these long 5' or 3' overhangs are complementary to each other, annealing at low DNA concentrations results in cyclization.

Although the cyclized product contains two nicks (noncovalently linked ends), it may be used directly to transform competent *E. coli*. Once inside the bacterial cell, the two nicks are covalently joined and the recombinant plasmid is replicated.

## 2. Materials

1. Plasmid vector.
2. Appropriate restriction endonuclease(s) and buffer(s) for linearization of the plasmid vector.
3. Oligonucleotide primers *(see* Note 1). Prepare the four oligonucleotide primers required for ligase-free subcloning by synthesizing chemically with the last dimethoxytrityl (DMT) group off. Cleave the oligonucleotide from the resin by treatment with concentrated ammonia at room temperature, and deprotect by heating to 55°C in concentrated ammonia overnight *(see* Chapter 1). The oligonucleotides may be used for PCR after desalting through a G-25 spin column (Boehringer Mannheim Biochemicals, Indianapolis, IN). Alternatively, the last DMT group may be left on and the oligonucleotides may be purified with NENsorb Prep columns (New England Nuclear, Boston, MA) according to the manufacturer's directions.
4. 10X GeneAmp PCR buffer: 500 m$M$ KCl, 100 m$M$ Tris-HCl, pH 8.3, at 25°C, 15 m$M$ MgCl$_2$, and 0.1 mg/mL of gelatin. GeneAmp is a trademark of Perkin-Elmer/Cetus (Norwalk, CT).
5. dNTPs: 10 m$M$ each.
6. *Taq* DNA polymerase.
7. Paraffin oil.
8. Autoclaved, distilled, deionized water.
9. Phenol equilibrated in TE buffer (10 m$M$ Tris-HCl, pH 8.0, at 25°C, 1 m$M$ EDTA).
0. Chloroform:isoamyl alcohol: 24:1, v/v.
1. tRNA: 10 mg/mL in sterile water.
2. Sterile, 5$M$ NaCl.
3. Autoclaved sterile Eppendorf tubes.
4. Vacuum dessicator.

15. Ultrafree-MC-100 and Ultrafree-MC-30 ultrafiltration devices (Millipore, Bedford, MA) (*see* Note 2).
16. 50 m*M* Tris-HCl buffer, pH 6.6, at 25°C.
17. DNA size marker: *Hae*III digest of φX174.
18. 2*M* NaOH.
19. Competent DH5-α *E. coli* (Bethesda Research Laboratories, Gaithersburg, MD; theoretical transformation efficiency >1 × 10⁹ transformants/μg PUC19).
21. LB agar plates with the appropriate antibiotic.
22. 10,000X ethidium bromide solution: 5 mg/mL in water. Keep in dark container at 4°C for up to 12 mo. Wear gloves when handling ethidium bromide.
23. High-melting-point agarose.
24. NuSeive GTG agarose (FMC Bioproducts, Rockland, ME).
25. 95% Ethanol.
26. 70% Ethanol.

## 3. Methods

### 3.1. Plasmid Linearization

1. Linearize 1–2 μg of the plasmid vector by digestion with the appropriate restriction endonuclease(s). Incubate the plasmid vector, the appropriate buffer, and restriction enzyme(s) in a final volume of 20 μL at the appropriate temperature for 2–3 h in an autoclaved Eppendorf tube.
2. Add 30 μL of sterile water and 1 μL of tRNA (10 mg/mL) to the reaction mixture.
3. Extract by adding 25 μL of phenol that has been pre-equilibrated with TE buffer. Vortex. Add 50 μL of chloroform:isoamyl alcohol. Vortex. Centrifuge in a microfuge at room temperature for 30 s to separate the phases. Remove the upper (aqueous) phase (approx 50 μL) and place in a clean autoclaved Eppendorf tube.
4. Repeat step 3 once.
5. Precipitate the linearized plasmid DNA by adding 12.5 μL of 5*M* NaCl, vortex briefly, then add 125 μL of cold 95% ethanol. Incubate at –20°C for at least 3 h.
6. Recover the linearized plasmid DNA by centrifugation in a microfuge (12,000g) at 4°C for 30 min. Pour off the supernatant and wash the pellet with 250 μL of cold 70% ethanol. Carefully decant the supernatant, remove excess liquid from the walls of the Eppendorf tube with a cotton swab, and vacuum-dry the pellet in a vacuum dessicator or Speed-Vac for 10 min.
7. Resuspend the pellet in 20 μL of sterile water. Confirm that the plasmid was linearized and determine the approximate DNA concentration by running 10% (2 μL) of the linearized plasmid DNA on a 1% agarose gel with 100 ng of uncut plasmid and a known amount of a *Hae*III-digest of φX174 DNA size marker (*see* Chapter 1). DNA should be visualized by staining in ethidium bromide solution for 10–15 min at room temperature and inspection (or photography) during UV transillumination using appropriate precautions (*see* Note 3). The intensity of staining of the sample DNA and the size marker DNA may be compared to each other so that the amount of sample DNA can be roughly quantified.

## 3.2. PCR Amplification and Ligation
## by Overlap Extension

1. Assemble the PCR reaction in a final volume of 100 μL containing 1X GeneAmp PCR buffer, 200 μM of each dNTP, 100 nM each of primers a and b, 1.5 U/tube of *Taq* polymerase, and DNA template (*see* Note 4). Overlay with 50 μL of paraffin oil if required.
2. Amplify by PCR using the following cycle profile (*see* Note 5):

   | | |
   |---|---|
   | Initial denaturation | 94°C, 5 min |
   | 25–35 main cycles of PCR | 55°C, 1 min (annealing) |
   | | 72°C, 1 min (extension) |
   | | 94°C, 1 min (denaturation) |
   | Final extension | 72°C, 10 min |

3. To assess the purity of the PCR product, remove 18 μL of the PCR reaction mixture, add 2 μL of 10X loading buffer, and perform electrophoresis on a composite gel consisting of 1% agarose and 2% NuSeive GTG agarose (*see* Chapter 1). Estimate the amount of DNA as described in step 7 in Section 3.1.
4. Add 320 μL of water to the remaining 82 μL of the PCR reaction mixture and filter through an Ultrafree-MC-100 device by centrifugation at 2000g at room temperature until the retentate volume is about 20–25 μL (approx 3–5 min), being careful not to overfilter (*see* Note 6).
5. After the first ultrafiltration, add 400 μL of water to the retentate and repeat ultrafiltration in the same device to a final retentate volume of approx 20–25 μL, again being careful not to overfilter.
6. Prepare a 100-μL PCR reaction mixture in two separate 0.5-mL Eppendorf tubes as follows (*see* Note 7):

   | | Tube 1 | Tube 2 |
   |---|---|---|
   | 10X GeneAmp PCR buffer | 10.0 μL | 10.0 μL |
   | dATP (10 m*M*) | 2.0 μL | 2.0 μL |
   | dTTP (10 m*M*) | 2.0 μL | 2.0 μL |
   | dCTP (10 m*M*) | 2.0 μL | 2.0 μL |
   | dGTP (10 m*M*) | 2.0 μL | 2.0 μL |
   | Primer a (5 μ*M*) | 2.0 μL | — |
   | Primer c (5 μ*M*) | 2.0 μL | — |
   | Primer b (5 μ*M*) | — | 2.0 μL |
   | Primer d (5 μ*M*) | — | 2.0 μL |
   | Linearized plasmid (50 ng/μL) | 2.0 μL | 2.0 μL |
   | PCR product (approx 50 ng/μL) | 2.0 μL | 2.0 μL |
   | Water | 73.6 μL | 73.6 μL |
   | *Taq* Polymerase (5 U/μL) | 0.4 μL | 0.4 μL |
   | Total volume | 100.0 μL | 100.0 μL |

   Overlay with approx 50 μL of paraffin oil if required.
7. Amplify by PCR (20 cycles) as described in step 2 of Section 3.2., except increase the extension time of the main cycles to 1.5 min.

### 3.3. Denaturation, Heterologous Annealing, and Cyclization

1. After thermocycling is completed, start water boiling in preparation for steps 3 and 4 *(see below)*.
2. Combine 15 μL of the PCR reaction mixture from tube 1 with 15 μL of the PCR reaction mixture from tube 2. Denature the DNA by adding 3.4 μL of 2*M* NaOH. Vortex and incubate for 5 min at room temperature.
3. During the incubation, transfer 400 μL of Tris-HCl buffer (50 m*M*, pH 6.6, at 25°C) to a 1.5-mL Eppendorf tube and place in the boiling water bath.
4. Add the denatured PCR products from step 2 (this section) to the heated Tris-HCl buffer. Vortex briefly and place into the boiling water bath for 5 min.
5. Cool to 60°C over approx 15 min *(see Note 8)* and incubate at 60°C for 3–24 h to accomplish heterologous annealing and cyclization.
6. Following heterologous annealing and cyclization, allow the reaction mixture from step 5 to cool to room temperature and place the entire contents in an Ultrafree-MC-30 device. Concentrate to a final volume of approx 10–20 μL by centrifugation at 2000*g* at room temperature (approx 3–5 min). Check the retentate volume frequently to avoid over-filtration and irreversible loss of DNA *(see Note 6)*.
7. Transfer the concentrated cyclized product into an autoclaved 0.5-mL Eppendorf tube. The concentrated cyclized product may be stored indefinitely at –20°C for transformation at a future time or may be used immediately. Transform 100 μL of competent DH5-α *E. coli* according to the manufacturer's directions with 1–5 μL of the cyclized product.
8. After transformation, plate the *E. coli* onto LB agar plates containing the appropriate antibiotic and incubate overnight at 37°C. If the plasmid encodes β-galactosidase (LacZ), colonies containing the appropriate recombinant plasmid may be chosen by blue-white selection on agar plates that also contain X-gal (50 μg/plate). Since this approach may be misleading *(see Note 9)*, we recommend that plasmid minipreparations be prepared from several white colonies *(12; see Chapter 22)*, and the presence and size of the DNA insert be confirmed by restriction endonuclease cleavage and gel electrophoresis. Alternatively, colony hybridization with the appropriate radiolabeled probe *(13)* or PCR of plasmid DNA with primers that flank the DNA insert *(14)* may be used to select colonies with plasmids containing the desired DNA insert.

## 4. Notes

1. Once the plasmid vector has been linearized with the appropriate restriction endonuclease(s), the oligonucleotide primers required for ligase-free subcloning may be designed. Primers a and b should be approx 39–49 bases in length with 20–25 bases at their 3' ends that are complementary to opposite strands of the target DNA at a predetermined distance from each other (Fig. 1). In addition, primers a and b must also contain 19–24 nucleotides at their 5' ends (the 5' addition sequences) that are identical to each of the 3' ends of the linearized plasmid vector (Figs. 1 and 2). In most of our experiments, 5' addition sequences were

arbitrarily chosen to be 24 nucleotides in length. However, 5' addition sequences 19 nucleotides in length may be used if the temperature of the second-stage PCR is decreased to 37°C. 5' addition sequences 14 nucleotides in length or less did not work reliably. Primers c and d should be approx 21 nucleotides in length, and be complementary to opposite strands of the plasmid vector internal to each of the 5' addition sequences (Figs. 1 and 2).

2. Centricon-100 and Centricon-30 ultrafiltration devices (Amicon, Danvers, MA) may be substituted for Ultrafree-MC-100 and Ultrafree-MC-30 devices, respectively. When Centricon devices are used, sodium azide, which is used as a preservative, must first be removed from the membrane by treatment with sodium hydroxide according to the manufacturer's directions. If the sodium azide is not removed, it will kill all of the bacteria during the transformation step.

3. On a 1% agarose gel, the linearized plasmid will run more slowly than the uncut (supercoiled) plasmid. If digestion is complete, it is generally not necessary to gel-purify the linearized plasmid.

4. When genomic DNA is used as starting template, 0.1–0.5 μg/tube will suffice. If plasmid DNA is being used as starting template, 0.1–1 ng/tube is sufficient. If RNA is being used as starting template, reverse transcription must first be performed followed by PCR as outlined. If the cloned PCR product will be used for applications that require precise nucleotide sequence integrity, substitution of *Taq* polymerase with a thermostable DNA polymerase with proofreading ability (i.e., Vent or Pyrococcus DNA polymerase) is preferable to minimize misincorporation errors during the PCR.

5. PCR conditions should be optimized for the specific application.

6. During centrifugation, frequently check the volume of the retentate since overfiltration can result in irreversible binding of DNA to the filter and loss of the PCR product.

7. Although never tested, use of *Taq* polymerase derivatives (or reagents) optimized for long extensions (i.e., TaKaRa LA Taq, Takara Shuzo Co., Ltd.; *Taq*/ Pyrococcus GB-D DNA polymerase mixture, Life Technologies [Gaithersburg, MD]; TaqPlus, Strategene) may be advantageous for this step.

8. This is most conveniently done by using a 500-mL beaker filled approximately halfway with water as the boiling water bath and transferring the entire beaker directly into a conventional water bath that has been preset at 60°C. This will result in cooling at the appropriate rate and automatic equilibration at 60°C.

9. In our hands, the percentage of antibiotic-resistant colonies that contain the desired DNA insert varies from 50–100%. If primer-dimers form during the initial PCR and are not removed prior to ligation by overlap extension, they contain the proper 5' addition sequences and are subcloned into the vector with high efficiency resulting in a higher background (i.e., colonies with recombinant plasmids that do not contain the desired DNA insert). If large quantities of primer-dimers are present, we recommend that the conditions of the PCR be modified to minimize these primer artifacts (i.e., decrease the primer concentration and/or decrease the amount of *Taq* polymerase). Alternatively, the PCR product may be gel-purified prior to ligation by overlap extension.

## Acknowledgments

We thank Jesse Roth, Charles T. Roberts, Jr., Steven Lasky, and Domenico Accili for their helpful comments on the manuscript. This research was supported in part by grants from the Diabetes Research and Education Foundation, the Juvenile Diabetes Foundation, and the Mallinckrodt Foundation. Dr. Shuldiner is the recipient of a Paul Beeson Physician Faculty Scholar Award from the American Federation of Aging Research.

## References

1. Saiki, R. K., Scharf, S., Faloona, F., Mullis, K. B., Horn, G. T., Erlich, H. A., and Arnheim, N. (1985) Enzymatic amplification of beta-globin genomic sequence and restriction site analysis for diagnosis of sickle cell anemia. *Science* **230,** 1350–1354.
2. Saiki, R. K., Gelfand, D. H., Stoffel, S., Scharf, S. J., Higuchi, R., Horn, G. T., Mullis, K. B., and Erlich, H. A. (1988) Primer-directed enzymatic amplification of DNA with a thermostable DNA polymerase. *Science* **239,** 487–491.
3. Scharf, S. J., Horn, G. T., and Erlich, H. A. (1986) Direct cloning and sequence analyses of enzymatically amplified genomic sequences. *Science* **233,** 1076–1078.
4. Lee, C. C., Wu, X., Gibbs, R. A., Cook, R. G., Muzny, D. M., and Caskey, C. T. (1988) Generation of cDNA probes directed by amino acid sequence: cloning of urate oxidase. *Science* **239,** 1288–1290.
5. Higuchi, R. (1989) Using PCR to engineer DNA, in *PCR Technology* (Erlich, H. A., ed.), Stockton, New York, pp. 61–70.
6. Scharf, S. J. (1990) Cloning with PCR, in *PCR Protocols* (Innis, M. A., Gelfand, D. H., Sninsky, J. J., and White, T. J., eds.), Academic, San Diego, pp. 84–91.
7. Kaufman, D. L. and Evans, G. A. (1990) Restriction endonuclease cleavage at the termini of PCR products. *BioTechniques* **9,** 304–306.
8. Shuldiner, A. R., Scott, L. A., and Roth, J. (1990) PCR induced subcloning polymerase chain reaction (PCR) products. *Nucleic Acid Res.* **18,** 1920.
9. Shuldiner, A. R., Tanner, K., Scott, L. A., and Roth, J. (1991) Ligase-free subcloning: a versatile method of subcloning polymerase chain reaction (PCR) products in a single day. *Anal. Biochem.* **194,** 9–15.
10. Higuchi, R., Krummel, B., and Saiki, R. K. (1988) A general method of *in vitro* preparation and specific mutagenesis of DNA fragments: study of protein and DNA interactions. *Nucleic Acid Res.* **16,** 7351–7367.
11. Horton, R. M., Hunt, H. D., Ho, S. N., Pullen, J. K., and Pease, L. R. (1989) Engineering hybrid genes without the use of restriction enzymes: gene splicing by overlap extension. *Gene* **77,** 61–68.
12. Del Sal, G., Manfioletti, G., and Schneider, C. (1989) A one-tube DNA minipreparation suitable for sequencing. *Nucleic Acids Res.* **16,** 9878.
13. Maas, R. (1983) An improved cloning hybridization method with significantly increased sensitivity for detection of single genes. *Plasmid* **10,** 296–298.
14. Gussow, D. and Clackson, T. (1989) Direct clone characterization from plaques and colonies by the polymerase chain reaction. *Nucleic Acid Res.* **17,** 4000.

# 9

## Cloning PCR Products
## Utilizing the T/A Overhang and a Kit

**Melissa Lail-Trecker**

### 1. Introduction

The ever expanding identification of new gene family members in recent years has depended in large part on the use of the polymerase chain reaction (PCR) technique. Direct cloning of PCR products into an appropriate vector allows identification of the product by sequencing and characterization of the product's transcript by Northern hybridization analysis or ribonuclease protection assay. Typically PCR products may be cloned into the vector by cohesive or blunt-end ligation. Cohesive-end ligation traditionally requires the addition of restriction sites to the PCR primers. This necessitates additional enzymatic manipulation and purification steps before the product may be cloned into the vector. Likewise, PCR products amplified with *Taq* polymerase must undergo further enzymatic manipulation before blunt-end ligation. In addition, blunt-end ligation is a less efficient process than cohesive-end ligation *(1)*.

The Invitrogen (San Diego, CA) TA Cloning® Kit combines the efficiency of cohesive-end ligation with the ease of direct cloning of PCR products. A sample may be removed from the completed PCR reaction and transferred directly to the ligation reaction mixture. This procedure is based on the fact that *Taq* polymerase has a nontemplate-dependent activity that adds a single deoxyadenosine to the 3' ends of double-stranded DNA *(2)*. The Invitrogen pCR™II vector is supplied as a linearized molecule with 3' T-overhangs on either side of the insert site. This allows cohesive-end ligation of the PCR product with the pCRII vector.

In this chapter, I describe the steps I followed in order to clone a 576-bp fragment of the rat N-cadherin gene into the pCRII vector. Although much of

From: *Methods in Molecular Biology, Vol. 67: PCR Cloning Protocols: From Molecular Cloning to Genetic Engineering* Edited by: B. A. White  Humana Press Inc., Totowa, NJ

the description of the ligation and transformation procedures is based on directions supplied by Invitrogen, some steps differ. It is also worth noting that subsequent variations have been introduced with more recent kits by Invitrogen. This plasmid was then used to generate RNA probes for use in a ribonuclease protection assay. The time involved for this procedure is a total of 4 d:

1. Set up PCR reaction in the morning of the first day. After completion, set up a ligation reaction that same afternoon; ligate overnight.
2. On the second day, transform bacteria with ligation reactions; grow transformed cells overnight on plates.
3. The third day pick colonies off plates and grow for minipreps. Plasmid miniprep DNA may be isolated on the third or fourth day.

On the fourth day restriction cut the miniprep DNA to determine which transformants contain the desired insert.

## 2. Materials

### 2.1. Preparation of PCR-Generated Insert

1. Source of insert sequence to be amplified: This may be genomic DNA, cDNA from a reverse transcriptase reaction or 100–200 ng of plasmid DNA. For this particular experiment, 10 μg of total GH₃ cell cytoplasmic RNA was reverse transcribed to cDNA with reverse transcriptase in a 20 μL reaction and subsequently used as the PCR template.
2. Oligonucleotide primers: The primers used for amplifying the N-cadherin fragment are 18-mers corresponding to amino acids 457–462 and 643–648 in the extracellular domain of mouse N-cadherin. Sequence for the upstream primer is 5' GGA TGT TTG TCC TTA CTG 3'. The downstream primer sequence is 5' TAG TCA CTG GAG ATA AGG 3'. Both primers have $T_m$ of 62.2°C as calculated by the %GC method.
3. 10X PCR buffer (Perkin-Elmer Cetus, Norwalk, CT). Alternatively, prepare by this recipe: 100 m$M$ Tris-HCl, pH 8.3, 500 m$M$ KCl, 15 m$M$ MgCl₂, 0.01% (w/v) gelatin. Use stock solutions that have been filter sterilized, autoclaved for 15 min on liquid cycle, and designated for PCR use only. For 1 mL of 10X PCR buffer combine: 100 μL of 1$M$ Tris-HCl, pH 8.3, 500 μL of 1$M$ KCl, 15 μL of 1$M$ MgCl₂, and 375 μL of UV-irradiated sterile water. Make up a 1% solution of gelatin in UV-irradiated sterile water. Heat at 60–70°C, mixing occasionally, to dissolve the gelatin. Filter the gelatin solution while it is still warm through a 0.2-μm filter, and add 10 μL of gelatin to each milliliter of 10X PCR buffer. Store 10X PCR buffer in 300–500 μL aliquots at –20°C. The buffer may be UV-irradiated before use.
4. Sterile UV-irradiated water: 0.2-μm filter deionized, distilled water into a sterile container. UV irradiate for 2 min in a Stratagene (La Jolla, CA) Stratalinker UV crosslinker (200 mJ/cm²) *(3)* or at 254 and 300 nm for 5 min *(4)*. Store at room temperature.

5. 10 m$M$ deoxynucleotide stocks (dATP, dCTP, dGTP, and dTTP). Supplied as 10 mg solids from Pharmacia (Piscataway, NJ). To make 10 m$M$ stocks, resuspend 10 mg of dNTP in 10% less sterile water than is required to give a 10 m$M$ solution. Adjust the pH to 7.0 using sterile NaOH and pH paper. Determine the exact concentration by OD, using the wavelength and molar extinction coefficient provided by the manufacturer for each dNTP. For example, the $A_m$ (259 nm) for dATP is $15.7 \times 10^3$; therefore a 1:100 dilution of a 10 m$M$ solution of dATP will have an $A_{259}$ of $(0.01M \times 15.7 \times 10^3$ OD U/M) $\times 1/100 = 1.57$. If the actual OD of a 1/100 dilution of the dATP is 1.3, the dATP concentration is $1.3/1.57 \times 10$ m$M$ = 8.3 m$M$. Store dNTPs at –20°C in 50–100-µL aliquots. Make a working stock containing 1.25 m$M$ of each dNTP in UV-irradiated sterile water for PCR. Unused working stock may be stored at –20°C for up to 2 wk. **Do not** UV-irradiate the dNTP solutions. Alternatively, dNTP solutions are available from several companies (e.g., Perkin-Elmer, Norwalk, CT).
6. Light mineral oil.
7. *Taq* DNA polymerase (Perkin-Elmer Cetus). If Vent or Pfu polymerase will be used to amplify the insert, *see* Note 1.
8. CHCl$_3$.
9. 7.5$M$ ammonium acetate. Filter through a 0.2-µm filter and store at room temperature.
10. 95% ethanol. Store at –20°C.
11. TE, pH 8.0: 10 m$M$ Tris-HCl, pH 8.0, 1 m$M$ EDTA. Sterilize by 0.2-µm filter and store at room temperature.

## 2.2. Ligation and Bacterial Transformation

1. pCRII vector at 25 ng/µL (supplied with Invitrogen TA Cloning Kit). Prepared insert. Determine approximate concentration of insert (*see* Note 2).
2. T4 DNA ligase and 10X ligation buffer (supplied with Invitrogen TA Cloning Kit). Store at –20°C.
3. Sterile water (supplied with Invitrogen TA Cloning Kit). Alternatively, 0.2-µm filter deionized, distilled water into a sterile container and store at room temperature.
4. One aliquot of TA One Shot™ Competent bacterial cells per ligation reaction, plus two aliquots for a "TE only" negative control transformation and a pUC18 positive control transformation (supplied with Invitrogen TA Cloning Kit). Store at –70°C.
5. 0.5$M$ β-mercaptoethanol (supplied with Invitrogen TA Cloning Kit). Store at –20°C.
6. pUC18 test plasmid (supplied with Invitrogen TA Cloning Kit).
7. TE (supplied with Invitrogen TA Cloning Kit). Alternatively, TE, pH 8.0, may be prepared by the investigator (*see* Section 2.1., item 11).
8. SOC Media (supplied with Invitrogen TA Cloning Kit). Store at room temperature.
9. Ampicillin or kanamycin. Prepare a stock solution of 50 mg/mL in water. Sterile filter. Store in aliquots at –20°C.

10. X-Gal (5-bromo-4-chloro-3–indolyl-β-D-galactoside). Prepare a stock solution of 40 mg/mL in *N,N* dimethylformamide under a fume hood *(5)*. Store at –20°C.

11. LB agar plates containing ampicillin or kanamycin. Prepare by dissolving 10 g of Bacto-tryptone, 5 g of yeast extract, 10 g of NaCl, and 15 g of agar in 1 L of $H_2O$ *(5)* in a 2.8-L Fernbach flask. Heat to near boiling while stirring to dissolve the agar. Autoclave for 20 min on liquid cycle. After solution has cooled to 50–55°C in a water bath, add ampicillin or kanamycin to 50 μg/mL (1 mL of a 50-mg/mL stock solution). Mix well. Pour molten agar into sterile Petri dishes, using 20–25 mL/plate. If plates are poured in a sterile hood, they may be allowed to cool partially uncovered until the agar has set. When solid, cover, invert, and store at 4°C. Antibiotic-containing agar has a shelf-life of about 1 mo. If longer storage is required, prepare agar plates without antibiotic and spread with ampicillin or kanamycin (25 μL of a 50 mg/mL stock) at least 1 h before use.

## 2.3. Plasmid Minipreps for Screening Transformant Colonies

1. 2X TY broth containing 50 μg/mL ampicillin or kanamycin. Prepare 2X TY broth by dissolving 16 g Bacto-tryptone, 10 g of yeast extract, and 5 g NaCl in 1 L of $H_2O$ in an autoclavable flask. Autoclave 20 min on liquid cycle. Cool and store at 4°C. Aliquot appropriate amount of 2X TY as needed for miniprep procedure and add antibiotic to 50 μg/mL.

2. 80% glycerol (v/v, in water), 0.2-μm sterile filter. Store at room temperature.

3. GTE buffer: 25 m*M* Tris-HCl, pH 8.0, 50 m*M* glucose, 10 m*M* EDTA. Sterilize by 0.2-μm filter and store at 4°C.

4. Lysis buffer: 0.2*M* NaOH, 1% SDS. Prepare fresh.

5. KOAc buffer: 3*M* potassium acetate, 2*M* acetic acid. Sterilize by 0.2-μm filter and store at room temperature.

6. TE, pH 8.0 *(see* Section 2.1., item 11).

7. Restriction enzyme, *Eco*RI, and the appropriate restriction digest buffer, React 3 (Life Technologies, Gaithersberg, MD). Store at –20°C.

## 3. Methods
### 3.1. Preparation of PCR-Generated Insert

1. Prepare four standard 100-μL PCR reactions, each containing: 200 μ*M* of each deoxynucleotide, 100 pmol of each primer, and 2.5 U of *Taq* polymerase in 1X PCR buffer. Two tubes are experimental, i.e., to amplify the target sequence; a third tube serves as a negative control, the fourth as a positive control. The two experimental tubes (reaction done in duplicate) and positive control tube will also contain 1/10 of the reverse transcriptase reaction as DNA template. The negative control tube should contain no cDNA template. This control tells you if there is contamination of the PCR solutions with DNA. If primers are available that amplify a known specific sequence in the cDNA, these may be substituted for the experimental primers in the fourth tube to check the integrity of the DNA template and the efficiency of the PCR reaction. For example, in this particular

experiment primers that amplify a portion of the prolactin gene were used in the positive control tube. *Taq* polymerase should be added last to the reaction mix and just prior to the cycling steps. Overlay the final reaction mix with a drop (approx 70 µL) of light mineral oil to prevent evaporation.

2. Amplify by PCR using the following cycle profile (*see* Note 3): 94°C for 1 min, 55°C for 2 min, 72°C for 1 min, for 17–25 main cycles; 72°C for 5 min (final extension).

3. Add 100 µL CHCl₃ to each PCR reaction tube to remove mineral oil. Spin for 1 min at approx 10,000$g$ in a microfuge.

4. Remove the top aqueous layer which contains the PCR reaction product, and transfer to a fresh microfuge tube. Discard the bottom layer as organic waste.

5. At this point the PCR product may be used directly in a TA ligation reaction (*see* Section 3.2., step 1), or held on ice until the amount of PCR product available for ligation can be determined (*see* Note 2 and the following procedure).

6. Precipitate the PCR product by adding 1/2 vol of 7.5$M$ ammonium acetate to the PCR reaction product. Mix briefly. Add 2.5 vol of 95% ethanol. Mix well.

7. Precipitate on dry ice for 1 h, or at –20°C overnight.

8. Spin out precipitate at 11,000–15,000$g$ for 30 min. Discard supernatant while ensuring DNA pellet remains in the tube. Dry pellet and resuspend in 10–20 µL of either TE or sterile water.

9. At this point the PCR product may be stored at –20°C or analyzed by gel electrophoresis to determine the size and amount of the amplified product (*see* Notes 2,4). If multiple sized products result from the PCR amplification, the band of interest may be purified out of agarose (*see* Note 5) and incubated with *Taq* polymerase to add the 3' A-overhangs required for TA cloning (*see* Note 1).

### *3.2. Ligation and Bacterial Transformation*

1. Combine 2 µL pCRII vector with PCR product insert in a 1:1 or 1:3 ratio. The formula for determining molar amounts of insert to use is as follows:

$X$ ng of PCR product to ligate = ($Y$ bp PCR product) (50 ng pCRII vector)/
(size in bp of the pCRII vector)

"$X$" ng is the amount of PCR product to be ligated for a 1:1 molar ratio. For the N-cadherin insert, $X$ = 7.3 ng ($X$ = [576 bp × 50 ng] / 3932 bp). Three times "$X$" ng would be used for a 1:3 molar ratio. If the PCR product will be used in a TA ligation without first determining the amount needed for a 1:1 and 1:3 molar ratio, set up ligation reactions that use various amounts of the PCR product. Under these circumstances we typically set up three ligation reactions per experimental PCR reaction using 1, 3, or 6 µL of the fresh PCR product in the ligation reaction.

2. Add the following to vector and insert to complete ligation reaction set-up: 1 µL 10X ligation buffer, $X$ µL sterile water (to bring reaction volume to 9 µL); and 1 µL T4 DNA ligase.

3. A self-ligation reaction may also be performed to test the stability of the pCRII vector. Carry out the reaction as above after substituting sterile water for the insert.

4. Incubate the ligation reactions overnight at 12°C.
5. When ligation reaction is complete, warm LB agar plates to 37°C. Use two plates for each transformation reaction.
6. Spin ligation reaction tubes briefly in a microfuge and place on ice.
7. Thaw 0.5*M* β-mercaptoethanol and OneShot competent cells on ice.
8. When cells have thawed, pipet 2 μL of 0.5*M* β-mercaptoethanol into each vial of competent cells. Tap gently to mix.
9. Pipet 1 μL of each ligation reaction (or 1 μL of pUC18 test plasmid or 1 μL TE) into a vial of competent cells. Tap gently to mix. Remaining ligation reaction may be stored at –20°C.
10. Incubate vials on ice for 30 min.
11. Incubate for 30 s at 42°C to heat shock bacterial cells (*see* Note 6).
12. Quickly place on ice for 2 min.
13. Add 450 μL of room temperature SOC medium to each vial of transformed cells. Place vials in a microcentrifuge rack and secure with tape.
14. Shake vials at 37°C for 1 h at 225 rpm in a rotary shaking incubator.
15. While vials are shaking, spread 25 μL of X-Gal stock solution (40 mg/mL) on top of LB agar plates with a L-shaped glass spreader. Use sterile technique. Let X-Gal diffuse into the agar for approx 1 h.
16. When incubation is complete, place transformation vials on ice. Spread 25 and 100 μL from each transformation vial on separate, labeled LB agar plates which contain either ampicillin or kanamycin and X-Gal.
17. Invert plates and incubate overnight at 37°C.

### *3.3. Plasmid Minipreps for Screening Transformant Colonies* (6)

1. The plates containing TE-transformed bacteria should be free of colonies. The presence of any colonies could indicate that the antibiotic is no longer working or the bacterial cells themselves are antibiotic resistant. Antibiotic resistance should not be a problem with fresh cells obtained from a commercially available kit.
2. The plates containing pUC18 control plasmid-transformed bacteria should have many colonies. This indicates the transformation worked.
3. The plates containing bacteria transformed with the self-ligation reaction should have mainly blue colonies. Any white colonies are caused by blunt-end self-ligation of the vector as a result of 3' T-overhang degradation. These colonies should be regarded as false positives. Knowing the percentage of false positives obtained for any given transformation allows an estimation of the number of colonies that can be expected to be true positives on the experimental plates.
4. The plates containing bacteria transformed with ligated constructs should contain a mix of blue and white colonies. Successful ligation of vector with insert will cause a disruption of the vector's *lacZ* gene and result in white colonies. However, if the insert size is 500-bp or less, the *lacZ* gene may not be disrupted as efficiently. This can result in light blue colonies that should then be treated as positives. Positive colonies may be screened by direct PCR sequencing or by preparation of plasmid DNA by the following miniprep procedure.

5. Touch a sterile loop or toothpick to each colony to be screened and inoculate each into a 2 mL aliquot of 2X TY broth containing 50 µg/mL kanamycin or ampicillin.

6. Incubate at 37°C with shaking (225 rpm) for 4–16 h, until broth is moderately turbid from bacterial growth.

7. Transfer 0.4 mL of each sample into an appropriately labeled 1.5-mL microfuge tube containing 0.1 mL of 80% glycerol. Mix and freeze on dry ice. Store these aliquots at −70°C as the future source of the desired transformants.

8. Transfer the remainder of each sample to another 1.5-mL tube, and centrifuge at 2600g for 2 min in a table-top microfuge. Aspirate the supernatant without disturbing the pellet. Leave the caps open for the next two steps.

9. Resuspend each pellet in 50 µL of GTE buffer by vortexing. Incubate at room temperature for 5 min.

10. Add 100 µL of lysis buffer to each tube while vortexing. Incubate on ice for 5 min.

11. Add 75 µL of KOAc buffer to each tube, close the caps, vortex to mix well, and incubate on ice for 5 min.

12. Centrifuge at approx 10,000g for 10 min in a table-top microfuge.

13. Transfer the supernatants to fresh 1.5-mL tubes containing 0.5 mL of 95% ethanol. Mix well and centrifuge again at top speed for 10 min.

14. Decant the supernatants and air-dry the DNA pellets for 20–30 min. Resuspend the DNA in 50 µL of TE, pH 8.0.

15. To determine which plasmids contain the desired insert, the miniprep DNA may be digested with restriction enzymes that cut within the pCRII vector's polylinker. For example, *Eco*RI restriction sites flank the TA cloning site. *Eco*RI may be used to cut out the insert (*see* Note 7). This leaves a minimal amount of vector DNA (6–10 bp) at each end of the insert. For each sample to be analyzed, combine 3–5 µL of miniprep DNA with 1 µL of 10X reaction buffer (React 3, Life Technologies), and 1 µL of *Eco*RI in a total volume of 10 µL. Incubate at 37°C for 30–60 min. Restriction digests were then treated with 1 µL of 10 µg/mL RNase A (Boehringer Mannheim, Indianapolis, IN) for 30 min at 37°C.

16. Analyze the digested DNA samples by agarose gel electrophoresis. Digested plasmids containing the N-cadherin insert will produce one approx 576-bp fragment in addition to the 3.9-kb linearized pCRII vector band (*see* Notes 5, 8). Digested plasmids that did not contain the insert will appear as only the 3.9-kb linearized band. If the restriction digest was incomplete, some plasmid may remain intact and supercoiled. Supercoiled plasmid will migrate faster in an agarose gel than the linearized plasmid.

## 4. Notes

1. If *Vent* or *Pfu* polymerase is used to amplify the PCR product, then an additional incubation with *Taq* polymerase is required to add 3' A-overhangs to the PCR product. A protocol for this procedure is available from Invitrogen.

2. The approximate concentration of insert may be determined by running one-fifth to one-tenth the PCR reaction product on an agarose gel with three different

amounts of λ DNA-*Bst*E II molecular weight markers (New England Biolabs, Beverly, MA). After ethidium bromide staining of the agarose gel, the approximate concentration of PCR product may be estimated by comparing the intensity of the product band to the λ DNA marker bands. Bands of the same intensity will have approx the same concentration of DNA. Because the percentage of each marker band present in the total amount of λ DNA preparation is known, the concentration of DNA present in a specific marker band is determined by multiplying the relative percentage for a specific marker band times the total amount of λ DNA loaded in that particular lane.

3. Since *Taq* polymerase has no 3'-5' exonuclease editing function *(7)*, it is important not to over-amplify, which may introduce mutations. We used 30 cycles in this particular experiment, but in the interest of minimizing mutations, one should begin with 17–20 cycles.

4. To ensure your PCR reaction amplified the correct-sized product, run a parallel PCR reaction that can be precipitated and analyzed by agarose gel electrophoresis or remove one-fifth to one-tenth of the PCR reaction product and analyze by agarose gel electrophoresis. In our experience the PCR reactions may be held at 15°C for up to 12 h without any noticeable decrease in ligation efficiency.

5. If the insert DNA is larger than 200-bp it may be purified out of the agarose gel after electrophoresis by using the Prep-a-Gene™ DNA purification kit (Bio-Rad, Hercules, CA). Alternatively, the insert may be purified out of agarose by electroelution *(8)*. After this point the purified insert may be labeled by random priming (e.g., Life Technologies) for use as a probe for Northern hybridization analysis.

6. Timing and temperature during transformation are critical. If the heat-shock step is too long or temperature is too high, then transformation efficiency may be reduced or cells may die. Also, if the heat shock temperature is too low, transformation efficiency may be reduced.

7. Choose an enzyme to cut out the insert that does not cut within the insert.

8. The pCRII vector contains T7 and Sp6 RNA promoters for production of sense and antisense RNA transcripts.

## Acknowledgments

This work was performed in the lab of Bruce A. White and supported by NIH grant DK43064.

## References

1. Ausubel, F. M., Brent, R., Kingston, R. E., Moore, D. D., Smith, J. A., Seidman, J. G., and Struhl, K. (eds.) (1987) *Current Protocols in Molecular Biology*. Wiley, New York.

2. Clark, J. M. (1988) Novel non-templated nucleotide addition reactions catalyzed by procaryotic and eucaryotic DNA polymerases. *Nucleic Acids Res.* **16,** 9677–9686.

3. Dycaico, M. and Mather, S. (1991) Reduce PCR false positives using the Stratalinker UV crosslinker. *Stratagene Strategies* **4(3),** 39,40.

4. Sarkar, G. and Sommer, S. S. (1990) Shedding light on PCR contamination. *Nature* **343,** 27.

5. Maniatis, T., Fritsch, E. F., and Sambrook, J. (1982) *Molecular Cloning. A Laboratory Manual.* Cold Spring Harbor Laboratory, Cold Spring Harbor, NY.

6. Sambrook, J., Fritsch, E. F., and Maniatis, T. (1989) *Molecular Cloning. A Laboratory Manual,* 2nd edition. Cold Spring Harbor Laboratory, Cold Spring Harbor, NY.

7. Eckert, K. A. and Kunkel, T. A. (1990) High fidelity DNA synthesis by the *Thermus aquaticus* DNA polymerase. *Nucleic Acids Res.* **18,** 3739–3744.

8. Zhen, L. and Swank, R. T. (1993) A simple and high yield method for recovering DNA from agarose gels. *BioTechniques* **14(6),** 894–898.

# 10

## Cloning Unmodified PCR Products Using Engineered *XcmI* Restriction Sites in a Portable Cassette

**Alessandro Testori and Paul Sollitti**

### 1. Introduction

The rapid amplification and isolation of specific DNA fragments made possible by the polymerase chain reaction *(1,2)* has become routine for a variety of molecular biology studies and applications. *Taq* DNA polymerase is still the most widely used enzyme for PCR amplifications, despite the fact that a number of other thermostable DNA polymerases are now available from commercial suppliers. *Taq* DNA polymerase is capable of catalyzing the addition of a single nucleotide, a deoxyadenosine (dA), to the 3' ends of amplified PCR products *(3)*, resulting in a single nucleotide 3' overhang. This renders blunt-end ligation of such PCR products to cloning vectors inefficient.

A number of strategies have been devised to clone PCR products. The incorporation of restriction sites into the primers used for PCR amplification is perhaps the most common way to clone PCR products. However, restriction endonucleases often fail to cleave these sites if they are located too close to the termini of the DNA molecules *(4)*.

More elaborate strategies involve ligation-independent cloning of the PCR products which require the synthesis of more than two primers. Additionally, 12 or more bases must be added to their 5' ends in order to allow the ligation of the insert to the vector and recircularization of the final product *(5,6);* this approach is effective but costly.

In recent years, T-vectors *(7)* have been widely used to clone PCR fragments. T-vectors are prepared by incubating a linearized, blunt-ended plasmid vector with *Taq* DNA polymerase and an excess of deoxythymidine triphosphate (dTTP) *(7)* or by using terminal deoxynucleotidyl transferase (TDT) to add

From: *Methods in Molecular Biology, Vol. 67: PCR Cloning Protocols: From Molecular Cloning to Genetic Engineering* Edited by: B. A. White  Humana Press Inc., Totowa, NJ

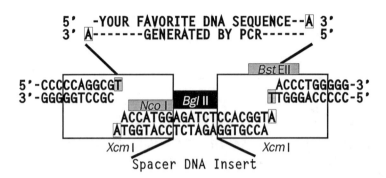

**Fig. 1.** Design of the oligodeoxynucleotide linker used in the construction of pES vector. All of the restriction sites included in this linker are unique to the pES vector. Locations for the insertion of PCR-generated DNA fragments and spacer DNA are indicated.

dideoxy thymidine triphosphate (ddTTP) to the 3' termini of the vectors *(8)*. In the absence of deoxyadenosine, the terminal transferase activity of *Taq* DNA polymerase will catalyze the addition of a single deoxythymidine at the 3' ends of the vector molecules. The resultant vectors may then be ligated to the PCR products bearing single, complementary, deoxyadenosine overhangs at their 3' ends *(7)*. The preparation of dT- or ddT-tailed vectors is not without problems. The addition of dT or ddT catalyzed by *Taq* DNA polymerase or TDT, respectively, is inefficient and not all treated vector molecules bear properly modified ends *(7)*.

One additional disadvantage of these T-vector systems is that, for some applications, subcloning into a second vector is required, for instance to prepare deletion constructs to study promoter activity in vectors that carry reporter genes. Thus, the use of traditional T-vectors may require an additional cloning step that could significantly lengthen the procedure and is limited by the available restriction sites that flank the insert in the T-vector.

In this report, we present an efficient protocol to prepare vectors, based on engineering *Xcm*I restriction sites, that contain single 3' overhanging dTs suitable for the rapid cloning of PCR products of any DNA fragments bearing single dA overhangs on their 3' ends *(9)*. The modified vectors can be easily propagated to generate large amounts of T-vector. A double-stranded oligodeoxyribonucleotide (oligos) linker carrying two *Xcm*I restriction sites arranged in tandem (Fig. 1) was generated by first synthesizing two complementary single-stranded oligos by the phosphoramidate method. These oligos were subsequently hybridized together and inserted into the *Eco*RV site of pBluescript SK II (+) to form the plasmid pES (Engineered Site) (Fig. 2). The orientation of the linker was confirmed by restriction mapping and by sequencing. The

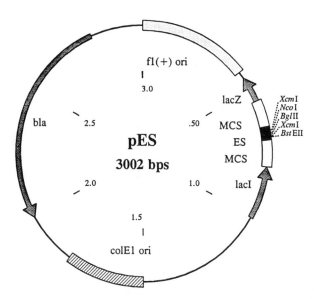

**Fig. 2.** Map of pES. ES, engineered restriction site linker containing the tandem *Xcm*I restriction sites without any inserted spacer DNA. MCS, multiple cloning site from pBluescript II SK (+) containing the ES region in the *Eco*RV restriction site. The *Eco*RV restriction site is disrupted in the pES vector. The orientation of the MCS in the pES vector is identical to pBluescript II SK (+).

*Xcm*I site is characterized by the CCA(N$_5$)/(N$_4$)TGG recognition sequence, in which the central nine nucleotides are not important for enzyme recognition; therefore, each of the tandem *Xcm*I sites can be engineered to produce 3' dT overhangs on digestion.

The tandem arrangement of the engineered *Xcm*I sites ensures the production of a single 3' dT overhang at each end of the vector following restriction by *Xcm*I. We chose to include the unique restriction sites *Bgl*II, *Bst*EII, and *Nco*I, originally absent from the sequence of pBluescript SK II (+), into the sequence of the oligodeoxyribonucleotide linker. These sites can be used for mapping purposes to verify that both *Xcm*I sites are present, or to confirm that both *Xcm*I sites have been cleaved, or as sites for routine cloning in this vector derivative. However, any other restriction site or otherwise random sequence that allows a dT to be present in the appropriate position to generate a 3'-dT overhang would be acceptable.

The *Xcm*I generated dT overhangs at each 3' end eliminates the possibility that the *Taq* DNA polymerase or TDT enzymes might fail to add dT or ddT overhangs to both 3' ends of the plasmid, giving rise to a vector incapable of ligation to the insert containing 3'-dA overhangs and subsequent recircularization.

## 2. Materials

1. Two complementary oligodeoxyribonucleotides that contain two *Xcm*I sites arranged in tandem (*see* Fig. 1).
2. T4 DNA ligase and buffer (containing ATP) supplied with enzyme.
3. *Xcm*I, *Bgl*II, *Bst*EII restriction endonucleases and a suitable restriction enzyme to be used to generate a blunt-end site on the target vector.
4. A plasmid vector of choice; pBluescript SK II (+) (Stratagene, La Jolla, CA) was used in these experiments.
5. Reagents for acrylamide gel electrophoresis, including acrylamide solution (19:1 ratio of acrylamide to *bis*-acrylamide), TBE buffer (10X: $1M$ Tris-HCl, pH 8.3, $0.83M$ boric acid, 10 m$M$ Na$_2$ EDTA-H$_2$O), TEMED, and 10% ammonium persulfate.
6. Tris-equilibrated phenol, pH 7.0–8.0.
7. Ethanol, 100 and 70%.
8. Reagents for bacterial culture, including *Escherichia coli* strain DH5α, LB agar and broth, and ampicillin, all from Life Technologies (Gibco-BRL, Gaithersburg, MD).
9. Reagents for agarose gel electrophoresis (*see* Chapter 1).
10 Salmon sperm DNA (Sigma, St. Louis, MO).
11. TE buffer: 10 m$M$ Tris-HCl, pH 8.0, 0.1 m$M$ EDTA.
12. CHCl$_3$.
13. PCR products amplified by *Taq* DNA polymerase.
14. Solutions for minialkaline lysis plasmid preparations:
    a. Solution I: 50 m$M$ glucose, 25 m$M$ Tris-HCl, pH 8.0, 10 m$M$ EDTA.
    b. Solution II: $0.2N$ NaOH, 1% SDS.
    c. Solution III: $5M$ postassium acetate, pH 4.8. (300 mL of $5M$ potassium acetate salt, 57.5 mL glacial acetic acid, and 145.5 mL of double-distilled H$_2$O).
15. Isopropanol.
16. $2M$ ammonium acetate.
17. RNase A.

## 3. Methods

### 3.1. Synthesis and Purification of Oligodeoxyribonucleotides Containing the Tandem XcmI Cassette

The first steps in constructing the tandem *Xcm*I vectors include the synthesis and purification of oligodeoxyribonucletides that have been designed to produce a single 3'-dT overhanging end after digestion with *Xcm*I. Although we suggest a particular sequence that includes several other restriction sites, many other sequence substitutions would also function well. Care must be used to ensure that thymidine is present in the proper place in the *Xcm*I recognition site so that a T-vector will be produced by digestion with that enzyme (*see* Note 1).

1. Synthesize two complementary oligodeoxyribonucleotides (oligos) by the phosphoramidate method according to the following sequences:

   5'-CCCCCAGGCGTACCATGGAGATCTCCACGGTAACCCTGGGGG -3'
   5'-CCCCCAGGGTTACCGTGGAGATCTCCATGGTACGCCTGGGGG -3'

   Alternatively, primers can be ordered from various vendors.

2. Dilute 10 µL of each oligo 100-fold in a quartz spectrophotometer cuvet and measure the absorbance at 260 nm ($A_{260}$). The concentration of the oligo is approx 25 µg/$A_{260}$ × $A_{260}$ × dilution factor. (The 27 µg/$A_{260}$ U applies to short, single-stranded oligodeoxyribonucleotides only.)

3. Based on the concentrations obtained by $A_{260}$, add 1 µg of each oligo to a micro-centrifuge tube in a total volume of 20–50 µL, heat the tube to 100°C in a 200-mL boiling water bath (in a beaker) for 5 min, and allow to cool to room temperature slowly by removing the water bath from the heat. This allows the two oligos to hybridize, forming a double-stranded DNA fragment.

4. Purify the annealed DNA by electrophoresis on a 10% polyacrylamide native gel (19 acrylamide: 1 *bis*-acrylamide) run in 1X Tris borate buffer *(10)*. Recover the DNA by soaking the excised gel fragment in TE buffer overnight at 4°C followed by precipitation with 100% ethanol and recovery by centrifugation.

5. The double-stranded DNA fragment resulting from the annealing of the oligos containing the tandem *Xcm*I sites is ready to clone into pBluescript SK II (+) by standard methods *(10; see* Section 3.2.).

## 3.2. Insertion of the Tandem XcmI-Containing Oligodeoxyribonucleotide into the Target Vector

Once the oligos containing the tandem *Xcm*I sites have been synthesized, annealed, and purified, they must be inserted into an appropriate vector that can be used as a recipient of PCR-generated fragments (*see* Notes 2,3). The pES construct described by Testori et al. *(9)* was based on pBluescript SK II (+) because it contained numerous restriction sites in its multiple cloning site region and because it is maintained at a high copy number in *E. coli,* thus facilitating its easy isolation. This protocol is adaptable to almost any plasmid for use in nearly any system. Be careful to choose plasmids containing appropriate replicons, markers, and restriction sites that will allow unhindered manipulation of the DNA inserted into the modified vector.

1. Prepare the pBluescript SK II (+) for cloning by digestion of 1 µg of the plasmid with *Eco*RV for 1 h. Extract the digested plasmid once with Tris-equilibrated phenol, precipitate with 2 vol of 100% ethanol, centrifuge 5 min at 13,000*g,* wash twice with 70% ethanol, dry, and redissolve in TE buffer. *Eco*RV produces a blunt cut in the DNA at its cognate site. However, any vector can be used once a blunt restriction site has been generated. Blunt sites can also be generated by filling in cohesive ends generated by many restriction enzymes *(10)*.

2. Ligate 3 pmol of the tandem *Xcm*I-containing double-stranded oligo to 0.5 pmol of the *Eco*RV digested pBluescript SK II (+) DNA using standard methods *(10)*.

A threefold molar excess of nonkinased insert double-stranded oligo and nondephosphorylated vector should be used to ensure insertion of only one copy of the insert. Remember that the number of pmol of the annealed, double-stranded oligo = ([pmol of oligo$_1$ + pmol of oligo$_2$]/2). Use 1 U of T4 DNA ligase in the ligation reaction and incubate at room temperature for 4 h.

3. Transform *E. coli* strain DH5α with 3 μL of the ligation reaction according to the suppliers' instructions provided with the "subcloning efficiency" competent cells. Transformants were plated onto LB plus 100 μg/mL of ampicillin (*see* Section 3.5. for more complete ligation and transformation conditions).

4. Since the inserted oligo was expected to maintain the reading frame of the *lacZ* gene in which the *Eco*RV site was located, screening for inserts by blue/white selection in the presence of IPTG plus X-gal was not possible. Screen inserts by growing them in LB broth plus 100 μg/mL of ampicillin and making minipreparations of plasmid DNA from randomly selected colonies (20–40) using the protocol outlined in Section 3.6.

5. Screen the plasmid minipreparations by separate digestions with *Bgl*II, *Nco*I, and *Bst*EII (this enzyme works at 65°C) followed by electrophoresis on a 1% agarose gel containing 0.3 μg/mL of ethidium bromide using TAE or TBE buffer *(10)*. These digestions will linearize the correct insert-containing plasmids. Additionally, since *Nco*I and *Bst*EII were engineered separately into the *Xcm*I sites (Fig. 1), digestion with these enzymes will confirm the presence of the tandem *Xcm*I sites.

## 3.3. Insertion of Spacer DNA into Tandem Xcm*I*-Generated T-Vectors

One problem with these constructions is the inability to easily determine whether both *Xcm*I sites are cleaved during digestion since the size of the region between the *Xcm*I sites is too small to cause an observable difference in migration on an agarose gel. Although *Xcm*I generally digests DNA efficiently, we considered this ambiguity a limitation in the ability to adequately purify the vector species that was cut at both *Xcm*I sites from the species that was only linearized by *Xcm*I digestion. To overcome this limitation, we inserted a piece of DNA into the *Bgl*II site as a spacer between the tandem *Xcm*I sites to allow easy separation of the various forms of cut and uncut T-vectors (Fig. 3).

1. Dissolve high molecular weight salmon sperm DNA in TE to a concentration of 10 mg/mL. One milliliter of this solution is a great excess since only a few microliters will be used for this procedure. Be careful not to shear the DNA by excessive rough pipeting and do not heat the DNA above 50°C.

2. Restriction digest 1 μg of the tandem *Xcm*I site T-vector with *Bgl*II for 1 h. Extract the restriction digest with an equal volume of phenol/chloroform *(10)*. Precipitate with 2 vol of 100% ethanol, collect the DNA by centrifugation, wash twice with 70% ethanol, dry, and redissolve in 20 μL of TE.

3. Once the salmon sperm DNA is completely dissolved in TE buffer, digest 5 μg of this DNA with both *Xcm*I and *Bgl*II for 2 h. (This can be performed sequentially

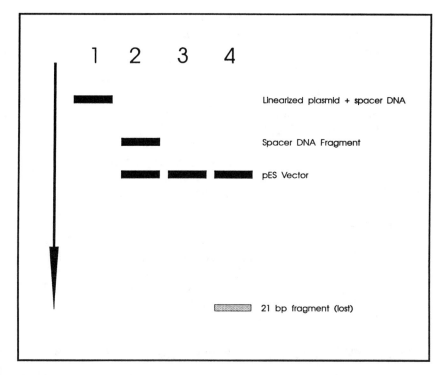

**Fig. 3**. Diagram showing how the insertion of a spacer DNA fragment into the central *Bgl*II site can facilitate the purification of the pES vector that has been digested completely with *Xcm*I and gel-purified. Lane 1, Linearized pES plasmid plus spacer DNA: Lane 2, completely digested pES plasmid plus spacer DNA showing two distinct bands; Lane 3, linearized pES plasmid without a spacer DNA insert; Lane 4, completely digested pES plasmid lacking spacer DNA. The 21-bp DNA excised from the linker is too small to be resolved on most agarose gels.

or concurrently.) After digestion, extract the restriction reaction with phenol/ chloroform, precipitate, wash, and redissolve in 20 μL of TE buffer as described in step 2.

4. Ligate together 5 μL each of the digested T-vector and digested salmon sperm DNA *(10)* at room temperature for 4 h to overnight. Any other ligation protocol should function equivalently, including those of "ultrafast" ligation kits. Transform the ligated DNA into DH5 α or some other suitable host strain. In the case of the pES vector, the inserted DNA was selected to be free of *Xcm*I sites and to contain *Bgl*II sites at their ends. Additionally, the insert was expected to disrupt expression of the *lacZ* gene; a function previously preserved after insertion of the DNA segment containing the tandem *Xcm*I sites into pBluescript SK II (+). Thus, those plasmids that contain the spacer inserts caused the host cells to be white, rather than blue, when plated on a medium containing IPTG and X-gal. This

strategy might not work with different vector constructs unless appropriate modifications are made to the original oligodeoxyribonucleotide sequence.

5. Pick white colonies and grow in LB plus 100 μg/mL of ampicillin for use in plasmid minipreparations described in Section 3.6.
6. Screen the plasmids by separate digestion with *Xcm*I (to ensure these sites remained intact) and *Bgl*II to ensure that the size is the same as that obtained by *Xcm*I digestion. These analyses will confirm that the insert does not contain an *Xcm*I site and that it is cloned into the *Bgl*II site between the tandem *Xcm*I sites. Analyze all restriction digests by electrophoresis using 1% agarose gels (horizontal, submerged) containing 0.3 μg/mL of ethidium bromide.
7. Choose a plasmid containing an insert that allows clean electrophoretic separation of the spacer insert, the uncut and linearized plasmids containing the insert and the plasmid from which the insert was excised (Fig. 3). Thus, the *Xcm*I digested plasmid, free from the spacer insert and definitely restricted at both *Xcm*I sites, can be isolated cleanly from the agarose gel using one of many available methods, such as a commercial kit, centrifugation through a 0.45 μm filter membranes, or electroelution. For the pES vector, we chose a plasmid that contained a 4 kb spacer insert after screening a small number of candidate plasmids.
8. The purified T-vector is now ready to be used for the direct cloning of PCR fragments.

## 3.4. Cloning PCR Products into Xcm I-Generated T-Vectors

We advise that the PCR fragments be purified before use in the ligation reaction. The Wizard PCR Preps DNA Purification System by Promega (Madison, WI) functions well for this purpose. This system allows purification of PCR fragments directly from the PCR mixture without running an agarose gel. Alternatively, the PCR-generated fragments can be cleaned using small centrifugal ultrafilters with a low nominal molecular weight cutoff (about 10–30 K), such as Millipore Ultrafree-MC (Bedford, MA) or Amicon Microcon 30 (Beverly, MA). These facilitate removal of unused primers and dNTPs as well as buffer exchanges.

Ligation of PCR products to the pES vector and their subsequent transformation into competent *E. coli* cells were performed according to established ligation and transformation protocols *(10)*. We chose T4 DNA ligase for use in our reactions. Recent evidence suggests that inactivation of the T4 DNA ligase after ligation is complete will dramatically increase the efficiency of transformation into *E. coli (11)*. The details of the ligation and transformation conditions are outlined below (*see* Note 4).

1. Mix 1 pmol of the redissolved vector DNA with 3 pmol of the PCR product and 2 μL of 10X ligation buffer. Adjust the volume to 19 μL with ddH$_2$O. Add 1 U/μL of T4 ligase to bring the total volume to 20 μL. Mix the reaction briefly by finger-vortexing, then incubate the mixture at room temperature for 4 h.

2. After incubation, heat the ligation reaction mixture to 65–70°C for 10 min to inactivate the ligase, then cool on ice.
3. Thaw on ice an aliquot (50 or 100 μL) of "subcloning efficiency" DH5α competent cells. The competent cells should be in a thin-wall microcentrifuge tube.
4. Add 3 μL of the ligation reaction to each 50 μL of competent DH5α cells and incubate on ice for 20–30 min.
5. Heat-shock the cells in a water bath for 20 s at 37°C. Cool on ice for 2 min then add 1 mL of 2X LB broth **(no antibiotic)**. Incubate the transformants 1 h at 37°C either shaking or on a cell rotating wheel. If shaking is used, it might be convenient to transfer the culture to a larger round bottom sterile tube so it can fit in a tube clamp on some shaker incubators or so it can be placed in a small flask or beaker that can be fit into a reciprocating or rotary shaker bath.
6. Vortex the cells briefly immediately before plating since they tend to aggregate. Spread 250 μL of the transformed culture evenly onto each LB agar plate containing 100 μg/mL of ampicillin. Incubate the plates overnight at 37°C.
7. Pick colonies for plasmid DNA minipreparations to screen for inserts (*see* Section 3.6.).

### 3.5. Mini-Alkaline Lysis Plasmid Preparation

Screening the recombinant plasmids at each step in the vector construction to final insertion of the PCR-generated fragment requires that the plasmid DNAs be isolated from the *E. coli* cells. Many kits are available from such companies as Nucleopore, Qiagen, Promega, Stratagene, BIO 101, and many others to whom we apologize for not mentioning, that will perform quite well for this purpose. We use a very abridged and inexpensive variation of the Birnboim and Doly *(12)* alkaline lysis minipreparation of plasmid DNA that provides good yields of plasmid that are suitable for restriction mapping, cloning, and sequencing. The solutions required and the complete protocol are outlined below.

1. Grow a 2 mL overnight culture of *E. coli* in LB-broth plus 20–100 μL/mL of ampicillin, shaking at 37°C.
2. Pour 1.5 mL of the culture into a microcentrifuge tube. Pellet the cells 15–20 s.
3. Resuspend cells in 100 μL of solution I and vortex to resuspend the pellet.
4. Add 200 μL of solution II, mix by inversion or gentle shaking. Cells lyse immediately.
5. Add 150 μL of solution III, cap, and invert tubes to mix or vortex briefly. Let stand about 2 min at room temperature.
6. Spin 5 min in a microcentrifuge to pellet the potassium dodecyl sulfate-protein complexes. Transfer 400 μL of the supernatant to a fresh tube. Add 200 μL of isopropanol, mix, and let stand 5 min at room temperature to precipitate the nucleic acids.
7. Pellet the precipitated nucleic acids 5 min in a microcentrifuge. Pour off supernatant. Redissolve the pellet in 200 μL of 2*M* ammonium acetate. Let stand 2–5 min at room temperature.

8. Add 1 mL of 100% ethanol. (More easily, just fill tube to the top with 100% ethanol from a squirt bottle). Mix, and let stand 1–2 min.

9. Pellet nucleic acids 5 min in a microcentrifuge. Decant the supernatant. Wash the nucleic acid-containing pellet twice with 1 mL of 70% cold ethanol, centrifuge 30 s and decant or aspirate (carefully) the supernatant. Dry pellets briefly under vacuum.

10. Redissolve the pellets in 45 μL of TE plus 100 μg/μL RNase A (final concentration). Use about 3 μL for restriction digest. Recovery of pUC-based plasmids is about 5–7 μg (*see* Note 5).

## 4. Notes

1. In the process of constructing the modified *Xcm*I cassette, we found that *Asp*I [5'GAC(N$_3$)/(N$_2$)GTC-3'] (Boehringer-Manneheim, Indianapolis, IN) or its isoschizomer *Eam* 1105I (New England Biolabs, Beverly, MA) would also be suitable for this purpose. In fact, while our original manuscript was being reviewed, Ichihara and Kurosawa *(13)* constructed such a vector. Vectors that used modified *Xcm*I sites were developed independently and concurrently by our group and others *(9,14,15)*. These reports were published within a few months of each other, but were reviewed simultaneously.

2. Once the oligos that contain the engineered *Xcm*I sites have been synthesized, they can be inserted into almost any vector and can be used in bacterial, yeast, or mammalian cloning systems. The sequences adjacent to the *Xcm*I cassette can be modified so that it can be used to make in-frame gene fusions in reporter systems by avoiding the incorporation of nucleotides that would form translational stop codons upstream of the fusion junctions.

3. Using the strategy of a portable, engineered tandem *Xcm*I PCR cloning cassette can be helpful, such as in deletion studies of promoters in which many constructs need to be generated. First, the PCR fragments are cloned into a T-vector then transferred to a reporter vector after choosing compatible internal restriction sites. In our approach the PCR fragment can be cloned directly into a reporter vector that has been modified by the addition of an engineered *Xcm*I oligodeoxyribonucleotide linker. To further expand the utility of this portable PCR cloning cassette, other modifications can be made in the sequence of the oligo. For example, promoter sequences for bacterial, yeast, insect, or mammalian systems can be included to facilitate in vivo expression of the PCR-generated inserts. Alternatively, if the tandem *Xcm*I linker is to be used in a vector other than pBluescript SK II (+), phage RNA polymerase recognition sites (T3, T7, SP6) can be inserted into the sequence for in vitro transcription of the inserts. Moreover, the entire customized tandem *Xcm*I cassette can be inserted into any number of vectors that might be useful to the researcher.

4. In practical terms, cloning unmodified PCR products is sometimes difficult. In our experience, the cloning efficiency was occasionally below the level of detection when conventional T-vectors were used. In addition, commercially available T-vectors are usually supplied in limited, expensive quantities whereas *Xcm*I T-

vectors can be generated at a modest cost in useful quantities. Since any vector can be engineered for cloning PCR fragments, this system is versatile and adaptable to a variety of applications.

5. Analysis of cloned PCR fragments:

a. The PCR fragments cloned into the pES T-vectors can be analyzed by restriction mapping or recovered easily by taking advantage of the numerous restriction sites that remain in the polylinker. Recall that the tandem *XcmI* cassette was inserted near the center of the pBluescript II SK (+) multiple cloning site (*EcoRV*). Other convenient restriction sites can be engineered into the tandem *XcmI* cassette and can be made to flank the actual insertion site of the PCR product. Some useful enzymes would be those that recognize 8 bp sequences, reducing the possibility of digesting the insert sequence as well.

b. The pES vector retains the T7 and T3 and *lac* promoter recognition sequences present on the parent pBluescript II SK (+) vector as well as the f1 replication origin. These features would facilitate:

   i. In vitro transcription of the PCR-generated insert for expression of fusion proteins.

   ii. Strand-specific RNA synthesis, useful for generating high specific-activity radiolabeled probes.

   iii. Fusion protein expression from the inducible *E. coli lac* promoter upstream from the *lacZ* gene. The tandem *XcmI* cassette in the pES vector was designed to maintain the reading frame of *lacZ*.

   iv. Generation of nested sequential deletions using exonuclease III and S1 or mung bean nuclease. The pBluescript II SK (+) plasmid contains restriction sites that leave 3' or 5' overhanging sequences. Targeting these restriction sites, in combination, facilitates generating unidirectional deletions since 3' overhanging sequences are resistant to exonuclease III digestion.

   v. The PCR inserts can be sequenced starting at their ends by using the flanking T7 and T3 promoters as priming sites for dideoxy sequencing. This is especially useful after sequential deletions have been generated. Sequencing can be performed directly on the double-strand template; alternatively, single-strand template can be generated using the f1 origin.

# References

1. Saiki, R. K., Gelfand, D. H., Stoffel, S., Scharf, S. J., Higuchi, R., Horn, G. R., Mullis, K. B., and Erlich, H. A. (1988) Primer-directed enzymatic amplification of DNA with a thermostable DNA polymerase. *Science* **239,** 489–491.

2. Saiki, R. K., Scharf, S., Faloona, F., Mullis, K. B., Horn, G. T., Ehrlich, H. A., and Arnheim, N. (1985) Enzymatic amplification of β globin genomic sequences and restriction site analysis for diagnosis of sickle cell anemia. *Science* **230,** 1350–1354.

3. Clark, J. M. (1988) Novel nontemplated nucleotide addition reactions catalyzed by procaryotic and eucaryotic DNA polymerases. *Nucleic Acids Res.* **16,** 9677–9686.

4. Kaufman, D. L. and Evans, G. A. (1990) Restriction endonuclease cleavage at the termini of PCR products. *BioTechniques* **9**, 305,306.
5. Shuldiner, A. R., Scott, L. A., and Roth, J. (1990) PCR-induced (ligase-free) subcloning: a rapid reliable method to subclone polymerase chain reaction (PCR) products. *Nucleic Acids Res.* **18**, 1920.
6. Aslanidis, C. and De Jong, P. J. (1990) Ligation-independent cloning of PCR products (LIC). *Nucleic Acids Res.* **18**, 6069–6074.
7. Marchuk, D., Drumm, M. M., Saulino, A., and Collins, F. S. (1991) Construction of T-vectors; a rapid and general system for direct cloning of unmodified PCR products. *Nucleic Acids Res.* **19**, 1154.
8. Holton, T. A. and Graham, M. W. (1991) A simple and efficient method for direct cloning of PCR products using ddT-tailed vectors. *Nucleic Acids Res.* **19**, 1156.
9. Testori, A., Listowsky, I., and Sollitti, P. (1994) Direct cloning of unmodified PCR products by exploiting an engineered restriction site. *Gene* **143**, 151,152.
10. Sambrook, J., Fritsch, E. F., and Maniatis, T. (1989) *Molecular Cloning: A Laboratory Manual,* 2nd ed., Cold Spring Harbor Laboratory, Cold Spring Harbor, NY.
11. Michelsen, B. K. (1995) Transformation of *Escherichia coli* increases 260-fold on inactivation of T4 DNA ligase. *Anal. Biochem.* **255**, 172–174.
12. Birnboim, H. C. and Doly, J. (1979) A rapid alkaline extraction procedure for screening recombinant plasmid DNA. *Nucleic Acids Res.* **7**, 1513–1523.
13. Ichihara, Y. and Kurosawa, Y. (1993) Construction of new T vectors for direct cloning of PCR products. *Gene* **130**, 153,154.
14. Cha, J., Bishai, W., and Chandrasegaran, S. (1993) New vectors for direct cloning of PCR products. *Gene* **136**, 369,370 and **141** (1994) 149 (Corrigendum).
15. Harrison, J., Molloy, P. L., and Clark, S. J. (1994) Direct cloning of polymerase chain reaction products in an *Xcm*I T-vector. *Anal. Biochem.* **216**, 235,236.

# 11

## A T-Linker Strategy for Modification and Directional Cloning of PCR Products

**Robert M. Horton, Raghavanpillai Raju, and Bianca M. Conti-Fine**

### 1. Introduction

The propensity of *Taq* polymerase to add 3'-A overhangs *(1,2)* to PCR-amplified DNA has made possible a simple method for cloning PCR products into a T-vector (Invitrogen [San Diego, CA]; *3–5*). Here we present a related strategy that uses T-linkers to add sequences, such as restriction sites, to the ends of PCR products (*see* Note 1). A single base T overhang at the end of a synthetic double-stranded oligonucleotide linker allows ligation of the linker to the unpolished ends of a PCR product. This avoids the expense of adding the "extra" sequences to sequence-specific primers.

#### 1.1. Examples

Two T-linker designs are presented here. In each case, the T-linker is a double-stranded synthetic oligonucleotide composed of complementary oligos (either TL-A and TL-B for the *Nde*T linker, or HisTL-A and HisTL-B for the HisT-linker) with a single 3' overhanging 't' at one end.

##### 1.1.1. Nde TL

The basic principles involved in using a T-linker are shown using the *Nde*-T-linker in Fig. 1. This T-linker contains complete *Eco*RI and *Not*I sites, and a third site (*Nde*I) is partially present, except for its final 'g'; the overhanging 't' is part of this site. The 5' end of TL-B is phosphorylated (indicated by an asterisk), so that it can be ligated. The other end of the linker contains a sticky *Hin*dIII-compatible end, which was not used in the approach described here. However, since this 5' overhang is filled in by the polymerase during PCR, these extra bases serve as a "clamp" or spacer, which permits the *Eco*RI site to be cut.

From: *Methods in Molecular Biology, Vol. 67: PCR Cloning Protocols: From Molecular Cloning to Genetic Engineering* Edited by: B. A. White Humana Press Inc., Totowa, NJ

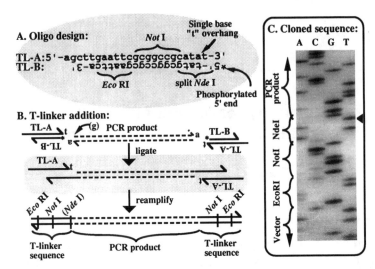

**Fig. 1.** Cloning of PCR products with a T-linker. (**A**) Oligonucleotide design. (**B**) Addition of the T-linker to a PCR product. The parenthetical 'g' and '*Nde*I' indicate that if a 'g' is the first base in the PCR product, the *Nde*I site will be completed. (**C**) Sequence from a pBluescript KS II+ plasmid containing a T-linker-reamplified insert cloned into its *Eco*RI site. This sequence was obtained using a *fmol* cycle sequencing kit (Promega, Madison, WI). The base representing the overhanging 't' of the T-linker is indicated by the solid triangle to the right of the sequence. Since the PCR product begins with a 'g', a complete *Nde*I site is formed.

The overhanging 't' of the linker matches the 'a' added by the polymerase and directs the ligation of TL-B, allowing reamplification of the sequence using TL-A as a primer. The resulting molecule contains the original PCR-amplified sequence flanked by inverted repeats of the T-linker.

If one of the sequence-specific primers has a 'g' at its 5' end, an *Nde*I site will be formed. This site is "split" between the T-linker and the PCR product to be cloned. In this way, very minor restraints in the PCR primer sequences (having one start with 'g', but not the other) can be used to complete the site at only one end of the reamplified product. One simple application of this idea is directional cloning (*see* Notes 2,3).

The use of this T-linker is illustrated in Fig. 1. A portion of a T-cell receptor V region was amplified from Jurkat tumor line cDNA using primers Vb8-1cpe5 and hpVbe3 (*see* Chapter X). The ligated products were reamplfied with TL-A and cloned into the *Eco*RI site of pBluescript KS II⁺. The sequence of the resulting product is shown in Fig. 1C. At the bottom of the gel is sequence from the polylinker of the vector to the *Eco*RI site, followed by the T-linker. The overhanging 't' is marked. The 'g' completing the *Nde*I site is the first base of Vb8–1cp.5.

## 1.1.2. HisTL

Our second example adds another level of sophistication, as shown in Fig. 2. Here an additional ability of the T-linker is brought into use. A new and useful sequence is added in addition to the restriction sites, namely a "histidine tag" sequence, which will be used for affinity purification of the expressed recombinant α3 protein on a $Ni^{2+}$ column (Qiagen, Chatsworth, CA). Note that the *Hin*dIII site is created only at the 3' end of the product in this case, because only the 3' primer begins with 't'. The histidine tag portion is removed from the 3' end by cutting with *Hin*dIII. No special sequences have been added to the sequence-specific primers hacpe5 and hacpe3, but their positions have been chosen to take advantage of the design of the HisT-linkers. The initial PCR-amplified sequence begins with a full codon, to put the sequence in the proper reading frame with the His tag. It ends with "ta"; the "a" added by the polymerase finishes the termination codon (taa) and is included in the *Hin*dIII site.

## 2. Materials

1. cDNA template: This was reverse transcribed from total RNA using Superscript RNase H-reverse transcriptase (Life Sciences) and an oligo (dT) primer using manufacturer's instructions.
2. Reagents for PCR: 10X buffer: 500 m*M* KCl, 100 m*M* Tris-HCl, pH 8.3; red sucrose is a PCR-compatible gel-loading dye consisting of ~1 m*M* cresol red in 60% sucrose *(6); Taq* DNA polymerase, 10 m*M* dNTP mix, 10 m*M* $MgCl_2$ solution (Perkin-Elmer Cetus, Norwalk, CT).
3. Reagents for agarose gel electrophoresis (*see* Chapter 1).
4. GeneClean (Bio 101, La Jolla, CA).
5. Ligation reagents: T4 DNA ligase (Stratagene, La Jolla, CA) 8U/μL, ligase buffer supplied by manufacturer, 10 m*M* ATP (Pharmacia, Piscataway, NJ), and PEG 8000 (Aldrich, Milwaukee, WI). Recently, we have used T4 ligase buffer from Life Technologies (Gibco-BRL, Gaithersburg, MD), which already contains ATP and PEG.
6. T-linker oligonucleotides:
   a. *Nde*-T-linkers: TL-A: (23-mer) 5'-agcttgaattcgcggccgcatat-3'; TL-B: (18-mer)* 5'-tatgcggccgcgaattca-3';
   b. His-T-linkers: HisTL-A: (55-mer) 5'-gcggccgcatatgggatcctcacatcatcat-caccatcactcgagtggccaagct-3'; HisTL-B: (21-mer) *5'-gcttggccactcgagtgatgg-3'
      The 5' end of each "B" oligonucleotide is phosphorylated (represented by the *) so that it can be ligated to the (nonphosphorylated) end of the PCR product. The 5' end of the "A" oligonucleotide is not phosphorylated. The "B" oligo only needs to be long enough to bind the "A" oligo during the ligation and in the annealing steps of the early rounds of reamplification. A 5' overhang on oligo "A" is not a problem, because this is filled in by the polymerase during reamplification.

# A

Oligo design:

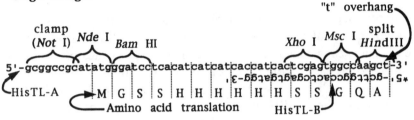

# B

Detail of junction at 5' end:

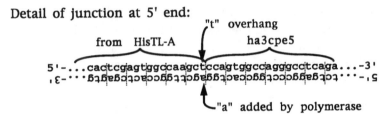

# C

Detail of junction at 3' end:

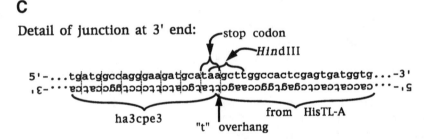

**Fig. 2.** Using a T-linker to add a "histidine tag" to one end of a PCR product. (A) Oligonucleotide design. The *Nde*I site allows the product to be cloned into an appropriate expression vector, in which translation begins at the ATG codon included in this site. The amino acids which make up the histidine tag are shown, and the reading frame is indicated by vertical bars between codons. The *Not*I site in HisTL-A is in parentheses because it would not be expected to cut so close to the end of a DNA molecule (although it should be possible to use it if the primer were phosphorylated and the products ligated to form concatamers). Its main purpose is to serve as a clamp to allow efficient cutting at the *Nde*I site. (B) Sequence at the junction between the *His*T-linker and the 5' end of the PCR-amplified sequence. The 5' primer begins at the first base in a codon, to put the sequence in frame with the histidine tag. (C) Sequence at the 3' junction. The 3' primer begins with 'ta'; the complementary 'ta' in the other strand is converted to a stop codon "taa" when the extra 'a' is added by *Taq* polymerase. Because the 5' primer begins with 'c', the *Hin*dIII site is completed only at the 3' end of the molecule. This allows directional cloning of the product, and removal of the T-linker sequences from the 3' end.

   c. Dissolve primers at a stock concentration of 10 µM, which is generally considered 20X for PCR. After mixing T-linker primers, they are at a final concentration of approx 5 µM for ligation reactions.
7. Ampligrease (*see* ref. 7). Plain petroleum jelly, Vaseline brand or generic. Apply quality control checks as described *(7)*.
8. Reagents for bacterial culture, including competent *Escherichia coli*, appropriate antibiotic, LB agar.
9. 95% ethanol containing 2% (w/v) potassium acetate.
10. 75% ethanol.

# 3. Methods
## 3.1. Polymerase Chain Reaction

The *His*T-linker was used to clone the coding region of the α3 subunit of the nicotinic acetylcholine receptor from human bronchial epithelium (manuscript in preparation; *see* Note 2) into the *E. coli* expression vector pT7–7 *(8)*. The following sequence-specific primers were used:

Vb8-1cpe5:    5'-ggagttatccagtcacc-3'
hpVbe3:       5'-gggaattcgtcgactgctggcrcagarrta-3'
ha3cpe5:      5'-ccagtggccagggcctcaga-3'
ha3cpe3:      5'-tatgcatcttccctggccatca-3'

1. Perform PCRs under fairly standard conditions. For example, for cloning the α3 subunit of the nicotinic receptor, the conditions for amplification were as follows: 3 µL 10X PCR buffer, 4.5 µL 10 mM MgCl$_2$, 3 µL dNTPs (2 mM each), 6 µL red sucrose, 16.5 µL H$_2$O, 0.75 µL ha3cpe5 primer (10 µM), 0.75 µL ha3cpe3 primer (10 µM), 1 µL cDNA template, and 0.25 µL *Taq* polymerase.
2. Perform PCR using 40 cycles at 94, 52, and 72°C, each for 0.5 min, in a programmable circulating air oven (ProOven, Integrated Separation Systems).
3. Products were purified by agarose gel electrophoresis and GeneClean (Bio 101) (*see* Chapter 1).

## 3.2. Ligation

*See* Note 4.

1. Assemble ligation reactions as follows: 2 µL purified PCR product, 7.5 µL H$_2$O, 2 µL 10X ligase buffer, 1 µL (0.5 mM final) 10 mM ATP (if not in buffer), 2.5% (5% final) 40% PEG 8000 (optional), 2 µL T-linkers (5 µM in 500 mM NaCl), and 1 µL ligase (1 U/µL dilution).
2. Incubate at room temperature for 20 min to overnight.

## 3.3. Reamplification with T-Linker Primer

1. After ligation of the T-linker, reamplfy the products as follows: 5 µL 10X PCR buffer, 7.5 µL 10 mM MgCl$_2$, 5 µL dNTPs, 10 µL Red sucrose, 16.5 µL H$_2$O, 5 µL TL-A (or HisTL-A) primer (10 µM), ligation mixture, and 0.35 µL *Taq* polymerase.

2. Cover the sample with mineral oil and amplify with 40 two-step cycles of 94°C for 0.5 min and 72°C for 2.5 min.

*See* Note 5.

## 3.4. Alternative Method: One-Tube Ligation/Reamplification

The whole T-linker reaction can be set up in one tube to make a "kit-like" product. This is done using a meltable barrier, as for hot-start PCR (we use AmpliGrease *[7]*). The "top mix" contains the ligation reaction, to which the PCR-amplified band is added. After a suitable incubation to allow ligation, the reaction is heated to melt the barrier, and the second PCR is begun:

1. Set up a 100-μL PCR reaction mix (*see* Section 3.1.) in a tube. Add only TL-A (or HisTL-A) as a primer, and no template.
2. Dispense approx 35 μL of petroleum jelly (AmpliGrease) onto the side of the tube with a syringe.
3. Heat the tube so that the grease melts to cover the bottom mix.
4. Allow to cool, so the grease resolidifies.
5. Add approx 30 μL of mineral oil on top of the grease.
6. Add 4 μL of ligation mix through the oil so the droplet rests above the grease barrier. The ligation mix is made as a master mix containing the following: 2 μL T-linkers (5 μM each in 50 mM NaCl), 2 μL 10X ligase buffer, 1 μL 10 mM ATP (if not in buffer), 0.5 μL ligase (8 U/μL), 10.5 μL H$_2$O.
7. Add 1 μL of the PCR product to be cloned into the droplet of top ligation mix.
8. Use the following reaction conditions: 25°C for 1 h (ligation); 94°C for 15 s, 55°C for 15 s, and 72°C for 45 s for 30 amplification cycles (*see* Note 6).

## 3.5. Digestion and Cloning

The reamplified PCR product now has the restriction sites from the T-linker at its ends, and may be cloned using standard procedures. The protocol we usually use is as follows:

1. Run a small portion of the reamplified product on a checking gel, to make sure you have enough of the correctly-sized product with which to work. About 5 μL should give a clear band. With care, you should be able to use this same gel to check your DNA at several stages of the process. Do not let it dry out!
2. Add ~1 μg of uncut (supercoiled) plasmid vector to the tube containing the reamplified material. This will both act as a carrier during the purification process, and help to make the effects of the restriction digests more obvious. You will probably not be able to see the slight size differences resulting from cutting off the ends of the T-linkers, but you should be able to see the difference as the vector is cut, and goes from being supercoiled to being linear.
3. Extract the DNA twice with phenol/chloroform, and once with chloroform.
4. Precipitate the DNA by adding 2 vol of 95% ethanol containing 2% potassium acetate to the aqueous supernatant and spin on high speed in a microcentrifuge a

4°C for 30 min. Discard the supernatant and carefully rinse the pellet with cold 75% ethanol.

5. Resuspend the DNA pellet in distilled water and add the appropriate 10X RE buffer. Add restriction enzymes and incubate until the vector is completely in the linear form on a checking gel. Directional cloning will require cutting with two enzymes. Manufacturers generally provide charts that indicate which buffer works reasonably well with both enzymes. Run a lane with uncut vector on the checking gel. Uncut vector should have three bands representing supercoiled, nicked circular, and (sometimes) linear forms. Cut vector should only have the linear form. Even small amounts of uncut vector will lead to high backgrounds on nonrecombinants.

6. Run the digested material on a preparative agarose gel, and cut out the vector and insert bands. Minimize exposure to UV light.

7. Recover the DNA separately from each band. Many protocols are available for this: We usually use GeneClean for inserts larger than 250 bp.

8. Run about one-tenth to one-fifth of the DNA extracted from the band on a checking gel to roughly estimate the amount of DNA recovered. The relative amounts of DNA in each band are crudely estimated by visually comparing the brightness of the bands to those with a known amount of DNA. The vector bands should contain more or less known quantities of DNA, if you know how much you started with, and assume about 70% was recovered from the preparative gel.

9. Mix vector and insert in at least a 2-to-1 molar ratio, and ligate. Use about 100 ng of vector per ligation. The ligation should resemble the following (for a larger insert, more DNA is needed for the same molar ratio): 100 ng vector (2.5 kb), 20 ng insert (250 bp), 2 μL 10X ligation buffer, 1 μL 10 m$M$ ATP (if not in buffer), 0.5 μL T4 DNA ligase, and up to 16.5 μL H$_2$O.

10. Dilute the ligation 1:5, and use 1 μL to transform 20 μL of competent *E. coli*. Plate on appropriate antibiotic medium.

11. Recombinant colonies can be screened using PCR. Depending on which restriction site was used, you may be able to screen with a T-linker primer. For example, with inserts cloned nondirectionally using the *Eco*R1 site of the *Nde*T-linker, *Nde*TL-A can be used for screening. Make a master mix for as many reactions as you need, in which a 10 μL reaction contains: 1 μL 10X PCR buffer, 1 μL dNTPs (2 m$M$ each), 1.5 μL MgCl$_2$ (10 m$M$), 2 μL red sucrose dye, 1 μL *Nde*TL-A (10 μ$M$), and 0.25 μL *Taq* polymerase. Cover these small reactions with oil while picking colonies to prevent evaporation.

12. Touch a recombinant colony with the tip of a sterile toothpick, dip the end of the toothpick through the oil into the reaction, and swirl. Do not add enough bacteria to the reaction to make it cloudy. Just a few bacteria are sufficient. Too much bacterial matter, or small amounts of agar, will inhibit the reaction. Be sure to include one reaction of a nonrecombinant ("blue") colony as a negative control.

13. Heat to 94°C for 1 min.

14. Amplify for 30 cycles using the following parameters: 94°C for 15 s, 55°C for 15 s, 72°C for 45 s.

## 4. Notes

1. Why not use a T-vector? The method of choice for routine cloning of PCR products is probably a T-vector. One of the more important considerations is that with
   a T-vector you do not need to digest the insert with a restriction enzyme, so you
   do not need to worry about whether or not the insert contains that site. Also,
   reamplification with the T-linker provides another set of opportunities for the
   polymerase to introduce errors. However, T-linkers have several differences that
   can provide advantage in certain circumstances.
   a. The efficiency of ligation can theoretically be increased because much higher
      concentrations of linkers can be achieved.
   b. The efficiency of ligation does not need to be as high because the ligated
      product can be reamplified with one of the linker oligonucleotides to give a
      product that has added restriction sites at the ends.
   c. Oligonucleotides, such as those used to construct the T-linker, are quite stable,
      and remain usable for many years. Stability has been a problem with some of
      the commercially available T-vectors.
   d. Because the ends of a PCR product can be precisely defined and/or modified,
      and because T-linkers can be custom-made, it is possible to "split" a DNA
      sequence, such as a restriction site, between the PCR product and the T-linker
      so that a complete site is formed only at one end of the final product, without
      having to include the entire site in the primer made for amplifying a specific
      gene. This can save on the cost of oligonucleotides but still allow directional
      cloning.
   e. T-linkers should be more flexible in terms of using a variety of restriction
      sites in a variety of vectors.
2. The directional cloning of the coding sequence for the α3 acetylcholine
   receptor subunit illustrated in Fig. 2 was accomplished by selecting the locations of the primers so that a HindIII site was created at only one end. However, by random chance, in one case out of four a PCR product made with
   primers not designed to complete the NdeI site will have a 'g' at a given end,
   and will thus end up with an NdeI site. Similarly, one out of 16 randomly
   chosen products will have NdeI sites at both ends. Thus, three out of 16 randomly chosen PCR products will have a 'g' at only one end, and could be
   cloned directionally using this T linker. A set of such linkers, with each
   depending on the presence of a different single nucleotide at the end of the
   PCR product, could therefore be used to directionally clone 75% of randomly
   chosen PCR products. Potential restriction enzymes for such complementable
   sites in T-linkers would be:

   | Site ends in | Restriction enzymes |
   | --- | --- |
   | ...t(g) | NdeI, PvuII (blunt), PmlI(blunt) |
   | ...t(c) | EcoRI, EcoRV (blunt), AatII, SacI |
   | ...t(a) | SnaBI (blunt) |
   | ...t(t) | HindIII, SspI (blunt) |

Three linkers would make up a complete set, i.e., you do not need an 'a' linker because any product that only has 'a' at one end automatically has one of the other three bases at the other end. Together with the T-linkers using split *Nde*I and *Hin*dIII sites described here, a T-linker that introduces a split *Sac*I site, for example, would complete a set.

3. Potentially, other DNA sequences, such as a promoter for in vitro transcription, could be split so that a functional site is completed at only one end of the product. If the majority of the DNA sequence of such sites can be added by ligating a T-linker, this could provide significant cost savings compared to synthesizing target-specific primers containing such sequences.

4. Blunt-ended ligation of linkers has been used to add primer sequences to DNA fragments *(9,10)*, but the T-linker application presented here is novel as far as we can tell. The T-linkers are more suitable for use with DNA fragments generated by PCR than blunt-ended linkers, because of the nontemplate derived 'a's added by the polymerase. Because the sequences at the ends of PCR products can be easily manipulated by incorporating changes in the primers, DNA sequences, such as restriction sites, can be split between the T-linker and the PCR product. In general, this sort of site splitting is not practical except with PCR-generated fragments. One exception is fragments tailed with a known homopolymer, such as the poly A tail on eukaryotic mRNAs *(11)*; such a linker is commercially available (Novagen, Madison, WI). Because the T-linker is added as an inverted repeat at the ends of the fragment, a single primer (TL-A) is used to reamplify after ligation. Single-primer amplification systems *(12)* have an advantage in that a "primer-dimer" cannot be formed from one primer, since this would produce an unamplifiable hairpin.

5. Because each product being cloned is subjected to an extra amplification, the overall frequency of errors should be increased, although our experience shows that the increase is not dramatic. In many cases, the risk of PCR errors is quite acceptable. For example, if one is producing an enzyme or other protein with a measurable function, the clones can be screened for activity. Deleterious mutations will thus be weeded out, and nondeleterious mutations may not matter (the $\alpha3$ subunit in our example will be used as a potential ligand-binding protein, and clones will be screened on this basis). Similarly, if a sequence is to be used as a hybridization probe, a low frequency of base substitutions is frequently acceptable. In situations where no mutations are acceptable, clones must be screened by careful sequencing.

6. The pH and the magnesium concentration of the PCR are different from those of the ligation. Using a small ligation volume and a large PCR volume helps to correct the conditions for the PCR. Alternatively, the pH of the bottom mix can be increased, and the added magnesium reduced, so that the final pH and ($Mg^{2+}$) are correct after mixing. This allows use of smaller volumes. The ability to make a quick, automated one-tube ligation/reamplification reaction makes this setup intriguing. However, it is easier to repeat parts of the experiment (such as the reamplification) if you have leftovers from a separate ligation reaction.

## Acknowledgments

This work was supported by research grants from the Muscular Dystrophy Association of America and from the Council for Tobacco Research, by the NIH grant N52319, and the NIDA Program Project grant DA05695. R. M. H. was the recipient of the Robert G. Sampson Neuromuscular Disease Research Fellowship from the Muscular Dystrophy Association.

## References

1. Clark, J. M. (1988) Novel nontemplated nucleotide addition reactions catalyzed by procaryotic and eucaryotic DNA polymerases. *Nucleic Acids Res.* **16,** 9677–9686.
2. Mole, S. E., Iggo, R. D., and Lane, D. P. (1989) Using the polymerase chain reaction to modify expression plasmids for epitope mapping. *Nucleic Acids Res.* **17,** 3319.
3. Mead, D. A., Pey, N. K., Herrnstadt, C., Marcil, R. A., and Smith, L. M. (1991) A universal method for the direct cloning of PCR amplified nucleic acid. *Bio/ Technology* **9,** 657–663.
4. Marchuk, D., Drumm, M., Saulino, A., and Collins, F. S. (1991) Construction of T-vectors, a rapid and general system for direct cloning of unmodified PCR products. *Nucleic Acids Res.* **19,** 1154.
5. Holton, T. A. and Graham, M. W. (1991) A simple and efficient method for direct cloning of PCR products using ddT-tailed vectors. *Nucleic Acids Res.* **19,** 1156.
6. Hoppe, B. L., Conti-Tronconi, B. M., and Horton, R. H. (1992) Gel loading dyes compatible with PCR. *BioTechniques* **12,** 679,680.
7. Horton, R. M., Hoppe, B. L., and Conti-Tronconi, B. M. (1994) AmpliGrease, "Hot Start" PCR using petroleum jelly. *BioTechniques* **16,** 42,43.
8. Tabor, S. and Richardson, C. C. (1985) A bacteriophage T7 RNA polymerase/ promoter system for controlled exclusive expression of specific genes. *Proc. Natl. Acad. Sci.* USA **82,** 1074–1078.
9. Ko, M. S. H., Takahashi, N., Nishiguchi, K., and Abe, K. (1990) Unbiased amplification of a highly complex mixture of DNA fragments by 'lone-linker'-tagged PCR. *Nucleic Acids Res.* **18,** 4293.
10. Bhat, G. J., Lodes, M. J., Myler, P. J., and Stuart, K. D. (1991) A simple method for cloning blunt ended DNA fragments. *Nucleic Acids Res.* **19,** 398.
11. Meissner, P. S. Sisk, W. P., and Berman, M. L. (1987) Bacteriophage lambda cloning system for the construction of directional cDNA libraries. *Proc. Natl. Acad. Sci.* USA **84,** 4171–4175.
12. Rich, J. J. and Willis, D. K. (1990) A single oligonucleotide can be used to rapidly isolate DNA sequences flanking a transposon Tn5 insertion by the polymerase chain reaction. *Nucleic Acids Res.* **18,** 6673–6676.

# 12

# Recovery of DNA Amplification Products from Silver-Stained Polyacrylamide Gels

*Applications in Nucleic Acid Fingerprinting and Genetic Mapping*

**Gustavo Caetano-Anollés and Robert N. Trigiano**

## 1. Introduction

The amplification of picogram quantities of DNA with the polymerase chain reaction (PCR) *(1,2)* targets specific nucleic acid sequences without the need for cloning, subcloning, and plasmid amplification *(3)*. However, although the purity of amplified DNA products can be enhanced by strategies that avoid false priming events during amplification, such as reamplification with nested primers *(4)* or "hot start" PCR *(3,5,6)*, ultimately it can only be achieved by the subsequent cloning of individual products.

DNA amplification fragments are routinely analyzed by agarose gel electrophoresis and staining with ethidium bromide *(3)*. This methodology is simple and quick, allows recovery of DNA for subsequent analysis, and has proved adequate in many PCR applications. However, it lacks resolution and sensitivity and is unable to produce a permanent record. Of the several alternatives available, the coupling of polyacrylamide gel electrophoresis (PAGE) and silver-staining has been particularly popular because of its simplicity, greater resolving power, and high sensitivity *(7)*. The overall procedure has been streamlined to a minimum number of steps, can take only about 2 h from gel casting to image development, is relatively inexpensive, and allows gels to be easily handled and permanently preserved. Silver-stained polyacrylamide gels have been used in a number of applications, including sequencing of DNA *(8)*, single-strand conformation polymorphism (SSCP) analysis *(9–14)*, DNA amplification fingerprinting (DAF) with arbitrary oligonucleotide primers *(15)*,

From: *Methods in Molecular Biology, Vol. 67: PCR Cloning Protocols: From Molecular Cloning to Genetic Engineering* Edited by: B. A. White  Humana Press Inc., Totowa, NJ

differential display (DD) of messenger RNA *(16),* and DNA profiling *(17,18).*
Fingerprinting techniques, such as DAF, produce a collection of amplification
products representing discrete portions of a genome with some products sharing the same molecular size *(15).* Confirmation of interesting monomorphic or
polymorphic products by Southern hybridization often requires their isolation
for subsequent use as DNA probes. Although silver-staining and PAGE offer
the ability to resolve complex DNA profiles, such as those routinely obtained
in DAF or DD analysis, the close proximity of bands in these profiles makes
DNA fragment isolation physically demanding.

The isolation of individual DNA fragments from an agarose or polyacrylamide gel matrix is the simplest approach to subcloning, but requires an effective procedure to elute the DNA molecules *(19).* Nucleic acids can be eluted
into liquid or solid supports by blotting *(20),* centrifugal filtration through filter membranes *(21,22)* or siliconized sterile glass wool *(23),* electroelution into
dialysis bags *(24)* or DEAE cellulose *(25,26),* use of low-melting point agarose
*(4,27,28),* gel compression by freeze-squeeze *(29,30),* mechanical gel extru-
sion *(31),* flush *(32)* or centrifugal *(33)* DNA extraction, and even passive overnight diffusion into buffer. These methods usually demand further purification
by phenol/chloroform extraction and sometimes concentration before subcloning. We have developed a simple procedure to isolate DNA amplification
products from silver-stained polyacrylamide gels *(34).* The method amplifies
DNA that diffuses passively from gel segments during thermal cycling in one
or more rounds of amplification, and is particularly suited to the isolation of
products from complex DNA profiles. In this chapter, we describe the separation of DNA amplification products in polyacrylamide gels, their detection
at the picogram level using a quick and simple silver-staining protocol *(35),*
and the isolation of these products by subsequent amplification. Isolated DNA
fragments can be used as probes for Southern hybridization without the need
for further subcloning, or can be cloned, sequenced, and converted into
sequence characterized amplified regions (SCAR) *(36),* a new class of landmark for genome mapping applications. SCARs are PCR-based sequence-
tagged sites originating from the amplification of DNA with arbitrary primers,
where two specific primers amplify a diagnostic PCR product detectable by
electrophoresis or presence of incorporated fluorophores or radioactive label.
Our method can also be used to isolate, clone, and analyze expressed sequences
amplified from RNA with arbitrary primers.

## 2. Materials
### 2.1. DNA Amplification Reagents
1. DNA amplification products, including complex mixtures from DAF, DD, or
   sequencing analysis.

2. 10X Reaction buffer: 100 m$M$ Tris-HCl, pH 8.3, 100 m$M$ KCl.
3. 10X PCR buffer: 100 m$M$ Tris-HCl, pH 8.3, 500 m$M$ KCl, 0.01% (w/v) gelatin.
4. Deoxynucleoside triphosphate (dNTP) (2 m$M$ of each dNTP).
5. 25 or 100 m$M$ MgCl$_2$ stock solution.
6. Thermostable DNA polymerase enzyme (preferably truncated derivates, such as AmpliTaq Stoffel DNA polymerase [Perkin-Elmer/Cetus, Norwalk, CT], when using nonstringent amplification conditions).
7. Oligonucleotide primers (3 or 30 µ$M$ stocks).

## 2.2. Polyacrylamide Gel Electrophoresis Reagents

1. Urea.
2. 10X Tris-borate-EDTA (TBE) buffer: 1$M$ Tris-HCl, pH 8.3, 0.83$M$ boric acid, 10 m$M$ Na$_2$ EDTA · H$_2$O.
3. Acrylamide-urea stock solution: 9.8% acrylamide, 0.2% piperazine diacrylamide, 10% urea, prepared in in 1X TBE.
4. 10% ammonium persulfate.
5. $N,N,N',N'$-tetramethylethylenediamine (TEMED).
6. Loading buffer: 1.2 g/mL of urea, 0.08% xylene cyanol FF.

## 2.3. Silver-Staining Reagents

Chemicals must be of high purity, analytical grade.

1. Fixative-stop solution: 7.5% (v/v) glacial acetic acid.
2. Silver impregnating solution: 1 g/L of silver nitrate and 1.5 mL/L of 37% formaldehyde.
3. Developer: 30 g/L of sodium carbonate, 3 mL/L of formaldehyde, and 2 mg/L of sodium thiosulfate.

## 2.4. Cloning Reagents

1. TA cloning kit with vector, pCRII (Invitrogen, San Diego, CA).
2. Competent *Escherichia coli* cells (INVαF').
3. T4 DNA ligase and supplied 10X ligation buffer.
4. TE buffer: 10 m$M$ Tris-HCl, pH 7.5, 1 m$M$ Na$_2$EDTA · H$_2$O.
5. β-mercaptoethanol.
6. X-Gal (5-bromo-4-chloro-3-indolyl-β-D-galactopyranoside).
7. Ampicillin.
8. SOC medium: 2% Bacto tryptone, 0.5% yeast extract, 10 m$M$ NaCl, 2.5 m$M$ KCl, 10 m$M$ MgCl$_2$, 10 m$M$ MgSO$_4$, and 20 m$M$ glucose.
9. LB agar (Difco, Detroit, MI).
10. Reverse and forward primers T3 (CGCAATTAACCCTCACTAAAGGG) and T7 (GTAATACGACTCACTATAGGGCG).

## 2.5. Other Supplies

1. Equipment: bench-top centrifuge, thermocycler (Bios, New Haven, CT), incubators, shaker, and electrophoresis apparatus.

2. Miscellaneous: heavy mineral oil, polyester gel-backing film (GelBond PAG, FMC Bioproducts, Rockland, ME; or other silanized plastic sheets), low-melting-point paraffin (Aldrich Chemical, Milwaukee, WI), membrane syringe filters, staining trays, and scalpel.

## 3. Methods

The isolation of DNA amplification fragments involves their initial separation by PAGE and silver-staining, and one or more cycles of gel band excision and subsequent amplification. Isolated fragments can be cloned and sequenced, such as in the covertion of DAF markers into SCAR landmarks.

### 3.1. DNA Separation

Amplification products can be separated in vertical or open-faced polyacrylamide slab gels using a variety of protocols (*see* Note 1). We here present a description of DNA separation by denaturing PAGE using 0.45-mm thick polyacrylamide gels (8 × 10 cm) backed on polyester film as modified from ref. *37*. Polyacrylamide gels are usually 4–10% T and 2% C.

1. Assemble the electrophoretic rig under running distilled water by placing the large glass plate on the clamp assembly, then the polyester backing sheet (hydrophilic side up), spacers, and small glass plate, in that order. Make sure that the backing sheet is in tight apposition to the large glass plate.
2. Allow to air dry in a dust-free area in the dark, overnight, or for 1 h in a 40–50°C oven.
3. Place gel rigs in casting stand.
4. Mix 10 mL of the acrylamide-urea stock solution with 150 µL ammonium persulfate and 15 µL TEMED and cast the polyacrylamide-urea gels immediately. If required, the acrylamide-urea stock solution can be made to contain 5% glycerol as mobility retardant.
5. Deliver the gel mix by injection through a 0.45 µm-pore size membrane syringe filter and insert Teflon comb. The filter will remove dust particles and the mix will begin to set in about 2 min and will fully polymerize in 30 min.
6. Attach gel rigs to electrode core, place into buffer tank, and fill upper and lower buffer reservoirs to the top with running buffer (1X TBE).
7. Remove combs and rinse well with running buffer using a 5-mL syringe.
8. Pre-electrophorese gels at 150 V for at least 5 min to equilibrate.
9. Rerinse wells and load 3 µL of a dilution of each amplification reaction (usually containing 30–40 ng DNA) mixed with 3 µL of loading buffer.
10. Electrophorese at 150 V for about 60–90 min.
11. Disassemble gel rigs and carefully remove backed-gels, preferably under running distilled water.

### 3.2. Silver-Staining

Polyester-backed gels can be silver-stained using the procedure of Bassam et al. *(35,38)* (*see* Notes 2,3). The stain is commercially available (Promega

Corp., Madison, WI). The following protocol is designed for 0.45-mm thick polyacrylamide minigels. Treatment times may need to be increased with larger electrophoretic setups or decreased when using nonbacked gels. Wear gloves when handling gels to avoid staining artifacts.

1. Place the gels in flat-bottomed, straight-sided staining trays. Clear plastic lids from 1000-μL pipet-tip racks can be used. Throughout subsequent steps (2–6) agitate the trays on a shaker at 50 rpm.
2. Immerse the gels in fixative-stop solution for 10 min.
2. Wash the gels three times with distilled water at least 2 min each.
3. Impregnate with silver solution for 20 min.
4. Rinse quickly (5–20 s) with water.
5. Develop the image with developer at 8–12°C until optimum image contrast is attained (approx 4 min).
6. Stop image development in fixative-stop solution for at least 1 min and wash extensively with water.
7. Preserve the gels permanently by drying at room temperature. Further details can be found in ref. *38*.

## 3.3. DNA Fragment Isolation

Selected DNA fragments embedded in polyacrylamide are isolated directly from the silver-stained gels by one or more cycles of isolation and amplification (*see* Notes 4–7).

1. Carefully excise a small piece of gel containing the desired DNA fragment with a flamed scalpel or dissection probe.
2. When using a preserved gel, clean its surface with 95% ethanol, use the scalpel to sharply delimit the segment of interest, and then rehydrate the excised band with a drop of sterile water.
3. Place the gel segment in 10–100 μL of standard amplification mixture. A typical amplification mixture contains 0.3–3 μ*M* of primer(s), 0.3 U/μL of AmpliTaq Stoffel fragment, 200 μ*M* of each dNTP, 1.5 m*M* $MgCl_2$, and buffer.
4. Heat the sample for a minimum of 5 min at 95°C (if the fragment to isolate is larger than 500 bp).
5. Amplify the eluting DNA in 35 cycles of 30 s at 96°C, 30 s at 50°C, and 30 s at 72°C in an oven thermocycler.
6. Alternatively, bands can be eluted from the gel segments at 95°C for 20–30 min and the recovered eluate used as template.
7. Separate the amplification products by PAGE and stain with silver.

## 3.4. SCAR Analysis

DNA polymorphisms detected by DAF *(15),* or other related techniques, can be isolated, cloned, sequenced, and converted into SCARs (*see* Note 8). Usually, a single research project may involve production of numerous SCARs.

To avoid confusion, SCARs should be named according to a clear nomenclature that defines template DNA, primer used (sequence and structural features), strategy (DAF, RAPD, and so forth), inheritance of marker, and molecular weight.

### 3.4.1. Detection and Isolation of Polymorphic DNA

1. Assemble amplification mixture in 10–20 µL total volume containing 0.1–1 ng/µL of template DNA, 3 µM of one or two arbitrary primer (typically 8 nucleotides in length), 0.3 U/µL of AmpliTaq Stoffel fragment, 200 µM of each dNTP, 1.5 mM MgCl$_2$, and buffer. Prepare a master mix with reagents that are common to avoid pipeting errors and contamination, and aliquot the mix into 200- or 500-µL microcentrifuge tubes. Then add the missing reagent to each tube individually.
2. Cover the amplification mixture with 1–2 drops of heavy mineral oil.
3. Amplify the DNA in an oven thermal cycler for 35 cycles of 30 s at 96°C, 30 s at 30°C, and 30 s at 72°C. When using other thermal cyclers adjust cycle parameters to the different ramp times, heating, and cooling rates.
4. Retrieve amplification mixtures by adding 100–200 µL of chloroform and pipeting out the aqueous droplet, or directly by using a long pipet tip or by rolling the amplification mixture over Parafilm.
5. Dilute aliquots of the samples 5–10-fold prior to electrophoresis.
6. Separate amplification products by PAGE (*see* Section 3.1.).
7. Silver-stain the polyester-backed polyacrylamide gels (Section 3.2.), and preserve gels by drying at room temperature.
8. Isolate interesting polymorphic or monomorphic bands directly from the silver-stained gels (Section 3.3.).
9. Confirm band recovery by PAGE and silver-staining.

### 3.4.2. Cloning of Isolated Fragments

Isolated fragments have 5' termini defined by the primer(s) used in the amplification reaction. SCARs can be constructed by using direct sequencing methods. However, cloning should precede DNA sequencing in those cases where the isolated DNA product defining the future SCAR was originally generated with a single primer or contains identical termini. In this case, the identity of recombinant clones should be confirmed by a variety of methods, including analysis of transformant DNA by PCR amplification, restriction analysis of cloned inserts, and amplification with arbitrary primers (*see* Note 6). The following is a cloning protocol based on the TA Cloning kit and using reagents supplied by the manufacturer.

1. Resuspend the lyophilized vector (pCRII) in TE buffer to a final concentration of 25 ng/µL. The resuspended vector is stable for 60–90 d.
2. Ligate vector and amplification fragment. Add 6 µL sterile water, 1 µL of 10X ligation buffer, 2 µL of resuspended vector, 1 µL of T4 DNA ligase, and 1 µL diluted amplification product (to give 1:3 vector-to-insert ratios). Incubate at 12°C for at least 4 h.

3. Retrieve a 50-μL aliquot of competent cells from the −70°C freezer, thaw, and gently add 2 μL of 0.5*M* β-mercaptoethanol and 1 μL of ligation mixture. Incubate on ice for 30 min.

4. Heat transformation reaction at 42°C for 1 min in a water bath, and then place on ice.

5. Add 450 μL of SOC medium. Incubate the mixture at 37°C for 1 h in a shaker at 225 rpm.

6. Spread 20 μL of 50 mg/mL X-Gal onto LB agar plates containing 50 μg/mL ampicillin (previously dried for 15 min in an incubator). Allow diffusion to occur for 1 h.

7. Plate 25 and 100 μL of transformed cells and incubate the plates inverted overnight at 37°C.

8. Choose the positive clones by picking white colonies, plating them in LB agar plates, and growing the bacteria for an additional 12–48 h. Using this protocol 90% of positive clones contain inserts. Note that small inserts (<500 bp) may not interrupt *lacZ* expression completely, rendering pale blue colonies.

9. Isolate inserts by PCR amplification with T3 and T7 primers (*see* amplification protocol described in Section 3.4.4.).

### 3.4.3. Sequencing and Primer Design

1. SCARs are defined by sequencing of isolated or cloned amplification products. Comprehensive sequencing protocols using dideoxy termination methods can be found in refs. *3,28*. Commercial sequencing kits are also available.

2. Several interactive software packages permit input, manipulation, and analysis of sequence data and, in some cases, primer design *(39)*. Such programs as Oligo (NBI, Plymouth, MN), Primer (Whitehead Inst., Cambridge, MA), and the GCG mainframe package (University of Wisconsin) can be used. For each amplification product, select more than one SCAR primer pair. This will provide alternative amplicons in the event that the selected SCAR fails to give clear and informative PCR products. Be aware that different software utilize algorithms that may select different primer sequences. Finally, evaluate structural features of primers and amplicons, including analysis of internal restriction sites, hairpin structures, and repetitive motifs.

### 3.4.4. Analysis, Validation, and Mapping of SCARs

Once a SCAR has been designed, it should be tested by PCR amplification (*see* Note 8). PCR protocols must be optimized for each application and primer pair. The following is a PCR protocol for the detection of a typical SCAR that uses the "hot start" *(5)* approach *(40)* to increase specificity of the reaction:

1. Assemble a PCR reaction (20 μL) containing 0.1 ng/μL of template, 0.2 μ*M* of each primer (20–25 nucleotides in length), 0.05 U/μL of AmpliTaq DNA polymerase, 200 μ*M* of each dNTP, 1.5 m*M* MgCl$_2$, and PCR buffer, using 500-μL microcentrifuge tubes.

2. Add 20 μL of melted paraffin wax to the side of the reaction tube containing the amplification mixture.

3. Remelt the wax by transferring the tubes to a block, oven or thermal cycler at 70°C, and then cool to form a hard seal.
4. Deposit a 2 μL drop containing the template DNA on top of the wax seal. Provide a final concentration of 1 ng/μL of template DNA.
5. Cover the amplification mixture with 1–2 drops of heavy white mineral oil.
6. Amplify in 30 cycles of 30 s at 96°C, 30 s at 60°C, and 30 s at 72°C.
7. Retrieve amplification mixtures through the wax-oil micellar layer directly by using a normal pipet tip.
8. Resolve amplification products using PAGE or agarose gel electrophoresis.

## 4. Notes

1. DNA separation: We have recovered silver-stained products separated using vertical and horizontal PAGE, discontinuous polyacrylamide isotachophoresis, and the PhastSystem (Pharmacia LKB, Piscataway, NJ), under both denaturing and nondenaturing conditions. When using open-faced discontinuous denaturing or nondenaturing polyacrylamide gels *(17,41)*, the gel mix is poured by capillary action between a glass plate and backing polyester film separated by 0.2–0.4-mm spacers, and the gel is washed and dried. Before electrophoresis, the gel is rehydrated with buffer containing a leading ion (50 m$M$ formate and 130 m$M$ Tris-HCl, pH 8.5), placed on the horizontal platen of an isothermal controlled electrophoresis apparatus (EC1001, E-C Apparatus Corp., St. Petersburg, FL), and electrodes placed on top of cathodal and anodal buffer strips containing 200 m$M$ of an appropriate trailing acid species (such as proline, glycine or tricine; *[41]*) in 100 m$M$ NaOH and leading ion buffer, respectively. DNA samples are applied directly to the gel surface with a sample applicator, and electrophoresed for about 1 h at 300 V. The isolation of DNA fragments from these gels is as described in Section 3.

   DNA fragments can also be separated using semiautomated miniaturized electrophoretic and staining devices, such as the PhastSystem. Amplification products are electrophoresed at 15°C using precast polyester-backed nondenaturing 5–15% polyacrylamide gels and under conditions modified from ref. *42*: prerun for 100 Vh (400 V, 10 mA, 2.5 W), sample application for 2 Vh (400 V, 1 mA, 2.5 W); DNA separation for 100 Vh (200 V, 10 mA, 2.5 W), and a slow hold for 100 Vh (50 V, 1 mA, 0.5 W). Gels are usually removed at the holding phase (about 200 Vh total). Despite the restricted resolving range (about 20 mm), loading constraints, and general loss of resolution of PhastSystem gels, products were recovered successfully from these miniature discontinuous plastic-backed matrices. This is important because the PhastSystem allows fast electrophoretic performance (30 samples/2 gels/30 min using one unit), high throughput (30–60 samples/h vs 15 samples/h using vertical PAGE), and reproducible, sensitive, and semiautomated DNA separation and staining.
2. Silver-staining: Silver-stained polyacrylamide gels are becoming the choice whenever a complex assortment of nucleic acids is to be separated or high band resolution required. Silver-staining avoids fluorophore or radioisotopic labeling

with their associated expense and handling limitations. However, it has been considered much of an "art" and in some cases cumbersome and time-consuming. Recent advances in the silver-staining technique allow for consistent detection of picogram quantities of nucleic acids separated on polyester-backed polyacrylamide gels *(7)*. We use a simple acidic silver-stain *(35)* that provides unsurpassed sensitivity, is highly reproducible, has few steps and reagents, is fast, and produces the least number of staining artifacts *(38)*. The silver-stained polyester-backed gels can be preserved for many years by drying, without suffering distortion or image loss. These gels are safe repositories of electrophoresed DNA amplification products, allowing their retrospective recovery and examination. Using our protocol, DNA fragments were recovered from gels that have been stored in a photographic album for 3 yr. Nonbacked gels can also be dried and preserved between plastic sheets, but will suffer distortion and sometimes damage from handling. Furthermore, the surface of these gels is not accesible for DNA recovery.

3. Automated silver-staining using the PhastSystem: Use the following staining program: step 1, 10 min fixation; steps 2–4, 2 min washes; step 5, 20 min silver impregnation; step 6, 0.1 min wash; step 7, 3.5 min image development; step 8, stop reaction with acetic acid (1 min); step 9, final wash for 4 min. Steps use four reagents delivered through six in-ports and discarded through two out-ports (silver discarded separately), and are done at 15°C, except for steps 7 and 8, which use ice-cold solutions. The use of ice-cold developer compensates for temperature shifts during image development resulting from time delays in the automated delivery and evacuation of developer solution, resulting in a highly reliable stain. Backed gels are preserved by drying at room temperature.

4. Recovery of DNA amplification products: Silver-stained DNA can be recovered from a polyacrylamide matrix by amplification *(34)*. The desired band, whether part of a simple or complex array of DNA fragments in a profile, can be directly excised and used as template for further amplification. One or more rounds of amplification, electrophoretic separation, and band isolation produce a single-strong band on a gel. The number of these "amplification cycles" varies with each fragment, but depends mostly on the complexity of the DNA pattern from where it is recovered. Recovery occurs in a single cycle when bands are isolated from PCR products, even if resulting from multiplex PCR reactions. In contrast, isolation from complex profiles, such as those generated by amplification with short oligonucleotide primers usually requires at least two cycles. During the past years, we have isolated many bands representing interesting monomorphic or polymorphic DAF products from bacterial, fungal, and plant fingerprints (cf. refs. *43–45*). Figure 1 shows a typical exercise of band isolation from relatively complex DAF profiles generated from two cultivars of flowering dogwood *(Cornus florida L.)*, an ornamental understory tree native of the eastern deciduous forest of North America. The dogwood has been threatened by the introduced fungal pathogen *Discula destructiva* Red., which causes anthacnose and devastation in native tree populations. DAF has allowed a

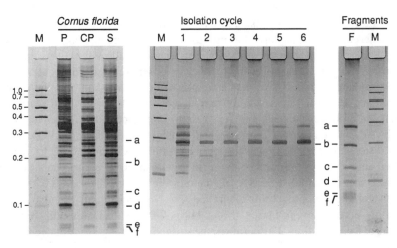

**Fig. 1.** Isolation of DNA products from dogwood *(Cornus florida L.)* DAF profiles. Six polymorphic fragments that were part of DNA patterns generated from cultivars 'Cherokee Princess' **(CP)** (fragments **a** and **f**) and dogwood anthracnose-resistant 'Santamour' **(S)** (**b–e**) by amplification with the octamer primer GATCGCAG (left panel), were subjected to up to six cycles of isolation (band excision and subsequent amplification) as described in Section 3. Aliquots of the isolated fragments were pooled and electrophoresed together (lane **F**) (right panel). In most cases few isolation cycles were required. For example, only two isolation cycles of fragment b were necessary to eliminate most amplification contaminants (middle panel). Molecular weight markers (lane **M**) are given in kilobases. Note that fragments present in 'Santamour' were also present in 'Presidential' **(P),** another tree putatively resistant to dogwood anthracnose.

detailed genetic analysis of fungal populations *(46)*. However, in the present study, DAF was used to find markers diagnostic of disease-resistant trees. We selected bands that were common between cultivars shown putatively resistant to the disease but polymorphic with those that were susceptible. Candidate bands generated from the anthracnose-resistant cultivars "Santamour" and the susceptible "Cherokee Princess" were recovered free of major contaminants in only 2–3 isolation cycles. Note that the majority of bands isolated in Fig. 1 constitute relatively minor amplification products that in some cases were very close in molecular weight to abundant products in the profile (cf. bands c and d). Nevertheless, the procedure was efficient and selective enough to remove the unwanted contaminants. In some cases, a few minor contaminating products persisted at very low level even after several isolation cycles. These contaminants are highlighted by the high sensitivity of our silver-staining technique, represent <5% of isolated DNA (as judged by staining of serially diluted amplifications), and probably depict the limits of our isolation method.

5. Factors affecting DNA recovery: Increasing polyacrylamide concentration appears to decrease the number of isolation cycles required for appropriate DNA recovery *(34)*, most probably by decreasing product carryover during electrophoresis. Band isolation from high concentration (20%) polyacrylamide gels in SSCP analysis necessitated thorough washing of the silver-stained gel prior to band excision and high number of temperature cycles during reamplification *(47)*. Although loading the gels with lower amounts of template did not alter the efficiency of band isolation *(34)*, the isolation of DNA fragments of more than 600 bp in length required that gel segments be incubated at 95°C for at least 20 min *(48)*. The DNA polymerase used also appears of importance. Variants of *Taq* polymerase, such as Klentaq LA (AB Peptides, St. Louis, MO), were found inefficient in their ability to eliminate contaminating products *(34)*. Finally, we found that higher annealing temperatures during band recovery from DAF gels help eliminate many amplification contaminants. For example, annealing at 50°C during amplification in the first cycle of isolation of product d *(see* Fig. 1) produced about half the number of contaminants than those produced at 30°C. In addition, "hot-start" PCR also can enhance relative enrichment and fragment recovery (J. Padilla, unpublished results).

   Calvert et al. *(47)* provide evidence that silver-staining components have inhibitory effect on PCR and that use of acetic acid during fixation may cause partial depurination and strand breakage with the subsequent heating of template molecules during amplification. Therefore, it is recommended that only a small gel segment be excised and that each round of isolation be given at least 25–30 temperature cycles to maximize amplification. In those cases where band isolation proves difficult, the use of weak acids (such as citric acid) may confer an advantage.

6. Cloning of isolated fragments: Isolated products from silver-stained gels have been used as probes for Southern hybridization *(34,44,45)*. These probes can be used to determine if the amplification product represents a single or multi-locus site in the genome, or to confirm its monomorphic or polymorphic nature in the fingerprint. Isolated DAF products that have and have not been subsequently cloned gave virtually the same genomic hybridization results *(34)*, and can therefore be used directly in hybridization studies. However, persistance of amplification contaminants during recovery, even if present at very low levels, may pose a problem for certain applications where purity of the selected fragment is required. In other cases, cloning may be the only way to allow sequencing of the isolated fragment, as when only one primer species is used in the amplification reaction.

   To varying extents, blunt double-stranded amplification products contain an extra 3' terminal nucleotide (usually dA) because of template-independent polymerization by DNA polymerase. Amplification products can be cloned into vector DNA using either "blunt-ended" or "overhanging-ended" cloning strategies *(49)*, as long as they guarantee appropriate and efficient ligation to the linearized vector. Several methods have been developed for PCR cloning, including:
   a. The use of T/A cloning vectors that rely on the adenosyl terminal deoxy-nucleotidyl-transferase activity of some DNA polymerases;

b. Blunt-ended DNA cloning of products amplified with kinased primers by ligation to blunt-cut phosphatased vector;

c. Blunt-ended ligation of vector and nonphosphorylated amplification products in the presence of a rarecutting restriction endonuclease;

d. Cloning of products amplified with primers containing built-in restriction endonuclease sites; and

e. The uracil-DNA glycosidase (UDG) method, which removes uracil from amplification products generated with dUMP-containing primers to expose clonable 3' overhangs.

Although most methods suffer from one or more drawbacks, the overhanging-ended approach is considered more efficient. For example, blunt-ended cloning is generally inefficient, and requires the use of large amounts of amplification product, phosphorylated primers, and dephosphorylated blunt-ended vectors. However, other strategies require the use of primers with nontarget-specific tags or expensive uracil substitutions and post-PCR manipulations. Despite limitations, a number of cloning kits that are commercially available guarantee efficient ligation to the linearized vector. Examples include blunt-end ligation into pBluescript IIsk or pCR-Script SK(+) vectors (Stratagene, La Jolla, CA) or into pNoTA/T7 of Prime PCR Cloner (5Prime-3Prime Inc., Boulder, CO), overhanging 3'A ligation to T/A cloning vector pCRII (Invitrogen), and the UDG-driven CloneAmp Systems (Life Technologies). In general, efficient cloning with high number of clones containing the desired insert requires amplification conditions or postamplification treatments that decrease artifactual amplification products, high fidelity of DNA polymerase to minimize mutation during amplification, and appropriate conditions for restriction endonuclease digestion and insert ligation.

Overhanging-ended cloning is considered the best strategy, but in some cases requires the use of primers encoding restriction endonuclease sites. In the absence of such primers, it is always possible to incorporate restriction sites to the end of isolated products (GC, unpublished). This is done by reamplification with tagged primers derived from those used during fragment recovery but containing restriction sites at the 5' terminus. Ultimately, recombinant plasmids are amplified by the PCR and cloned fragments radioactively labeled for hybridization or subjected to sequencing with forward and reverse primers complementary to sites in the cloning vector *(3)*.

7. Applications: Retrospective examination of silver-stained DNA may be especially important in forensic applications, parentage testing, plant variety rights enforcement, and whenever genetic evidence is entered in a court of law. Because bands can be easily dissected from dry gels, reamplified, and used as hybridization probes, our procedure becomes valuable in molecular ecology and evolution, genetic mapping, marker-assisted selection, and general fingerprinting. Isolated DNA fragments can also be cloned, sequenced, and converted into SCARs *(36)*. Finally, in the study of differential gene expression, DD fragments representing differentially expressed genes can be cloned directly from amplification-recovered products isolated from silver-stained cDNA fingerprints.

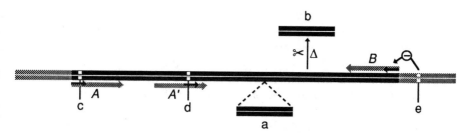

**Fig. 2.** Origins of polymorphic DNA in nucleic acid scanning and design of PCR primers defining SCARs. The sequencing of polymorphic fragments amplified with short arbitrary primers (black arrows) permits the search for longer primer annealing sites, the synthesis of PCR primers (shaded arrows), and SCAR amplification. If the DNA polymorphism is caused by an insertion **(a)** or a deletion **(b)** within the amplicon, both DAF and SCAR markers will be codominant. Loss of an annealing site **(c)** results in a dominant DAF and SCAR marker but the SCAR will amplify alleles from both parents in a segregating population. The gain of an annealing site **(d)** probably results in a codominant DAF marker but its SCAR will be dominant and will amplify alleles from both parents (primer A) or perhaps from only one (primer A'). Finally, a change in sequence within or outside **(e)** the amplicon can condition the annealing of the arbitrary primer resulting in dominant DAF and SCAR markers with SCARs amplifying alleles from either one or both parents. Other origins of polymorphic DNA are possible but are not shown.

8. SCAR analysis: DNA amplification markers generated using arbitrary primers are versatile and universal *(50)*. However, the use of arbitrary primers is poised with a number of limitations, including the generation of artifactual and nonparental bands, marker dominance, low allele number, existence of repetitive sequences, and band comigration. A way to overcome some of these limitations is the conversion of these markers into sequence-characterized amplified regions (SCARs) *(36)*. Diagnostic polymorphic or monomorphic fragments can be cloned directly from the silver-stained gels and used as hybridization probes. Alternatively, partial sequencing of the isolated product can define a PCR amplicon within the selected locus detectable by electrophoresis or presence of incorporated fluorophores and radioactive labels. These SCARs become diagnostic PCR markers of a dominant or codominant nature and represent single, genetically defined loci.

SCARs are usually generated by cloning and sequencing of interesting amplification products. Primers complementary to the ends of an isolated DAF fragment are used in a PCR reaction to amplify single loci from the originating DNA template. These SCARs retain the original dominant segregation behavior of the DAF marker or become codominant. DNA polymorphisms can be caused by changes in nucleotide sequence of the primer annealing sites, by structural rearrangements within the amplicon, or by secondary structure (Fig. 2). Depending on the nature of the polymorphisms and selected PCR primers, SCARs will be dominant or codominant. Usually, most SCARs are dominant but show amplification of alleles from both parents. In this case, originating polymorphisms are probably caused by mismatches in primer annealing sites. However, the longer SCAR priming sites tolerate these mismatches and therefore sustain amplification. Other SCARs are codominant and most probably result from amplicon sequence rearrangements, such as insertions or deletions. Sequence divergence causing gain, loss, or altered orientation of annealing sites generate either dominant or codominant markers, but derived SCARs often behave as dominant markers with only one allele being amplified. By selecting appropriate SCAR primers one can take advantage of these dominant markers (Fig. 2). Alternatively, dominant SCAR markers can be converted to codominant SCARs by digestion with restriction endonucleases *(36),* or fingerprinting with minihairpin or microsatellite primers *(51,52).*

A SCAR is only useful as genetic marker or physical landmark if it characterizes only a single locus. In some cases, the identity of cloned amplification products and derived SCARs should be verified by Southern hybridization to DAF profiles or genomic DNA. If possible, this study should be extended to a segregating population. Hybridization to DAF patterns will confirm the polymorphic nature of the isolated fragment. Hybridization to digested genomic DNA will show if the SCAR contains repetitive DNA sequences and may detect codominant restriction fragment length polymorphisms (RFLPs). To avoid selecting PCR products that are amplified from multiple loci, SCARs must be polymorphic in a mapping population and defined genetically as a single locus by segregation analysis. The SCAR should cosegregate with the DAF marker from which it was derived. This confirms that the SCAR represents a single locus identical to that revealed by the original DAF marker.

SCARs are widely applicable and offer several advantages. They are similar to sequence tagged sites (STS) *(53)* and behave as landmarks for physical mapping and as anchoring points between physical and genetic maps. Genomic libraries can be screened with these markers to construct contigs (contiguous segments) or identify overlapping clones in chromosome walking. SCARs containing interspersed repetitive sequences can still be used as reliable PCR markers in map-based cloning. In contrast, arbitrary amplification markers containing repetitive DNA are useless as hybridization probes and as physical landmarks. In genetic mapping applications, codominant SCARs are much more informative than dominant arbitrary amplification markers and constitute reliable tags of specific genomic regions. Finally, SCARs are reliable because of stringent PCR conditions and adequate for comparative mapping of related species.

# References

1. Mullis, K. B. and Faloona, F. A. (1987) Specific synthesis of DNA in vitro via a polymerase catalized reaction. *Methods Enzymol.* **255,** 335–350.
2. Erlich, H. A., Gelfand, D., and Sninsky, J. J. (1991) Recent advances in the polymerase chain reaction. *Science* **252,** 1643–1651.

3. Sambrook, J., Fritsch, E. F., and Maniatis, T. (1989) *Molecular Cloning: A Laboratory Manual,* 2nd ed., Cold Spring Harbor Laboratory, Cold Spring Harbor, NY.

4. Zintz, C. A. and Beebe, D. C. (1991) Rapid reamplification of PCR products purified in low melting point agarose gels. *BioTechniques* **11,** 158–162.

5. D'Aquila, R. T., Bechtel, L. J., Videler, J. A., Eron, J. J., Gorczyca, P., and Kaplan, J. C. (1991) Maximizing sensitivity and specificity of PCR by preamplification heating. *Nucleic Acids Res.* **19,** 3749.

6. Mullis, K. B. (1991) The polymerase chain reaction in an anemic mode, how to avoid cold oligodeoxyribonuclear fusion. *PCR Methods Applic.* **1,** 1–4.

7. Caetano-Anollés, G. and Gresshoff, P. M. (1994) Staining nucleic acids with silver, an alternative to radioisotopic and fluorescent labeling. *Promega Notes* **45,** 13–18.

8. Storts, D. R., Wu, L. C., Mendoza, L., and Oler, J. K. (1993) Silver-staining, a new approach to nonradioactive DNA sequencing. *Promega Notes* **42,** 10–14.

9. Ainsworth, P. J., Surh, L. C., and Coultier-Mackie, M. B. (1991) Diagnostic single-strand conformational polymorphism (SSCP): a simplified nonradioisotopic method as applied to a Tay-Sachs B1 variant. *Nucleic Acids Res.* **19,** 405,406.

10. Dockhorn-Dworniczak, B., Dworniczak, B., Brommelkamp, L., Bulles, J., Horst, J., and Bocker, W. W. (1991) Non-isotopic detection of single-strand conformation polymorphism (PCR-SSCP): a rapid and sensitive technique in diagnosis of phenylketonuria. *Nucleic Acids Res.* **19,** 2500.

11. Mohabeer, A. J., Hiti, A. L., and Martin, W. J. (1991) Nonradioactive single-strand conformation polymorphism (SSCP) using the Pharmacia PhastSystem. *Nucleic Acids Res.* **19,** 3154.

12. Maekawa, M., Sudo, K., Kitajima, M., Matsuura, Y., Li, S., and Kanno, T. (1993) Detection and characterization of new genetic mutations in individuals heterozygous for lactate dehydrogenase-B(H) deficiency using DNA conformation polymorphism analysis and silver-staining. *Hum. Genet.* **91,** 163–168.

13. Sugano, K., Kyogoku, A., Fukayama, N., Ohkura, H., Shimosato, Y., Sekiya, T., and Hayashi, K. (1993) Rapid and simple detection of c-Ki-*ras*-2 gene codon 12 mutations by nonradioisotopic single-strand conformation polymorphism analysis. *Lab. Invest.* **68,** 361–366.

14. Ainsworth, P. J., Rodenhiser, D., and Costa, M. (1993) Identification and characterization of sporadic and inherited mutations in exon 31 of the neurofibromatosis (NF1) gene. *Hum. Genet.* **91,** 151–156.

15. Caetano-Anollés, G., Bassam, B. J., and Gresshoff, P. M. (1991) DNA amplification fingerprinting using short arbitrary oligonucleotide primers. *Bio/Technology* **9,** 553–557.

16. Lohmann, J., Schickle, H., and Bosch, T. C. G. (1995) REN display, a rapid and efficient method for nonradioactive differential display and mRNA isolation. *Biotechniques* **18,** 200–202.

17. Allen, R. C., Graves, G., and Budowle, B. (1989) Polymerase chain reaction amplification products separated on rehydratable polyacrylamide gels and stained with silver. *Biotechniques* **7,** 736–744.

18. Budowle, B., Chakraborty, R., Giusti, A. M., Eisenberg, A. J., and Allen, R. C. (1991) Analysis of VNTR locus DIS80 by the PCR followed by high-resolution PAGE. *Am. J. Hum. Genet.* **48,** 137–144.

19. Smith, H. O. (1980) Recovery of DNA from gels. *Methods Enzymol.* **65,** 371–380.
20. Southern, E. M. (1975) Detection of specific sequences among DNA fragments detected by gel electrophoresis. *J. Mol. Biol.* **98,** 503–517.
21. Zhu, J., Kempenaers, W., Van der Straeten, D., Contreras R., and Fiers, W. (1985) A method for fast and pure DNA elution from agarose gels by centrifugal filtration. *Bio/Technology* **3,** 1014–1016.
22. Krowczynska, A. M., Donoghue, K., and Hughes, L. (1995) Recovery of DNA, RNA and protein from gels with microconcentrators. *Biotechniques* **18,** 698–703.
23. Heery, D. M., Gannon, F., and Powell, R. (1990) A simple method for subcloning DNA fragments from gel slices. *Trends Genet.* **6,** 173.
24. McDonell, M. W., Simon, M. N., and Studier, F. W. (1977) Analysis of restriction fragments of T7 DNA and determination of molecular weights by electrophoresis in neutral and alkaline gels. *J. Mol. Biol.* **110,** 119–146.
25. Dretzen, G., Bellard, M., Sassone-Corsi, P., and Chambon, P. (1981) A reliable method for the recovery of DNA fragments from agarose and acrylamide gels. *Anal. Biochem.* **112,** 295–298.
26. Sylvers, L. A. and Beresten, S. (1993) A rapid automated method for simultaneous elution and purification of RNA from polyacrylamide gels. *BioTechniques* **14,** 378–380.
27. Wieslander, L. (1979) A simple method to recover intact high molecular weight RNA and DNA after electrophoretic separation in low gelling temperature agarose gels. *Anal. Biochem.* **98,** 305–309.
28. Ausubel, F. M., Brent, R., Kingston, R. E., Moore, D. D., Seidman, J. G., Smith, J. A., and Struhl, K. (1992) *Current Protocols in Molecular Biology* (2nd ed.), (Unit 2.6), Wiley, New York.
29. Thuring, R. W., Sanders, J. B., and Borst, P. A. (1975) Freeze-squeeze method for recovering long DNA from agarose gels. *Anal. Biochem.* **66,** 213–220.
30. Tautz, D. and Renz, M. (1983) An optimized freeze-squeeze method for the recovery of DNA fragments from agarose gels. *Anal. Biochem.* **132,** 14–19.
31. Li, Q. and Ownby, C. L. (1993) A rapid method for extraction of DNA from agarose gels using a syringe. *Biotechniques* **15,** 976–978.
32. Grey, M. and Brendel, M. (1992) Rapid and simple isolation of DNA from agarose gels. *Curr. Genet.* **22,** 83,84.
33. Schwarz, H. and Whitton, J. L. (1992) A rapid, inexpensive method for eluting DNA from agarose or acrylamide gel slices without using toxic or chaotropic materials. *Biotechniques* **13,** 205,206.
34. Weaver, K. R., Caetano-Anollés, G., Gresshoff, P. M., and Callahan, L. M. (1994) Isolation and cloning of DNA amplification products from silver-stained polyacrylamide gels. *Biotechniques* **16,** 226,227.
35. Bassam, B. J., Caetano-Anollés, G., and Gresshoff, P. M. (1991) Fast and sensitive silver-staining of DNA in polyacrylamide gels. *Anal. Biochem.* **196,** 80–83.
36. Paran, I. and Michelmore, R. W. (1993) Development of reliable PCR-based markers linked to downy mildew resistance genes in lettuce. *Theor. Appl. Genet.* **85,** 985–993.

37. Caetano-Anollés, G. and Bassam, B. J. (1993) DNA amplification fingerprinting using arbitrary oligonucleotide primers. *Appl. Biochem. Biotechnol.* **42,** 189–200.
38. Bassam, B. J. and Caetano-Anollés, G. (1993) Silver-staining of DNA in polyacrylamide gels. *Appl. Biochem. Biotechnol.* **42,** 181–188.
39. Dieffenbach, C. W., Lowe, T. M. J., and Dveksler, G. S. (1993) General concepts for PCR primer design. *PCR Methods Applic.* **3,** S30–S37.
40. Bassam, B. J. and Caetano-Anollés, G. (1993) Automated "hot start" PCR using mineral oil and paraffin wax. *Biotechniques* **14,** 30–34.
41. Doktycz, M. J. (1993) Discontinuous electrophoresis of DNA, adjusting DNA mobility by trailing ion net mobility. *Anal. Biochem.* **213,** 400–406.
42. Baum, T. J., Gresshoff, P. M., Lewis, S. A., and Dean, R. A. (1994) Characterization and phylogenetic analysis of four root-knot nematode species using DNA amplification fingerprinting and automated polyacrylamide gel electrophoresis. *Mol. Plant-Microbe Interact.* **7,** 39–47.
43. Caetano-Anollés, G., Bassam, B. J., and Gresshoff, P. M. (1992) Primer-template interactions during DNA amplification fingerprinting with single arbitrary oligonucleotide. *Mol. Gen. Genet.* **235,** 157–165.
44. Caetano-Anollés, G., Bassam, B. J., and Gresshoff, P. M. (1993) Enhanced detection of polymorphic DNA by multiple arbitrary amplicon profiling of endonuclease digested DNA: identification of markers tightly linked to the supernodulation locus in soybean. *Mol. Gen. Genet.* **241,** 57–64.
45. Weaver, K. R., Callahan, L. M., Caetano-Anollés, G., and Gresshoff, P. M. (1995) DNA amplification fingerprinting and hybridization analysis of centipedegrass. *Crop Sci.* **35,** 881–885.
46. Trigiano, R. N., Caetano-Anollés, G., Bassam, B. J., and Windham, M. T. (1995) DNA amplification fingerprinting provides evidence that *Discula destructiva,* the cause of dogwood anthracnose in North America, is an introduced pathogen. *Mycologia* **87,** 490–500.
47. Calvert, R. J., Weghorst, C. M., and Buzard, G. S. (1995) PCR amplification of silver-stained bands from cold SSCP gels. *Biotechniques* **18,** 782–786.
48. Sanguinetti, C. J., Diaz-Neto, E., and Simpson, A. J. G. (1994) Rapid silver-staining and recovery of PCR products separated on polyacrylamide gels. *Biotechniques* **17,** 15–19.
49. Frohman, M. A. (1994) Cloning PCR products. In *The Polymerase Chain Reaction* (Mullis, K. B., Ferré, F., Gibbs, R. A., eds.), Birkhäuser, Boston, pp. 14–37.
50. Caetano-Anollés, G. (1994) MAAP, a versatile and universal tool for genome analysis. *Plant Mol. Biol.* **25,** 1011–1026.
51. Caetano-Anollés, G. and Gresshoff, P. M. (1994) DNA amplification fingerprinting using minihairpin oligonucleotide primers. *Bio/Technology* **12,** 619–623.
52. Caetano-Anollés, G., and Gresshoff, P. M. (1996) Generation of sequence signatures from DNA amplification fingerprints with minihairpin and microsatellite primers. *Biotechniques* **20,** 1044–1056.
53. Olson, M., Hood, L., Cantor, C., and Botstein, D. (1989) A common language for physical mapping of the human genome. *Science* **245,** 1434,1435.

# III

## MUTAGENESIS, RECOMBINATION, AND IN VITRO SELECTION

# 13

# Recombination and Site-Directed Mutagenesis Using Recombination PCR

## Douglas H. Jones and Stanley C. Winistorfer

## 1. Introduction

The polymerase chain reaction (PCR) *(1)* provides a rapid means for the recombination and site-directed mutagenesis of DNA *(2)*. DNA modification can occur during PCR because the primers are incorporated into the ends of the PCR product. The simplest PCR-based method for site-directed mutagenesis and DNA recombination is recombination PCR.

Recombination PCR uses in vivo recombination in *Escherichia coli* to generate site-directed mutants and recombinant constructs *(3,4)*. In the recombination PCR method, PCR adds homologous ends to DNA. These homologous ends mediate recombination between these linear PCR products in *E. coli*, resulting in the formation of DNA joints in vivo. If two PCR products have homologous ends that can recombine to form a circle, and if this circle constitutes a selectable plasmid, *E. coli* can be readily transformed by the linear PCR products. Recombination PCR has almost no steps apart from PCR amplification and transformation of *E. coli*, and this method works well in Rec A minus *E. coli* strains used routinely in cloning. Since the introduction of this method in 1991, it has been used by numerous investigators *(5–9)*. One example of DNA recombination using recombination PCR is illustrated in Fig. 1. This figure illustrates a protocol for amplifying a portion of a donor plasmid and inserting it in a recipient plasmid at a defined position and orientation. The donor plasmid is shown on the left side and the recipient plasmid is on the right side. The steps corresponding to this figure are briefly outlined below:

1. The DNA segment that is to be inserted into the recipient plasmid is amplified from the donor plasmid using primers 1 and 2. In a separate PCR amplification, the recipient plasmid is amplified with primers 3 and 4. The 5' regions of primers

From: *Methods in Molecular Biology, Vol. 67: PCR Cloning Protocols: From Molecular Cloning to Genetic Engineering* Edited by: B. A. White  Humana Press Inc., Totowa, NJ

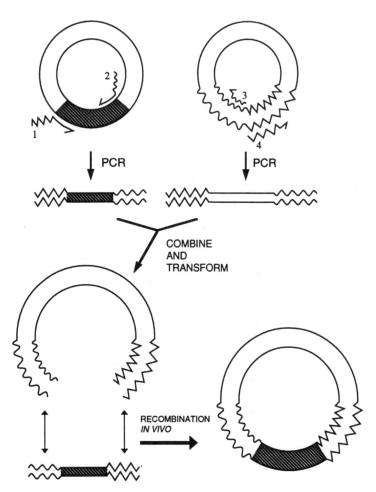

**Fig. 1.** Diagram illustrating DNA recombination using recombination PCR. The primers are numbered hemiarrows. The insert is the cross-hatched region. Smooth circles represent the DNA strands of the donor plasmid. Circles with wavy and jagged portions represent DNA strands of the recipient plasmid. Reprinted by permission from *BioTechniques* **10,** 62–66.

1 and 2 that do not anneal to the donor plasmid are complementary to primers 4 and 3, respectively (or 3 and 4, depending on the orientation of the insert desired in the recombinant construct). Frequently, a plasmid template can be linearized outside the region to be amplified by restriction endonuclease digestion prior to PCR amplification. When this can be done, the PCR product does not need to be purified, because linearized plasmids transform *E. coli* inefficiently. If a plasmid template cannot be linearized by restriction endonuclease digestion outside the region to be amplified prior to PCR amplification, the PCR product must be

removed from the plasmid prior to transformation in order to prevent background transformants arising from the supercoiled plasmid template. PCR product purification is accomplished either by agarose gel purification followed by glass bead extraction or by adding the restriction endonuclease *Dpn*I to the PCR mixture. *Dpn*I is a restriction endonuclease that digests methylated GATC sites. These sites are methylated in the plasmid by strains of *E. coli* used routinely in cloning (by *dam* methylase), but are not methylated in the PCR products, permitting *Dpn*I to digest the plasmid without cutting the PCR product *(10)*.

2. The two PCR products are combined and used to transform MAX efficiency competent *E. coli* (BRL, Life Technologies, Gaithersburg, MD). If each plasmid template is restriction endonuclease digested outside the region to be amplified prior to PCR amplification, the two crude PCR products can simply be combined, and the resulting mixture used to transform *E. coli*.

In a simple variation of this recombination PCR strategy, the inserted segment can be an unmodified PCR product. In that case, primers 3 and 4 have 5' ends that are homologous to the ends of the PCR fragment to be inserted, and the recipient plasmid is linearized by restriction endonuclease digestion prior to PCR amplification. We routinely use this approach to clone any PCR product *(4)*.

In recombination PCR, the sum goal of the two amplifications is to yield two PCR products where each end of one product is homologous to a distinct end of the other PCR product. Since the amplifying primer sequences are incorporated into the ends of a PCR product, so long as primers 1 and 2 contain regions that are complementary to regions of primers 3 and 4 (or 4 and 3), the PCR products will contain ends that are homologous to each other, and these primer-determined DNA ends do not need to be determined by the original donor or recipient templates. The only requirement of this recombination PCR strategy is that primers 1 and 2 must have regions of complementarity to primers 3 and 4. Therefore, recombination PCR can be used not only to generate recombinant constructs, such as gene chimeras, but also for the site-directed mutagenesis of two distal sites concurrently (Fig. 2), or for the rapid site-directed mutagenesis of single sites (Fig. 3) *(11)*. In the point mutagenesis protocol illustrated in Fig. 3, the plasmid is linearized by restriction endonuclease digestion prior to each PCR amplification. In each of the two amplifications, the mutating primers (primers 1 and 3) mutate the identical base pair so that the mutated ends of each product are homologous to each other and the nonmutating primers (primers 2 and 4) are also designed to produce ends that are homologous to each other. Both unpurified PCR products are combined to transform *E. coli*, generating clones with the mutation of interest.

## 2. Materials

1. *Taq* DNA polymerase (AmpliTaq 5 U/mL; Perkin-Elmer, Norwalk CT) *(see Note 1)*.
2. 10X PCR buffer II: 500 m$M$ KCl, 100 m$M$ Tris-HCl, pH 8.3.
3. 25 m$M$ MgCl$_2$ solution.

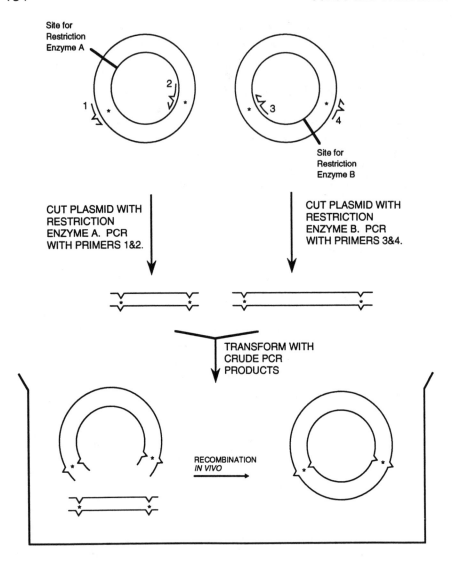

**Fig. 2.** Diagram illustrating site-directed mutagenesis of two distal sites using recombination PCR. The primers are numbered hemiarrows. Asterisks designate the mutagenesis sites. There is no purification of the PCR products. Notches designate point mismatches in the primers and resulting mutations in the PCR products. Reprinted by permission from *Technique* **2**, 273–278.

4. 10 m*M* stocks of each dATP, dCTP, dTTP, and dGTP, neutralized to pH 7.0 with NaOH.
5. Restriction endonucleases (New England BioLabs, Beverly, MA).

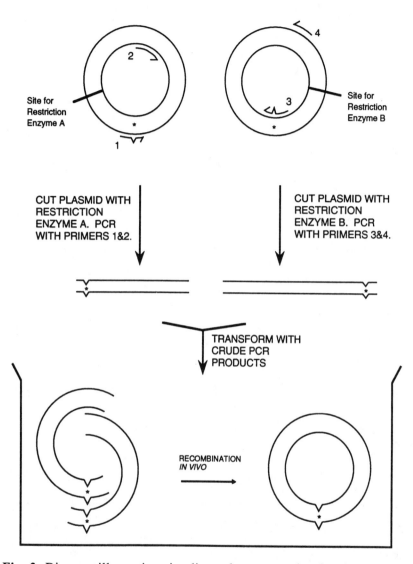

**Fig. 3.** Diagram illustrating site-directed mutagenesis of a single site using recombination PCR with 4 primers. The primers are numbered hemiarrows. The asterisk designates the mutagenesis site. Primer 2 is complementary to primer 4. Restriction endonuclease sites A and B bracket the insert. Notches designate point mismatches in the primers and resulting mutations in the PCR products. There is no purification of the PCR products. For each additional single site-directed mutagenesis reaction, only a new primer 1 and 3 need to be synthesized, and the same cut templates can be used. Reprinted by permission from *Technique* **2,** 273–278.

6. PCR primers. In Fig. 1, PCR amplification with primers 1 and 2 results in a product with 24–30 basepairs of homology with the products of primers 3 and 4. For PCR primers that introduce mutations, *see* Note 2.

7. Long thin micropipet tips (gel loader tips T-010; Phenix Research Products, Hayward, CA).

8. Agarose.

9. Ethidium bromide.

10. TAE buffer: 40 m*M* Tris-acetate, 2 m*M* EDTA, pH 8.5 *(12)*.

11. Geneclean (Bio 101, La Jolla, CA).

12. TE buffer: 10 m*M* Tris-HCl, pH 8.0, 1 m*M* EDTA.

13. MAX Efficiency DH5α Competent *E. coli* (BRL, Life Technologies). Once a tube is thawed it should not be reused (*see* Note 3).

14. SOC Media *(13)*.

15. Top agar *(14)*.

16. LB plates with 100 µg/mL Ampicillin *(14)*.

17. Luria-Bertani medium (LB broth) *(15)*.

18. Qiagen (Chatsworth, CA) midi-columns.

19. Sequenase 2.0 sequencing kit (US Biochemical, Cleveland, OH).

## 3. Methods

1. Linearize the plasmid template by restriction endonuclease digestion outside the region to be amplified, if possible. The plasmid digest does not need to be purified prior to its use as a PCR template (*see* Notes 4 and 5).

2. Assemble a PCR in a total volume of 50 µL containing the following: 2 ng of plasmid template, 25 pmol of each primer, 200 µ*M* each dNTP, 1X PCR buffer, 2.5 m*M* MgCl$_2$, and 1.25 U DNA *Taq* DNA polymerase. (*See* Note 6.)

3. Pipet 50 µL of mineral oil on top of each reaction mix prior to amplification.

4. Perform PCR amplification using the following parameters (*see* Note 6): 94°C for 1 min (initial denaturation), 94°C for 30 s (denaturation), 50°C for 30 s (anneal), 72°C for 1 min/kb of PCR product (extension), 14–20 amplification cycles, and 72°C for 7 min (final extension step).

5. Remove each PCR product by inserting a long thin micropipet tip through the mineral oil layer and drawing up the sample. Visualize the PCR product on an agarose minigel. If 5 µL of the PCR product can be clearly seen following ethidium bromide staining (>15 ng/5 µL), there is enough product.

6. Withdraw 2.5 µL from each PCR tube (typically 10–60 ng/2.5 µL) by inserting a long thin micropipet tip through the mineral oil, and then combine the two samples. If the PCR template is linearized by restriction endonuclease digestion outside the region to be amplified, no purification of PCR products is necessary (*see* Note 5).

7. Transform *E. coli* using a mixture containing 2.5 µL of each PCR product. Maintaining an even molar ratio of one product to another is not necessary. Then, transform MAX efficiency competent *E. coli* (BRL) with the 5-µL sample containing the two PCR products. Transformation is carried out following the manufacturer's instructions with the following modifications:

a. Use 50 μL of *E. coli* for each sample transformed, because this is effective and less expensive than the 100 μL recommended.

b. After incubation at 37°C in a shaker for 1 h, plate the entire sample onto an LB plate containing 100 μg/mL ampicillin. In order to keep the sample on the plate, add 2 mL of top agar, prewarmed to 42°C, to each sample immediately prior to pouring it onto the plate.

c. Once an aliquot of bacteria is thawed, do not use it again.

Set up the following transformations:

Plate A: 2.5 μL of PCR 1 + 2.5 μL of PCR 2.

Plate B: 2.5 μL of PCR 1 + 2.5 μL of TE.

Plate C: 2.5 μL of PCR 2 + 2.5 μL of TE.

Plate D: 0.5 ng of a supercoiled template in 5 μL of TE.

Plate E: 5 μL of TE

The yield of colonies from plate A is >2× that from plate B + C, confirming a high percentage of recombinants in plate A. Plate D is a transformation control, and should yield a thick lawn of colonies. Plate E is an antibiotic control, and should yield no colonies since the bacteria that have not been transformed are sensitive to ampicillin. Only 25 μL of cells are used for the control plates D and plate E, so that only one BRL tube, which contains 200 μL of bacteria, needs to be used.

8. Screen plasmids by placing individual colonies in 2 mL of LB broth containing 100 μg/mL of ampicillin. Grow the colonies at 37°C for 6–24 h.

9. Screen the plasmids by removing 2 μL of the LB broth, place it directly in a PCR tube, and amplify for 25 cycles (*see* steps 2–4) using primers that flank the mutated site or insert (e.g., M13 primers) *(16)*.

10. In a mutagenesis protocol, a base pair can be mutated to either create or remove a restriction endonuclease site. In particular, the degenerate amino acid code allows one to create or eliminate a restriction endonuclease recognition site without altering the amino acid encoded at that site. Screen for the mutation by adding 3 U of the restriction endonuclease and 1 μL of the appropriate 10X restriction buffer directly to 5 μL of the unpurified PCR product in a total volume of 10 μL.

11. Assess cutting by minigel analysis. Typically, 50–100% of the clones contain the recombinant of interest. Then, purify the plasmid and sequence the mutated region (*see* Note 7) using Qiagen columns for plasmid purification and Sequenase 2.0 for sequencing following the manufacturers' instructions.

## 4. Notes

1. Other investigators have used Pfu DNA polymerase instead of *Taq* DNA polymerase in recombination PCR *(6)*. Pfu DNA polymerase has better fidelity than *Taq* DNA polymerase *(17)*.

2. The primers that introduce point mutations (primers 1–4 in Fig. 2 and primers 1 and 3 in Fig. 3) are designed to generate 15–45 bp of homology between each end of one PCR product relative to the other PCR product. In all recombination

PCR protocols, 24 bp of homology works very well, and alterations that generate long regions of homology do not work noticeably better. Decreasing the length of homology from 25–12 bp in an early protocol did decrease the transformation efficiency 4–5-fold. Single point mismatches lie no closer than six nucleotides from the 3' end of a primer and are frequently placed toward the middle. Placing point mutations near the 5' end of each mutating primer will generate two PCR products whose mutated ends have <24-bp of homology. Multiple point mismatches should be placed in the middle or toward the 5' end of a primer, with primer lengths long enough to create 24-bp of homology between the mutated ends of the two PCR products. Primers that are nonmutating are generally 24–30 nucleotides long. These nonmutating primers can be designed to anneal to the β-lactamase gene so that they can be used with a variety of different plasmids. Frequently, the mutating and nonmutating primers are designed to be perfect complements to each other.

For site-directed mutagenesis, since unique restriction endonuclease recognition sites almost always bracket the insert, the same linearized templates can be used for the mutagenesis of any single site in the insert. Primers 2 and 4 are nonmutating (*see* Fig. 3), and are conserved for each new site targeted for mutagenesis, so that only two new primers need to be generated for each site targeted for mutagenesis (via primers 1 and 3). Furthermore, only approximately one half of the length of the entire template needs to be amplified in each of the two PCR amplifications, facilitating the mutagenesis of large constructs and permitting considerable flexibility in the primer design and sequence. Recombination PCR has been used to mutate constructs up to 7.1 kb *(18)*.

3. Since the transformation efficiency is low, highly competent bacteria (transfection efficiency $>1 \times 10^9/\mu g$ of monomer pUC19) should be used. Using restriction endonuclease digested templates, the transformation efficiency is about 10 colonies with the mutation/ng total DNA used to transform *E. coli*.

4. Following 14 amplification cycles, the PCR product yield is much higher when using a linear template than when using a supercoiled template.

5. If a plasmid template cannot be linearized outside the region to be amplified prior to PCR amplification, the PCR product must be removed from the supercoiled plasmid template. This can be accomplished either by agarose gel electrophoresis and extraction using GeneClean or by digestion with the restriction endonuclease *Dpn*I. When agarose gel resolution and GeneClean extraction are used, the entire PCR product should be gel purified and reconstituted in 25–30 μL TE, and 2.5 μL is combined with the 2.5 μL of the other PCR product prior to transformation. If *Dpn*I is used, remove the PCR product from under the mineral oil, add 20 U of *Dpn*I directly to the 25 μL of the PCR sample using the recommended 10X *Dpn*I buffer in a final total volume of 30 μL, and incubate the mixture at 37°C for 1 h. No further purification of the PCR product is necessary.

6. The exact buffer components and conditions for PCR vary with different primers and template (*see* Chapters 1–3 for a discussion of PCR optimization).

7. There is always the possibility of a sequence error in a single clone following PCR amplification. The altered region should be sequenced, and one may choose to clone a restriction fragment containing the mutated or recombined region of interest into a construct that has not undergone PCR amplification.

## Acknowledgments

This work was supported by the Roy J. Carver Charitable Trust, the University of Iowa through funds generated by the Children's Miracle Network Telethon, and by National Institutes of Health grant R01 HG00569. We thank Jim Hartley for suggesting use of the restriction endonuclease *Dpn*I in order to remove supercoiled template from the PCR mixture.

## References

1. Mullis, K., Faloona, F., Scharf, S., Saiki, R., Horn, G., and Erlich, H. (1986) Specific enzymatic amplification of DNA in vitro: the polymerase chain reaction, in *Cold Spring Harbor Symposia on Quantitative Biology,* vol. LI, Cold Spring Harbor Laboratory, Cold Spring Harbor, NY, pp. 263–273.
2. White, B. (1993) *Methods in Molecular Biology,* vol. 15, *PCR Protocols: Current Methods and Applications,* Humana, Totowa, NJ.
3. Jones, D. H. and Howard, B. H. (1991) A rapid method for recombination and site-specific mutagenesis by placing homologous ends on DNA using polymerase chain reaction. *BioTechniques* **10,** 62–66.
4. Jones, D. H. (1994) PCR mutagenesis and recombination in vivo. *PCR Methods Appl.* 3, S141–S148.
5. Coco, W. M., Rothmel, R. K., Henikoff, S., and Chakrabarty, A. M. (1993) Nucleotide sequence and initial functional characterization of the *clc*R gene encoding a *Lys*R family activator of the *clc*ABD chlorocatechol operon in *Pseudomonas putida. J. Bacteriol.* **175,** 417–427.
6. Fridovich-Keil, J. L. and Jinks-Robertson, S. (1993) A yeast expression system for human galactose-1-phosphate uridylyltransferase. *Proc. Natl. Acad. Sci. USA* **90,** 398–402.
7. Goulden, M. G., Kohm, B. A., Santa Cruz, S., Kavanagh, T. A., and Baulcombe, D. C. (1993) A feature of the coat protein of potato virus X affects both induced virus resistance in potato and viral fitness. *Virology* **197,** 293–302.
8. Gibbs, J. S., Regier, D. A., and Desrosiers, R. C. (1994) Construction and in vitro properties of HIV-1 mutants with deletions in "nonessential" genes. *AIDS Res. Hum. Retrovir.* **10,** 343–350.
9. Singh, K. K., Small, G. M., and Lewin, A. S. (1992) Alternative topogenic signals in peroxisomal citrate synthase of *Saccharomyces cerevisiae. Mol. Cell. Biol.* **12,** 5593–5599.
0. Weiner, M. P., Costa, G. L., Schoettlin, W., Cline, J., Mathur, E., and Bauer, J. C. (1994) Site-directed mutagenesis of double-stranded DNA by the polymerase chain reaction. *Gene* **151,** 119–123.

11. Jones, D. H. and Winistorfer, S. C. (1992) Recombinant circle PCR and recombination PCR for site-specific mutagenesis without PCR product purification. *BioTechniques* **12,** 528–534.
12. Sambrook, J., Fritsch, E. F., and Maniatis, T. (1989) *Molecular Cloning: A Laboratory Manual,* 2nd ed., Cold Spring Harbor Laboratory, Cold Spring Harbor, NY, p. B. 23.
13. Sambrook, J., Fritsch, E. F., and Maniatis, T. (1989) *Molecular Cloning: A Laboratory Manual,* 2nd ed., Cold Spring Harbor Laboratory, Cold Spring Harbor, NY, p. A. 2.
14. Sambrook, J., Fritsch, E. F., and Maniatis, T. (1989) *Molecular Cloning: A Laboratory Manual,* 2nd ed., Cold Spring Harbor Laboratory, Cold Spring Harbor, NY, p. A. 4.
15. Sambrook, J., Fritsch, E. F., and Maniatis, T. (1989) *Molecular Cloning: A Laboratory Manual,* 2nd ed., Cold Spring Harbor Laboratory, Cold Spring Harbor, NY, p. A. 1.
16. Liang, W. and Johnson, J. P. (1988) Rapid plasmid insert amplification with polymerase chain reaction. *Nucleic Acids Res.* **16,** 3579.
17. Lundberg, K. S., Shoemaker, D. D., Adams, M. W. W., Short, J. M., Sorge, J. A., and Mathur, E. J. (1991) High-fidelity amplification using a thermostable DNA polymerase isolated from *Pyrococcus furiosus. Gene* **108,** 1–6.
18. Yao, Z., Jones, D. H., and Grose, C. (1992) Site-directed mutagenesis of herpesvirus glycoprotein phosphorylation sites by recombination polymerase chain reaction. *PCR Methods Appl.* **1,** 205–207.

# 14

## In Vitro Recombination and Mutagenesis of DNA

### SOEing Together Tailor-Made Genes

**Robert M. Horton**

## 1. Introduction

Gene splicing by overlap extension (gene SOEing) provides a powerful method of recombining sequences without depending on restriction sites or ligase, and a simple, generally applicable way of using polymerase chain reaction (PCR) to perform site-directed mutagenesis in vitro. This technique allows even those with minimal molecular biology expertise to generate quickly genetic constructs that might otherwise have been impractical using only the older (restriction enzyme-based) technology.

This method is made possible by the fact that the ends of a PCR-generated DNA fragment can incorporate new, custom-designed sequences that were not present in the original template. The primers must match their template sequence well enough to prime, but they do not have to match exactly, especially toward the 5' end. Any mismatches will be incorporated into the product and will represent changes in the original sequence. This idea, originally called "mispriming" by Mullis et al. *(1,2)*, provides a simple way to perform site-directed mutagenesis *(3,4)*. However, simple mispriming is a limited way of creating site-directed mutants because the changes can only be made at the ends of a PCR product.

Overlap extension was originally devised as a way of introducing mutations in the center of PCR-generated sequence segments in order to make PCR mutagenesis more universally applicable *(5,6)*. The concept is illustrated in Fig. 1. First, two separate fragments are amplified from the target gene. The first is amplified with primers a and b (product AB): Primer b introduces a sequence change at the right end of product AB. The second fragment (product CD) is

From: *Methods in Molecular Biology, Vol. 67: PCR Cloning Protocols: From Molecular Cloning to Genetic Engineering* Edited by: B. A. White Humana Press Inc., Totowa, NJ

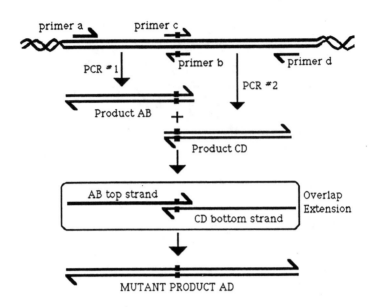

**Fig. 1.** Mutagenesis by overlap extension. The segment of the "wild-type" gene to be mutated is shown as straight lines with the flanking regions coiling away on both sides *ad infinitum*. Synthetic oligonucleotide primers are named with lower case letters. The PCR product made with two primers is named by capital letters of the primers used to make it; for example, primers a and b make product AB. PCR products are shown as two paired strands, with half arrowheads indicating the direction in which each strand can act as a primer for DNA polymerase (the 5'–3' direction). The mutagenic primers, b and c, are complementary to one another, with the mutation shown as a black rectangle. When these intermediate PCR products are mixed, denatured, and reannealed, the top strand of AB and the bottom strand of CD can pair with this overlap at their 3' ends, and act as primers on one another to make the mutant product. The other strands, with the overlap at their 5' ends, cannot prime, and are not shown.

amplified with primers c and d, with primer c introducing the same mutation, but into the left end of product CD. These two products now share a segment of identical sequence called the overlap region. When these intermediate products are mixed together, melted, and reannealed, the top strand of AB can anneal to the bottom strand of CD in such a way that the two strands act as primers on one another. Extension of this overlap by DNA polymerase creates the full-length mutant molecule AD, which has the mutation at an arbitrary distance from either end.

In the final overlap extension step, two separate PCR-generated sequences (AB and CD) are joined together. If AB and CD are made from different genes, then product AD is a recombinant molecule *(7,8)*. Figure 2 illustrates the concept of PCR-mediated recombination, or gene SOEing. Here, extra sequence is added to the 5' end of primer b, which results in a short segment of gene II

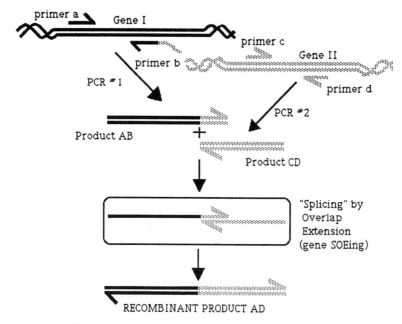

**Fig. 2.** Gene splicing by overlap extension (SOEing). Here, products AB and CD are derived by two different genes. SOEing primer b has sequences added to its 5' end such that the right end of AB is made complementary to the left end of CD. This allows their sequences to be joined by overlap extension, or "SOEn" together.

being added to the right-hand end of the PCR product amplified from gene I (product AB). This causes the two intermediate products to have an overlap region of common sequence, so that they can be joined together by overlap extension. The two strands having the overlap at their 3' ends (the "productive" strands, shown in the boxes in Figs. 1 and 2) each act as both a primer and a template to produce a giant "primer dimer," which is the recombinant molecule.

When should you use gene SOEing? For site-directed mutagenesis, overlap extension is simple, and has advantages over other methods in that

1. Recombination and mutagenesis can be performed simultaneously *(7)*;
2. Essentially all of the product molecules are mutated (i.e., 100% mutation efficiency *(6)*; and
3. The product is produced in vitro (without having to grow it up in a plasmid or phage) and may be used directly in experiments *(5,9)*.

As a method of DNA recombination, gene SOEing is tremendously useful in situations where no leeway can be given to use nearby restriction sites. Protein engineering projects provide excellent examples *(7,8,10,11)*. Two major drawbacks of recombination by SOEing are the cost of the primers and the potential for introducing random errors with PCR (*see* Notes 3, 5, and 6). Therefore, gene SOEing is most practical in "complicated" constructions where there are no convenient restriction sites.

## 2. Materials

1. Thermal cycling ("PCR") machine.
2. *Taq* DNA polymerase.
3. 10X PCR buffer: 500 m*M* KCl, 100 m*M* Tris-HCl, pH 8.3.
4. 10 m*M* MgCl$_2$.
5. dNTPs.
6. Primers (*see* Section 3.1. and Note 1).
7. DNA templates (*see* Note 2).
8. Agarose gel electrophoresis supplies and equipment (*see* Chapter 1).
9. GeneClean (Bio101, La Jolla, CA), or your favorite method for purifying DNA from agarose gels (*see* Chapter 1).

## 3. Methods

### *3.1. Primer Design (see Note 3)*

1. Unlike a regular PCR primer, a SOEing primer (such as b in Fig. 2) has to include two sequence regions capable of hybridizing to a template well enough to act as primers. The first is the priming region at the 3' end of the oligonucleotides, which allows it to act as a PCR primer. The second is the overlap region at the 5' end of the oligonucleotide; this allows the strand of the PCR product that is complementary to the oligonucleotide to act as a primer in the overlap extension reaction. As shown in Fig. 2, the priming region of primer b contains a sequence from gene I, whereas the overlap region contains a sequence from gene II which is complementary to primer c.

   For example, some primers from published overlap extension projects *(8)* are shown in the following (*see* Note 4):

   ```
                priming
        |—region—||-overlap region-|
   3'-gggtccgggtgtgcttctgctgtaactccg-5'  (SOEing primer b)
              5'-gaagacgacattgaggc-3'  (primer c)
                |-priming region-|
   ```

2. In mutagenesis, there is only one template, so the priming and overlap regions may completely coincide *(6)*. The mismatches are placed in the center of the oligo so that both its 3' end and the 3' end of the product made with it can act as primers.
3. My colleagues and I have adopted the convention of designing both the priming and overlap regions of SOEing primers to have an estimated $T_m$ of around 50°C using the following formula *(12)*:

$$T_m = [(G + C) \times 4] + [(A + T) \times 2] \quad (\text{in } °C) \qquad (1$$

Using SOEing primer b as an example, the $T_m$ for its overlap region is 52°C ([3 + 6] × 4] + [(2 + 6) × 2]) and for its priming region is 46°C ([8 + 2) × 4] + [(0 + 3) × 2]). The mismatches in a mutagenic oligo do not count when estimating $T_m$. These $T_m$ estimates, although crude, are generally conservative approximations of the annealing temperature needed for an oligo to be used in PCR (e.g., a 50°C annealing step will work with these primers). They lead to priming and overlap regions being somewhere around 13–20 bases long, depending on the GC content. The "~50°C rule" has resulted in quite reliable primers, but it does not represent a systematic effort to determine the absolute minimum lengths that SOEing primers can be (*see* Note 5).

4. The normal considerations in primer design also apply, such as avoiding complementarity between or within primers, and so forth (*see* Chapter 1).
5. The flanking primers a and d should also be capable of priming at 50°C, and may include restriction sites at the ends to facilitate cloning of the product.
6. "Megaprimer" reactions. Modifications of the gene-SOEing protocol make it possible to use only one primer at the recombinant joint *(13,14)*. Rather than using the symmetrical approach shown in Fig. 2, in which the recombination event is formation of a giant primer-dimer, the top strand of AB can be used as a megaprimer in place of primer c directly on template gene II to make the final product AD. For recombination of two sufficiently different genes, only the recombinant product can be amplified using primers a and d, since each of these primers will match one of the template genes but not the other. Therefore, a and d can be included in the reaction to drive the synthesis of the productive strands, and to amplify the product as soon as it forms *(13,15)*. For megaprimer mutagenesis, however, primer a cannot be included in the final reaction containing primer d because a and d would simply amplify the wild-type sequence from the template. This means that product AB must be added in sufficient quantity to supply all of the megaprimer strand necessary to generate AD *(14)*. For further discussion of megaprimer mutagenesis, *see* Chapter 17.

## 3.2. PCR Generation of Intermediate Products AB and CD (Fig. 2)

1. Set up two separate PCRs as follows (*see* Note 6):

| Ingredients | PCR 1 | PCR 2 |
|---|---|---|
| 10X buffer II | 10 µL | 10 µL |
| 10 m$M$ MgCl$_2$ | ~15 µL | ~15 µL |
| 10X dNTPs | 10 µL | 10 µL |
| 5' primer (10 µ$M$) a | 10 µL | c, 10 µL |
| 3' primer (10 µ$M$) b | 10 µL | d, 10 µL |
| Template | gene I, ~0.5 µg | gene II, ~0.5 µg |
| H$_2$O | to 100 µL | to 100 µL |
| *Taq* polymerase | ~3 U | ~3 U |
| Product | AB | CD |

2. Amplify by PCR using the following cycle profile: 20–25 main cycles: 94°C for 1 min (denaturation), 50°C for 1 min (annealing), and 72°C for 1 min (extension).

## 3.3. Purification of Intermediates (see Note 7)

1. Run the PCR products on an agarose gel to size-purify them (*see* Note 8).
2. The method you use to extract the DNA from the agarose also depends on the size of the piece of DNA with which you are working. Fragments larger than ~300 bp can be purified using silica adsorption, e.g., GeneClean.
3. Smaller fragments are not recovered efficiently by this method, so you should use another procedure, such as electroelution. A simple procedure (described in ref. *16*) is to cut a well in front of the band, run the band into the well, and remove the DNA from the well with a pipet. Recover the DNA from the buffer by alcohol precipitation.

## 3.4. Generation of the Recombinant Product AD

1. Set up the overlap extension (SOE) reaction as follows (*see* Note 9): 10 µL 10X buffer II, ~15 µL 10 m*M* MgCl$_2$, 10 µL 10X dNTPs, 10 µL 5' primer (10 µ*M*) a, 10 µL 3' primer (10 µ*M*) d, intermediate product AB, intermediate product CD, H$_2$O to 100 µL, and ~3 U polymerase.
2. Amplify by PCR using the following cycle profile (*see* Note 10): 20–25 main cycles: 94°C for 1 min (denaturation), 50°C for 1 min (annealing), and 72°C for 1 min (extension).

## 4. Notes

1. Extensive (and expensive!) primer purification schemes, such as acrylamide gel electrophoresis or HPLC, are not necessary. Desalting by gel filtration over Sephadex should be sufficient.
2. The starting templates contain the gene sequences that you want to recombine into a tailor-made molecule. Any template suitable for PCR, such as reverse-transcribed RNA, can be used for gene SOEing (e.g., *see* ref. *10*). However, because high template concentrations minimize the probability of the polymerase introducing errors into the sequences (*see* Note 6), your starting templates will probably be cloned genes in plasmids.
3. The single simplest and most complete way to mess up a SOEing reaction is to have mistakes in the primer sequences (R. H., personal experience!). As PCR reactions go, amplifying an insert from a plasmid that is present in microgram amounts is like falling off the proverbial log. If this reaction cannot be made to work after the normal titrations of Mg$^{2+}$, template, and so forth, then something is drastically wrong, and you should recheck the design of your primers. Similarly, if products AB and CD, when mixed together in near-microgram amounts in a SOE reaction, completely fail to produce a recombinant product, the sequences of the primers in the overlap region should be immediately suspect. Writing out the sequence of the desired product and making sure that the SOEing primers each match one strand of the desired product at the recombination point is helpful.
4. The overlap region does not all have to be added to one primer. If, for example, instead of adding a 17-bp overlap region to primer b we had added 9 bases

complementary to primer c to the 5' end of b and 8 bases complementary to b to the 5' end of c, the overlap between AB and CD would still have been 17 bp (*see* ref. *7*). This approach avoids using very long primers. However, since making a 40- or 50-mer is now routine, there is usually no need to split the overlap region.

5. The optimal length to design the overlap region has not been settled. Regions as long as 164 bp (*15*) or as short as 12 bp (*17*) have been reported. Although the "50°C rule" reliably produces overlaps that are long enough to work, it does not indicate the minimum workable length. For example, the 12-bp-long overlap region reported in ref. *17* has an estimated $T_m$ of 34°C, although it was used at an annealing temperature of 50°C. On theoretical grounds, however, merely producing the recombinant molecule is not the only consideration to bear in mind. A very short overlap region might lead to an inefficient SOEing reaction, requiring the final product to be amplified through more rounds of PCR. This in turn may increase the error frequency.

6. Errors introduced by polymerase. Studies have demonstrated that *Taq* polymerase is capable of high fidelity DNA synthesis under PCR conditions (*18*). Clones produced by a single round of overlap extension have error frequencies of around 1 in 4000 bases (*6*), whereas more complicated constructs involving tandem SOEing reactions lead to slightly higher frequencies (~1 in 1800; ref. *7*). Because of the possibility that random mutations will be introduced by polymerase errors, several precautions are in order. First, the highest concentration of template plasmid consistent with amplification should be employed (this will probably be around 500 ng of plasmid in 100 μL). This minimizes the number of rounds of replication required to produce enough product with which to work, and gives the polymerase fewer opportunities to make errors. Since the reaction plateaus after producing a certain amount of product, it is probably not necessary to minimize the number of heating/cooling cycles to which the samples are subjected. Second, the lowest concentration of magnesium compatible with amplification should be used because error rates increase with increasing [$Mg^{2+}$] (*18*). Finally, for many applications, it is advisable to sequence the final product to ensure that it is free of errors. For this reason, a "cassette" approach, in which PCR manipulations are performed on a small recombinant segment, which is then ligated into a vector containing the remaining portions of the construct, may in some cases be preferable to SOEing directly into a vector (*see* Chapter 15). Around 300–500 bp is a convenient size for a cassette because it is large enough to handle conveniently yet small enough to sequence quickly. It should be pointed out that other thermostable DNA polymerases (i.e., Vent polymerase, New England Biolabs, Beverly, MA) are reportedly capable of significantly higher fidelity synthesis than *Taq*, and can be used for overlap extension (*19*).

7. Ho et al. (*6*) found that gel purification of the intermediate products AB and CD led to a cleaner reaction and increased product yield. Purification of intermediates may be most important when the initial template concentration is high. Gel purification removes not only the template plasmids, but also open-ended primer extension products, which, coming from the template, may be longer than the

PCR product. These open-ended products may not be obvious on an ethidium-stained gel because they are of indeterminate length and possibly single-stranded, but they have the potential to generate unwanted side products. Nevertheless, other workers have successfully used less extensive purification schemes *(5)* or none at all *(10,13)*.

8. The percent agarose you use depends on the size of the products you are isolating; smaller fragments need a higher percentage of agarose. For up to 1% agarose, "regular" agarose is fine, but for higher percentages, NuSieve (FMC BioProducts, Rockland, ME) gives better resolution. Please note that NuSieve is used to supplement regular agarose: The first 1% agarose is the regular variety, and only the additional percentage is NuSieve (up to a total of 4%).

9. Using large quantities of the intermediates should minimize polymerase errors (*see* Note 6). About 25% of what you recover from the gel should be plenty if the PCRs worked well, and this will leave you some extra in case you have to repeat it.

10. Related applications. Although the concept is simple, overlap extension is a tremendously powerful technology, and the reader is encouraged to spend some time contemplating modifications and applications. A general theoretical review of the subject is given in ref. *15*. Some of the more important and thought-provoking technical developments related to synthetic uses of PCR are given in refs. *14,20–22; see also* Chapters 13, 15–22.

## Acknowledgments

I am deeply grateful to my colleagues Steffan Ho, Jeff Pullen, Henry Hunt, Zeling Cai, and Larry Pease for making it possible for me to participate in the development of this technology.

## References

1. Mullis, K. B. and Faloona, F. A. (1987) Specific synthesis of DNA in vitro via a polymerase-catalysed chain reaction. *Methods Enzymol.* **155,** 335–350.

2. Mullis, K., Faloona, F., Scharf, S., Saiki, R., Horn, G., and Erlich, H. (1986) Specific enzymatic amplification of DNA in vitro: the polymerase chain reaction. *Cold Spring Harbor Symp. Quant. Biol.* **51,** 263–273.

3. Kadowaki, H., Kadowaki, T., Wondisford, F. E., and Taylor, S. I. (1989) Use of polymerase chain reaction catalysed by Taq DNA polymerase for site-specific mutagenesis. *Gene* **76,** 161–166.

4. Vallette, F., Mege, E., Reiss, A., and Milton, A. (1989) Constuction of mutant and chimeric genes using the polymerase chain reaction. *Nucleic Acids Res.* **17,** 723–733.

5. Higuchi, R., Krummel, B., and Saiki, R. K. (1988) A general method of in vitro preparation and specific mutagenesis of DNA fragments: study of protein and DNA interactions. *Nucleic Acids Res.* **16,** 7351–7367.

6. Ho, S. N., Hunt, H. D., Horton, R. M., Pullen, J. K., and Pease, L. R. (1989) Site-directed mutagenesis by overlap extension using the polymerase chain reaction. *Gene* **77,** 51–59.

7. Horton, R. M., Hunt, H. D., Ho, S. N., Pullen, J. K., and Pease, L. R. (1989) Engineering hybrid genes without the use of restriction enzymes: gene splicing by overlap extension. *Gene* **77**, 61–68.

8. Horton, R. M., Cai, Z., Ho, S. N., and Pease, L. R. (1990) Gene splicing by overlap extension: tailor-made genes using the polymerase chain reaction. *BioTechniques* **8**, 528–535.

9. Kain, K. C., Orlandi, P. A., and Lanar, D. E. (1991) Universal promoter for gene expression without cloning: expression-PCR. *BioTechniques* **10**, 366.

10. Davis, G. T., Bedzyk, W. D., Voss, E. W., and Jacobs, T. W. (1991) Single chain antibody (SCA) encoding genes: one-step construction and expression in eukaryotic cells. *Biotechnology* **9**, 165–169.

11. Daughtery, B. L., DeMartino, J. A., Law, M.-F., Kawka, D. W., Singer, I. I., and Mark, G. E. (1991) Polymerase chain reaction facilitates the cloning, CDR-grafting, and rapid expression of a murine monoclonal antibody directed against the CD18 component of leukocyte integrins. *Nucleic Acids Res.* **19**, 2471–2476.

12. Suggs, S.V., Hirose, T., Miyake, T., Kawashima, E. H., Johnson, M. J., Itakura, K., and Wallace, R. B. (1981) Use of synthetic oligo-deoxyribonucleotides for the isolation of cloned DNA sequences, in *Developmental Biology Using Purified Genes* (Brown, D. D. and Fow, C. F., eds.), Academic, New York, pp. 683–693.

13. Yon, J. and Fried, M. (1989) Precise gene fusion by PCR. *Nucleic Acids Res.* **17**, 4895.

14. Sarkar, G. and Sommer, S. S. (1990) The "megaprimer" method of site-directed mutagenesis. *BioTechniques* **8**, 404–407.

15. Horton, R. M. and Pease, L. R. (1991) Recombination and mutagenesis of DNA sequences using PCR, in *Directed Mutagenesis: A Practical Approach* (McPherson, M. J., ed.), IRL, Oxford, pp. 217–247.

16. Sambrook, J., Fritsch, E. F., and Maniatis, T. (1989) *Molecular Cloning: A Laboratory Manual*, 2nd ed., Cold Spring Harbor Laboratory, Cold Spring Harbor, NY.

17. Yolov, A. A. and Shaborova, Z. A. (1990) Constructing DNA by polymerase recombination. *Nucleic Acids Res.* **18**, 3983–3986.

18. Eckert, K. A. and Kunkel, T. A. (1990) High fidelity DNA synthesis by the Thermus aquaticus DNA polymerase. *Nucleic Acids Res.* **18**, 3739–3744.

19. Hanes, S. D. and Brent, R. (1991) A genetic model for interaction of the homeodomain recognition helix with DNA. *Science* **251**, 426–430.

20. Rudert, R. A. and Trucco, M. (1990) DNA polymers of protein binding sequences generated by PCR. *Nucleic Acids Res.* **18**, 6460.

21. Shuldiner, A. R., Scott, L. A., and Roth, J. (1990) PCR-induced (ligase-free) subcloning: a rapid reliable method to subclone polymerase chain reaction (PCR) products. *Nucleic Acids Res.* **18**, 1920.

22. Jones, D. H. and Howard, B. H. (1990) A rapid method for site-specific mutagenesis and directional subcloning by using the polymerase chain reaction to generate recombinant circles. *BioTechniques* **8**, 178–183.

# 15

## In-Frame Cloning of Synthetic Genes Using PCR Inserts

### James C. Pierce

### 1. Introduction

Since many genes of biological interest are larger than the maximum size that current synthetic oligonucleotide synthesizers can produce, there is a need for methods that allow rapid production and expression of genes constructed from multiple synthetic DNA fragments. The cloning method described here follows a series of steps in which multiple PCR products or synthetic duplex oligonucleotides are positionally cloned into a plasmid vector *(1)*. A synthetic gene of practically any sequence or length can be built using the in-frame cloning method. Genes are assembled such that open reading frames are maintained by linking DNA fragments through the use of six base pair blunt-end restriction sites. Each cloning step uses an anchored sticky-end restriction site and a variable blunt-end restriction site that result in specific insert orientation and high cloning efficiencies. The overall strategy of in-frame cloning allows the researcher total control over nucleotide sequence, codon usage, promoter and other regulatory elements, and the placement of unique restriction sites throughout the recombinant construct. One advantage of the in-frame cloning method described here is that it allows for flexible yet precise construction of synthetic genes using standard recombinant techniques. Another advantage is that it employs inexpensive, readily available materials.

The in-frame cloning method is based on the observation that standard plasmid cloning vectors, such as pUC or pGEM, contain very few six base pair blunt-end restriction recognition sites *(2)*. The amino acid sequence of a protein, whose gene is to be cloned, is scanned for those amino acids encoded by the blunt-end restriction sites listed in Table 1. Each of these "signpost" amino

From: *Methods in Molecular Biology, Vol. 67: PCR Cloning Protocols: From Molecular Cloning to Genetic Engineering* Edited by: B. A. White Humana Press Inc., Totowa, NJ

**Table 1
Blunt-End Restriction Sites that Can Be Used
for In-Frame Cloning**

| Enzyme | Recognition site | Amino acids |
| --- | --- | --- |
| *Bsa*I | CAC-GTG | His-Val |
| *Bst*1107I | GTA-TAC | Val-Tyr |
| *Cac*8I | GGN-NCC | Gly-Ala |
| *Eco*47III | AGC-GCT | Ser-Ala |
| *Eco*RV | GAT-ATC | Asp-Ile |
| *Ehe*I | GGC-GCC | Gly-Ala |
| *Hinc*II | GTC-GAC | Val-Asp |
| *Hpa*I | GTT-AAC | Val-Asn |
| *Msc*I | TGG-CCA | Trp-Pro |
| *Nae*I | GCC-GGC | Ala-Gly |
| *Pml*I | CAC-GTG | His-Val |
| *Sma*I | CCC-GGG | Pro-Gly |
| *Sna*BI | TAC-GTA | Tyr-Val |
| *Stu*I | AGG-CCT | Arg-Pro |

acids can then be used to fragment a protein sequence into sections that are easily encoded by synthetic oligonucleotides. When two contiguous DNA fragments are joined in the plasmid vector by blunt-end ligation the open reading frame of the synthetic gene is maintained. As described below, a combination of relatively simple cloning techniques and electroporation give high overall cloning efficiencies. The polymerase chain reaction (PCR) is used for both synthesis of duplex DNA from oligonucleotides and for the rapid screening of intermediate and final synthetic plasmid constructs by direct amplification of plasmid DNA from transformed host bacteria (*Escherichia coli*) colonies. By monitoring the final plasmid-synthetic gene construct for the presence or absence of diagnostic restriction sites, one can have good confidence that the correct synthetic gene has indeed been cloned before confirmation by DNA sequencing.

The in-frame cloning method was initially developed to make synthetic genes from complementary synthetic oligonucleotides that were annealed to generate a duplex DNA molecule and then directly cloned into the plasmid vector *(1)*. One advantage of using synthetic oligonucleotides as DNA fragments for direct cloning is the ability to use them without further purification. Also, since oligonucleotides do not contain a terminal phosphate group, no problems are encountered with multiple tandem copies of the DNA insert following ligation reactions in which the insert is in significant molar excess relative to the plasmid scaffold. The protocol described here is based on the work of a number of researchers in which duplex insert DNA is first made by the PCR using overlapping oligonucleotide primers *(3–6)*. This modification allows for larger blocks of insert DNA to be made for each cloning event and will decrease the overall cost of oligonucleotide primers. Two problems associated with PCR-derived templates are poor DNA sequence fidelity and difficulty in obtaining flush blunt-ends. With the availability of thermostable DNA polymerase with high replication fidelity and 3'-5' exonuclease activity that removes terminal sequences, many of the problems associated with poor quality PCR-derived templates can be overcome.

## 2. Materials

1. Plasmid vector DNA (e.g., pUC or pGEM).
2. Synthetic oligonucleotides designed using the format in Fig. 1. Resuspend deprotected oligonucleotides in sterile water or Tris-EDTA buffer, pH 8.0, at a final concentration of about 2 mg/mL. There is no need to purify oligonucleotides as long as the synthesis was relatively efficient. The quality of the oligonucleotides can be checked by simple agarose or acrylamide gel electrophoresis and staining with ethidium bromide.
3. *E. coli* host strain for plasmid transformation (e.g., strain TB-1, New England Biolabs, Beverly, MA).
4. Restriction enzymes and buffers.
5. Reagents for agarose gel electrophoresis, including low-melting-point agarose for in-gel cloning (e.g., GTG SeaPlaque agarose, FMC BioProducts, Rockland, ME).
6. DNA ligase and buffer.
7. Microdialysis membrane, such as Millipore VSP 0.025-mm filters (Bedford, MA).
8. A high efficiency method to transform *E. coli* cells with plasmid DNA (e.g., Bio-Rad Gene Pulser, electroporator, Hercules, CA).
9. Media, Petri plates and ampicillin for growth and selection of transformed bacteria.
10. PCR primers that border the cloning site of the plasmid vector (e.g., Sp6 and T7 primers are used for pGEM vectors).
11. Reagents and thermocycler for PCR.
12. Reagents needed for alkaline lysis, miniplasmid purification procedure *(7)*.
13. DNA sequencing reagents and equipment for final confirmation of synthetic gene sequence.
14. TE buffer: 10 m$M$ Tris-HCl, pH 8.0, 1 m$M$ EDTA.
15. TBE buffer: 90 m$M$ Tris, pH 8.3, 90 m$M$ borate, 2 m$M$ EDTA.
16. TE-saturated phenol.
17. Chloroform:isoamyl alcohol (24:1).
18. 95% Ethanol.
19. 70% Ethanol.

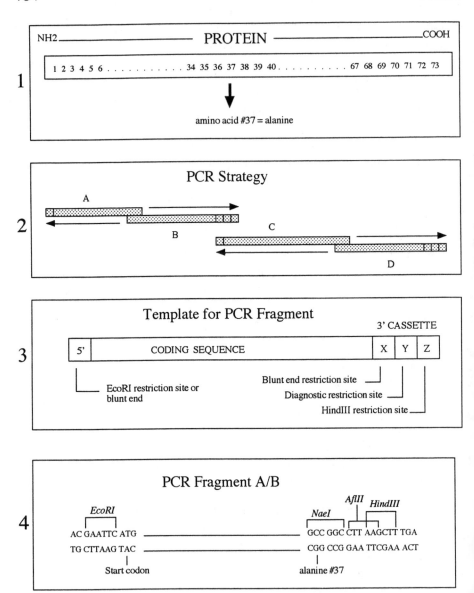

**Fig. 1.** Design strategy for the in-frame cloning method using PCR-generated DNA inserts. The design strategy is broken into four parts. Part **1** illustrates the analysis portion of the in-frame cloning procedure. The primary amino acid sequence is scanned for amino acids that are present in Table 1. The protein sequence is divided into segments that are encoded by overlapping synthetic oligonucleotides (Part **2**). In this example, amino acid 37 is an alanine that defines the endpoint between segment 1 (encoded by oligos A/B) and segment 2 (encoded by oligo C/D). The PCR is used to

## 3. Methods

### 3.1. Organizational Strategy for Oligonucleotide Design

1. The in-frame cloning method assumes that the amino acid sequence of the protein is known for the synthetic gene under construction. Plasmid vectors, such as pUC and pGEM are used as the scaffold to build the synthetic gene. These plasmids have relatively few six base pair blunt-end restriction recognition sites (listed in Table 1). The in-frame cloning method uses this knowledge to design a cloning strategy that allows the forced orientation cloning of insert DNA such that the open reading frame is maintained by the assembly of DNA fragments at one endpoint by blunt-end ligation and at the other endpoint by sticky-end ligation (for overview *see* Figs. 1 and 2).

2. A number of symmetrical six base pair blunt-end restriction sites, listed in Table 1, can be used for in-frame cloning. The protein is broken into a number of fragments that contain one of the amino acids listed in Table 1. These selected amino acids define the last amino acid of a protein segment in each of the cloning steps. It is possible to build a gene starting from either the amino terminus or the carboxyl terminus. This gives more flexibility in the choice of amino acids that are encoded by the blunt-end restriction sites. Once the protein sequence has been segmented into specific insert blocks (Fig. 1, parts 1 and 2), a PCR strategy is utilized that will result in DNA fragments that contain all the elements needed for in-frame cloning.

3. Figure 1 illustrates the design strategy for generating PCR fragments that are used in the in-frame cloning method. The protein sequence is segmented into blocks which are easily covered by moderate sized oligonucleotides of about 50–90 bases. Part 1 of Fig. 1 shows a hypothetical 73-amino-acid protein that will be broken into two parts between amino acids 37 and 38. Note that amino acid 37 is alanine, which is encoded by one or more of the blunt-end restriction enzymes listed in Table 1. The oligonucleotides are designed such that a 20-base-pair region of complementary DNA sequence overlap will result when the two cognate oligonucleotides are annealed together (e.g., oligos A and B in part 2 of Fig. 1). As an example, if two 80-base oligos were annealed together with a 20 base overlap, a protein coding sequence of up to 46 amino acids could be generated. Once annealed, duplex DNA is made by extension of the 3' end of each oligo using PCR for a limited number of cycles (*see* Note 1).

---

extend the 3' end of each oligonucleotide. Part **3** demonstrates the template from which to design the synthetic oligonucleotides. Each PCR-generated fragment contains three main components: the 5' end, the coding sequence, and the 3' cassette. The 3' cassette is further divided into three sections. Section *X* represents the blunt-end restriction site picked from Table 1, section *Y* represents the diagnostic restriction site for segment identification, and section *Z* represents the 3' cloning site, usually a *Hind*III sticky-end. Part **4** shows the partial sequence of PCR fragment A/B.

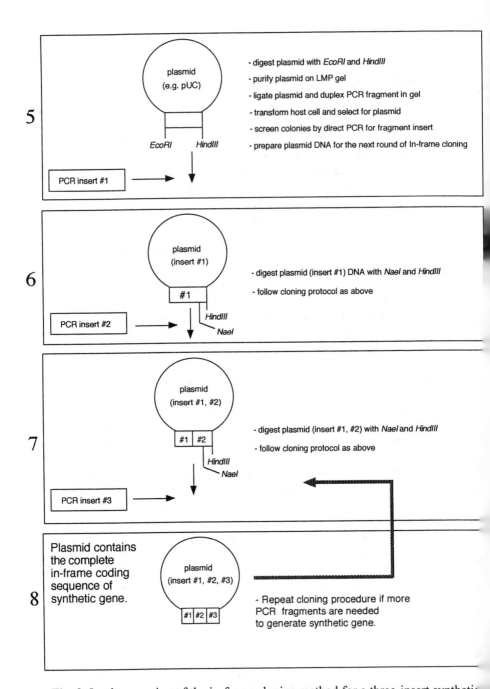

**Fig. 2.** Implementation of the in-frame cloning method for a three-insert synthetic gene. The execution of the cloning strategy is broken into four parts, **5–8**. Part 5 shows the cloning of PCR insert 1 into a standard plasmid vector (e.g., pUC). This duplex PCR-generated segment is cloned as an *Eco*RI/*Hin*dIII fragment. The steps used in the in-frame cloning protocol are listed in sequential order next to the plasmid diagram.

4. The PCR fragment must contain a number of elements as illustrated in Section 3 of Fig. 1. The 5' end contains either an *Eco*RI restriction site or a blunt-end. The *Eco*RI site is used during the first round of cloning. Subsequent rounds of cloning use a 5' blunt-end that on ligation to the 3' blunt-end generated during the cloning process will result in the correct translational reading frame. The 5' blunt end can be made by either of two (different) methods. If a DNA polymerase is used that generates flush blunt ends during the amplification process, then oligonucleotides can be designed such that the 5' end of the fragment (first three nucleotides) will contain the next codon in the amino acid sequence. If there is uncertainty concerning the sequence (e.g., terminal synthetase activity) at the 5' (left) end, then it would be necessary to generate an oligonucleotide that contains an internal blunt-end recognition site that on restriction digestion of the PCR fragment with the appropriate enzyme will leave a suitable blunt-end. This second scenario is somewhat restrictive in that the amino acid must be encoded by a symmetrical six base blunt-end restriction enzyme (not necessarily from Table 1) and makes the design of oligonucleotides more difficult. The internal part of the oligonucleotide contains the nucleotide sequence information that encodes the amino acids of the protein. The 3' end contains a number of restriction sites and is termed the 3' cassette. As shown in Fig. 1, part 3, the 3' cassette contains three sites labeled *X, Y,* and *Z.* Site *X* contains the blunt-end restriction site which will align the amino acids of the open reading frame on blunt-end ligation. Site *Y* contains a diagnostic restriction site used to monitor placement of the insert when screening plasmid clones. Site *Z* contains the *Hin*dIII restriction site that is used to anchor the 3' DNA fragment to the plasmid during each ligation step of the cloning process.

5. Figure 1, part 4 illustrates the partial sequence of PCR fragment A/B that has been generated using the template from Fig. 1, part 3. This fragment contains a 5' *Eco*RI cloning site, a translational start codon, a *Nae*I blunt-end site, an *Afl*II diagnostic site, and a 3' *Hin*dIII cloning site. After restriction digestion with *Eco*RI and *Hin*dIII, this fragment can be cloned into the corresponding sites of the plasmid vector. Note that after cloning this fragment into the plasmid vector, cleavage with the restriction enzyme *Nae*I will generate a blunt-end that encodes the GCC codon of alanine (amino acid 37) at the 3' terminus. As shown below (*see* Fig. 2), subsequent rounds of in-frame cloning will insert the 5' blunt-end of the incoming fragment (C/D) next to the 3' *Nae*I-generated blunt-end of the first insert (A/B). Assuming the first three base pairs of fragment C/D encode amino acid 38, the translational reading frame of the synthetic gene will be maintained.

---

Part 6 illustrates the cloning of PCR insert 2 (blunt-end/*Hin*dIII fragment). Part 7 illustrates the cloning of PCR insert 3 (blunt-end/*Hin*dIII fragment). Part 8 shows the completed plasmid construct containing the synthetic gene. The bold arrow represents the ability to continue to clone as many DNA inserts as are needed to complete an open reading frame for any synthetic gene.

6. The design of primers for PCR fragment synthesis must be carefully monitored before oligonucleotide synthesis. Three points are worth noting: First, make sure that the inadvertent engineering of necessary unique restriction sites in other parts of the fragment does not occur. Screen the oligonucleotide sequence for all six base restriction recognition sites before synthesis. It is important that the *Eco*RI, *Hin*dIII, blunt-end, and diagnostic restriction sites all remain unique to the engineered cloning vector. Second, make sure that the restriction sites needed for moving the synthetic gene from one vector to another have been correctly engineered. In most cases the synthetic gene will be engineered in a standard plasmid vector (e.g., pGEM or pUC) and then subcloned into an expression vector (e.g., pMAL, baculovirus). Finally, remember to include all regulatory elements, such as ribosome binding sites, transcription terminator sequences, and so forth, if these elements are not present in the expression vector into which the synthetic gene is to be cloned.

## 3.2. Construction of a Synthetic Gene Using In-Frame Cloning

1. Figure 2 illustrates the in-frame cloning process once an insert fragment has been generated. The protocol described in this Chapter uses PCR-generated fragments but synthetic duplex oligonucleotides or other cloned plasmid-derived fragments could also be used. The first round of in-frame cloning uses a PCR insert that has *Eco*RI and *Hin*dIII sticky-ends at the 5' (left) and 3' (right) ends, respectively. All subsequent rounds use a 5' blunt-end and 3' *Hin*dIII sticky-end duplex DNA fragments as inserts. By a reiterative process of cutting the plasmid vector with *Eco*RI/blunt-end and *Hin*dIII restriction enzymes followed by agarose gel purification and ligation, a synthetic gene of almost any length can be constructed. The only limiting factor is the availability of appropriately spaced single amino acids in the protein sequence that are encoded in the blunt-end restriction sites listed in Table 1. Screening for plasmids with successful ligation of insert DNA is done by direct PCR analysis of antibiotic-resistant bacterial colonies. This method greatly simplifies the isolation of positive clones and allows for the rapid construction of moderate to large-sized synthetic genes.
2. Assemble the following for a standard 40 µL PCR: 0.5–1.5 µg primer A (approx 30 pmol of a 60–80 base oligo), 0.5–1.5 µg primer B (approx 30 pmol of a 60–80 base oligo), 4 µL 10X buffer, 3.2 µL 25 m$M$ MgCl$_2$, 3.2 µL 2.5 m$M$ dNTPs, 1 U *Taq* polymerase, and water to 40 µL total volume.
3. Perform PCR using the following conditions (*see* Note 1): 94°C for 1 min (denaturation); 55°C for 2 min (annealing); 72°C for 2 min (extension) for 5 cycles; and 72°C for 10 min (final extension).
4. If a mineral oil overlay was used, add an equal volume of chloroform, mix, and transfer the aqueous layer to a sterile microcentrifuge tube. Incubate the tube with the lid open at 37°C for 15 min or until there is no trace odor of chloroform.
5. While the sample is drying, fill a 100-mm sterile Petri plate with 35 mL of 1X TE buffer (*see* Note 2). Float a Millipore VS$M$ 0.025-µm filter (shiny side up) on the TE buffer.

6. Place the PCR amplified DNA sample (up to 100 μL) on the filter and allow 30–60 min for dialysis. Remove the sample and place it in a sterile microcentrifuge tube.

7. Digest an aliquot of the PCR amplified DNA sample by adding the following reagents to a microcentrifuge tube and incubating at 37°C for 1 h: 16 μL DNA sample, 2 μL 10X buffer, 1 μL *Hind*III, and 1 μL *Eco*RI.

8. After the digestion is complete, purify the PCR DNA fragment away from the released cut ends (*see* Note 3). The protocol listed here is based on the Qiagen QIAquick-spin PCR purification columns. For more details see manufacturers' product description.

9. Following restriction digestion add 0.5 mL of binding buffer to the DNA sample. Apply the sample to the QIAquick-spin column and centrifuge the column in a 2-mL centrifuge tube for 60 s at maximum speed. Remove flowthrough and wash the column with 0.75 mL of wash buffer as above. Remove flowthrough and centrifuge again for 60 s to remove residual wash buffer. Elute DNA by adding 50 μL of TE buffer to the spin column, placing the spin column in a sterile 1.5-mL microcentrifuge tube, and centrifuging for 60 s at maximum speed. The DNA fragment is now ready to be cloned into the plasmid cloning vector.

10. Plasmid vector DNA is prepared by cutting with restriction enzymes in the following manner: 1–2 μg plasmid DNA (e.g., pGEM), 2 μL 10X buffer, 1 μL *Eco*RI, 1 μL *Hind*III, and water to 20 μL total volume. Incubate restriction digest at 37°C for 1 h.

11. The cut plasmid DNA is purified away from the small DNA fragment released during the double digest using low-melting-point agarose gel electrophoresis (*see* Note 4). The purified DNA fragment is then ligated to the PCR insert fragment that was purified in step 9. Add to a microcentrifuge tube the following components: 5 μL purified PCR fragment (~0.5 μg), 10 μL plasmid DNA fragment in agarose plug (50–100 ng), 4 μL 5X ligation buffer, and 1 μL DNA ligase. The ligation reaction is incubated overnight at room temperature (*see* Note 5).

12. Transform appropriate *E. coli* host cell (e.g., strain TB-1 from New England Biolabs) by electroporation or other high efficiency transformation protocol. Before electroporation, remove the salt from the ligation reaction by dialysis against 0.1X TE buffer for 1 h at room temperature (*see* Note 2).

13. Electroporation using a Bio-Rad *E. coli* Gene Pulser (Hercules, CA) is performed as follows. The agarose plug containing the ligation reaction is melted at 70°C for 10 min and 5 μL of the molten reaction is added to 80 μL of electrocompetent host cells (*E. coli* strain TB-1). The DNA/host cell mixture is then placed in a 0.1-cm cuvet and electroporated at a setting of 1.80 kV. Add 1 mL of liquid media (e.g., L Broth) and incubate with shaking at 37°C for 1 h. One-tenth of the transformed cells are then spread on a LB agar selector plate containing 50 mg/mL ampicillin. The plates are then incubated overnight at 37°C.

14. Screening for positive PCR fragment inserts is performed by direct PCR analysis of colonies from the ampicillin selector plates. An agar plug (*see* Note 6) of

an individual ampicillin resistant colony is resuspended in 50 μL of sterile water and mixed well. To each 40 μL PCR sample tube add the following reagents: 5 μL agar plug colony mixture (DNA template), 100 ng (ca. 15 pmol) each forward and reverse pUC/M13 primers, 4 μL 10X buffer, 3.2 μL 25 m$M$ MgCl$_2$, 3.2 μL 2.5 m$M$ dNTPs, 1 U *Taq* polymerase, and water to 40 μL final volume.

15. Perform PCR using the following conditions: 94°C for 1 min (denaturation), 55°C for 2 min (annealing), and 72°C for 2 min (extension) for 30 cycles.

16. Analysis of PCR samples is performed by agarose gel electrophoresis using 0.5X TBE buffer. To observe DNA fragment sizes in the range of 100–300 base pairs, a 3% agarose gel fractionates well. Bacteria that contain clones that are positive for the correctly ligated DNA insert will contain a PCR DNA fragment that is larger than those PCR DNA fragments amplified from bacteria that contain only vector DNA (*see* Note 8).

17. Once positive colonies have been identified, plasmid DNA is prepared by the alkaline lysis or other DNA miniprep procedures *(7,8)*.

18. To confirm that the PCR fragment has been cloned into the plasmid vector, a restriction digest is performed. The choice of restriction enzyme depends on which diagnostic site was engineered into the DNA insert (*see* Fig. 1, part 3). In addition to the insert diagnostic restriction enzyme, use restriction enzymes that will reveal correct insert ligation and the removal of plasmid vector multiple cloning region (*see* Note 9). To a microcentrifuge tube add the following reagents: 5 μL plasmid DNA (~0.2 μg), 2 μL 10X restriction buffer, 1 μL restriction enzyme (e.g., *Afl*II), and water to 12 μL total volume. Incubate the restriction digest(s) at the appropriate temperature for 1 h.

19. Fractionate the plasmid DNA from the restriction digests by agarose gel electrophoresis using a 0.8% gel and 0.5X TBE buffer. A plasmid clone that is positive for the correct DNA insert will have a restriction pattern that indicates the presence of the diagnostic restriction site (e.g., *Afl*II). Once a positive clone is identified by PCR and then confirmed by restriction mapping, the next round of in-frame cloning is initiated. The next round of in-frame cloning (*see* Note 10) proceeds with a ligation reaction using the blunt-end and *Hin*dIII cut plasmid (insert 1) that was engineered in the first round of in-frame cloning (Fig. 2, part 5), and the PCR fragment generated from oligonucleotides C and D (Fig. 1, part 2).

20. Using the plasmid DNA prepared in the miniprep procedure (step 17) from one of the positive clones, perform a restriction enzyme double digest. One restriction enzyme will be *Hin*dIII. The second restriction enzyme will be a six base pair, blunt-end recognition site enzyme listed in Table 1. If the buffer and/or temperature conditions are incompatible for both enzymes together in one reaction, then perform the digest sequentially. To a microcentrifuge tube add the following: 1 μg plasmid DNA (containing insert 1), 2 μL 10X buffer, 1 μL *Hin*dIII, 1 μL blunt-end restriction enzyme from Table 1, and water to 20 μL total volume. Incubate restriction enzyme at the appropriate temperature for 1 h.

21. Purify the cut plasmid (insert 1) away from the small DNA fragment released during the double restriction digest using low-melting agarose gel electrophoresis exactly as described in step 11 and Note 4.

22. Prepare the second DNA insert (insert 2) by following the protocol outlined in steps 1–9. It is generally more efficient to prepare all of the DNA inserts (e.g., insert 1, insert 2, and so forth) at the same time. After PCR extension to generate the double-stranded DNA fragment, the fragment must be digested with *Hin*dIII. As mentioned in Section 3.1., part 4, if blunt-ends are difficult to achieve during the PCR reaction, then a second restriction digest must be performed to generate the correct fragment ends needed for ligation. Perform the *Hin*dIII digest as follows: 17 μL DNA sample, 2 μL 10X buffer, and 1 μL *Hin*dIII. Incubate the reaction at 37°C for 1 h. Purify the DNA fragment away from small released fragment(s) by column chromatography as described in steps 8 and 9.

23. A ligation reaction is prepared exactly as described in step 11. After ligation and transformation (steps 11–13), positive clones are identified as described in steps 14–19. In-frame cloning can be reiterated as many times as is necessary to engineer a synthetic gene of the desired length (Fig. 2, part 8).

24. When all of the necessary rounds of in-frame cloning have been completed, the insert nucleotide sequence must be confirmed by DNA sequence analysis. Candidate plasmid constructs should first be confirmed by analysis of the correct insert size (as determined by PCR amplification and comparison to vector with no insert and intermediate plasmid constructs) and by restriction mapping to indicate that the correct diagnostic restriction sites are present and that certain vector restriction sites have been removed. Sequence analysis can be performed by double-stranded PCR sequencing using M13/pUC universal primers *(8)*. Single-stranded DNA can be made and sequenced using vectors, such as the pGEM-f1 ori series.

25. A rough timeline is presented below to serve as a guide for experiment design. It is assumed that all oligonucleotide design and synthesis has been completed. It also assumes that a preliminary PCR experiment has shown that the oligonucleotide PCR fragments are of the correct size and of decent quality.
    a. Day 1: PCR amplification of oligonucleotide insert fragments. Gel electrophoretic analysis of PCR samples
    b. Day 2: Cut PCR fragments with restriction enzyme(s). Cut plasmid vector with restriction enzymes. Purify PCR fragments by spin chromatography. Purify plasmid vector by low-melting point agarose gel electrophoresis. Set up ligation reactions; incubate overnight.
    c. Day 3: Transform *E. coli* host cells. Grow colonies overnight on ampicillin selector plates.
    d. Day 4: Screen selected colonies by direct PCR. Gel electrophoresis of PCR/colony samples. Overnight culture of tentatively positive colonies.
    e. Day 5: Miniprep plasmid DNA isolation. Restriction mapping using diagnostic site analysis. Gel electrophoresis of plasmid restriction digests. Prepare vector for the next round of in-frame cloning.

## 4. Notes

1. The generation of the duplex DNA fragments in this protocol is really just a DNA polymerase extension reaction. The polymerase chain reaction is used since it is simple and the reagents are generally available. A small number of amplification cycles (e.g., 5) should generate sufficient product. If priming artifact occurs try using fewer cycles. One cycle may be enough to generate sufficient duplex DNA for cloning. If PCR product artifacts are a problem, the use of 5' outside primers to amplify the correct size duplex fragment from the initial PCR amplification may be needed. The use of synthetic oligonucleotide duplex DNA for direct insert cloning is always an option.

2. I use microdialysis routinely to prepare DNA substrates that must undergo a number of different and separate enzymatic treatments. Although many DNA/enzyme reactions work adequately under "universal" buffer conditions, complex protocols that require a number of linked reactions often show low overall efficiency. It has been my experience that if each DNA modification reaction is treated individually, the success of the entire experiment is greatly enhanced. Microdialysis is a technique that allows the efficient exchange of buffers for any DNA sample. Once mastered it is often more convenient and efficient than ethanol precipitation or spin chromatography. The following protocol works well for DNA fragments 100 bases or larger.

    a. Fill a 100-mm sterile Petri plate with 35 mL of 1X TE buffer. Float a Millipore VSM 0.025-mm filter (shiny side up) on the TE buffer. Make sure no buffer wets the top side of the filter membrane.

    b. Place the DNA sample on the filter. Volumes from 10–100 mL can be used but care must be taken with larger volumes. Multiple samples can be placed on one filter as long as they do not contact one another. As described below, sample volumes often increase, so beware!

    c. Place the lid on the Petri plate and incubate for about 1 h at room temperature. Make sure the Petri plate is in a safe place on the bench since if it is disturbed the sample may be lost. Length of time will depend on how dramatic the change in buffer conditions will be. Simple desalting can usually be accomplished in less than an hour.

    d. Remove the sample using a pipeting device and place it in a sterile tube. The volume of the DNA sample can change significantly because of osmotic imbalances so it is often useful to record the pre- and post-microdialysis volumes. It has been my experience that very little DNA is lost by nonspecific binding to the membrane in this procedure.

    e. This protocol works well with agarose plugs when it is necessary to remove the electrophoresis buffer before enzymatic manipulation. The agarose plug is melted at 70°C and then placed directly on the filter that is floating on the TE buffer. The molten agarose will form a semisolid plug on the filter but this does not interfere with buffer exchange. To remove the agarose plug containing the DNA sample, first lift the entire filter membrane off the buffer using forceps. Then scrape the agarose plug off the filter directly into a microcentrifuge tube using a sterile scalpel. The sample is then incubated at 70°C and the agarose–DNA sample can then be added to the next reaction mixture.

3. Ligation reactions are not always logical. Often the products recovered from a ligation reaction are a mixture of intended constructs and a collection of obscure and unlikely side-products. The source of unintended DNA constructs is often insert–insert ligation events. Therefore, it is necessary to think critically about just what DNA substrates are being placed in the ligation reaction. When synthetic duplex oligonucleotides are used as DNA inserts, there is not much concern about insert ligating to itself since the 5' ends are not phosphorylated. With the PCR generated DNA fragment approach described in this protocol, self ligation is a potential problem. Although the forced orientation cloning approach minimizes ligation artifacts, there is still the possibility of multiple inserts being cloned into the vector during a ligation reaction.

The first round of in-frame cloning uses an *Eco*RI/*Hin*dIII sticky-ended fragment. There is significant possibility for selfligation under the conditions used here (*see* Note 5). Subsequent rounds of in-frame cloning using blunt-end/*Hin*dIII substrates pose less of a problem because of the lower efficiency of blunt-end ligation. A tripartite *Eco*RI/*Hin*dIII fragment will insert into the plasmid vector giving an incorrect construct. Since the in-frame cloning method screens for positive DNA inserts using direct PCR amplification of selected colonies followed by fragment size analysis using gel electrophoresis, this multiple insert problem should not interfere with the engineering experiment.

4. The double-digested plasmid is purified on a 0.6% low-melting-point agarose gel using 0.5X TBE buffer. The gel is stained with ethidium bromide and the linearized plasmid cut out and then placed in 1 mL of sterile water to remove excess TBE buffer. After 30 min remove the water and heat the agarose gel plug at 70°C for 10 min. It is important that the components of the next reaction have already been prepared before the agarose gel fragment is melted. After heating, the molten agarose is added directly to the ligation reaction before it solidifies. Using the pipet tip, gently mix the molten agarose with the ligation reaction components and allow the tube to sit at room temperature. Do not centrifuge the sample once the agarose has been added because this will separate the plug from the other reaction components. Depending on the reaction buffer and the amount of agarose added, the reaction mixture may or may not solidify at room temperature.

5. The molar ratio of insert to vector in the ligation reaction described in this protocol is approximately 100:1. This is calculated by assuming the plasmid vector to be 3000 bp of which 0.1 ng is added to the ligation reaction. This would equal about 0.05 pmol of DNA (1 bp = 660 daltons, 1 pmol of a base pair of DNA = 660 pg). If the average insert fragment is about 150-bp and 0.5 ng were used in the ligation reaction then about 5 pmol would be present in the sample. Thus, a 5–0.05 pmol DNA ratio is achieved. This ratio is somewhat high for a standard sticky-end ligation reaction but is used to drive the ligation reaction in the agarose plug. If problems with multiple inserts occur, then decrease the concentration of fragment insert. It is difficult to modulate the concentration of vector DNA

since an agarose plug strategy is being used and low cloning efficiencies are a problem if too little plasmid vector is used during transformation. In general, the conditions described here work well and the majority of plasmids contain the correct insert. Forced orientation and removal of restriction cleavage ends "push" the ligation reaction in the direction of correct insertion.

6. Screening for positive vector inserts is best accomplished by direct PCR amplification of bacterial colonies taken from the ampicillin selector plates. A sterile 50-mL capillary pipet (or 1-mL pipet) is used to isolate or "plug" well separated colonies from the agar plate. The agar plug is then placed in a sterile 1.5-mL microcentrifuge tube containing 50 mL of sterile water and mixed well using a vortex mixer. A 5-μL sample of this bacterial colony/agar plug/water mixture is then added to a standard 40-μL PCR tube as described in the protocol. I have found it convenient to work up 18 samples at a time and then perform gel electrophoretic analysis using a 20-well comb with two lanes used as molecular weight marker and control PCR reaction (vector no-insert) respectively. The tubes containing the bacterial colony/agar plug are stored at 4°C until the PCR and electrophoresis results are obtained. Once positive colonies are tentatively identified, an overnight culture is made using 3 mL of Luria broth, 50 mg/mL of ampicillin and 25 mL of the bacterial colony/agar plug sample. This is grown overnight at 37°C with moderate shaking. The overnight culture can then be used to make an alkaline lysis miniprep DNA sample that is then used for restriction mapping and/or prepared for the next round of in-frame cloning. Once positive constructs are identified, the remaining overnight culture can be kept for long term storage by adding glycerol to 30% of total volume, mixing well, and then storing at –80°C.

7. These primers flank the multiple cloning site of many pUC-based cloning vectors. If another vector is being used for in-frame cloning you may need a different set of primers. A control reaction should be performed from a cell containing vector DNA that has not been manipulated (usually taken from the selector plate used to monitor transformation efficiency, e.g., uncut pGEM DNA).

8. The difference in fragment size between the control PCR sample (colonies with vector/no insert) and experimental samples (colonies with vector plus insert) will allow you to determine which host cells contained plasmid DNA with the PCR fragment insert cloned into the *Eco*RI/*Hin*dIII site. Remember to subtract the length of the multiple cloning region (~50-bp) of uncut pGEM or pUC-based vectors when comparing fragment sizes.

9. Analysis of the diagnostic restriction site (e.g., *Afl*II) as an identifier for the correct insert fragment, and loss of restriction sites (e.g., *Pst*I) from the multiple cloning region (MCR) of the vector, ensures that this PCR fragment was cloned and the vector MCR removed. You can also check for the regeneration of the *Eco*RI and *Hin*dIII sites. This restriction digest screen will allow you to have good confidence that the correct fragment has indeed been cloned. During this restriction analysis the plasmid DNA can be prepared for the next step of in-frame cloning by cutting with *Hin*dIII and the blunt end recognition restriction enzyme. All of these samples can be run on a low-melting-point agarose gel and the *Hin*dIII/blunt end cut plasmid purified as described above (*see* Note 2).

10. This process can be repeated as many times as is necessary to clone a moderate to large sized synthetic gene (200–600-bp). If a very large gene is going to be made by the synthetic in-frame cloning method it may be more practical to use two or more plasmids as the templates for constructing different portions of the gene. Just duplicate each of the cloning and processing steps using different DNA inserts. The assembled gene fragments are then isolated by blunt-end/blunt-end or blunt-end/*Hin*dIII restriction digestion and then linked to the primary plasmid template by the in-frame cloning process.

## References

1. Pierce, J. C. (1994) In-frame cloning of large synthetic genes using moderate-size oligonucleotides. *BioTechniques* **16**, 708–715.
2. Yanisch-Perron, C., Vieira, J., and Messing, J. (1985) Improved M13 phage cloning vectors and host strains: nucleotide sequences of the M13mp18 and pUC19 vectors. *Gene* **33**, 103–119.
3. Barnett, R. W. and Erflel, H. (1990) Rapid generation of DNA fragments by PCR amplification of crude, synthetic oligonucleotides. *Nucleic Acids Res.* **18**, 3094.
4. Dillion, P. J. and Rosen, C. A. (1990) A rapid method for the construction of synthetic genes using the polymerase chain reaction. *BioTechniques* **9**, 298–300.
5. Prodromou, C. and Pearl, L. H. (1992) Recursive PCR: a novel technique for total gene synthesis. *Protein Eng.* **5**, 827–829.
6. Sandhu, G. S., Aleff, R. A., and Kline, B. C. (1992) Dual asymmetric PCR: one-step construction of synthetic genes. *BioTechniques* **12**, 14–16.
7. Birnboim, H. C. and Doly, J. (1979) A rapid alkaline extraction procedure for screening recombinant plasmid DNA. *Nucleic Acids Res.* **7**, 1513–1523.
8. Sambrook, J., Fritsch, E. F., and Maniatis, T. (1989) *Molecular Cloning: A Laboratory Manual*. Cold Spring Harbor Laboratory, Cold Spring Harbor, NY.

# 16

## Creation of Chimeric Junctions, Deletions, and Insertions by PCR

### Geneviève Pont-Kingdon

## 1. Introduction

Recombinant PCR *(1)* is the method of choice if one wants to modify a cloned DNA. It is a versatile technique that allows operations as different as creation of deletions, addition of small insertions, site directed mutagenesis, and construction of chimeric molecules at any chosen location in the molecule of interest (*see* Note 1). This chapter describes in detail a simplification of the original recombinant PCR method. This fast and efficient method has been successful in fusing two different sequences with precision *(2–4)*. It can also be used to create deletions or insert small fragments of DNA.

The method (*see* Fig. 1) relies on a "chimeric primer" (C) and two outside primers (A and B). The final product can be obtained in one or two rounds of PCR. The figure illustrates the construction of a chimeric molecule in which two different templates are joined. The creation of a deletion, or the introduction of a small insertion within a given template, would follow the same pathway (*see* Note 1). In all cases, the new junction is designed in the chimeric primer; the 3' half of the chimeric primer pairs with one of the templates (or one side of the deletion/insertion point), and its 5' half has homology with the other template (or the other side of the deletion/insertion point).

Both templates and the three primers (A–C) are placed in a reaction tube (step 1) and one PCR reaction is performed. During the first cycles, only the primers A and C can prime exponential amplification (step 2). This amplification reaction gives rise to an "intermediate fragment" (step 3), that can itself act as a primer. One of its 3'-ending strands anneals to the second template and is extended (step 4). This extension provides a template (step 5) for exponen-

From: *Methods in Molecular Biology, Vol. 67: PCR Cloning Protocols: From Molecular Cloning to Genetic Engineering* Edited by: B. A. White  Humana Press Inc., Totowa, NJ

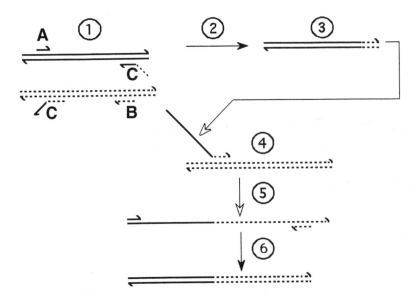

**Fig. 1.** Construction of a chimeric product. See text for explanation of steps 1–6. The dsDNA templates are the plain and dotted double lines. The outside primers are short, plain (**A**) or dotted (**B**) single lines. The drawing of the chimeric primer (half plain, half dotted) reflects its homologies to the templates. Half arrowheads indicate 3' ends. Closed arrows indicate amplification steps (2 and 6).

tial amplification of the final product using the primers A and B (step 6). In order to limit the amplification of the intermediate fragment to the first cycles of PCR, the chimeric primer is used at a lower concentration than the two outside primers *(3,4)*.

## 2. Materials

1. Primers (*see* Section 3.1.).
2. DNA templates, linearized at a site outside the region to be amplified.
3. GeneAmp PCR Core Reagent Kit (Perkin-Elmer, Foster City, CA) containing: AmpliTaq DNA polymerase (5 U/μL), Gene Amp dNTPs (10 m$M$ solutions of dATP, dGTP, dTTP, dCTP), GeneAMP 10X PCR buffer II (100 m$M$ Tris-HCl, pH 8.3, 500 m$M$ KCl), and 25 m$M$ MgCl$_2$ solution.
4. Mineral oil.
5. Restriction endonucleases and buffers.
6. Agarose gel electrophoresis reagents and equipment.
7. Phenol/CHCl$_3$/isoamyl alchohol (25:24:1 v:v:v).
8. 7.5$M$ Ammonium acetate.
9. 100% Ethanol.
10. 70% Ethanol.

## 3. Methods

### 3.1. Design of Chimeric Primer

The chimeric primer is crucial because it contains the new junction and because sequences on each side of the junction serve as primers in different steps (2 and 4 in Fig. 1) of the reaction. To allow priming by the nucleotides found on each side of the junction, the new junction should be placed in the middle of an oligonucleotide of sufficient length. Otherwise, the design of a chimeric primer should follow the classical rules of primer-design *(5)*. Our chimeric primer was a 34-mer, with 17 bases homologous to one template and 17 bases homologous to the other. Some slightly longer chimeric primers (36- and 40-mer) have been used *(3,4)*. *See* Note 2 for more information on the design of the chimeric primer.

### 3.2. Design of Outside Primers

Fewer constraints apply to the design of the two outside primers, and following the general rules of primer design should be adequate. In principle, the outside primers can be kilobase pairs away (within the limit of PCR feasibility) from the new junction. However, because their location determines the size of the product that will be cloned, the choice of their position is important. For several reasons, it is advantageous to plan the cloning of a fragment of few hundred nucleotides instead of a longer one: First, the smaller the fragment is, the less chance there is to find PCR-induced mutations in the final clone. Second, a smaller piece of DNA has to be sequenced in order to verify the integrity of the newly cloned DNA. This can be achieved by using primers that anneal close to the junction and have "built in" restriction sites at their 5' ends. Another approach is to choose primers that anneal further from the junction and clone with restriction sites that closely surround the chimeric junction in the final product. The choice between these two possibilities depends on the cloning strategy and the availability of cloning sites in the amplification product. If the chimeric fragment has to be cloned into a new vector, restriction sites unique in both vector and chimeric fragment can be engineered at the 5' ends of the two outside primers. If the chimeric junction has to be replaced by cloning in one of the original templates, restriction sites that exist in this template and in the chimeric junction have to be used.

The size of the outside primers can be different than the size of the chimeric primer. We have been successful with a 34-mer chimeric primer, and two outside primers of 23 and 20 nucleotides, respectively. If the size of the two outside primers is very different than half the size of the chimeric primer, series of cycles with different annealing temperatures can be performed (*see* Note 3).

### *3.3. Procedure*

1. Assemble the components in one 50 μL reaction containing: both templates (10 fmol each), both outside primers (25 pmol each), chimeric primer (1 pmol), 50 μ*M* each dNTP (*see* Note 4), 1X AmpliTaq DNA polymerase buffer, 1.5 m*M* MgCl$_2$ (*see* Note 5), and 2.5 U of AmpliTaq DNA polymerase.
2. Top PCR mix with 50 μL of mineral oil.
3. Perform PCR as follows (*see* Note 3):
   a. Three to five initial cycles to allow initiation and amplification (steps 2–4 in Fig. 1) using the chimeric primer: 95°C for 30 s, T1 (*see* Note 3) for 30 s, and 72°C for 1 min for each kilobase of intermediate fragment.
   b. Twenty to thirty cycles to amplify the final chimeric product: 95°C for 30 s, T2 (*see* Note 3) for 30 s, and 72°C for 1 min for each kilobase of chimeric product.
4. Directly analyze 10 μL of the amplified DNA by standard procedures (restriction enzyme digests and electrophoresis, *see* Note 6).
5. In order to clone the new junction, clean the PCR product by extraction with an equal volume of phenol:chloroform:isoamyl alcohol (25:24:1).
6. Transfer the aqueous phase to a sterile microcentrifuge tube. Remove the excess dNTP (it can inhibit T4 ligase) by ethanol precipitation. Add 1/2 vol of 7.5*M* ammonium acetate and 2.5 vol of 100% ethanol.
7. Mix well and incubate at room temperature for 10 min.
8. Spin in a microfuge at maximum speed for 5 min.
9. Invert tube and allow to drain.
10. Wash pellet with 70% ethanol.
11. Resuspend the DNA in 20–50 μL of TE. Quantify the DNA by UV spectrophotometry and use for cloning.

## 4. Notes

1. The applications of this method are diverse; it allows the creation of:
   a. Chimeric molecules: This method is well suited to cases in which the two molecules to be joined are unrelated. In the case where the two templates are fused in a region of homology, another PCR technique *(6)* might be preferred.
   b. Deletions: A set of different deletions can be easily obtained from the same template by using a set of different chimeric primers and only one set of outside primers.
   c. Insertions: The size of potential insertion using this technique is limited to the size of the chimeric primer. A restriction site sequence, sandwiched into a chimeric primer, could be introduced at will into any DNA.
2. As in the "Megaprimer" method *(7)*, the technique described here uses a PCR product as a primer. It has been mentioned for the Megaprimer method *(8)* that mutations can be found in the final product because of the tendency of *Taq* polymerase to add nontemplated nucleotides at the 3' end of newly synthethized DNA strands. The frequency of these mutations should be low since 3'-ending DNA strands carrying nontemplated nucleotides should not prime well for the synthesis of the final

product. Although we did not observe such mutations in the two final clones that we obtained and sequenced, this phenomenon could apply here, and we encourage the reader to refer to Note 1 in ref. *8* for a complete discussion.

3. In this technique, each half of the chimeric primer must anneal with its target. We have limited the size of our chimeric primer to 34 nucleotides, giving us 17 nucleotides for each half. Because our outside primers are 20 and 23 nucleotides in length, we felt that it was necessary to plan a first set of cycles with a lower annealing temperature to allow the stable annealing of each half of the chimeric primer in the steps 2 and 4. This precaution is not necessary if all the sequences with a "priming" role (each half of the chimeric primer and both outside primers) are close in length and in G+C content.

    The temperature T1 is the approximate "annealing temperature" ([number of G + C × 4°C] + [number of A + T × 2°C] − 10°C) of the less stable half of the chimeric primer. The temperature T2 is the approximate annealing temperature of the less stable of the two outside primers.

4. In order to limit the number of PCR-induced mutations in the final chimeric product, a low concentration of each dNTP (50 μ*M*) is used.

5. The optimal MgCl$_2$ concentration can vary among different pairs of primers. It is wise to define the best MgCl$_2$ concentration in a test experiment *(9)*. In fact, it is possible that the two consecutive PCRs that occur in the tube have incompatible requirements for MgCl$_2$. If this is the case, the chimeric product can be obtained by a two-step method *(2)*. In this alternative, the intermediate fragment obtained during the first reaction is extracted with phenol/chloroform/isoamyl alcohol and purified from excess primers by precipitation with isopropanol from 2*M* NH$_4$OAc. The purified DNA is then used as a primer in a second reaction that contains the second template and the two outside primers.

6. It is often stated that the oil that tops the PCR has to be removed in order to properly load the DNA into the well of an electrophoresis gel. We found that this step is not needed if the pipet tip is cleaned with tissue-paper (Kimwipe) just after it has been filled with a DNA sample. In fact we found that the oil left in the tube allows for longer conservation of the sample at 4°C.

## Acknowledgment

I want to thank Janet E. Lindsley and Dana Carroll for helpful comments.

## References

1. Higuchi, R. (1989) Using PCR to engineer DNA, in *PCR Technology, Principles and Applications for DNA Amplification*. Stockton, New York, pp. 61–70.
2. Pont-Kingdon, G. (1994) Construction of chimeric molecules by a two-step recombinant PCR method. *Biotechniques* **16,** 1010,1011.
3. Cao, Y. (1990) Direct cloning of a chimeric gene fused by the polymerase chain reaction. *Technique* **2,** 109–111.
4. Yon, J. and Fried, M. (1989) Precise gene fusion by PCR. *Nucleic Acids Res.* **17,** 4895.

5. Sharrocks, A. D. (1994) The design of primers for PCR, in *PCR Technology: Current Innovations* (Griffin, G. G. and Griffin A. M., eds.), CRC, Boca Raton, FL, pp. 5–11.
6. Klug, J., Wolf, M., and Beato, M. (1991) Creating chimeric molecules by PCR directed homologous DNA recombination. *Nucleic Acids Res.* **19,** 2793.
7. Sarkar, G. and Sommer, S. S. (1990) The "megaprimer" method of site-directed mutagenesis. *BioTechniques* **8,** 404–407.
8. Barik, S. (1993) Site-directed mutagenesis by double polymerase chain reaction, in *Methods in Molecular Biology, vol. 15: PCR Protocols: Current Methods and Applications* (White, B. A., ed.), Humana, Totowa, NJ, pp. 277–286.
9. Saiki, R. K. (1989) The design and optimization of the PCR, in *PCR Technology: Principles and Applications for DNA Amplification* (Erlich, H. A., ed.), Stockton, New York, pp. 7–16.

# 17

# Mutagenesis and Gene Fusion by Megaprimer PCR

## Sailen Barik

## 1. Introduction

Among the many variations of polymerase chain reaction (PCR)-based mutagenesis procedures, the "megaprimer" method *(1–5)* is probably the simplest and most versatile. The method utilizes three oligonucleotide primers and two rounds of PCR performed on a DNA template containing the cloned gene which is to be mutated. The rationale of the basic method is shown schematically in Fig. 1A, where A and B represent the "flanking" primers that can map either within the cloned gene or outside the gene (i.e., within the vector sequence) and M represents the internal "mutant" primer containing the desired base change. The first round of PCR is performed using the mutant primer (e.g., M1 in Fig. 1) and one of the flanking primers (e.g., A). The double-stranded product is purified and used as one of the primers (hence the name "megaprimer") in the second round of PCR along with the other flanking primer (B). The wild-type cloned gene is used as template in both PCR reactions. The final PCR product containing the mutation can be used in a variety of standard applications, such as cloning in expression vectors and sequencing, or in more specialized applications, such as production of the gene message in vitro if primer A (or the template sequence downstream of primer A) also contains a transcriptional promoter (e.g., that of SP6 or T7 phage). This basic procedure can be adopted to create site-specific insertions (Fig. 1B), deletions (Fig. 1C), and gene fusions (Fig. 1D). Successful and error-free PCR in the second round often requires special considerations. The reader is, therefore, strongly urged to go through the whole chapter, including Section 4., before proceeding with the actual experiment.

It is also to be noted that the double-stranded megaprimer is directly used in the second round of PCR; in other words, prior separation of the two strands is unnecessary. Melting of the megaprimer is essentially achieved in the denaturation

From: *Methods in Molecular Biology, Vol. 67: PCR Cloning Protocols: From Molecular Cloning to Genetic Engineering* Edited by: B. A. White  Humana Press Inc., Totowa, NJ

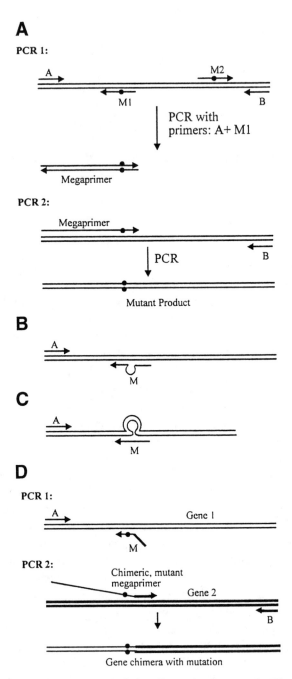

**Fig. 1**. The megaprimer method of site-directed mutagenesis. The general procedure is shown in **(A)**. For insertion **(B)** and deletion **(C)** mutagenesis, only PCR 1 is
shown; the corresponding PCR 2 will use the megaprimer produced in PCR 1 and
primer B. The primers A, B, M1, and M2 (as well as the priming strand of the
megaprimer, AM1) are indicated by single lines with arrowhead, while the doubl

steps of the second PCR. Although both strands of the megaprimer have the potential to anneal to the respective complementary strands of the template, the basic rules of PCR amplification automatically ensure that only the correct strand (one that extends to the other primer, B, in Fig. 1) will be amplified into the double-stranded product. Under some conditions, particularly with large megaprimers (1-kb and above), self-annealing of the megaprimer tends to reduce the yield of the product *(2)*. In order to avoid this, use of higher amounts of template (in microgram range, as opposed to nanogram quantities used in standard PCR) in the second PCR is recommended. This often improves the yield *(2)*, presumably because extra template strands anneal to both strands of the megaprimer, thereby preventing renaturation of the megaprimer. The other choice is to reamplify the final product in a third PCR, as recommended in Section 3.

## 2. Materials

1. DNA template containing the cloned gene (e.g., in pUC or pGEM vector) to be mutated, approx 1 μg.
2. Oligonucleotide primers A and B: the "upstream" primer A in the message sense, and the "downstream" primer B in the anti-message sense. Include restriction sites, preferably unique, in these primers so that the final product can be restricted and cloned (*see* Section 3.1. and Note 1).
3. Standard PCR reagents (*see* Chapter 1).
4. An electroelutor (IBI, New Haven, CT).
5. Standard reagents and apparatus for agarose gel electrophoresis.
6. 8*M* ammonium acetate, 0.01% bromophenol blue.
7. TE buffer: 10 m*M* Tris-HCl, pH 8.0, 1 m*M* EDTA.
8. TE-saturated phenol.
9. CHCl₃.
10. 95% Ethanol at −20°C.
11. 70% Ethanol at −20°C.
12. Speed-Vac (Savant Instruments, Farmingdale, NY).

## 3. Methods

### 3.1. Primer Design

1. For technical reasons described below, avoid making megaprimers that approach the size of the final, full-length product (gene) AB (Fig. 1); in brief, if M1 is too

---

nes represent the template. For gene fusion **(D)**, the two different genes are shown as arrow and thick double lines. The dots shown on some M primers indicate the desired mutations (base changes) to be introduced into the product via the megaprimer. The rimer in (B) (Insertion) and the template in (C) (Deletion) are shown as looped-out, lely to indicate the region to be inserted or deleted; such structures may not necessarily form in actual PCR.

close to B, it will make separation of AB and AM1 (leftover megaprimer) difficult after the second round of PCR. Ideally, the megaprimer should be shorter
than the full-length gene by more than 200–500 bp, depending on the exact length
of the gene. (Example: If the gene [AB] is 2 kb, megaprimer [AM1] can be up to
~1.5 kb long, since 2 and 1.5 kb can be separated reasonably well in agarose gels.
However, if the gene is 8 kb, the megaprimer should not be bigger than, say, 7 kb,
since 8- and 7-kb fragments would migrate too close to each other.)

2. When the mutation is to be created near B (Fig. 1), one should make an M primer
of the opposite polarity, e.g., M2, and synthesize BM2 megaprimer (rather than
AM1 or AM2).

3. When the mutation is very near one terminus of the gene (say, within 1–50 nucleotides), there is no need to use the megaprimer method! One can simply incorporate the mutation in either A or B primer and do a straightforward PCR. In
borderline situations, such as when the mutation is, say, 120 nucleotides away
from the 5' end of the gene, incorporation of the mutation in primer A may make
the primer too big to synthesize, or else it will make the megaprimer AM1 too
short to handle conveniently. In such a case, simply back up primer A a few
hundred bases further into the vector sequence in order to make AM1 megaprimer
longer. In general, remember that primers A and B can be located anywhere on
either side of the mutant region and try to utilize that flexibility as an advantage
when designing these primers.

4. In addition to the standard rules of primer design (such as GC-content matching
the template, sequence specificity, extra "clamp" sequence for restriction, absence
of self-complementarity, and so forth) *(7)*, attention should be paid to the following aspects. As stated before, primers A and B should contain unique restriction
sites for ease of cloning. The minimum length of extra sequence needed for efficient restriction varies from one restriction enzyme to another; a list for some
selected enzymes has recently been published *(8)*. Regarding the M primer, two
additional considerations are important. First, the mutational mismatch should
not be too close to the 3' end of the primer. Mismatch at the very 3' nucleotide of
the primer will virtually abolish amplification by *Taq* polymerase. For best results,
the mismatch should be at least four bases away from the 3' end of the primer.
Second, as described in detail in Note 1, the 5' end of the M primer should preferably be located such that there is at least one (two or more is better) T residue in
the template strand of the same sense just upstream of this end of the primer *(3–5)*.
If a T is not available, try to have the "wobble" base of a codon just upstream of
the 5' end of the M primer so that substitution of this base with A will code for the
same amino acid *(see* Note 1).

5. Essentially the same considerations apply in designing primers (M in Fig. 1B–D)
for insertion, deletion, and fusion of two gene sequences. Using variations of the
megaprimer method as depicted in Fig. 1, we have inserted, deleted, and fused
hundreds of base-pairs. Thus, for all practical purposes, there appears to be no
upper limit for gene lengths that can be manipulated, so long as about 15–18
nucleotides at each end of the M primer anneal to the template, allowing enough

stringency in both PCR 1 and 2 (*9;* S. Barik and R. Honkanen, unpublished). More importantly, one can combine gene fusion with mutagenesis of either or both genes by incorporating the desired mutation(s) in the M primer (Fig. 1D).

### 3.2. PCR 1: Synthesis of the Megaprimer

1. Assemble a standard 100-μL PCR *(6)* containing 5–50 ng of template DNA (plasmid containing the cloned gene, for example), 0.2–0.4 μg each of primers A and M1 (assuming a primer length of 25 nt), using standard PCR buffer and temperature cycles (*see* Chapter 1; ref. *6*). *See* Notes 2,3.

2. PCR amplify using the following cycle parameters: 94°C for 3 min (initial denaturation); 94°C for 2 min (denaturation); 55°C for 2 min (annealing); 72°C for appropriate time, *n* min (elongation) (*see* Note 2) for 30–35 cycles; and 72°C for 1.5 × *n* min (final extension). If your thermal cycler does not have a heated lid (bonnet), overlay your reaction with mineral oil (80–100 μL).

3. Following PCR, remove oil overlay as follows. Freeze PCR tubes, then thaw just enough so that the oil overlay melts, but the aqueous reaction stays frozen. Remove as much oil as possible. Then use a drawn-out round tip to transfer about 80 μL of the lower aqueous layer carefully to a clean Eppendorf tube, wiping the outside of the tip to remove any adhering oil. It is important to remove the oil completely, otherwise the sample will float up when loaded in horizontal agarose gels in step 4. Reduce volume to about 20 μL in SpeedVac concentrator or by other means of evaporation or lyophilization.

4. Gel-purify megaprimer: Any method that can purify the megaprimer from the small primers (A and M1) will work (*see* Notes 4,5). Gel electrophoresis is routinely used in our laboratory and is described here. Electrophorese PCR 1 in a standard agarose gel made in TBE in the presence of ethidium bromide. Use an appropriate concentration of agarose (0.7–1.2%) depending on the length of the megaprimer product to be purified. Perform electrophoresis until a good separation of the megaprimer and the small primer has been achieved.

5. Locate the megaprimer band by UV light and cut out the gel slice. The DNA can be extracted from the gel slice by electroelution as described here or by using commercial kits (such as the Qiaex Gel Extraction Kit, Qiagen, Chatsworth, CA). For electroelution, place the gel slice containing megaprimer in one slot of the electroelutor that is already filled with TE or TBE (*see* Instruction manual of the electroelutor). Make sure that the apparatus is leveled, that buffers in the two chambers are connected, and that all buffer flow has stopped. Clear out any air bubbles that may be trapped in the V-shaped grooves and put 75 μL of 8*M* ammonium acetate containing 0.01% BPB in each groove. Electrophorese at 150 V for required period (approx 15–30 min), as judged by the disappearance of the DNA band from the gel slice into the V groove (monitored by a hand-held UV light). Do not disturb the apparatus during elution.

6. Discontinue electrophoresis, carefully drain buffer out of both chambers, collect 400 μL ammonium acetate solution containing DNA from each V groove. Add 2–4 μg carrier tRNA at this point to improve recovery.

7. Clean the DNA by the usual phenol/chloroform treatment, precipitate with 2.5 vol (1 mL) prechilled (–20°C) ethanol (do not add extra salt), followed by centrifugation at maximum speed in a microcentrifuge for 15 min.
8. Wash DNA pellet with prechilled 70% ethanol and dissolve in 20 µL of water. Expect 30–60% recovery of DNA irrespective of size.

## *3.3. PCR 2*

1. Assemble a second PCR (100-µL) as follows: 0.5–1 µg (note the higher than usual amount; *see* Note 6) of plasmid template (the same template that was used in step 2 above), 0.2–0.4 µg primer B, 20–30 pmol of gel-purified megaprimer (3–5 µg for a 250-bp DNA), and standard concentration of buffer and nucleotides *(6)*. Generally, expect to use up all the megaprimer (recovered in step 4, Section 3.2.) in this PCR in order to achieve a good yield of the product! Use the maximum allowable temperature for annealing in the thermal cycle, as dictated by the smaller primer B. Ignore the megaprimer for annealing considerations, since its $T_m$ will be too high for the smaller primer.
2. Amplify by PCR using the cycle profiles described in step 2 of Section 3.2., using 30–35 main cycles.
3. Purify the mutant PCR product as described in Section 3.2., steps 4–8 above. It is now ready for restriction, ligation, and the like by the use of standard procedures. If the yield is poor, reamplify the mutant product using primers A and B using standard PCR (*see* Notes 7 and 8).

## 4. Notes

1. The problem of nontemplated insertions and its solution: This issue has been discussed previously *(2–5);* some of the solutions have been subsequently confirmed in specific applications. In brief, *Taq* polymerase has a natural tendency to incorporate nontemplated residues, particularly A, at the 3' end of the daughter polynucleotide strand at a certain frequency *(3,10)*. In the subsequent cycles of the PCR, these are copied and amplified into the final double-stranded product. This is generally not a problem in standard PCR where the termini of the product are usually cleaved off by restriction enzymes for cloning purposes. However, in the megaprimer method, the 3' end of the megaprimer is directly incorporated into the final product. Therefore, nontemplated A residues in the megaprimer will eventually show up in a certain percentage of the final product and cause a mutation that may be undesirable. The frequency of such "error" is usually low and megaprimers with a mismatch at the 3' end will not prime well; however, the frequency may be appreciable in some cases *(3)*. There are two kinds of solutions to this problem; one kind, exemplified by (a) below, does not introduce the nontemplated base or remove it; the other kind, described in (b), does not remove the nontemplated base but tolerates the alteration.
   a. Use a thermostable DNA polymerase that has a 3'-exonuclease activity, such as *Pfu* (Stratagene Cloning Systems, La Jolla, CA) or Vent™ polymerases (New England Biolabs, Beverly, MA). *Pfu* polymerase can use the standard

PCR conditions described in Section 3. For Vent, follow the manufacturer's recipe: A typical 100-μL PCR will contain 20 m$M$ Tris-HCl, pH 8.8 (at 25°C), 10 m$M$ KCl, 10 m$M$ (NH$_4$)$_2$SO$_4$, 2 m$M$ MgCl$_2$, 0.1% Triton X-100, 100 μg/mL acetylated BSA, 200 μ$M$ of each dNTP, 2 U of Vent polymerase, and standard amounts of template and primers. The temperature and time values of the thermal cycles are identical to those for *Taq* polymerase. Before use, make sure that these products are licensed for use in PCR.

b. Tolerate the alteration: This is the method of choice. It relies on clever primer design and does not require any extra step *(3)*. There are two ways of achieving this. As an example, suppose the relevant region of the wild-type sequence is (the amino acids are shown at the bottom in single-letter codes):

5'—AAA CTG CCA ACT CCG TCA TAT CTG CAG—3'

3'—TTT GAC GGT TGA GGC AGT ATA GAC GTC—5'

    K    L    P    T    P    S    Y    L    Q

and the Ser (TCA) is to be mutated to Ala (GCA). A mutant primer in the message sense (like M2 in Fig. 1) may have the sequence 5' CA ACT CCG GCA TAT CTG CAG 3' (the boldface G being the mutant base). However, when this M2 primer and primer B is used in PCR, the nontemplated A incorporated at the M2 end of the product (megaprimer) will result in the sequence:

5'<u>T</u>CA ACT CCG GCA TAT CTG CAG—

3'<u>A</u>GT TGA GGC CGT ATA GAC GTC—

(the nontemplated A/T is underlined). When incorporated into the final product, this megaprimer will produce the following mutant (the underlined amino acids are altered from the wild-type sequence):

5'—AAA CTG TCA ACT CCG GCA TAT CTG CAG—3'

    K    L   <u>S</u>   T    P   <u>A</u>   Y    L    Q

resulting in an undesired Ala—> Ser change (boldface). To avoid this, make the following M$_2$ primer: 5' G CCA ACT CCG GCA TAT CTG CAG 3' so that there is a T residue upstream of the 5' end of M2 in the template sequence; any extratemplated T in this strand of the megaprimer will therefore match with the T residue in the wild-type sequence and will not cause any mutation.

When no T residues are available, use the wobble base of a codon. This is possible when the primary purpose of the clone is to produce a protein product; thus, substitution of a codon with another, synonymous codon is permissible (make sure that the resultant change in the nucleotide sequence is acceptable in terms of introduction or loss of restriction sites, and so forth). Now, make the following M2 primer: 5' ACT CCG GCA TAT CTG CAG 3', so that the codon upstream of it is CCA. The nontemplated T will change this codon to CCT; however, since they both code for proline, the protein will remain unaltered.

2. As a rule, elongation time (at 72°C) in a PCR cycle should be proportional to the length of the product. An approximate guideline is 1 min of elongation/kb, i.e., 100 nt = 10 s; 500 nt = 40 s; 1 kb = 1 min 10 s; 2 kb = 2 min 20 s, and so forth *(6)*.

3. Annealing temperature is primarily governed by the base composition of the primers. A golden rule is to calculate the $T_m$ of the primer as follows: add 2°C for each A or T, and 4°C for each G or C, then deduct 4°C. For example, for a 22-nt primer with 10 G + C and 12 A + T, $T_m$ is ($[10 \times 4] + [12 \times 2]$) = 64°C; therefore, anneal at 60°C. However, the upper limit of the annealing temperature for any primer is 72°C for since it is the elongation temperature of the *Taq* polymerase (Chapters 1 and 2). More detailed thermodynamic criteria of primer selection can be found elsewhere *(7)*.

4. The electroelution procedure for purification of PCR products described here has been chosen as the single most versatile method for such purpose. Several workers may prefer other methods for more specialized and routine applications, a few of which are as follows:

   a. Centricon-100 spin filtration column (Amicon, Beverly, MA) may be used directly with the PCR reaction to separate the megaprimer product from the small PCR primers provided the megaprimer is sufficiently bigger than the smaller primers. Megaprimer purified by this method will also contain the template; however, this is of no concern since the same template is going to be used in the second round of PCR. Follow the recommended procedure *(11)*.

   b. Freeze-squeeze method *(12)*: The method works best for fragments smaller than 500-bp. The agarose gel slice containing the DNA fragment is taken in an Eppendorf tube, frozen in a dry ice-ethanol bath (10 min) or in a –70°C freezer (20 min) and then spun at room temperature in a microcentrifuge for 15 min. The recovered (30–70%) megaprimer DNA can be used directly in PCR.

   c. The GeneClean method (Bio 101, La Jolla, CA) utilizes the property of DNA to bind to glass (specially prepared, fine glass powder is supplied in the kit), however, it works best for DNA segments larger than 500-bp. It is a rather elaborate procedure and is outside the scope of this chapter. Follow the detailed instructions that come with the kit. A somewhat modified version has been published *(1)*.

5. In a recent method (13), DNA fragments in low-melting-point agarose slices have been directly used in PCR reaction apparently without any problem. Since a good quantity of the megaprimer is important for the second PCR, it might be worth trying to use it in a similar manner. This will bypass the need to recover it from the gel. Further optimization of the method is advised.

6. Higher product yield: Note that higher amounts of template have been used in the second PCR. As stated in Section 1., this tends to improve product yield *(2)*. If the yield is still poor, the best choice is to analyze PCR 2 by electrophoresis in standard agarose gels, purify the final (mutant) product (based on its size) by electroelution, and use it as the template in a third PCR with A and B as primers.

7. As in any cloning procedure, the final mutants obtained by the megaprimer method must be confirmed by DNA sequencing. This can be done either by directly

sequencing the PCR product *(14,15)* or after cloning the mutant product in plasmid vectors *(16)*. When using the dideoxy method, PCR primer A or B can be used as sequencing primers as well.

8. For gene fusion only, one can combine PCR 1 and 2 in a single PCR containing both templates, and primers A, M (the chimeric primer), and B *(4)*, provided that primers A and B do not anneal to their noncognate templates. Clearly, combining PCR 1 and 2 is not feasible for substitution, deletion, or insertion mutagenesis (i.e., those procedures that use the same template for both PCR 1 and 2), since primers A and B will simply amplify the wild-type template and primer M will be essentially ignored!

## Acknowledgments

Research in my laboratory was supported by a National Institutes of Health Research Grant AI37938 and by an Intramural Research Grant Award from the College of Medicine, University of South Alabama.

## References

1. Sarkar, G. and Sommer, S. S. (1990) The "megaprimer" method of site-directed mutagenesis. *BioTechniques* **8,** 404–407.
2. Barik, S. and Galinski, M. (1991) "Megaprimer" method of PCR: increased template concentration improves yield. *BioTechniques* **10,** 489,490.
3. Barik, S. (1993) Site-directed mutagenesis by double polymerase chain reaction: megaprimer method. *Meth. Mol. Biol.* **15,** 277–286.
4. Barik, S. (1995) Site-directed mutagenesis by PCR: substitution, insertion, deletion, and gene fusion. *Meth. Neurosci.* **26,** 309–323.
5. Barik, S. (1995) Site-directed mutagenesis in vitro by megaprimer PCR. *Meth. Mol. Biol.,* in press.
6. Delidow, B., Lynch, J. P., Peluso, J. J., and White, B. A. (1993) Polymerase chain reaction: basic protocols. *Meth. Mol. Biol.* **15,** 1–29.
7. Rychlik, W. (1993) Selection of primers for polymerase chain reaction. *Meth. Mol. Biol.* **15,** 31–40.
8. New England BioLabs (1995, Catalog), Cleavage close to the end of DNA fragments, pp. 208,209.
9. Barik, S. (1993) Cloning, expression, and biochemical characterization of a novel protein serine/threonine phosphatase encoded by bacteriophage λ. *Proc. Natl. Acad. Sci. USA* **15,** 10,633–10,637.
10. Clark, J. M. (1988) Novel nontemplated nucleotide addition reactions catalyzed by procaryotic and eucaryotic DNA polymerases. *Nucleic Acids Res.* **16,** 9677–9686.
11. Krowczynska, A. M. and Henderson, M. B. (1992) Efficient purification of PCR products using ultrafiltration. *BioTechniques* **13,** 286–289.
12. Stoflet, E. S., Koeberl, D. D., Sarkar, G., and Sommer, S. S. (1988) Genomic amplification with transcript sequencing. *Science* **239,** 491–494.

13. Zintz, C. B. and Beebe, D. C. (1991) Rapid reamplification of PCR products purified in low-melting-point agarose gels. *BioTechniques* **11,** 158–162.
14. Meltzer, S. J. (1993) Direct sequencing of polymerase chain reaction products. *Meth. Mol. Biol.* **15,** 137–142.
15. Dicker, A. P., Volkenandt, M., and Bertino, J. R. (1993) Manual and automated direct sequencing of product generated by polymerase chain reaction. *Meth. Mol. Biol.* **15,** 143–152.
16. Maniatis, T., Fritsch, E. F., and Sambrook, J. (1982) *Molecular Cloning: A Laboratory Manual.* Cold Spring Harbor Laboratory, Cold Spring Harbor, NY, pp. 113–119.

# 18

## Rapid and Efficient
## One-Tube PCR-Based Mutagenesis Method

### Veronique Picard and Susan Clark Bock

### 1. Introduction

Oligonucleotide-directed mutagenesis techniques are extensively used for studying gene regulation and DNA and protein structure/function relationships. A number of PCR-based mutagenesis methods have been developed recently *(1–8)*. In general, these PCR-based approaches are simpler and faster than classical mutagenesis methods employing in vivo DNA synthesis, and several of them are now widely practiced.

This chapter describes a one-tube megaprimer PCR mutagenesis approach that is fast, efficient (>50% mutants), and not labor intensive. The protocol described here is a simplified version of a method we reported previously *(8)*. The principle of the method is outlined in Fig. 1. It takes 4.5 h from start to finish, and "human work" is required for <15 min of this time: 10 min at the beginning of the procedure for adding reaction components to the tubes and putting them in the PCR machine, and 5 min between steps 1 and 2 for adding additional primer to the sample. A further advantage of this method is the production of DNA with a low rate of background sequence errors. The misincorporation rate is minimized by utilizing *Pfu* DNA polymerase, which has a proofreading 3'→5' exonuclease activity *(9)* and by limiting the number of amplification cycles.

### 2. Materials

1. DNA template (e.g., plasmid carrying DNA sequence to be mutagenized).
2. Oligonucleotide primers: two flanking primers (U and D) and one mutagenic primer (M).
3. *Pfu* polymerase (*see* Note 1) and 10X reaction buffer (Stratagene, La Jolla, CA).
4. dNTP mix (e.g., 100 m$M$ [25 m$M$ of each dNTP] mix; Stratagene).

From: *Methods in Molecular Biology, Vol. 67: PCR Cloning Protocols: From Molecular Cloning to Genetic Engineering* Edited by: B. A. White Humana Press Inc., Totowa, NJ

STEP 1   (15 cycles)

Synthesis of megaprimer from:
    3.6 - 36 fmol template
    10 pmol mutagenic primer (M)
    10 pmol downstream primer (D)

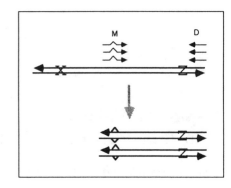

STEP 2    (15 cycles)

Addition of 50 pmol upstream primer (U)

• Megaprimer synthesis continues
• Mutant DNA synthesis from
  U and megaprimer
• Parental DNA synthesis from
  U and residual D primer

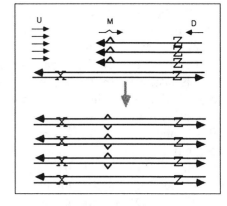

**Fig. 1.** $x$ and $z$ are restriction sites that can be used for subcloning of PCR products.

5. PCR thermal cycler.
6. Mineral oil.
7. Equipment and materials for DNA electrophoresis, electroelution, phenol/chloroform extraction, and ethanol precipitation.

# 3. Methods
## 3.1. Primer Design
### 3.1.1. Flanking Primers

Upstream (U) and downstream (D) primers are respectively located 5' and 3' to the mutation site. Locating the U and D primers distal to restriction sites flanking the mutation site (*x* and *z* in Fig. 1) will facilitate subcloning of the mutated region. A common set of flanking primers can be used for many different mutagenesis projects. It is often convenient to employ "universal" primers corresponding to plasmid sequences flanking the vector polylinker cloning site as the U and D primers.

### 3.1.2. Mutagenic Primer

We recommend maintaining at least ten perfect base pairs on each side of the mismatched region in the mutagenic (M) primer sequence (*see* Section 4.4.). The M primer may correspond to either strand of the DNA template. Choice of which strand to select for the M primer depends on the location of the flanking U and D primers and associated subcloning sites. It is useful to choose primers so that the size of the megaprimer is minimized. The length of the megaprimer is an important parameter in this protocol, with shorter megaprimers leading to higher yields. In practice we have been successful with megaprimers in the 100–500 base-pair range.

For protein mutagenesis projects, it is usually possible to introduce a translationally silent restriction site change via the mutagenic primer. This is highly recommended since it greatly simplifies identification of the subclones containing mutant sequences.

When creating several different mutations at the same site, we have found it helpful and cost-efficient to utilize oligonucleotide pools of up to four different sequences for the M primer. However, pools containing greater than four different sequences are not recommended since, in our experience, these more complex pools do not contain component oligonucleotides at equal frequencies and rare sequences may be difficult to recover.

## 3.2. PCR Mutagenesis Reaction

### 3.2.1. Synthesis of the Megaprimer

1. Assemble the initial 100 µL PCR reaction containing the following: 3.6–36 fmol template DNA (e.g., 10–100 ng of a 4.2-kb plasmid; *see* Note 2), 10 pmol (100 n*M*) mutagenic primer (M), 10 pmol (100 n*M*) downstream primer (D), 20 nmol each (0.2 m*M* each) dNTPs, 10 µL 10X reaction buffer, 2.5 U *Pfu* polymerase (add enzyme last, after thoroughly mixing the other components), and water to 100 µL final volume.
2. Overlay reaction with mineral oil.
3. Perform 15 cycles of amplification according to the following profile: 94°C for 2 min (denaturation), 50°C for 2 min (annealing; *see* Note 3), and 75°C for 2 min (extension; *see* Note 4).
4. Follow the fifteenth cycle with 5 min additional extension at 75°C then hold at 4°C.

### 3.2.2. Step 2: Synthesis of Full-Length DNA

1. Following completion of Section 3.2.1., add 50 pmol of upstream primer to the 100-µL sample by pipeting 2.5 µL of 20 µ$M$ U primer under the oil overlay.
2. Perform PCR using repeat of program for Step 3.2.1. (15 cycles plus additional 5 min extension step).

## 3.3. Analysis of Mutagenesis Reaction Products and Subcloning

### 3.3.1. Characterization of PCR Mutagenesis Reaction Products

Verify the size of and determine the approximate amount of full-length DNA produced in the mutagenesis reaction by running a 5-µL aliquot of the 100-µL reaction on a polyacrylamide gel (*see* Chapter 20 for electrophoresis conditions). If a restriction site was introduced or removed via the mutagenic primer, a second 5-µL aliquot of the PCR reaction should be digested with the appropriate enzyme and run beside the untreated sample. This allows one to estimate the mutant:parental DNA ratio, which is useful to know when deciding how many subclones to miniprep (*see* next section).

### 3.3.2. Subcloning

Prepare insert by digesting 50 µL of the remaining sample with appropriate enzymes (*x* and *z* in Fig. 1). Gel purify insert and ligate with appropriate vector. Transform ligated DNA into *E. coli* to obtain subcloned mutagenesis reaction products. For mutagenesis reactions producing approximately equal amounts of mutant and parental DNA according to the restriction digest results described in Section 3.3.1., we find it is almost always more than sufficient to prepare plasmid DNA minipreps from six subclones. If the restriction analysis shows a mutant:parental ratio of substantially more or substantially less than 1:1, decrease or increase the number of minipreps respectively.

### 3.3.3. Identification and Verification of Mutant Subclones

If the M oligonucleotide changed a restriction site, mutant subclones may be distinguished from parental subclones by restriction analysis of the miniprep DNAs. One of the mutant subclones is then sequenced to verify introduction of intended mutation(s) and absence of unplanned sequence changes in background DNA.

For cases where no restriction site change is associated with the *M* oligonucleotide, we sequence 4–6 subclones and usually obtain >1 mutant. Gel space and material and labor requirements can be reduced (and the number of subclones screened increased) by running "tracking" reactions of one base that will give distinctive mutant and parental ladders, instead of full four-base sequencing reactions. Once mutant(s) are identified, one is fully sequenced as above.

## 4. Notes

1. Based on several favorable properties, we chose to develop this procedure with *Pfu* DNA polymerase rather than *Taq* DNA polymerase. Use of the *Pfu* enzyme reduces the error rate caused by nucleotide misincorporation. *Pfu* polymerase is reported to have a 12-fold increased fidelity of DNA synthesis compared to *Taq* polymerase caused by the presence of a 3'→5' exonuclease activity ("proofreading activity") *(9)*. However, because of this 3'→5' exonuclease activity of *Pfu* polymerase, it is important to design mutagenic oligonucleotides with at least ten perfectly matching bases on each side of the introduced mismatch(es). Another favorable property of *Pfu* polymerase is its lack of terminal transferase activity; i.e., in contrast to *Taq* and Vent polymerases, it does not catalyze addition of a nontemplated nucleotide to the 3' ends of DNA fragments *(10)*. Therefore, concerns about the sequence of the template immediately following the 5' end of the mutagenic primer (which becomes the 3' end of the megaprimer) *(11)* are not relevant here. Finally, it is noted that we chose 75°C as the extension temperature since this has been reported to be optimal for *Pfu* polymerase.

2. Template DNA should be kept in the indicated range because a too small amount will decrease the yields, whereas excess template may increase the recovery of the parental sequences, instead of mutant sequences, during the second step.

3. Annealing temperature should be selected according to standard guidelines; we have found that 50°C is usually suitable for reactions using 18-mer flanking primers and a 20–25-mer mutagenic primer of average base composition (50% G + C).

4. Duration of the PCR extension step can be optimized based on the length of the fragments to be generated. For short megaprimers (<500 base pairs) extension time during step 1 can be reduced to 1 min.

5. In an earlier version of this procedure *(8)* mutant yields ranged from 25–100% of subcloned PCR products (>50% mutants in six out of nine reactions). Sequencing of 16.5 kb from 37 subcloned mutant DNAs revealed four unplanned mutations, corresponding to an error rate of 0.024%/base sequenced (1/4125).

## References

1. Higuchi, R., Krummel, B. K., and Saiki, R. K. (1988) A general method of in vitro preparation and specific mutagenesis of DNA fragments: study of DNA and protein interactions. *Nucleic Acids Res.*, **16**, 7351–7367.
2. Ho, S. N., Hunt, H. D., Horton, R. M., Pullen, J. K., and Pease, L. R. (1989) Site directed mutagenesis by overlap extension using the polymerase chain reaction. *Gene* **77**, 51–59.
3. Landt, O., Grunert, H.-P., and Hahn, U. (1990) A general method for rapid site-directed mutagenesis using the polymerase chain reaction. *Gene* **96**, 125–128.
4. Sarkar, G. and Sommer, S. S. (1990) The "megaprimer" method of site-directed mutagenesis. *BioTechniques*, **8**, 404–407.
5. Marini III, F., Naeem, A., and Lapeyre, J.-N. (1993) An efficient 1-tube PCR method for internal site-directed mutagenesis of large amplified molecules. *Nucleic Acids Res.* **21**, 2277,2278.

6. Barik, S. (1993) Site-directed mutagenesis by double polymerase chain reaction megaprimer method, in *Methods in Molecular Biology, vol. 15: PCR Protocols: Current Methods and Applications* (White, B. A., ed.), Humana, Totowa, NJ, pp. 277–286.

7. Barettino, D., Feigenbutz, M., Valcarcel, R., and Stunnenberg, H. G. (1994) Improved method for PCR-mediated site-directed mutagenesis. *Nucleic Acids Res.* **22,** 541,542.

8. Picard, V., Ersdal-Badju, E., Lu, A., and Bock, S.C. (1994) A rapid and efficient one-tube PCR based mutagenesis technique using *Pfu* DNA polymerase. *Nucleic Acids. Res.* **22,** 2587–2591.

9. Lundberg, K. S., Shoemaker, D. D., Adams, M. W. W., Short, J., Sorge, J. A., and Mathur, E. J. (1991) High-fidelity amplification using a thermostable DNA polymerase isolated from *Pyrococcus furiosus. Gene* **108,** 1–6.

10. Hu, G. (1993) DNA polymerase-catalyzed addition of nontemplated extra nucleotides to the 3' end of a DNA fragment. *DNA Cell Biol.* **12,** 763–770.

11. Kuipers, O. P., Boot, H. J., and de Vos, W. M. (1991) Improved site-directed mutagenesis method using PCR. *Nucleic Acids Res.* **19,** 4558.

# 19

## Thermostable Ligase-Mediated Incorporation of Mutagenic Oligonucleotides During PCR Amplification

### Scott F. Michael

## 1. Introduction

PCR amplification has become an extremely powerful and universally used technique for cloning and manipulating DNA segments. It is especially useful for the abillity to alter the terminal sequences of an amplified product simply by using primers containing the desired changes. However, the inability to alter easily regions of a product between the amplification primers has limited the use of PCR for more general mutagenesis. The protocol presented in this chapter employs a simple method that removes this limitation.

The combined use of thermostable ligase and thermostable polymerase during a thermocycling reaction allows the incorporation of additional phosphorylated oligonucleotides into any position of an amplified product *(1)*. The basis for this reaction lies in the lack of strand displacement activity shown by many thermostable polymerases, which allows the efficient annealing of an extra primer internally to a set of amplification primers. If the internally annealing primer is phosphorylated at the 5' end it can be ligated into the final product during thermal cycling by the activity of a thermostable ligase *(2)* (*see* Fig. 1). This combined reaction can be used to introduce site-specific mutations, deletions, or insertions into any position of an amplified product in a single amplification reaction, using a single mutagenic oligo. The ability to generate mutations in a product that can be cloned or used directly eliminates the need to construct and use special plasmids designed for mutagenesis *(3)*. In addition, there are no restrictions on the placement of the outer amplification primers, so these oligos can be designed to anneal to regions convenient for cloning

From: *Methods in Molecular Biology, Vol. 67: PCR Cloning Protocols: From Molecular Cloning to Genetic Engineering* Edited by: B. A. White  Humana Press Inc., Totowa, NJ

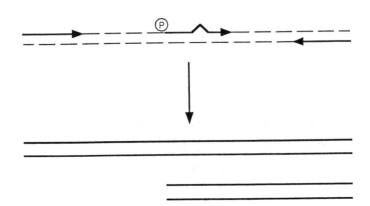

**Fig. 1.** Schematic representation of mutagenesis procedure and resulting products. Primers are drawn as solid arrows, extension products as dashed lines. The phosphorylated, mutagenic oligo serves as a third primer and is incorporated into the full-length product by the activity of the thermostable ligase. Two different sized fragments are produced from the reaction; a full-length product that potentially contains the mutation of interest, and a smaller fragment resulting from amplification from the mutagenic primer to the outer primer annealing to the complementary strand.

or other uses. This method is also faster and more economical than other PCR-based mutagenesis techniques in that it requires only a single amplification reaction and a single mutagenic oligo. Other methods require either multiple reactions or multiple mutagenic oligos *(4–8)*.

Successful incorporation of the mutagenic oligonucleotide primer relies on optimizing the efficiency of the ligation reaction relative to the competing extension reactions. There are two competing reactions: extension from the outer primers without incorporation of the mutagenic primer, and extension from the mutagenic primer to the outer primer that binds to the opposite strand. This second reaction generates a truncated product that should be separated from the full-length product before cloning (*see* Fig. 2). In order to increase the level of mutagenic primer incorporation into the final, full-length product, the reactions are run using an excess of mutagenic oligo, and buffer and thermal conditions designed to optimize the thermostable ligase activity. Theoretically, if the mutagenic oligo is incorporated with 100% efficiency, only nine rounds of amplification would be sufficient to yield greater than 99% mutation-containing product. This is because the mutagenic oligo is not only directly incorporated into newly synthesized DNA strands, but previously synthesized strands containing the mutagenic oligo are also used as template in subsequent rounds, increasing the proportion of mutant product with each cycle. Although

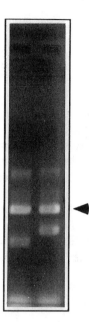

**Fig. 2.** Ethidium bromide-stained agarose gel showing typical reaction products. Two different mutagenic primers, annealing to different locations, were used in amplification/mutagenesis reactions with identical outer primer sets. The resulting products consisted of identically sized full-length fragments (indicated by arrow) and differently sized smaller fragments resulting from amplification between the mutagenic primer and the outer primer annealing to the opposite strand. Both reactions also produced larger nonspecific products.

100% efficiency is generally not achieved, in practice many reactions produce a >50% mutation rate.

## 2. Materials

1. Template DNA containing the target sequence to be mutated.
2. Oligonucleotide amplification primers that will produce the entire DNA segment of interest.
3. Mutagenic oligonucleotide primer.
4. Polynucleotide kinase and 10X buffer.
5. Thermostable polymerase and 10X buffer (*see* Note 1).
6. Thermostable ligase and 10X buffer (*see* Note 1).
7. dNTP mix.
8. Thermocycler.
9. GeneClean (Bio 101, La Jolla, CA).
10. Appropriate restriction enzymes and buffers.
11. Agarose gel apparatus and reagents.

## 3. Methods

### 3.1. Primer Design

1. Synthesize two amplification primers of approx 15–25 bp, which will generate the desired full-length product. There are no limitations on the placement of these primers and they can include convenient restriction sites for cloning. In general, the full-length product should not be too large, because amplification may be difficult and the incidence of unwanted, secondary mutations will increase with size. Care should be taken in designing these primers to avoid sequences that will either self-hybridize or hybridize to the other primers (*see* Note 2).
2. Synthesize a mutagenic oligonucleotide primer that contains the mutation(s) of interest and anneals to the appropriate region of the template, somewhere between the two amplification primers. Again, care should be taken to avoid primer sequences that will self-anneal or anneal to the outer amplification primers (*see* Note 2). In addition, to maximize incorporation, this primer should be designed to anneal better than either outer primer. This can be accomplished easily by including GC-rich sequences in the mutagenic primer and/or by making it somewhat longer than the outer primers. This oligo can be either chemically phosphorylated during synthesis or enzymatically phosphorylated as detailed in Section 3.2.

### 3.2. Mutagenic Primer Phosphorylation

1. Assemble an oligonucleotide phosphorylation reaction using 5 nmol of mutagenic primer, 10 U of T4 polynucleotide kinase, and 100 nmol of ATP in a total volume of 100 μL (*see* Note 3).
2. Incubate the reaction at 37°C for the time recommended by the enzyme manufacturer (generally 1 h). There is no need to heat inactivate the reaction after completion, and the phosphorylated oligo can be added directly to the subsequent amplification/mutagenesis reaction without further purification (*see* Note 4).

### 3.3. Amplification/Mutagenesis Reaction

1. Set up a 100-μL reaction consisting of: 100 pmol (1 μ$M$) of each outer primer; 1 nmol (10 μ$M$) of the phosphorylated mutagenic primer (20 μL of the phosphorylation reaction described in Section 3.2.); 40 nmol (400 μ$M$) each dNTP; 10 μL 10X thermostable ligase buffer; 1 μL thermostable polymerase (*see* Note 1); 1 μL thermostable ligase (*see* Note 1); and $10^8$–$10^9$ copies of template DNA (equivalent to 1–10 ng of a 10-kbp plasmid containing the target sequence).
2. Add a mineral oil overlay if needed. For a final product size of approx 1 kbp, perform the reaction using the following profile: 94°C for 1 min, 40°C for 1 min, and 65°C for 4 min for 30 cycles. (*See* Note 5.)

### 3.4. Product Restriction and Isolation

1. Purify the amplified products away from remaining primers, primer dimers, and other reactants. For products >500 bp, use the GeneClean method as described by the manufacturer (Bio 101, La Jolla, CA). Alternatively, or for smaller products,

use any of several spin-filtration devices, such as the 30,000 mol wt cut-off Ultrafree-MC filters (Millipore, Bedford, MA). Other DNA fragment purification methods can also be used (*see* Note 6).

2. Digest the purified products with the appropriate restriction enzyme(s) following the enzyme manufacturer's instructions. Digestion times should be increased to account for the reduced activity many restriction enzymes have for sites near the ends of linear DNA (*see* Note 2). Overnight digests are often necessary.

3. Separate the full-length product from the smaller, partial product by ethidium bromide-stained agarose gel electrophoresis. Use a highly sieving agarose, such as 3 or 4% Nu-Sieve GTG low-melting-temperature agarose (FMC Bioproducts, Rockland, ME), because many agarose formulations do not have sufficient resolving power to give sufficient separation of small products. Low-melting-temperature agarose has the added advantage that the full-length product can be isolated by gel slice excision, melted, and used directly in a ligation reaction. Alternatively, the full-length product can be isolated from the agarose gel by some other method, including electroelution, spin-filtration, or matrix adsorption.

## 4. Notes

1. There are numerous thermostable polymerases currently commercially available, and several different thermostable ligases as well. This protocol has been written using *Taq* polymerase and *Taq* ligase, however, several other enzymes can also be used. The combination of *Taq* polymerase (Promega, Madison, WI) and *Taq* ligase (New England Biolabs, Beverly, MA), AMPLIGASE (Epicenter, Madison, WI), *Tth* ligase (Stratagene, La Jolla, CA), or *Pfu* ligase (Stratagene) all produced mutant product when used in this protocol. Unfortunately, none of the "proof-reading" thermostable polymerases that have been tested function extremely well. Vent and Deep Vent polymerases (New England Biolabs) showed some activity in experiments using the ligases and ligase buffers mentioned above, but in general produced little more than the small, truncated product. *Pfu* polymerase (Stratagene, La Jolla, CA) showed no activity at all. Apparently these polymerases are less tolerant of altered buffer conditions than *Taq* polymerase. However, it is likely that some buffer conditions exist that will allow the use of proofreading polymerases for this method.

2. It is extremely important to avoid primer sequences that will either self-hybridize, or hybridize to the other primers. This is especially true for the mutagenic oligo because unwanted annealing will reduce the oligo concentration available to bind to the desired template site and directly reduce the efficiency of mutagenesis. To increase the efficiency of restriction digestion and subsequent cloning of the amplified product, design the outer primers so that restriction sites are not at the very end. (An additional GCG sequence directly 5' to a terminal restriction site generally works well.) Alternatively, if the product will be amplified from, and cloned back into a construct already containing convenient sites, the primers can be placed to include these sites internally in the product.

3. T4 kinase may be used from any of various sources and will generally be supplied with an appropriate 10X buffer (that may or may not contain ATP) and instructions.

4. Complete phosphorylation of the mutagenic primer is essential for the successful generation of a high percentage of mutant-containing clones. In general, it may be desirable to use mutagenic primers that have been chemically phosphorylated during synthesis, because the efficiency of chemical phosphorylation is near 100%. However, enzymatic phosphorylation is simple and efficient if the reactants are handled appropriately. Probably the most significant source of problems is ATP degradation. To avoid this, ATP-containing solutions, whether prepared or purchased as part of a phosphorylation kit, should be aliquoted, stored frozen, and each tube used only once.

5. The amplification/mutagenesis reaction will take significantly longer to run than a standard amplification reaction with the same number of cycles because of the longer extension times. Also, the amplification/mutagenesis reaction is carried out using buffer conditions optimized for the ligase. These different thermal and chemical conditions enhance the activity of the ligase, but reduce the activity of the polymerase. Consequently, the overall product yield is often low. One adjustment that partially offsets this is to use larger amounts of starting template. As indicated above, 1–10 ng of a 10-kbp template-containing plasmid is generally required to produce a significant product. If your product yield is poor, try increasing the amount of starting template used in the reaction.

Another parameter that may need to be adjusted for some primers and templates is the annealing temperature. The suggested temperature is rather low in order to facilitate annealing and generate larger yields. In general, amplification of nonspecific products resulting from low annealing temperatures is not a problem since the final product will be separated by electrophoresis from other undesired products. However, it may be necessary to increase the annealing temperature if nonspecific amplification products are reducing the yield of the desired product, or are interfering with product purification. Increasing the annealing temperature to some optimum might also increase the efficiency of mutagenesis when the mutagenic primer has a higher $T_m$ than the outer primers. In this case, an annealing temperature that allowed complete annealing of the mutagenic primer while only allowing partial annealing of the outer primers should result in the generation of a higher proportion of final product containing the mutagenic primer. However, in a small set of experimental trials designed to test this, no evidence for increased efficiency could be found by increasing the annealing temperature.

Increased extension times may be necessary for the amplification of longer sequences under these conditions. Keep in mind that longer sequences have increased chances of producing an amplified product containing unwanted, secondary mutations caused by polymerase error.

6. There are probably nearly as many methods to isolate and clone PCR fragments as there are molecular biologists using PCR. I have presented the methods that I have found useful, however, other techniques will almost certainly work as well.

Whatever techniques are most comfortable for you, it is important to digest and clone your amplified products right away. PCR products are notorious for being difficult to ligate because of degradation of the termini, and successfully digested products will continue to degrade if not used immediately.

## References

1. Michael, S. F. (1994) Mutagenesis by incorporation of a phosphorylated oligo during PCR amplification. *Biotechniques* **16,** 410–412.
2. Baraney, F. (1991) Genetic disease detection and DNA amplification using cloned thermostable ligase. *Proc. Natl. Acad. Sci. USA* **88,** 189–193.
3. Kunkel, T. A., Bebenek, K., and McClary, J. (1991) Efficient site-directed mutagenesis using uracil-containing DNA. *Methods Enzymol.* **204,** 125–139.
4. Sarkar, G. and Sommer, S. S. (1990) The "megaprimer" method of site directed mutagenesis. *Biotechniques* **8,** 404–407.
5. Higuchi, R., Krummel, B., and Saiki, R. K. (1988) A general method of in vitro preparation and specific mutagenesis of DNA fragments: study of protein and DNA interactions. *Nucleic Acids Res.* **16,** 7351–7367.
6. Ho, S. N., Hunt, H. D., Horton, R. M., Pullen, J. K., and Pease, L. R. (1989) Site-directed mutagenesis by overlap extension using the polymerase chain reaction. *Gene* **77,** 51–59.
7. Gustin, K. E. and Burk, R. D. (1993) A rapid method for generating linker scanning mutants utilizing PCR. *Biotechniques* **14,** 22,23.
8. Rouwendal, G. J. A., Wolbert, E. J. H., Zwiers, L.-H., and Springer, J. (1993) Simultaneous mutagenesis of multiple sites: application of the ligase chain reaction using PCR products instead of oligonucleotides. *Biotechniques* **15,** 68–75.

# 20

# Linker Scanning Mutagenesis by Three-Step PCR

## Judith T. Schanke

## 1. Introduction

A number of mutagenesis methods allow the systematic survey of a region of transcriptional regulatory sequence for identifying functional elements. In these methods, clusters or blocks of point mutations are introduced at discrete locations that span a suspected regulatory region of a gene. The individual mutants are then examined for retention of transcriptional activity typically in a reporter gene assay. Traditional methods, including linker-scanning mutagenesis *(1)* and microscale "shot-gun" gene synthesis *(2)*, are either tedious and time-consuming or relatively inefficient at producing the desired mutated sequence. A modification of the PCR-based mutagenesis method originally described by Li and Shapiro *(3)* is much simpler, faster, and more efficient than the traditional methods.

Creating systematic blocks of mutations by PCR requires three sequential amplifications. Mutations are introduced by internal primers in the first round of amplification *(see* Fig. 1). The mutation primer (C), containing a central cluster of bases mismatched to the template, is paired with the 3' external primer to produce a partial length product containing the mutated sequence. The second amplification is asymmetric. Using the first round product as template, a single-stranded product containing the cluster of mutations is produced using only the 3' external primer (primer C in Fig. 1). This single-stranded product is then used as a primer and is paired with the 5' external primer (A) in the third amplification. This final round of amplification, which uses the original wild-type template, produces the full-length regulatory region containing the block mutation.

From: *Methods in Molecular Biology, Vol. 67: PCR Cloning Protocols: From Molecular Cloning to Genetic Engineering* Edited by: B. A. White  Humana Press Inc., Totowa, NJ

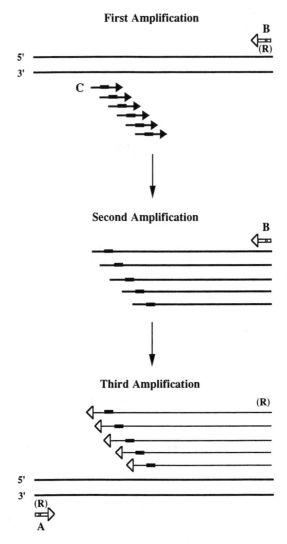

**Fig. 1**. Diagram of the three-step PCR method for generating scanning block mutations. The template or target regulatory sequence is depicted with bold lines. Oligonucleotide primers are depicted with arrows at their annealing sites along the template. Mismatches in the external primers (**A,B**) create restriction sites (R) in the final product to facilitate subcloning. The first amplification requires the mutation primer (**C**) and the 3' external primer (B). The double-stranded product of this reaction is then used as the template in the second amplification with the 3' external primer alone (B). This asymmetric PCR results in the production of single-stranded products that are purified and serve as primers in the third round of amplification. Amplification with the single-stranded products, when paired with the 5' external primer (A) and the original wild-type template, results in the production of full-length double-stranded products containing the block of mutations.

## 2. Materials

1. Template DNA, typically a plasmid subclone of the regulatory region to be mutated.
2. Two external primers flanking the region to be analyzed by block mutagenesis (Fig. 1A,B). Primer A should be in the message sense and primer B in the antisense. These primers can be designed with mismatches to allow endonuclease restriction for ease in cloning the mutant product *(4)*.
3. One message sense internal mutation primer is required for each mutation (Fig. 1C). Primer design is described in Section 3.
4. PCR reagents including dNTPs, thermal stable polymerase, buffer containing $MgCl_2$, and any cosolvents *(see* Note 1).
5. Reagents for either native polyacrylamide or Nusieve agarose gel purification, including TBE buffer and glass wool (for spin columns).

## 3. Methods

### 3.1. Primer Design

1. The standard rules of primer design apply to both the external primers and the mutation primers. Briefly this includes avoiding self complementarity, direct repeats, and single nucleotide runs at either end of the primer *(see* ref. *5)*.
2. Flanking primers A and B may include mismatches to allow restriction endonuclease cleavage of the final product. These mismatches should be anchored by complementary sequence as previously described *(4)*. Alternatively, the final product can be directly subcloned into a PCR cloning vector, such as pCRII (TA cloning kit, Invitrogen, San Diego, CA) or pGEM-T (Promega, Madison, WI). The PCR cloning vector can be used as a shuttle vector if subcloning into a reporter gene vector is the desired end point.
3. The design of the mutation primers depends on the number of clustered mismatches. A block of at least ten bases can be mutated at once as long as the mutation primers contain sufficient sequence, complementary to the template, flanking the mismatches to allow annealing of the primer. For a 5–10-base block mutation, ten complementary bases on both sides of the block is typically sufficient for annealing and extension in both the first and third amplifications.
4. The annealing temperature of the mutation primers should be calculated in order to adjust the standard cycling temperatures for the first and third amplifications *(see* Notes 2,3). Use the standard formula (2°C for each A and T, plus 4°C for each G and C, minus 4°C), but only count the bases which are complementary to the template.
5. When possible, a polymerase with 3' exonuclease activity should be used or the mutation primer should be designed so that the next base in the same sense template sequence is a T. The nontemplated A additions by some thermal stable polymerases may result in a mutation or insertion at the end of the mutation primer sequence in the first round of amplification. If neither of the above approaches is possible then treatment of the first round products with T4 DNA polymerase or Klenow is suggested.

## 3.2. First Amplification

1. Perform the first round of amplification with 20 pmol of mutation primer (C), 20 pmol of the 3' external primer (B), 200 ng of template DNA (containing the regulatory region of interest), 50 μ*M* each dNTP, 1–2.5 U of thermal stable polymerase and the appropriate buffer containing from 1–3 m*M* MgCl$_2$ (*see* Note 1 and Chapters 1–3) in a 50-μL reaction.
2. Amplify using the hot start technique (*see* Note 1 and ref. *9*) and the following cycle profiles: 95°C for 3 min (initial denaturation); 40 cycles: 95°C for 50 s, 45°C for 60 s (*see* Note 2), 72°C for 60 s; and 72°C for 7 min (final extension).
3. Use the products directly in the next amplification.

## 3.3. Second Amplification

1. Setup reactions with 10 μL of the first PCR product, 20 pmol 3' external primer (B), and standard buffer, enzyme and nucleotide concentrations (*see* Chapters 1–3) in 50-μL reactions.
2. Amplify using the hot start technique and the following cycle profiles: 95°C for 3 min (initial denaturation); 30 cycles: 95°C for 50 s, 60°C for 60 s (*see* Note 3), 72°C for 60 s; and 72°C for 7 min (final extension).
3. Resolve the single-stranded products resulting from this asymmetric amplification on a native polyacrylamide gel (*see* Chapter 20) or Nusieve agarose gel. Stain the gel with ethidium bromide and isolate gel slices containing the single-stranded products (*see* Note 4).
4. Crush the acrylamide gel slices and soak in 200 μL of dH$_2$O for approx 2 h with shaking.
5. Transfer the gel slurry to a glass wool spin column (*see* Note 5) and centrifuge for 10 s at 5000 rpm (about 2000 rpm) in a microfuge.
6. Use the collected product directly in the next amplification.

## 3.4. Third Amplification

1. Approximately 20 μL of purified PCR 2 product is used as the 3' primer in the final amplification (*see* Note 6). The 50-μL reactions also contain 20 pmol 5' primer, 200 ng template and standard enzyme, buffer and nucleotide concentrations.
2. Amplify using the hot start technique and the following cycle profiles: 95°C for 3 min (initial denaturation); 40 cycles: 95°C for 50 s, 45°C for 60 s, 72°C for 60 s (*see* Note 6); and 72°C for 7 min (final extension).
3. Gel purify the full-length PCR products as performed for the products of the second amplification.
4. The purified products are then either cloned directly into a PCR cloning vector or are restricted within the external primer sequence and subcloned into a suitable reporter gene vector by standard methods.

## 4. Notes

1. The inclusion of cosolvents, such as glycerol or DMSO *(7)*, optimization of MgCl$_2$ concentration *(8)* and the use of hot start procedures (adding the final

reaction components to reactions already heated to 95°C *[9]*), may all improve product amounts and decrease nonspecific annealing and amplification.

2. The annealing temperature could be increased if the calculated annealing sequence for the specific mutation primer allows. This annealing temperature should be determined using only the complementary bases of the primer. This would increase the specificity of the annealing and may allow for increased product formation.

3. The annealing temperature is increased to 60°C or to as high a temperature as the sequence of the specific 3' flanking primer allows.

4. Single-stranded products run slightly faster than double-stranded products in agarose, but may run slightly slower in acrylamide.

5. Glass wool spin columns are constructed by puncturing the bottom of a 500-mL Eppendorf tube with a 23-gage needle, inserting glass wool to cover the bottom one-fourth of the tube, and placing it in a 1.5-mL collection Eppendorf tube.

6. If difficulties arise in the third amplification, phenol/chloroform extraction and precipitation of the gel slice eluate may help to cleanup and concentrate PCR 2 products. Maximizing the amount of PCR 2 product in the third round of amplification can help increase product formation. Therefore, either concentrating the PCR 2 product or increasing the volume included in the PCR 3 reaction may improve efficiency.

7. The extension time for full-length products should be proportional to the length of the product. Therefore, if the final product is larger than 1 kb, the extension time should be increased.

## Acknowledgments

This is publication 35-003 of the Wisconsin Regional Primate Research Center.

## References

1. McKnight, S. L. and Kingsbury, R. (1982) Transcriptional control signals of a eukaryotic protein-coding gene. *Science* **217**, 316–324.

2. Grundstrom, T., Zenke, M. W., Wintzerith, M., Matthes, H. W., Staub, A., and Chambon, P. (1985) Oligonucleotide-directed mutagenesis by microscale 'shotgun' gene synthesis. *Nucleic Acids Res.* **13**, 3305–3316.

3. Li, X.-M. and Shapiro, L. J. (1993) Three-step PCR mutagenesis for "linker scanning." *Nucleic Acids Res.* **16**, 3745–3748.

4. Vallette, F., Mege, E., Reiss, A., and Adesnik, M. (1989) Construction of mutant and chimeric genes using the polymerase chain reaction. *Nucleic Acids Res.* **17**, 723–733.

5. Dieffenbach, C. W., Lowe, T. M., and Dveksler, G. S. (1993) General concepts for PCR primer design. *PCR Methods Appl.* **3**, S30–S37.

6. Smith, K. D., Valenzuela, A., Vigna, J. L., Aalbers, K., and Lutz, C. T. (1993) Unwanted mutations in PCR mutagenesis: avoiding the predictable. *PCR Meth. Appl.* **2**, 253–257.

7. Varadaraj, K. and Skinner, D. M. (1994) Denaturants or cosolvents improve the specificity of PCR amplification of a G + C-rich DNA using genetically engineered DNA polymerases. *Gene* **140,** 1–5.

8. Park, Y. H. and Kohel, R. J. (1994) Effect of concentration of $MgCl_2$ on random-amplified DNA polymorphism. *Biotechniques* **16,** 652–656.

9. Chou, Q., Russell, M., Birch, D. E., Raymond, J., and Bloch, W. (1992) Prevention of pre-PCR mis-priming and primer dimerization improves low-copy-number amplifications. *Nucleic Acids Res.* **20,** 1717–1723.

# 21

# Sequence Inversion by Flip-PCR

## Judith T. Schanke

## 1. Introduction

Flip-PCR was designed to allow the sequence inversion of small regions of DNA within a regulatory region in order to examine the importance of cis-acting elements on transcriptional activity *(1)*. Flip-PCR, a modification of the two-step overlap extension method of site-specific mutagenesis *(2,3)*, requires two rounds of amplification utilizing two internal inversion primers and two wild-type external or flanking primers. The method is diagrammed in Fig. 1. Primers A and B are the flanking primers and primers C and D are the inversion primers. The first round of amplification, which separately utilizes sets of one flanking primer (Fig. 1A,B) and one inversion primer (Fig. 1C,D), results in the formation of two overlapping products. The internal primers are designed to include a region that is complementary to the template as well as a region of inverted sequence. The two internal primers (C and D) are complementary in the inverted regions and will anneal, extend, and be amplified by a second round of PCR using only the external primers (A and B). The resulting product can then be cloned directly into a PCR cloning vector or restricted within the external primers and cloned into the vector of choice. The products of the first round of amplification serve as the only template in the second round. Therefore, there is no contaminating amplification of the wild-type template to decrease cloning efficiency.

The fidelity of this method depends on the design of the inversion primers. Factors to consider with inversion primer design include the location of the primers within the template in relation to external primers and the length of the regions of complementarity and inversion, as well as other factors typically considered when designing a primer, including specificity to a unique template sequence and the absence of self complementarity *(see* ref. *5)*. Primer design is described in detail in Section 3.

From: *Methods in Molecular Biology, Vol. 67: PCR Cloning Protocols: From Molecular Cloning to Genetic Engineering* Edited by: B. A. White  Humana Press Inc., Totowa, NJ

**First Round Amplifications**

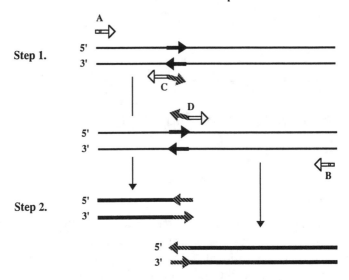

**Second Round Amplification**

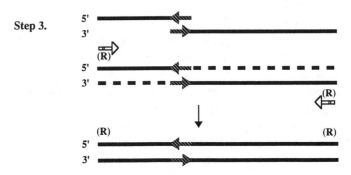

**Fig. 1.** Diagram of Flip-PCR procedure. The wild-type orientation of the inversion sequence within the template is indicated on the template by solid black arrows. Oligonucleotide primers are depicted with arrows at their annealing sites along the template. The two inversion primers (C and D) anneal to opposite strands of the DNA template. The inversion portion of the inversion primers (C and D, indicated by hatched arrows) are complementary to each other, and do not anneal to the wild-type template. They contain the template sequence (black arrows) inverted (hatched arrows). The first round of PCR involves two separate amplifications using A and C, and B and D (step 1). The products are separated from primers and template by gel purification (step 2). The purified products are mixed, denatured and allowed to reanneal. Extension of the recombinants with recessed 3' ends results in the full length inversion product which is then amplified further with the external primers A and B (step 3). Mismatches in the external primers create restriction sites (R) in the final product to facilitate subcloning.

Final product yield can be relatively low compared to wild-type amplifications because of the number of annealing possibilities of the first round products. Therefore, other known sources of competition for the second amplification should be minimized. Gel purification is used to isolate the first round PCR products from both the original template and inversion primers. This purification step decreases competition for extension and amplification of the first round products with the external primers and results in maximal production of full-length inversion products.

## 2. Materials

1. DNA template containing the sequence to be inverted (typically a subclone of a regulatory sequence in a plasmid vector).
2. Oligonucleotide primers flanking the region to be inverted. Primer A (upstream of the sequence to be inverted) in the message sense and primer B (downstream of the sequence to be inverted) in the antisense (Fig. 1). Restriction sites can be engineered into these primers in order to facilitate direct cloning of the final product into the vector of choice *(4)*.
3. Mutant internal primers containing the entire region to be inverted as well as flanking sequence complementary to the template (described in detail in Section 3.).
4. PCR reagents, including dNTP mix, reaction buffer containing $MgCl_2$, and thermal stable polymerase (*see* Chapter 1).
5. Reagents for either acrylamide or agarose gel electrophoresis, including acrylamide soution (20:1 acrylamide to *bis* mixture) and TBE buffer (89 m$M$ Tris-borate, 2 m$M$ EDTA) (*see* Note 1).
6. A method for purification of PCR products from either acrylamide or agarose gel slices, such as glass wool (*see* Note 1).

## 3. Methods

### 3.1. Primer Design

1. The standard rules of primer design apply to both the inversion primers and the flanking primers. These rules include lack of self complementarity or direct repeats and lack of single nucleotide runs at either end of the primer among others (*see* ref. *5*).
2. Flanking primers A and B may include mismatches to allow restriction endonuclease cleavage for ease of subcloning after the final amplification. These mismatches should be anchored by complementary sequence as previously described *(4)*. Alternatively, the final product can be directly subcloned into a PCR cloning vector, such as pCRII (TA cloning kit, Invitrogen, San Diego, CA) or pGEM-T (Promega, Madison, WI).
3. The design of the inversion primers is critical to the success of the method. Annealing temperatures should be determined for each portion of the primer, both the complementary portion and the inversion portion. The first round ampli-

fications rely only on the annealing temperature of the complementary portions of the internal primers. Therefore, design the complementary portion so that the annealing temperature is as high as possible (include many G and C residues); 12–20 bases of complementarity should be sufficient.

4. The inversion portion of the internal primers (C and D) should be complementary to each other (*see* Fig. 1). The second round of amplification relies on the annealing temperature of the inversion portion only, therefore this region should be long enough to provide a sufficiently high annealing temperature for specific amplification. Fifteen to twenty nucleotide-long inversions have been performed (*1*) but technically the length could be significantly extended and possibly shortened. The inversion portion of the primers should also be examined for sequence complementary to the template, after inversion. Mis-annealing of a long inversion primer could interfere with amplification.

5. When possible, the 5' end of each inversion primer should be located so that a T residue is the next base in the template sequence. If a T is not available, use a polymerase with 3' exonuclease activity. These strategies will prevent undesirable A insertions at the end of the first round of amplification (*6*) that could then be incorporated into the final product. Alternatively, the first round amplification products can be treated with T4 DNA polymerase or Klenow to remove the terminal A addition before the final amplification.

## 3.2. PCR 1: Sequence Inversion

1. Setup two to four replicates of a standard 50-µL reaction containing 100–500 ng template DNA (containing the target sequence for inversion), 20 pmol of each primer (A and C in one set of reactions and B and D in the other set), 50 µM of each dNTP, thermal stable polymerase and the appropriate buffer with the empirically determined amount of $MgCl_2$, typically 1–3 mM (*see* Note 2).

2. Amplify using the following profiles and the hot start technique; warm samples to 94°C before the addition of the final reaction component (*see* Note 2): 94°C for 3 min (initial denaturation); 5 cycles: 94°C for 30 s, 45°C for 45 s (*see* Note 3), and 72°C for 60 s (*see* Note 4); 25 cycles: 94°C for 30 s, 55°C for 45 s, and 72°C for 60 s; and 72°C for 7 min (final extension).

3. Pool the replicate reaction products from the first round of amplification and gel purify from both primers and template by gel electrophoresis (*see* Note 1) Resolve products on a native polyacrylamide gel in 1X TBE and stained with ethidium bromide. Use an 8% gel for products in the range of 60–400 nt and a 4% gel for products from 400–1000 nt in length. Run the gel at 200 V for approx 2 on 16-cm plates.

4. Stain the gel on one of the plates in 0.5 µg/mL of ethidium bromide for 10–2 min. Localize the products with UV light and isolate gel slices containing th product.

5. Crush and soak the gel slices in 200 µL $dH_2O$ for about 2 h with shaking.

6. Spin through glass wool to remove gel fragments. Construct glass wool spin col umns by puncturing the bottom of a 500-mL Eppendorf tube with a 23-G needle

Insert glass wool to cover the bottom one-fourth of the tube and place the tube into a 1.5-mL Eppendorf collection tube. Place the crushed gel slice in the glasswool containing tube and centrifuge the column for 10 s at 5000 rpm (about 2000$g$).

7. Collect the flow-through in the lower tube and use directly.

## 3.3. PCR 2: Overlap Extension

1. Set up four or five standard reactions with 5 μL (of 200 μL) of each first round product as template and 20 pmol of each flanking primer (Fig. 1A,D).
2. Amplify using the following profiles and the hot start technique: (initial denaturation) 94°C for 3 min; 30 cycles: 94°C for 30 s, and 55°C for 45 s (*see* Note 3) and 72°C for 60 s (*see* Note 4); and 72°C for 7 min (final extension).
3. The final product is now ready for direct TA cloning, restriction and ligation or gel purification followed by restriction and ligation (*see* chapters in Part II).

## 4. Notes

1. Gel purification: A number of methods exist to isolate DNA from agarose or polyacrylamide gel slices, including electroelution using DEAE paper, glass milk kits, such as GeneClean II (Bio 101, Vista, CA), or by quick freeze or crush and soak methods followed by spin columns. The efficiency of each method varies by the size of PCR product being purified, small products (<500 bp) are poorly recovered from agarose, therefore product length should be considered when choosing a method.
2. Reactions may need to be optimized for specific product amplification, especially if low annealing temperatures are used. The inclusion of cosolvents, such as glycerol or DMSO *(7)*, optimization of MgCl$_2$ concentration *(8)*, and the use of hot start procedures *(9)* may all improve product amounts and decrease non-specific annealing and amplification (*see* Chapter 1).
3. Amplification annealing temperatures used should be determined by the actual primer sequences used (although the conditions described have been successfully used with a number of different primers *[1]*). The annealing temperature used for the 25 cycles should be the lowest primer annealing temperature calculated in the standard manner (2°C for each A and T, and 4°C for each G and C, then deduct 4°C). The only nucleotides used to calculate the annealing temperature are those complementary to the template or first round product. The annealing temperature used for the initial five cycles of the first round of amplification should be lower than the calculated annealing temperature. This has been demonstrated to increase first round product amount without sacrificing final sequence integrity *(1)*.
4. The extension times noted have been used for the amplification of products within the 50–500 bp range. As a rule, elongation times should be proportional to product length, therefore if the product is larger than 1 kb this extension time should be increased.

## Acknowledgments

I gratefully acknowledge the assistance of Lynn M. Quam and Brian G. Van Ness in the development of this method. This is publication 35-002 of the Wisconsin Regional Primate Research Center.

## References

1. Schanke, J. T., Quam, L. M., and Van Ness, B. G. (1994) Flip-PCR for DNA sequence motif inversion. *Gene* **16,** 414, 415.
2. Ho, S. N., Hunt, H. D., Horton, R. M., Pullen, J. K., and Pease, L. R. (1989) Site-directed mutagenesis by overlap extension using the polymerase chain reaction. *Gene* **77,** 51–59.
3. Higuchi, R., Krummel, B., and Saiki, R. (1988) A general method of in vitro preparation and specific mutagenesis of DNA fragments: study of protein and DNA interactions. *Nucleic Acids Res.* **16,** 7351–7367.
4. Vallette, F., Mege, E., Reiss, A., and Adesnik, M. (1989) Construction of mutant and chimeric genes using the polymerase chain reaction. *Nucleic Acids Res.* **17,** 723–733.
5. Dieffenbach, C. W., Lowe, T. M., and Dveksler, G. S. (1993) General concepts for PCR primer design. *PCR Methods Appl.* **3,** S30–S37.
6. Smith, K. D., Valenzuela, A., Vigna, J. L., Aalbers, K., and Lutz, C. T. (1993) Unwanted mutations in PCR mutagenesis: avoiding the predictable. *PCR Methods Appl.* **2,** 253–257.
7. Varadaraj, K. and Skinner, D. M. (1994) Denaturants or cosolvents improve the specificity of PCR amplification of a G + C-rich DNA using genetically engineered DNA polymerases. *Gene* **140,** 1–5.
8. Park, Y. H. and Kohel, R. J. (1994) Effect of concentration of $MgCl_2$ on random-amplified DNA polymorphism. *Biotechniques* **16,** 652–656.
9. Chou, Q., Russell, M., Birch, D. E., Raymond, J., and Bloch, W. (1992) Prevention of pre-PCR mis-priming and primer dimerization improves low-copy-number amplifications. *Nucleic Acids Res.* **20,** 1717–1723.

# 22

## PCR Site-Directed Mutagenesis Using *Pyrococcus* sp GB-D Polymerase Coupled to a Rapid Screening Procedure

*Application to a β-Glucanase Gene*

**Jaume Pons, Antoni Planas, Miquel Juncosa, and Enrique Querol**

### 1. Introduction

PCR methodology is one of the fastest available procedures for site-directed mutagenesis *(1,2)*. However, it has been criticized for a lack of reliability because of unwanted mismatches produced during the PCR reaction *(3,4)*. In the present protocol, we describe an improvement on the efficiency of site-directed mutagenesis by PCR using the *Pyrococcus* species GB-D polymerase instead of the commonly used *Thermus aquatiqus (Taq)* polymerase. *Taq* polymerase lacks a 3'–>5' proofreading exonuclease activity that is not crucial for several PCR applications, but is advisable for site-directed mutagenesis experiments. Some thermophilic DNA polymerases have this activity, among them the *Thermococcus litoralis* and the *Pyrococcus* species GB-D enzymes. A 10-fold higher efficiency has been reported for these enzymes over that observed for *Taq* polymerase *(5)*. PCR site-directed mutagenesis is specially suitable for protein engineers when it is coupled to a screening procedure directly performed on the transformant plates. In such cases the procedure is rapid (3 d from mutagenic primers to selection of clones) and efficient (98–100% of successful mutagenesis).

In this chapter, we describe the application of PCR site-directed mutagenesis using *Pyrococcus* sp GB-D polymerase coupled to a rapid screening procedure on a 1,3–1,4-β-D-glucan 4-glucanohydrolase (β-glucanase,

From: *Methods in Molecular Biology, Vol. 67: PCR Cloning Protocols: From Molecular Cloning to Genetic Engineering* Edited by: B. A. White  Humana Press Inc., Totowa, NJ

EC3.2.1.73). β-glucanase is a glycosidase that hydrolyses β-glucans containing mixed b-1,3 and b-1,4 linkages as lichenin and barley β-glucan. We have characterized the enzymatic activity and cloned the corresponding gene from *Bacillus licheniformis (6)*. In order to gain insight into the structure/function of the enzyme by protein engineering, a search of the essential catalytic residues was performed by a screening mutagenesis strategy. It is generally accepted that the acid-base catalytic mechanism of glycohydrolases lays on acidic residues: Asp or Glu *(7)*. Thus, the 14 individual Asp and Glu residues of the *B. licheniformis* enzyme conserved among the different *Bacillus* β-glucanases were mutated to their respectively isosteric and uncharged Asn and Gln residues, with the aim of finding those single mutants with completely abolished glycosidase activity *(8)* in a rapid screening method. β-glucanase is secreted to the extracellular medium by the host expressing the gene; those transformant colonies expressing an active enzyme will degrade the β-glucan polysaccharide contained in the plate that, after staining with Congo red, will appear as a colorless halos where β-glucan has been depolymerized. The assay allows rapid selection of clones and evaluation of the mutagenesis efficiency before DNA sequencing and biochemical characterization of the mutant enzyme. The plate screening and enzyme analysis of the mutants indicated that only one mutant, E138Q, yielded an inactive enzyme, and E134Q and D136N retained <0.5% of the wild-type activity. Together with the three-dimensional analysis reported for another glucanase and epoxyalkyl inibitors *(9,10)*, we assigned Glu138 as the general acid catalyst and Glu 134 as the catalytic nucleophile *(8,11)*.

Figure 1 illustrates the site-directed mutagenesis procedure in two PCR amplifications. The gene is flanked by two single restriction sites and two primers defining the region to be amplified by PCR (one upstream and the other downstream of the gene). In the first PCR reaction this template is amplified by one flanking primer and the mutagenic primer (*see* Note 1), each one matching in a different strand to yield a fragment containing the desired mutation. In the second PCR amplification the first PCR product is used as a primer, together with the other flanking primer, to yield the whole insert containing the desired mutation. The mutated insert is digested in the flanking restriction sites, subcloned in a fresh expression vector and transformed again to the expression host to proceed with the screening method.

The coupling of the PCR site-directed mutagenesis procedure with a rapid screening assay on plates is one advantage of this methodology, allowing the preparation of several mutants in parallel. It is of general use for the search of essential or important residues for enzyme activity in glycosidases that catalyze the hydrolysis of polysaccharide substrates, such as cellulases, cellobiohydrolases, amylases, and glucanases of different specificity, or in general for any enzyme for which a colorimetric assay on plates can be used.

**SCHEME**

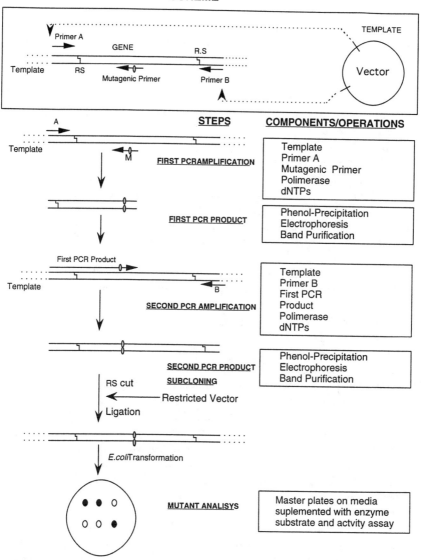

**Fig. 1.** General scheme of the procedure. Site-directed mutagenesis strategy. A and B are the gene-flanking primers and M is the mutagenic primer. RS: restriction enzyme sites.

Briefly, site-directed mutagenesis in two PCR amplifications coupled with a rapid screening method comprises the following steps:

1. Site-directed mutagenesis in two PCR amplifications as described above.

2. Digestion of the flanking restriction sites and subcloning to a fresh expression vector.

3. Transformation of the previous construct on the expression host, to proceed with the screening method directly on transformation plates, or in master-replica plates.

## 2. Materials

1. DNA template (gene to be mutated). It should be clean and well quantified. Prepare a stock solution at 25 ng/μL and use in reaction mixture at 0.5 ng/μL.
2. Deep Vent® polymerase (New England Biolabs, Beverly, MA).
3. Deep Vent Buffer (provided in 10X presentation by New England Biolabs) 1X: 10 m$M$ KCl, 10 m$M$ (NH$_4$)$_2$SO$_4$, 20 m$M$ Tris-HCl, pH 8.8, 2 m$M$ MgSO$_4$, and 0.1% Triton X-100. 100 m$M$ MgSO$_4$ solution is provided in case a higher concentration of magnesium is needed in the reaction (*see* Note 2).
4. dNTPs: Prepare a stock solution 10 m$M$ in each dNTP (Boehringer Mannheim, Indianapolis, IN) in deionized H$_2$O. Store at –20°C. Use in reaction mixture at 200 μ$M$ (*see* Note 3).
5. Oligonucleotide primers (synthesized by Clontech, Palo Alto, CA), in a stock solution of 5 pmol/μL. Store at –20°C. Use in reaction mixture at 0.2 pmol/μL (*see* Note 1).
6. Phenol/chloroform/isoamylalcohol (25:24:1) (for preparation *see* ref. *12,* pp. B.5).
7. 3$M$ Sodium acetate, pH 5.2.
8. Ethanol: 100 and 70% in water.
9. Polyacrylamide 0.25% (*see* Note 4).
10. Agarose gel components (ref. *12,* pp. 6.3, and Chapter 1).
11. Ethidium bromide. Prepare a 10 mg/mL stock solution and use at 0.5 μg/mL.
12. GeneClean kit (Bio101, La Jolla, CA).
13. Restriction endonucleases, T4 DNA ligase, and deoxyribonucleoside triphosphates (dNTPs). All are from Boehringer Mannheim (Indianapolis, IN).
14. *E. coli* TG1 (*supE hsd*Δ5 *thi* Δ(*lac-proAB*) F'[*traD36 proAB*⁺ *laclq lacZ*ΔM15]) competent cells, prepared by the calcium chloride method (ref. *12,* pp. 1.82) and stored at –80°C in 15% glycerol as a cryoprotector. Once a tube is thawed it should not be reused.
15. Ampicilline (Boehringer Mannheim). Prepare a 100 mg/mL stock solution in deionized H$_2$O and sterilize by 0.22-μm filter (Millex from Millipore, Bedford, MA). Store at –20°C. Use at 50 μg/mL in liquid media and plates after autoclaving (allow the medium to cool to 50°C before adding the antibiotic). Add to agar before plating.
16. Barley β-glucan (Sigma, St. Louis, MO). Prepare a 5 mg/mL stock solution in deionized H$_2$O. Store at +4°C. Use in plates at 0.04% (w/v).
17. IPTG (Isopropylthio-β-ᴅ-galactoside) (Boehringer Mannheim). Prepare a 200 mg/mL stock solution in deionized H$_2$O and sterilize by 0.22-μm filter (Millex from Millipore). Store at –20°C. Add at 0.5 m$M$ in plates (allow the medium to cool to 50°C before adding it).

18. X-gal (5-bromo-4-chloro-3-indolyl-β-D-galactoside) (Boehringer Mannheim). Prepare a 20 mg/mL stock solution in dimethylformamide. It is not necessary to sterilize the solution. Store in the dark (wrapped in aluminum foil) at –20°C. Add at 120 μg/mL in plates (allow the medium to cool to 50°C before adding it).

19. LB medium (ref. *12*, pp. A.1). For culture in liquid, add the following to 950 mL of $H_2O$: 10 g of Bacto-tryptone, 5 g of bacto-yeast extract, and 10 g of NaCl. Adjust the pH to 7.0 with $5N$ NaOH and the volume to 1 L with deionized $H_2O$. Autoclave for 20 min at 15 lb/sq. in. on liquid cycle. For culture on plates: Add 15 g/L of bacto-agar to the media described above just before autoclaving. For activity plates add β-glucan to 0.04% after autoclaving. For blue-white selection plates, add IPTG (to 0.5 m$M$) and X-gal (to 120 μg/mL) after autoclaving. Store the plates at +4°C.

20. Congo red (Merck, Rahway, NJ) is prepared and used at 1 mg/mL in deionized $H_2O$.

## 3. Methods

### 3.1. Site-Directed Mutagenesis

Mutate the template in a series of two PCR amplifications as described in Fig. 1.

### 3.1.1. First PCR Amplification

1. Prepare the reaction mixture by adding the following: 10 μL 10X DeepVent buffer, 2 μL dNTP 10 m$M$ solution, 2 μL DNA template (25 ng/mL), 4 μL mutagenic primer (5 pmol/μL), 4 μL flanking primer 1 (5 pmol/μL), and autoclaved $H_2O$ to a vol of 99 μL. Keep all the components and the mixture on ice (*see* Note 5).

2. Prewarm the thermocycler at 94°C.

3. Add 1 μL of DeepVent polymerase to the reaction mixture and overlay with 50 μL of mineral oil.

4. Amplify by PCR using the following cycle profile: 94°C for 1 min (initial denaturation); 30 main cycles: 94°C for 1 min 15 s (denaturation), 55°C for 1 min (annealing) (*see* Note 6), 72°C for 1 min/kb (elongation); and 72°C for 5 min (final extension).

5. Recover each PCR product by inserting a standard 200 μL micropipet tip through the mineral oil layer and draw up the sample. It is important to avoid any oil extraction with the sample. The volume removed will be about 90 μL.

6. Purify the DNA by adding 90 μL of phenol/chloroform/isoamylalcohol to 90 μL of sample. Mix by vortexing and spin at approx 10,000$g$ for 2 min in a microfuge, and transfer the aqueous (upper) layer to a fresh tube.

7. Precipitate the 90-μL sample of DNA by adding 9 μL of 3$M$ sodium acetate, 8 μL of 0.25% polyacrylamide (as a carrier, *see* Note 4), and 270 μL of 100% ethanol (stored at –20°C). Mix by inverting the tube, and leave 10 min at –80°C.

8. Spin 10 min at approx 10,000$g$ and remove the supernatant by gentle aspiration. Rinse the DNA pellet with 500 μL of 70% ethanol (stored at room temperature), spin 10 min at 10,000$g$, and remove the supernatant by gentle aspiration. Stand

the tube in a inverted position on a paper towel to allow all the fluid to drain away, and remove any drops of fluid adhering to the walls of the tube. Allow the pellet to air dry for 10 min.

9. Redissolve the pellet in 10 µL of autoclaved $H_2O$.
10. Agarose electrophoresis of the sample (*see* ref. *12,* pp. 6.9): Dissolve 1.2% (*see* Note 7) standard high-melting-point agarose in 1X TAE buffer by boiling, allow to cool to 55°C, and add 0.5 µg/mL of ethidium bromide. Pour the gel and let it solidify. Mix each entire PCR product with an electrophoresis loading buffer and electrophorese in TAE buffer with 0.5 µg/mL of ethidium bromide. Electrophoresis should be carried out until each PCR product has traveled at least 4 cm, in order to separate adequately the PCR products from the supercoiled plasmid template. Each PCR product will typically contain more than 200 ng of DNA, and therefore will be easily visualized by UV light.
11. Remove the band of interest with a razor blade and extract each PCR product from the agarose using GeneClean. Suspend each PCR product in 30 µL of $H_2O$. Since GeneClean recovery is 25–90%, it is recommended to quantify the final concentration by an agarose gel electrophoresis of 2 µL of each PCR product.

### 3.1.2. Second PCR Amplification

1. Prepare the reaction mixture by adding the following: 10 µL 10X DeepVent buffer, 10 µL dNTPs (10 m*M*), 2 µL DNA template (25 ng/µL), 20 pmol (approx 50% of total product) first PCR product, 4 µL flanking primer 2 (5 pmol/µL), *x* µL $Mg(SO_4)_2$ (*see* Note 2), and autoclaved $H_2O$ to a volume of 99 µL. Keep all components and the mixture on ice (*see* Note 5).
2. Prewarm the thermocycler at 94°C.
3. Add to the reaction mixture 1 µL of DeepVent polymerase and overlay with 50 µL of mineral oil.
4. Amplify by PCR using the following cycle profile: 94°C for 1 min (initial denaturation); 30 main cycles: 94°C for 1 min 15 s (denturation), 45°C for 1 min 30 s (annealing) (*see* Note 6), 72°C for 1 min/kb (elongation); and 72°C for 10 min (final extension).
5. Repeat steps 5–11 (Section 3.1.1.) with the minor modifications described below:
   a. Step 7: Polyacrylamide is usually not required for DNA precipitation after the second PCR amplification because of the length of the DNA fragment.
   b. Step 10: Use an adequate concentration of agarose for the length of the desired band (*see* Note 7). Since the amount of the desired band after the second PCR amplification is lower than that obtained after the first PCR, we recommend destaining by soaking the gel in water for 20 min to reduce the background fluorescence caused by unbound ethidium bromide.
   c. Step 11: Suspend each PCR product after GeneClean in 9.5–17 µL (*see* Section 3.1.2.). Since the amount of DNA is low, do not quantify it.
6. Restriction enzyme modification of the PCR products:
   a. One step restriction enzyme modification (for restriction enzymes that use the same or compatible buffers): Add 2 µL of restriction buffer and 0.5 µL of each

restriction enzyme to 17 μL of each PCR product. Incubate at the correct temperature (usually 37°C) for 2 h.

   b. Two step restriction enzyme modification (for restriction enzymes that do not use the same or compatible buffers): Add 1 μL of restriction buffer and 0.5 μL of restriction enzyme to the 9.5 μL of each PCR amplification and incubate at the correct temperature for 1 h. Add $H_2O$ to 90 μL, repeat steps 6–9 (Section 3.1.1.), and repeat digestion with the second restriction enzyme as described above.

7. Similarly, cut the vector with the appropriate enzymes.
8. Purify the band of restricted PCR product and fresh vector by agarose gel electrophoresis as described above. Resuspend each band after GeneClean in 10 μL.
9. Mix an equimolar amount of the vector with the insert in a final vol of 17 μL. Add 2 μL of ligation buffer (which must contain the appropriate concentration of ATP) and 1 U of T4 ligase (for more information about the ligation reaction *see* ref. *12*, pp. 1.63). Ligate overnight at 16°C.
10. Tranform 100 μL of chilled *E. coli* competent cells (prepared by calcium chloride method, ref. *12*, pp. 1.83) by adding 10 μL of ligation mixture and store the tubes on ice for 30 min. Transfer the tubes to a rack placed in a water bath that has been prewarmed to 42°C. Leave the tubes in the rack for exactly 90 s. Do not shake the tubes. Rapidly transfer the tubes to an ice bath. Allow the cells to chill for 1–2 min. Add 300 μL of 2YT medium to each tube and leave for 30 min at 37°C for plasmid resistance expression (shaking is not required).
11. Plate 100 μL and 300 μL of the transformation mixture on agar plates supplemented with the appropriate antibiotic. Leave the plates at room temperature until the liquid has been absorbed. Invert the plates and incubate at 37°C. Colonies should appear in 12–16 h.

### 3.2. Screening Method

This part of the protocol will change for any particular enzyme, but is applicable to any enzyme for which a colorimetric assay on plates can be used. Here, as an example, the screening procedure for β-1,3-1,4-glucanases is described. As the efficiency of the mutagenic method is very high *(14)*, the phenotype of the majority of colonies will be the phenotype of the mutant. This method is very convenient when the researcher is looking for essential residues of the enzyme, which will inactivate it.

1. Pick up about 50 individual colonies and transfer to duplicate master plates containing the appropriate antibiotic. One of the plates should contain the substrate of the enzyme (in this case Barley β-glucan). Leave colonies to grow for 8 h at 37°C.
2. Develop the master plate by staining the plate containing β-glucan by pouring on it 20 mL of Congo red at 1 mg/mL *(13)*. Leave 15 min in an orbital shaker and remove the supernatant. The active colonies will show clear halos on a red background of β-glucan-Congo red complex (*see* Fig. 2).

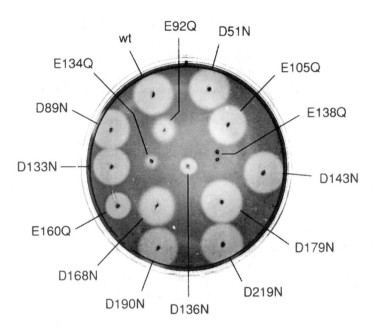

Fig. 2. Activity on plate of the mutant clones. Congo red activity halos on plate of the different recombinant *E. coli* cells expressing the wild-type and mutant β-glucanase genes. The plate contained 0.04% barley β-glucan.

## 4. Notes

1. Primer design:
   a. Primers were designed according to the literature reported elsewhere for PCR mutagenesis *(15,16)*.
   b. Mutagenic primer: To avoid corrections of the desired mutations by the 3'→5' proofreading exonuclease activity of DeepVent polymerase, 13–15 matched bases flanking the mismatch in both sides are recommended.
   c. Orientation of the mutagenic primer (going upstream or going downstream). It is important to get a first PCR product long enough for an easy purification (longer than 200 bp) but as short as possible to avoid annealing problems in the second PCR amplification.
   d. Flanking primers: a 17–20-mer primer with 0% mismatch is enough. General sequencing primer of most cloning vectors are appropriate.
2. MgSO$_4$ concentration: The 1X DeepVent buffer is 2 m$M$ MgSO$_4$, which appears to be appropriate for all the first PCR amplifications and most of the second PCR reactions. If the desired band is not observed after the second PCR amplification, increasing the concentration to 4, 6, and 8 m$M$ usually solves the problem.
3. Use 200 μ$M$ dNTPs in 1X reaction buffer for elongation until 3 kb, for longer elongation increase the dNTPs concentration to 400 μ$M$.

**Table 1**
**Agarose Gel Electrophoresis**

| Agarose concentration, % W/V | Efficient range of separation of linear DNA molecules, kb |
|---|---|
| 0.3 | 5–60 |
| 0.6 | 1–20 |
| 0.7 | 0.8–10 |
| 0.9 | 0.5–7 |
| 1.2 | 0.4–6 |
| 1.5 | 0.2–3 |
| 2.0 | 0.1–2 |

4. 0.25% polyacrylamide is used as a carrier for precipitation of short length DNA fragments (below 1 kb). Prepare it by adding 1/100 vol of 10% ammonium persulfate and 1/1000 vol of TEMED to a solution of 5% acrylamide in 40 m$M$ Tris, 20 m$M$ sodium acetate, and 2 m$M$ EDTA, pH 7.8. Leave 30 min at room temperature for polymerization. Add sodium acetate to 0.3$M$ and precipitate by adding 2.5 vol of 100% ethanol. Spin at 12,000 rpm and dry under vacuum. Resuspend the pellet in water to a concentration of 0.25%. Store it at –20°C.

5. Because DeepVent polymerase has 3'–>5' proofreading exonuclease activity, it may correct the desired mutation. To prevent primer degradation and correction of the mutation during the mixing time, it is important to keep the reaction tube chilled (DeepVent polymerase has no appreciable activity at 4°C).

6. The annealing temperature for the first PCR amplification must be 2°C below the lowest melting temperature (thermodynamic) of the primers (usually the flanking primer). Using the Forward and the Reverse universal primers as flanking primers, 55°C seems to be an appropriate temperature. The annealing temperature for the second PCR amplification is difficult to set. A good starting point is 10°C below the annealing temperature of the first PCR amplification. If no band is observed, even after increasing the MgSO$_4$ concentration (*see* Note 2), try to find a better annealing temperature increasing and decreasing it by 2°C steps.

7. The amount of agarose in DNA gels is chosen according to the length of the desired band (Table 1).

# References

1. Landt, O., Grunert, H. P., and Hahn, U. (1990) A general method for rapid site-directed mutagenesis using the polymerase chain reaction. *Gene* **96,** 125–128.
2. Barik, S. (1993) Site-directed mutagenesis by double polymerase chain reaction, in *PCR Protocols: Current Methods and Applications* (White, B. A., ed.), Humana, Totowa, NJ, pp. 277–286.
3. Karlovsky, P. (1990) Misuse of PCR. *Trends Biochem. Sci.* **15,** 419.

4. Sayers, J. R., Krekel, C., and Eckstein, F. (1992) Rapid high-efficiency site-directed mutagenesis by the phosphorothioate approach. *Biotechniques* **14,** 592–596.
5. Tindall, K. R. and Kunkel, T. A. (1988) Fidelity of DNA synthesis by the thermus aquaticus DNA polymerase. *Biochemistry* **27,** 6008–6013.
6. Lloberas, J., Pérez-Pons, J. A., and Querol, E. (1991). Molecular cloning, expression and nucleotide sequence of the endo-β-1,3-1,4–D-glucanase gene from *Bacillus licheniformis*. Predictive structural analyses of the encoded polypeptide. *Eur. J. Biochem.* **197,** 337–343.
7. Svensson, B. and Søgaard, M. (1993) Mutational analysis of glycolase function. *J. Biotechnol.* **29,** 1–37.
8. Juncosa, M., Pons, J., Dot, T., Querol, E., and Planas, A. (1994) Identification of active site carboxylic residues in *Bacillus licheniformis* endo-1,3-1,4-D-glucan 4-glucanohydrolase by site-directed mutagenesis. *J. Biol. Chem.* **269,** 14,530–14,535.
9. Keitel, T., Simon, O., Borriss, R., and Heinemann, U. (1993) Molecular and active-site structure of a *Bacillus* 1,3-1,4-β-glucanase. *Proc. Natl. Acad. Sci. USA* **90,** 5287–5291.
10. Høj, P. B., Condron, R., Traeger, J. C., McAuliffe, J. C., and Stone, B. A. (1992) Identification of glutamic acid 105 at the active site of *Bacillus amyloliquefaciens* 1,3-1,4-β-D-glucan 4-glucanohydrolase using epoxide-based inhibitors. *J. Biol. Chem.* **267,** 25,059–25,066.
11. Planas, A., Juncosa, M., Lloberas, J., and Querol, E. (1992) Essential catalytic role of Glu134 in endo-β-1,3-1,4-D-glucan 4-glucanohydrolase from *B. licheniformis* as determined by site-directed mutagenesis. *FEBS Lett.* **308,** 141–145.
12. Sambrook, J., Fritsch, E. F., and Maniatis, D. F. (1989) *Molecular Cloning: A Laboratory Manual.* Cold Spring Harbor Laboratory, Cold Spring Harbor, N.Y.
13. Teather, R. M. and Wood, P. J. (1982) Use of a Congo Red-polysaccharide interaction in enumeration and characterization of cellulolytic bacteria from bovine rumen. *Appl. Environ. Microbiol.* **43,** 777–780.
14. Juncosa, M., Pons, J., Planas, A., and Querol, E. (1994) Improved efficiency in site-directed mutagenesis by PCR using a *Pyrococcus* sp GB-D polymerase. *Biotechniques* **16,** 820,821.
15. Kuipers, O. P., Boot, H. J., and de Vos, W. M. (1991) Improved site-directed mutagenesis method using PCR. *Nucleic Acids Res.* **19,** 4558.
16. Sharrocks, A. and Shaw, P. E. (1992) Improved primer design for PCR-based, site-directed mutagenesis. *Nucleic Acids Res.* **20,** 1147.

# 23

## Using the SELEX Combinatorial Chemistry Process to Find High Affinity Nucleic Acid Ligands to Target Molecules

**Craig Tuerk**

### 1. Introduction

The SELEX (Systematic Evolution of Ligands by EXponential enrichments) combinatorial chemistry process is a procedure by which nucleic acid ligands of high affinity can be isolated against a molecular target *(1)*. A wide variety of target molecules have been used successfully in the SELEX process, including nucleic acid binding proteins *(1–19)*, nucleic acid enzymes *(1,20–25)*, proteins with no known function involving nucleic acids *(26–31)*, a peptide *(32)*, an RNA hairpin *(33)*, and small molecules, such as organic dyes *(34)*, amino acids *(35,36)*, cyanocobalamin *(37)*, theophylline *(38)*, and ATP *(39,40)*. The procedure involves:

1. The synthesis of a template containing a variable region flanked by 5' and 3' fixed regions that enable amplification by PCR;
2. Conversion of the template to an amplified library of candidate ligands;
3. A binding reaction in which a fraction of the candidate ligands are bound by the target molecule;
4. Separation of target molecule/ligand complexes from unbound candidate ligands;
5. Purification of the nucleic acid ligands from the separated complexes; and
6. Replication of the nucleic acid ligand population.

The steps of selection and amplification are repeated until the pool of nucleic acids is dominated by sequences of high affinity for the target molecule. Individuals are cloned from this population and sequenced. The sequences can be analyzed to extract a consensus description of nucleic acid ligands for the target molecule. Figure 1 depicts a cycle of the SELEX process for RNA libraries for steps 1–3, and 6.

From: *Methods in Molecular Biology, Vol. 67: PCR Cloning Protocols: From Molecular Cloning to Genetic Engineering* Edited by: B. A. White  Humana Press Inc., Totowa, NJ

One cycle of SELEX

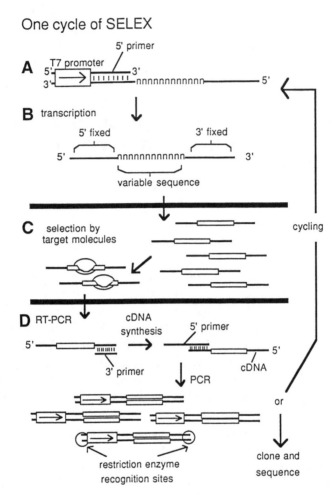

**Fig. 1.** Schematic representation of SELEX. **(A)** A dsDNA variable template is transcribed in vitro to produce RNAs with 5' and 3' fixed regions flanking a variable region. **(B)** The population is subjected to selection for binding to a protein by partitioning RNA-protein complexes. **(C)** The 3' primer is annealed to selected RNAs at the 3' fixed region sequence and a cDNA is synthesized from the RNA template using reverse transcriptase. **(D)** During the first found of PCR the 5' primer anneals at the complement of the 5' fixed region sequence in the cDNA so that a complement to the cDNA strand can be synthesized. An important feature of the 5' primer is the T7 promoter sequence at its 5' end. Ther is no complement to which it can anneal on the cDNA, as shown, but it must be present to begin the next transcription in SELEX. (This cDNA as illustrated can also serve as a model for the original variable oligonucleotide to be chemically synthesized that van be PCR'd to provide the input variable template appearing in A as described in text.) The cycle is completed by PCR of the cDNA to provide the dsDNA template for step A.

## 2. Materials

### 2.1. Reagents

1. T7 RNA polymerase (US Biochemical, Cleveland, OH).
2. *Taq* DNA polymerase (Perkin-Elmer/Cetus, Norwalk, CT).
3. AMV reverse transcriptase (Life Technologies [BRL], Gaithersburg, MD).
4. RNase-free DNase I (Pharmacia, Piscataway, NJ).
5. Nucleotide triphosphates (Pharmacia).
6. $\alpha$-[$^{32}$P]-ATP, 3000 Ci/mmol (New England Nuclear, Boston, MA).
7. Thermocycler (Perkin Elmer/Cetus).
8. Filtration manifold (Millipore, Bedford, MA).
9. Nitrocellulose filters, 25 mm, 0.45 µm pore size.
10. Fluorescent labels (Stratagene, La Jolla, CA).
11. Template oligonucleotide (*see* Note 1). 5'-CCGGATCCGTTGATATAAAATTC-[50n]TTGATTGTGGTGTTGGCTCCC-3'.
12. 5' primer oligonucleotide-5'-CCGAAGCTTCTAATACGACTCACTATAGGGA-GCCAACACCACAATCAA-3'.
13. 3' primer-5'-CCGGATCCGTTGATATAAAATTC-3'.

### 2.2. Buffers and Solutions

1. Binding buffer: 200 m$M$ potassium acetate, 50 m$M$ Tris-HCl, pH 7.7, 10 m$M$ DTT.
2. T7 buffer: 10X: 0.5$M$ Tris-HCl, pH 8.3, 0.6$M$ NaCl, 0.06$M$ magnesium acetate, 0.1$M$ DTT.
3. RT buffer: 10X: 0.5$M$ Tris-HCl, pH 8.3, 0.6$M$ NaCl, 0.06$M$ magnesium acetate, 0.1$M$ DTT.
4. *Taq* buffer: 10X: 0.5$M$ KCl, 0.1$M$ Tris-HCl, pH 8.4, 75 m$M$ MgCl$_2$, 0.5 mg/mL BSA.
5. dNTPs: 10X: 5 m$M$ each of dA-, dC-, dG-, and dTTP.
6. NTPs. 10X: 5 m$M$ each of A-, C-, G-, and UTP.
7. Carrier tRNA (Sigma, St. Louis, MO) at 20 mg/mL in H$_2$O (extract twice with phenol and once with ether).
8. 50:50 v/v Ethanol/isopropanol.
9. 2 m$M$ EDTA.
10. 3$M$ Na acetate, pH 5.2 (with glacial acetic acid) and pH 7.0.
11. 50 m$M$ Tris-acetate, pH 7.7.
12. Phenol (equilibrated with 0.1$M$ NaOAc, pH 5.2).
13. 7$M$ urea: Prepare fresh (<2 wk old). Keep this mixture at room temperature to avoid recrystallization of the urea.
14. Chloroform.
15. Reagents for ethanol precipitation of nucleic acid.

## 3. Methods

### 3.1. Template Design

The templates used in the SELEX process must have two regions of fixed sequence (which are labeled arbitrarily as the 5' fixed region and the 3' fixed region in Fig. 1B) flanking a variable sequence. During synthesis of the oligonucleotide template population, the variable region (illustrated as a run of $n$s in Fig. 1A,B) is created by mixing the phosphoramidite A, G, C, and T reagents in equal amounts during synthesis to effect library diversity as a function of $4^n$, where $n$ is the number of variable positions (*see* Note 1). The fixed regions serve as 3' fixed region or encode 5' fixed region primer annealing sites during the steps of PCR and RT-PCR amplification. If T7 RNA polymerase is used to generate a single-stranded "RNA" library, a T7 promoter sequence is added to the 5' fixed sequence. (The quotation marks refer to the inadequate description for 2'-modified nucleotides that can be incorporated by T7 RNA polymerase.) The fixed region length can vary from 7 *(25)* to 30 *(1)* nucleotides depending on the application, but in general 18 nucleotides is a workable average. The addition of restriction enzyme recognition sites with 3–4 additional nucleotides at the ends of the template (as shown for one of the double-stranded templates in Fig. 1F facilitates the necessary cloning of ligand sequences once a population of desired or acceptable affinity for the target has been isolated.

## 3.2. Initial Pool Setup

1. Synthesize the single-stranded oligonucleotides (template, 5' primer, 3' primer). Dissolve in a final vol of approx 200 pmol/µL.
2. Do pilot PCRs with varying numbers of cycles to produce dsDNA template with few aberrant products. The aim of these pilots is to find conditions for scale-up of the reaction for PCR of approx 500 pmol of input template. A typical condition for the pilot is: 15 pmol input template, 0.5 µL (100 pmol) of the 5' primer, 0.5 µL (100 pmol) of the 3' primer, 10 µL 10X *Taq* buffer, 10 µL 10X dNTPs, and 0.3 µL *Taq* DNA polymerase (2.5 U). Bring to 100 µL total volume with $H_2O$.
3. Perform PCR for 0, 2, 5, 7, 10, and 15 cycles of 93°C, 0.5 min; 55°C, 10 s; 72°C, 1 min.
4. Separate the PCR products on nondenaturing 8% polyacrylamide gels and visualize by ethidium bromide staining and UV illumination. The 0 cycle tube should show the position of the primers. As the number of cycles are increased a PCR product will become visible. With longer cycles, as the amount of free primer diminishes, this PCR band may also disappear to be replaced by a diffuse band of apparently higher molecular weight. This probably results from undesirable and opportunistic mispairing of individual strands within the variable region sequences between the members of the population.
5. Select the optimum number of cycles (that which produces the cleanest single band). Scale up the PCR reaction to accommodate 500 pmol of input template. Pool the large-scale reaction (typically 3 mL distributed over 30 0.5-mL tubes) extract with phenol, extract with ether, and ethanol precipitate. Resuspend the PCR product in 500 µL of TE buffer.

6. Set up the following pilot in vitro transcription reactions: 0, 1, 5, 10, and 25 μL PCR product; 10 μL 5X T7 buffer; 5 μL 10X NTPs; 5 μL α-[$^{32}$P]-ATP; 20, 19, 15, 10, 0 μL H$_2$O; and 5 μL (100 U) T7 RNA polymerase to a total volume of 50 μL.

7. Incubate at 37°C for 45 min.

8. Separate by electrophoresis on 7$M$ urea/8% polyacrylamide gel.

9. Gel purification of labeled RNA: Before exposing the gel to X-ray film attach a fluorescent label (from Stratagene) and briefly flash with UV light. After exposing the gel to X-ray film for an appropriate amount of time, the gel can be aligned with the exposure from the fluorescent label so that the exact position of the band can be determined and cut out. There is no need to dice the gel slice into small pieces. Put the gel slice into a screw-cap Eppendorf tube and freeze in a dry-ice/ethanol bath for about 15 min. Add to the gel slice (frozen or thawed) 300 μL of 2 m$M$ EDTA. Crush with the plunger of a 1-mL syringe until it is a slurry with no visible pieces. (Be careful; during crushing some of the solution can squirt past the plunger head out of the tube.) Add 50 μL of 3$M$ NaOAc (pH 7.0 for DNA and pH 5.2 for RNA). Pass through a 0.2 μm filter (Micro Filtration Systems, MFS 25) with a 5-mL syringe and rinse with 1 mL of 50:50 v/v ethanol/isopropanol. Alternatively, pass through silanized glass wool. Freeze in a dry-ice/ethanol bath, spin, and wash the pellet with 70% ethanol. Dry and resuspend in 20 μL of H$_2$O or TE. This also works well for PCR products purified from nondenaturing acrylamide gels.

10. Defining affinity parameters for the target protein: Generate a protein concentration-dependent binding curve with the nitrocellulose filter binding assay for the variable RNA population (with $^{32}$P labeled RNA with varied protein concentration, *see* ref. *1*). Such plots usually exhibit a sigmoid curve when the logarithmic values for protein concentration are plotted on the $x$-axis with a plateau at the higher concentrations indicating saturated binding. For selection, use protein concentrations that will yield 5–10% of this saturated binding (as interpolated from the binding curve) with RNA concentrations approx 10-fold higher than protein.

11. Scale up the in vitro transcription reaction conditions that were established to be optimum in step 5. Transcribe one half of the PCR reaction created in step 4 omitting the (α-$^{32}$P)-ATP, if desired.

## 3.3. Selection on Nitrocellulose Filters

1. Negative selection (*see* Note 2): Preheat the Binding buffer (1–50 mL) to 37°C for 5 min. Add the variable RNA pool to the binding buffer and syringe or suction through a nitrocellulose filter, collecting the flow-through. (This step subtracts from the pool potential binders of the support matrix independent of target protein.)

2. Selection: Add the target protein to the RNA solution and incubate for 5 min at 37°C (*see* Note 3). Place a nitrocellulose filter on a **clean** filtration apparatus and wash with 5 mL of 50 m$M$ Tris-acetate, pH 7.7, with applied suction. Suction the binding reaction through the filter and wash with 10 mL of 50 m$M$ Tris-acetate, pH 7.7. (Crosscontamination between experiments occurs very easily; therefore,

a soak of the filtration apparatus parts in concentrated nitric acid followed by $5M$ NaOH and ultrapure $H_2O$ is advised.)

3. Transfer the nitrocellulose filter to a clean glass plate and slice into four pieces with a fresh scalpel blade. Place the pieces into a screw-cap tube that contains 400 µL of phenol (equilibrated with $0.1M$ NaOAc, pH 5.2) and 400 µL of $7M$ urea. Shake, get all of the filter pieces back down into the phenol by tapping on the benchtop, and incubate at 95°C for 3 min. Allow to cool at room temperature and add 400 µL of chloroform, cap, and mix.

4. Centrifuge at full speed (preferably at room temperature) for 1 min in a microfuge. Use a pipet tip to squeeze the filter pieces into the bottom of the tube and respin for 5 min. Remove the aqueous phase (top) into a clean tube, extract with ether, and add 20 µg of carrier tRNA (20 mg/mL), 30 µL of $3M$ NaOAc, pH 5.2, and 1 mL of 50% ethanol:50% isopropanol. Cool in a dry-ice/ethanol bath for 15 min and spin full speed for 15 min in a microfuge. Resuspend the washed (80% ethanol) and dried pellet in 20 µL $H_2O$.

5. Reverse transcription: Set up a large-scale cDNA synthesis with 50 pmol of 3' primer in 50 µL volume with about 4 U of AMV reverse transcriptase (approx 0.2 µL/reaction): 50 pmol 3' primer, 5 µL 10X RT buffer, 5 µL 10X dNTPs, 20 µL of recovered RNA (all of it), 0.2 µL AMV reverse transcriptase, and 20 µL of $H_2O$ for a total volume of 50 µL.

6. Incubate at 37°C for 60 min.

7. PCR amplification: To the cDNA reaction (on ice) add: 50 pmol of the 5' primer (If using gel-purified cDNA, then add 50 pmol of 3' primer as well.), 10 µL 10X *Taq* buffer, 10 µL 10X dNTPs, and 0.3 µL *Taq* DNA polymerase (2.5 U) for a total volume of 100 µL (including cDNA reaction).

8. Conduct PCR amplification for 10–20 cycles. (The optimum number of cycles is best determined empirically by conducting 10-µL pilot PCRs with 5 µL cDNA reaction each with varying numbers of cycles, such as 10, 15, and 20 each (*see* Note 4). Phenol extract, ether extract, and ethanol precipitate the PCR product. Resuspend the washed (80% ethanol) and dried template in 20 µL $H_2O$.

9. In vitro transcription: Do the following in vitro transcription reaction: 10 µL PCR product, 10 µL 5X T7 buffer, 5 µL 20 m$M$ NTPs, and 5 µL (100 U) T7 RNA polymerase. Bring to 50 µL total volume with $H_2O$

10. Incubate at 37°C for 45 min.

11. Add 5 U of RNase-free DNase I and incubate for an extra 15 min at 37°C.

12. Phenol and ether extract, ethanol precipitate. Resuspend washed pellet in 50 µL of TE.

13. Repeat steps 1–11 for each round of the SELEX process.

### *3.4. Evaluating Progress*

The fraction of the population bound to target molecule, or the measured affinity, should increase over a number of rounds. During every third round of the SELEX process, conduct the in vitro transcription reaction with ($\alpha$-$^{32}$P)-ATP, purify the labeled RNA as in step 6, and redetermine the affinity of the target protein for labeled RNA. The purification step is especially recommended when PCR pilots indicate heterogeneity of template sizes (*see* Note 4). After each affinity determination lower the concentration of target molecule to maintain stringency in selection at the level of 5–10% of the RNA pool bound (*see* Note 3). Once the increase in binding affinity has begun during subsequent rounds of the SELEX process, it should continue until a plateau is reached (that is, subsequent affinity determinations that result in no increase in affinity). At this plateau of affinity, the number of ligands in the population may be significantly reduced, which can be confirmed by sequencing the population by dideoxynucleotide chain termination procedures.

Once a population of adequate affinity has been isolated, the population should be converted to double-stranded DNA, cut with restriction enzymes, and cloned into a suitable plasmid vector. The sequences of the cloned individuals are grouped according to primary sequence and secondary structure similarity. The judgement of any particular ligand's sequence significance should be weighed by its affinity for the target molecule. Model ligands of high affinity to the target molecule may be further analyzed by "sizing" the sequence region of optimal affinity by determining the 5' and 3' boundaries of the ligand sequence required for specific binding *(20)*. Such establishment of boundaries greatly simplifies sequence analysis and consensus description by limiting the field of considered sequence. As in the case of a natural binding sequence, the winners of a SELEX procedure may be rerandomized and subjected to further rounds of selection and amplification to determine (by observing the degree of conservation) nucleotides crucial to interaction with target molecule *(21,35,41)*. Other avenues of investigation of SELEX products include testing for effects on the function of target molecules *(21,22,24,26–31)* and chemical modification to analyze potential atomic groups involved in the interaction with target *(41,44)*.

## 4. Notes

1. The lengths of the variable region used in such libraries have ranged from 3 *(2)* to 120 *(39,40)*. The upper limit is set by the length of oligonucleotide that can be synthesized; the lower limit decreases the diversity of the library. The diversity is also constrained by the amount of nucleic acid in the first selection reaction. As a hypothetical illustration, if the size of the first selection reaction is 50 mL with a nucleic acid concentration of $10^{-6}M$, then the total number of molecules used is about $3 \times 10^{16}$. There is more potential variety in a library with a $28n$ variable region ($4^{28} = 7.2 \times 10^{16}$). If a transcription reaction has been used to convert a random DNA template library to an RNA library, then multiple copies (75–100) of individual sequences exist in the population, further reducing the diversity of the starting pool of candidate ligands so that in the above example there is actually about $10^{14}$ unique sequences in the first selection. Connel et al. *(35)* argue,

based on Poisson distribution calculations, that the number of sequences available for sampling by the target molecule is even lower.

Increases in length of the variable region beyond $28n$ can be rationalized by the observation that many ligands have complicated structures that require long nucleic acid chains within which some specific sequences are conserved at the primary sequence level. An illustrative case is that in which Sassefran and Szostak *(39)* used a $120n$ random region to get RNA ligands of ATP. One of the ligand sequences could be greatly reduced to a 54 nt bulged loop ligand containing 12 consensus positions without loss of binding activity. Based on the consensus, a 40-mer was designed that had optimal binding activity. Because the reduced yield during oligonucleotide synthesis is exponential with the increase in nucleotide length, such a large random region probably decreased the functional diversity of the pool. A $50n$ library may have more thoroughly described a consensus for RNAs that bind ATP. An argument for larger sized random regions is that Schneider et al. *(9)* used a $30n$ library to find two families of consensus sequences that bound to *E. coli rho* transcription termination protein less well than a natural *rho* binding site. They argued that the *rho* hexamer binding site on RNA was about 70 nts in length and that perhaps the SELEX derived sequences were too small. In isolation of RNA binders of arginine, Connell et al. *(35)* used a $25n$ library whose winning candidates' binding motifs involved a significant amount of fixed sequence region, enough to bias the resultant consensus; a rerandomization revealed a more general consensus for arginine binding RNA sequences. For most applications a random region of 40–50 nts is adequate.

2. No matter what the separation scheme, the nucleic acid pool contains sequences that will ultimately be selected by the separation mechanism independent of the presence of target molecules. Such unintended products of the selection process can at least significantly dilute the actual ligands of target molecules *(20)* and at worst (especially when target molecule concentration or supply is low) prevent success in obtaining target-specific ligands. The surest way of eliminating or reducing such problems is to subtract potential "background binders" from the initial population by running them through a targetless cycle and collecting what comes through as the starting pool for selection. This negative selection should be performed regularly, at best before each round of selection.

3. Calculations based on computer modeling of the SELEX process have suggested that binding reactions that yield 6% of the total RNA bound to target would be optimal *(42)*. A series of pilot binding reactions can be conducted to determine the ratios of nucleic acid to target molecule that produce the desired fraction bound, or the actual binding reaction for selection can be chosen from such pilots to fix the level of stringency. Schneider et al. *(9)* showed that when reactions were arranged to select 10% of the RNA population bound to *rho* protein (low stringency) or 1% (high stringency) during SELEX, ligands of similar sequence were produced by both selection strategies, although the speed of affinity enhancement of the pool was increased somewhat for the high stringency selections. This illustrates that precise fine-tuning of ligand-target ratios is not crucial

for success in obtaining optimal binding ligands. I would suggest that the first binding reaction be conducted in large volumes (about 50 mL) with target molecule concentrations at 0.2 times the $K_d$ of the original nucleic acid pool-target interaction and all of the starting nucleic acid (preferably near in concentration to the starting $K_d$). This would yield about 9% of the original nucleic acid population bound to a large number of target molecules, decreasing the chance that low abundance, unique sequences of high affinity are lost in the first selection. Subsequent to this first selection, optimized binding reactions can be used to speed enrichment. Of course, the binding reaction conditions, such as pH, ion concentrations, and temperature, should be chosen to suit the optimal conditions for target molecule structure and function or to match the site of intended application.

4. There are special problems associated with cyclic enzymatic manipulation and replication of nucleic acid libraries with large random regions. The most evident is that an initial nucleic acid population that appears as a discrete band on a separating gel may after three or more rounds of the SELEX process smear into a multisized population. The most likely explanation is that there is mispriming of the 3' ends of DNAs in the random regions of other molecules in the population during reverse transcription and/or PCR. Indeed, Crameri and Stemmer *(43)* found that the addition of *E. coli* SSB (which decreases the level of mispairing of nucleic acids) prevents such smearing after many rounds of PCR in random library amplification. They do not report whether the addition of SSB would decrease the rate of mispairing in RT steps, which are done at lower temperatures than PCR and would also be subject to such artifactual size problems. In lieu of adding SSB, these size heterogeneities can be dealt with by periodic gel band purification of the nucleic acid population *(1)*.

# References

1. Tuerk, C. and Gold, L. (1990) Systematic evolution of ligands by exponential enrichment, RNA ligands to bacteriophage T4 DNA polymerase. *Science* **249,** 505–510.
2. Blackwell, T. K. and Weintraub, H. (1990) Differences and similarities in DNA-binding preferences of MyoD and E2A protein complexes revealed by binding site selection. *Science* **250,** 1104–1110.
3. Tsai, D. E., Harper, D. S., and Keene, J. D. (1991) U1 snRNP-A protein selects a 10 nucleotide consensus sequence from a degenerate RNA pool presented in various structural contexts. *Nucleic Acids Res.* **19,** 4931–4936.
4. Bartel, D., Zapp, M. L., Green, M. R., and Szostak, J. W. (1991) HIV-1 rev regulation involves recognition of non-Watson-Crick base pairs in viral RNA. *Cell* **67,** 529–536.
5. Schneider, D., Tuerk, C., and Gold, L. (1992) Selection of high affinity RNA ligands to the bacteriophage R17 coat protein. *J. Mol. Biol.* **228,** 862–869.
6. Christianson, A. M., King, D. L., Hatzivassiliou, E., Casas, J. E., Hallenbeck, P. L., Nikodem, V. M., Mitsialis, S. A., and Kafatos, F. C. (1992) DNA binding and

heteromerization of the Drosophila transcription factor chorion factor 1/ultra-spiracle. *Proc. Natl. Acad. Sci. USA* **89,** 11,503–11,507.

7.  Tsai, D. E., Kenan, D. J., and Keene, J. D. (1992) In vitro selection of an RNA epitope immunologically crossreactive with a peptide. *Proc. Natl. Acad. Sci. USA* **89,** 8864–8868.

8.  Tuerk, C. and MacDougal-Waugh, S. (1993) In vitro evolution of functional nucleic acids: high-affinity RNA ligands of HIV-1 protein. *Gene* **137,** 33–39.

9.  Schneider, D., Gold, L., and Platt, T. (1993) Selective enrichment of RNA species for tight binding to *Escherichia coli rho* factor. *FASEB J.* **7,** 201–207.

10. de Jong, R., van der Heijden, J., and Meijlink, F. (1993) DNA-binding specificity of the S8 homeodomain. *Nucleic Acids Res.* **21,** 4711–4720.

11. Tsai, D. E. and Keene, J. D. (1993) In vitro selection of RNA epitopes using autoimmune patient serum. *J. Immunol.* **150,** 1137–1145.

12. Tuerk, C., MacDougal-Waugh, S., Hertz, G. Z., and Gold, L. (1994) In vitro evolution of functional nucleic acids: high-affinity RNA ligands of the HIV-1 rev protein, in *The Polymerase Chain Reaction* (Mullis, K. B., Ferre, F., and Gibbs, R. A., eds.), Birkhauser, Boston, MA, pp. 233–243.

13. Sakashita, E. and Sakamoto, H. (1994) Characterization of RNA binding specificity of the Drosophila sex-lethal protein by in vitro ligand selection. *Nucleic Acids Res.* **22,** 4082–4086.

14. Paca-Uccaralertkun, S., Zhao, L. J., Adya, N., Cross, J. V., Cullen, B. R., Boros, I. M., and Giam, C. Z. (1994) In vitro selection of DNA elements highly responsive to the human T-cell lymphotropic virus type I transcriptional activator, *Tax. Mol. Cell Biol.* **14,** 456–462.

15. Gorlach, M., Burd, C. G., and Dreyfuss, G. (1994) The determinants of RNA-binding specificity of the heterogeneous nuclear ribonucleoprotein C proteins. *J. Biol. Chem.* **269,** 23,074–23,078.

16. Gorlach, M., Burd, C. G., and Dreyfuss, G. (1994) The mRNA poly (A)-binding protein, localization, abundance, and RNA-binding specificity. *Exp. Cell Res.* **211,** 400–407.

17. Henderson, B. R., Menotti, E., Bonnard, C., and Kuhn, L. C. (1994) Optimal sequence and structure of iron-responsive elements. Selection of RNA stem-loops with high affinity for iron regulatory factor. *J. Biol. Chem.* **269,** 17,481–17,489.

18. Blackwell, T. K., Bowerman, B., Priess, J. R., and Weintraub, H. (1994) Formation of a monomeric DNA binding domain by Skn-1 bZIP and homeodomain elements. *Science* **266,** 621–628.

19. Andres, V., Chiara, M. D., and Mahdavi, V. (1994) A new bipartite DNA-binding domain: cooperative interaction between the cut repeat and homeo domain of the cut homeo proteins. *Genes Dev.* **8,** 245–257.

20. Tuerk, C., MacDougal, S., and Gold, L. (1992) RNA pseudoknots that inhibit human immunodeficiency virus type I reverse transcriptase. *Proc. Natl. Acad. Sci. USA* **89,** 6988–6992.

21. Peterson, E. T., Blank, J., Sprinzl, M., and Uhlenbeck, O. C. (1993) Selection for active *E. coli* tRNA(Phe) variants from a randomized library using two proteins. *EMBO J.* **12,** 2959–2967.

22. Chen, H. and Gold, L. (1994) Selection of high-affinity RNA ligands to reverse transcriptase, inhibition of cDNA synthesis and RNase H activity. *Biochemistry* **33,** 8746–8756.
23. Peterson, E. T., Pan, T., Coleman, J., and Uhlenbeck, O. C. (1994) In vitro selection of small RNAs that bind to *Escherichia coli* phenylalanyl-tRNA synthetase. *J. Mol. Biol.* **242,** 186–192.
24. Conrad, R., Keranen, L. M., Ellington, A. D., and Newton, A. C. (1994) Isozyme-specific inhibition of protein kinase C by RNA aptamers. *J. Biol. Chem.* **269,** 32,051–32,054.
25. Nazarenko, I. A. and Uhlenbeck, O. C. (1995) Defining a smaller RNA substrate for elongation factor *Tu*. *Biochemistry* **34,** 2545–2552
26. Bock, L. C., Griffin, L. C., Latham, J. A., Vermaas, E. H., and Toole, J. J. (1992) Selection of single-stranded DNA molecules that bind and inhibit human thrombin. *Nature* **355,** 564,565.
27. Tsai, D. E., Kenan, D. J., and Keene, J. D. (1992) In vitro selection of an RNA epitope immunologically crossreactive with a peptide. *Proc. Natl. Acad. Sci. USA* **89,** 8864–8868.
28. Jellinek, D., Lynott, C. K., Rifkin, D. B., and Janjic, N. (1993) High-affinity RNA ligands to basic fibroblast growth factor inhibit receptor binding. *Proc. Natl. Acad. Sci. USA* **90,** 11,227–11,231.
29. Lin, Y., Qui, Q., Gill, S. C., and Jayasena, S. D. (1994) Modified RNA sequence pools for in vitro selection. *Nucleic Acids Res.* **22,** 5229–5234.
30. Kubik, M. F., Stephens, A. W., Schneider, D., Marlar, R. A., and Tasset, D. (1994) *Nucleic Acids Res.* **22,** 2619–2626.
31. Jellinek, D., Green, L. S., Bell, C., and Janjic, N. (1994) Inhibition of receptor binding by high-affinity RNA ligands to vascular endothelial growth factor. *Biochemistry* **33,** 10,450–10,456.
32. Nieuwlandt, D., Wecker, M., and Gold, L. (1995) In vitro selection of RNA ligands to substance P. *Biochemistry* **34,** 5651–5659.
33. Mishra, R. K. and Toulme, J. J. (1994) In vitro selection of antisense oligonucleotides targeted to a hairpin structure. *CR Acad. Sci. III* **317,** 977–982.
34. Ellington, A. D. and Szostak, J. W. (1990) In vitro selection of RNA molecules that bind specific ligands. *Nature* **346,** 818–822.
35. Connell, G. J., Illangesekare, M., and Yarus, M. (1993) Three small ribooligonucleotides with specific arginine sites. *Biochemistry* **32,** 5497–5502.
36. Connell, G. J. and Yarus, M. (1994) RNAs with dual specificity and dual RNAs with similar specificity. *Science* **264,** 1137–1141.
37. Lorsch, J. R. and Szostak, J. W. (1994) In vitro selection of RNA aptamers specific for cyanocobalamin. *Biochemistry* **33,** 973–982.
38. Jenison, R. D., Gill, S. C., Pardi, A., and Polisky, B. (1994) High-resolution molecular discrimination by RNA. *Science* **263,** 1425–1429.
39. Sassanfar, M. and Szostak, J. W. (1993) An RNA motif that binds ATP. *Nature* **364,** 550–553.
40. Huizenga, D. E. and Szostak, J. W. (1995) A DNA aptamer that binds adensosine and ATP. *Biochemistry* **34,** 656–665.

41. Jensen, K. B., Green, L., MacDougal-Waugh, S., and Tuerk, C. (1994) Characterization of an in vitro-selected RNA ligand to the HIV-1 *Rev* protein. *J. Mol. Biol.* **235,** 237–247.
42. Irvine, D., Tuerk, C., and Gold, L. (1990) SELEX with integrated optimization by nonlinear analysis. *J. Mol. Biol.* **222,** 739–761.
43. Crameri, A. and Stemmer, W. P. (1993) 10(20)-fold aptamer library amplification without gel purification. *Nucleic Acids Res.* **21,** 4410.
44. Green, L., Waugh, S., Binkley, J. P., Hostomska, Z., Hostomsky, Z., and Tuerk, C. (1995) Comprehensive chemical modification interference and nucleotide substitution analysis of an RNA pseudoknot inhibitor to HIV-1 reverse transcriptase. *J. Mol. Biol.* **247,** 60–68.

# IV

# CLONING UNKNOWN NEIGHBORING DNA

# 24

# Rapid Amplification of cDNA Ends

## David Bertioli

## 1. Introduction

Rapid amplification of cDNA ends (RACE) allows the amplification of either the 5' or 3' end of a specific cDNA starting from a mRNA population (5'-RACE and 3'-RACE respectively: refs. *1,2*). The method allows the amplification of regions with unknown sequence and only requires the knowledge of a single short sequence within the mRNA of interest. The method is often used for cloning the remainder of a cDNA using a sequence from an incomplete cDNA. Incomplete cDNAs are frequently obtained by, for instance, the screening of a library or from PCR using degenerate primers.

A schematic representation of RACE is shown in Fig. 1. The key feature of the method is the use in PCR of a "specific primer" designed using the known sequence within the mRNA together with a "general primer" complementary either to the mRNA poly A tail (for 3'-RACE) or to a homopolymer tail that has been added to the 3' end of the cDNA (for 5'-RACE). Since homopolymers do not make good PCR primers, and to facilitate the cloning of RACE products, the general primers contain a short sequence with a restriction endonuclease site at their 5' end. The cDNA template for the PCR may be produced either using an oligo-dT primer (for 3' or 5'-RACE) or using a primer complementary to the known sequence within the mRNA (for 5'-RACE only). Where the RACE PCR gives a complex mix of products, an aliquot of this mix may be used as template for another PCR using a second specific primer (nested within the first) and the original general primer (nested RACE).

Several modifications and variants of RACE have been published *(3–6)*. The method that follows has been published *(7)* and was designed to be simple and inexpensive. Some examples of 5'-RACE products made using this method

From: *Methods in Molecular Biology, Vol. 67: PCR Cloning Protocols: From Molecular Cloning to Genetic Engineering* Edited by: B. A. White  Humana Press Inc., Totowa, NJ

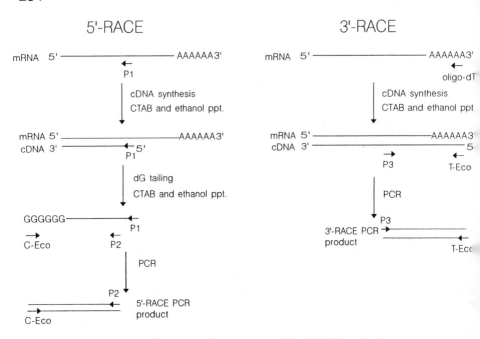

**Fig. 1.** A diagrammatic summary of the RACE method.

are shown in Fig. 2. This RACE method differs from that of Frohman et al. *(1)* in several ways including:

1. The use of dG instead of dA for the homopolymer tail. This enables the use of a higher annealing temperature in 5'-RACE and avoids nonspecific amplification of total cDNA that occurs when the "nonspecific" dT-based primer is complementary to both ends of the cDNAs.
2. The use of Hexadecyltrimethylammonium bromide (CTAB) precipitation to simultaneously purify, size select, and protect nucleic acids during the manipulative procedures.
3. The use of total RNA as the usual substrate, although mRNA may also be used. When total RNA is used, the structural RNAs act as carriers for the much less abundant mRNAs and cDNAs. Although the structural RNAs are substrates for terminal transferase, their presence does not seem to have adverse effects.

## 2. Materials

1. RNA from the tissue of interest: Many extraction methods could be used, but it is important that the RNA used is intact and essentially free of DNA. Our usual method is as described by Logemann et al. *(8)* (*see* Note 1).
2. Moloney murine leukemia virus RNase H⁻ reverse transcriptase (*see* Note 2; Superscript™; Life Technologies, Gaithersburg, MD).

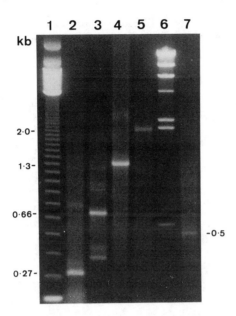

**Fig. 2.** Some RACE products separated by agarose gel electrophoresis (from ref. *7*). Lane 1, 123-bp ladder; lane 6, Lambda DNA digested with *Hin*dIII. Other samples are RACE reactions designed to amplify the cDNA end corresponding to the 5' end of the mRNA of: lane 2, a "purple pod" specific mRNA from *P. sativum.;* lane 3, β-subunit of pyrophosphate-dependent phosphofructokinase from potato; lane 4, RNA-2 of arabis mosaic virus; lane 5, acteylcholinesterase from house fly; lane 7, as lane 4 but cDNA synthesised by avian myoblastosis virus reverse transcriptase (AMV RT). Note the smaller RACE product in lane 7 derived from cDNA synthesised by AMV RT compared to the equivalent product in lane 4 from cDNA synthesised by superscript (RNase H). Bars indicate amplified cDNA. Numbers next to bars indicate approximate size of RACE products in kilobases.

3. 5X reverse transcription buffer: 250 m$M$ Tris-HCl, pH 8.3, 375 m$M$ KCl, 15 m$M$ MgCl$_2$ and 0.1$M$ DTT, as supplied by Life Technologies.
4. dNTPs: an aqueous solution of 20 m$M$ of each dNTP (dGTP, dATP, dTTP, and dCTP).
5. Oligonucleotide primer for cDNA synthesis: This may be dT$_{12-18}$ (for 5' or 3'-RACE) or a 15–25-mer primer designed to be complementary to the target mRNA (for 5'-RACE only). The use of the latter is more likely to give a single RACE product (*see* Note 3).
6. Reagents for homopolymer tailing of first strand cDNA: terminal transferase (LifeTechnologies), 5X tailing buffer (0.5$M$ potassium cacodylate, pH 7.2, 10 m$M$ CoCl$_2$ 1 m$M$ DTT) as supplied by Life Technologies and an aqueous solution of 100 µ$M$ dGTP.
7. Reagents for PCR, including 10X reaction buffer: 100 m$M$ Tris-HCl, pH 8.8, at 25°C, 500 m$M$ KCl, 15 m$M$ MgCl$_2$, 1% Triton X-100 (*see* Chapter 1).

8. Oligonucleotide primers for PCR: The sequences of the general primers contain a homopolymer at the 3' end and a restriction endonuclease site at the 5' end, for instance:

   For 5'-RACE, CAAGGAATT(dC)$_{14}$ (C-*Eco*RI)

   3'-RACE, GCGCCAGGAATTCGC(dT)$_{17}$ (T-*Eco*RI)

   The specific primers are designed using the known sequence from the target mRNA. For 5'-RACE the PCR primer must be nested to the 3' side (cDNA sense) of the primer used for cDNA synthesis. The PCR primer should be homologous to about 20 bases of the target mRNA/cDNA. To facilitate cloning, a non-homologous region with a restriction site may be included 5' to these 20 homologous bases.
9. 10% (w/v) CTAB (hexadecyltrimethylammonium bromide). It may be necessary to warm this to keep the CTAB in solution.
10. 3.5 and 1.2$M$ NaCl.
11. Absolute and 75% (v/v) ethanol.
12. Reagents for agarose gel electrophoresis.

## 3. Methods

1. Perform first-strand cDNA synthesis using a 20-μL reaction volume. The quantity of RNA used is not critical and we have used anything from approx 40–0.1 μg total RNA (although, if at all possible, avoid using small quantities). It is preferable to do two or three reverse transcriptions with tenfold different quantities of total RNA. Add the following to a microcentrifuge tube: cDNA synthesis primer: 500 ng oligo dT$_{12-18}$ or 15 pmol of a primer complementary to the mRNA (approx 200 ng of a 20-mer) total RNA; and RNase free water to 12.5 μL.
2. Heat at 70°C for 5 min and place on ice. Centrifuge to collect condensate.
3. Add the following: 4 μL of 5X reverse transcriptase buffer; 2 μL of 0.1$M$ DTT; 0.5 μL of 20m$M$ dNTPs; and 1 μL (200U) of Superscript reverse transcriptase.
4. Mix well and incubate at 37°C for 1 h.
5. Purify and size select nucleic acids by CTAB and ethanol precipitations (*see* Note 4). To the reverse transcription reaction add: 2.5 μL of 3.5$M$ NaCl and 2.5 μL of 10% CTAB (mix after each addition).
6. Recover CTAB:nucleic acid complexes by centrifugation in a microfuge at room temperature for 30 min.
7. Discard supernatant and redissolve pellet in 100 μL of 1.2$M$ NaCl. Precipitate nucleic acids by adding 2.7 vol of ethanol. Recover nucleic acids by centrifugation in microfuge, wash with 70% ethanol, dry and resuspend in 15.25 μL of water. This may be used directly for 3'-RACE PCR (go to step 12). For 5'-RACE continue.
8. To the microfuge tube add: 5 μL of 5X tailing buffer, 3.75 μL of 100 μ$M$ dGTP, and 1 μL of terminal transferase 15 U/μL.
9. Mix and incubate for 1 h.
10. To the tailing reaction add: 3.1 μL of 3.5$M$ NaCl and 3.1 μL of 10% CTAB (mix after each addition).

11. Recover CTAB:nucleic acid complex and ethanol precipitate as described in steps 6 and 7.
12. Use 1 µL of the cDNA mix as a template in PCR. To a microcentrifuge tube add the following: 5 µL of 10X reaction buffer, 0.5 µL of 20 m$M$ solution of each dNTP, 0.5 µL of *Taq* polymerase (5 U/µL), 1.5 µL of general oligonucleotide primer (50 ng/µL), 1.5 µL of specific primer (50 ng/µL), and water to 50 µL.
13. Use the following conditions for PCR before modifying them if necessary: 94°C for 3 min (initial denaturation); 30 cycles: 96°C for 1 min, 72°C for 3 min, 55°C for 2 min; and 72°C for 10 min (final extension).
14. Analyze a 10-µL aliquot of the PCR products by agarose gel electrophoresis, and if possible by Southern blotting before attempting to clone the products. (*See* Notes 5,6).

## 4. Notes

1. To check the integrity of RNA it is unnecessary to use denaturing agarose gel electrophoresis. Instead use formamide sequencing buffer *(9)* as 1X gel loading buffer and heat the sample with buffer to 85°C before analysis by electrophoresis in 1% agarose buffered in 0.5X TBE with ethidium bromide. Since the quantity of RNA used for this RACE protocol is not critical, it is unnecessary to quantitate the RNA accurately.
2. The use of the reverse transcriptase Superscript which lacks RNase H activity, gave larger RACE products than Avian myoblastosis virus reverse transcriptase.
3. For 5'-RACE it is important not to use the same primer for cDNA synthesis as for the specific primer in PCR because the annealing of the primer for cDNA synthesis is at 37°C and all the products of reverse transcription will have this primer at their 5' end. Therefore, if the same primer is used for both cDNA synthesis and PCR undesired products are very likely to be produced during PCR.
4. CTAB precipitation provides a convenient method for the simultaneous purification and size selection of nucleic acids. CTAB binds nucleic acids to form an insoluble complex, the formation of which is influenced by salt concentration. When the concentration of NaCl is above 1$M$ no complex formation occurs, when below 0.2$M$ all nucleic acids are included in the complex, and when between 0.3 and 0.4$M$ the incorporation of small single-stranded nucleic acids, such as oligonucleotide primers, is very inefficient *(10)*. The use of CTAB has the added benefit that it protects nucleic acids (particularly the structural RNAs that act as carrier) from degradation *(11)*.
5. For the cloning of RACE products it is preferable to restriction digest the ends and then purify by gel elution before ligation.
6. If the PCR gives many nonspecific products then it may be necessary to use nested RACE. For this, use aliquots (three or four tenfold dilutions from 1 mL downward) of the first PCR as template for a further PCR, which uses as primers the same general primer and a second specific primer nested to the 3' of the first specific primer.
7. Some workers maintain that the removal of RNA by alkaline hydrolysis or RNase treatment before PCR improves the quantity of PCR product obtained.

## References

1. Frohman, M. A., Dush, M. K., and Martin, G. R. (1988) Rapid production of full-length cDNAs from rare transcripts: amplification using a single gene-specific oligonucleotide primer. *Proc. Natl. Acad. Sci. USA* **85,** 8998–9002.
2. Ohara, O., Dorit, R. L., and Gilbert, W. (1989) One-sided polymerase chain reaction: the amplification of cDNA. *Proc. Natl. Acad. Sci. USA* **86,** 5673–5677.
3. Harvey, R. J. and Darlison, M. G. (1991) Random-primed cDNA synthesis facilitates the isolation of multiple 5'-cDNA ends by RACE. *Nucleic Acids Res.* **19,** 4002.
4. Borson, N. D., Salo, W. L., and Drewes, L. R. (1992) A lock-docking oligo(dT) primer for 5' and 3' RACE PCR. *PCR Methods Appl.* **2,** 144–148.
5. Jain, R., Gomer, R. H., and Murtagh, J. J., (1992) Increasing specificity from the PCR-RACE technique. *Biotechniques* **12,** 58,59.
6. Troutt, A. B., McHeyzer-Williams, M. G., Pulendran, B., and Nossal, G. J. V. (1992) Ligation-anchored PCR: a simple amplification technique with single-sided specificity. *Proc. Natl. Acad. Sci. USA* **89,** 9823–9825.
7. Bertioli, D. J. and Burrows, P. R. (1995) A simple RACE method based on CTAB precipitation. *Methods Mol. Cell. Biol.* (in press).
8. Logemann, J., Schell, J., and Lothar, W. (1987) Improved method for the isolation of RNA from plant tissues. *Anal. Biochem.* **163,** 16–20.
9. Sambrook, J., Fritsch, E. F., and Maniatis, T. (1989) *Molecular Cloning: A Laboratory Manual,* 2nd ed., Cold Spring Harbor Laboratory, Cold Spring Harbor, NY.
10. Belyavsky, A., Vinogradova, T., and Rajewsky, K. (1989) PCR-based cDNA library construction: general cDNA libraries at the level of a few cells. *Nucleic Acids Res.* **17,** 2919–2923.
11. Macfarlane, D. E. and Dahle, C. E. (1993) Isolating RNA from whole blood—the dawn of RNA-based diagnosis? *Nature* **362,** 186–188.

# 25

# Amplification of Gene-Regulating Regions with Single-Sided Specificity

## Mei-Zhong Luo and Rino Cella

## 1. Introduction

The isolation of genomic 5' regulating regions of genes starting from a suitable sequence (either a cDNA or an amino acid derived oligonucleotide) can be achieved in several ways. One of these, although long and labor-intensive, is the screening of a genomic library, followed by the isolation, subcloning, and sequencing of putative positive clones. A more direct approach is offered by inverse polymerase chain reaction (IPCR) using selfcircularized genomic DNA (1). Yet, PCR performed on circular DNA sometimes fails to produce an amount of amplification product sufficient to be visualized by EtBr-staining (2), thus hampering further isolation and cloning of the amplification product. The problem is usually overcome by relinearizing the circular DNA with a suitable restriction enzyme that has to cut in the region between the diverging amplification primers; such restriction site must not be present in the unknown flanking region, otherwise amplification will not take place.

To overcome most of the problems mentioned above, a relatively simple procedure, shown schematically in Fig. 1, has been developed; this is based on the ligation of size-selected DNA (containing the fragment of interest) to a plasmid vector possessing standard sequencing primers, followed by the transformation of *Escherichia coli* cells and the use of plasmid DNA extracted from the resulting minilibrary for PCR.

The developed procedure appears to have the following advantages over IPCR:

1. The DNA sequence needed as the specific primer can be relatively short (20–25 nucleotides);
2. The DNA to be used for PCR is available in high amount, thus facilitating all manipulations;

From: *Methods in Molecular Biology, Vol. 67: PCR Cloning Protocols: From Molecular Cloning to Genetic Engineering* Edited by: B. A. White  Humana Press Inc., Totowa, NJ

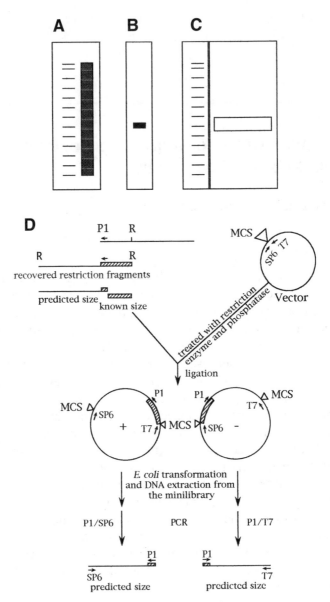

**Fig. 1.** Strategy for the isolation of 5' gene-regulating regions. **(A)** Genomic DNA (5–10 mg) is digested with a suitable restriction enzyme (e.g., *Eco*RI), separated by agarose gel electrophoresis and visualized by EtBr staining. In order to locate precisely the position of the gel corresponding to the desired band it is important to use molecular weight markers of appropriate size. **(B)** Identification of the band of interest by Southern blot hybridization. **(C)** Digestion and separation on a preparative scale of genomic DNA. Excision of the gel slice corresponding to the recognized band. **(D)** Schematic diagram of the PCR procedure. DNA fragments obtained from the gel slice are ligated to the linearized and dephosphorylated plasmid vector. The ligation mix

3. Several restriction sites are available in the vector Multiple Cloning Site (MCS) if relinearization of the DNA is required before PCR;

4. The minilibrary construction step, although apparently burdensome, allows one to obtain a DNA, which is a better template for PCR and possibly contributes to the elimination of PCR inhibitors that have been found to be present in genomic DNA of some plant species *(3)*.

The method could also be applied to the isolation of 3' unknown flanking regions or as an alternative to *de novo* PCR *(4)* for obtaining a full-length cDNA from a cDNA library.

## 2. Materials

1. Genomic DNA extracted from the species of interest. The extraction procedure of choice will depend on the type of explant used with the only requirement that the resulting DNA be readily digestible with a restriction enzyme that yields DNA fragments of suitable length. Depending on the starting tissue, the purification by CsCl *(5)* could be necessary.

2. Reagents for agarose gel electrophoresis *(5,6)*: Depending on the method chosen for the recovery of DNA from the agarose slice, normal or low-melting-point agarose will be used *(see* point 5 and Section 3.1.2., point 6).

3. Hybond N+ filters (Amersham, Buckinghamshire, UK).

4. Prime-a-Gene labeling system (to be used with $^{32}$P-dCTP; Promega, Madison, WI).

5. Reagents and materials for DNA hybridization and autoradiography *(5,6)*.

6. UV transilluminator and a razor blade for the excision from the agarose gel of the slice containing the wanted DNA fragment.

7. Reagents for the extraction of DNA fragments from the agarose gel slice. These will depend on the agarose used for electrophoresis *(see* Section 3.1.2., step 6).

8. A plasmid vector; any cloning vector with characteristics similar to those of pGEM-4Z (Promega).

9. Reagents for ligation *(5,6)*.

10. Bacterial cells for plasmid transformation (e.g., *E. coli* TOP-10 or DH5α).

11. Primers for PCR. A gene-specific primer *(see also* Section 3.3. and Note 1) and a vector-specific primer (oligonucleotides from either Sp6 or T7 promoters present in pGEM-4Z that are commercially available).

12. Reagents for PCR *(5,6) (see* Chapter 1).

---

ture is then used to transform *E. coli* cells obtaining a minilibrary. Extract and purify plasmid DNA from an aliquot of the minilibrary. P1 represents the gene specific primer, SP6 and T7 are the standard sequencing primers of the plasmid vector (in the case of pGEM-4Z), and MCS is the multiple cloning site. Signs "+" and "-" indicate the possible cloning orientations of the target fragment. The hatched box corresponds to the known 5' region of the cDNA, which overlaps the 3' region of the restriction fragment containing the unknown regulating sequence. *See text* for details.

## 3. Methods

### 3.1. Isolation of the Genomic DNA Fragment that Hybridizes with the 5' Region of a cDNA

#### 3.1.1. Identification of the Genomic Fragment that Overlaps with the 5' End of a cDNA

1. Digest 5–10 µg samples of genomic DNA with several restriction enzymes (*see* Section 3.1.2., step 1) and proceed to agarose gel electrophoresis; run DNA size markers in parallel. Stain the gel with EtBr and take a picture with a ruler at the side.
2. Transfer DNA to Hybond N+ filters according to manufacturer's instructions.
3. Hybridize the filter to a radiolabeled probe corresponding to the 5' region of the cDNA using a standard procedure *(5)*. Carefully locate the position of the autoradiographic band with respect to DNA size markers on the picture of the agarose gel.

#### 3.1.2. Isolation of the DNA Fragment of Interest

1. Choose a restriction enzyme able to produce DNA fragments of the desired size. This should depend on the length of the unknown region one wants to reveal and size limitation for subsequent cloning. Since the cloning efficiency could depend on fragment length, it is suggested to produce fragments not longer than 4–5 kbp. To achieve this goal, it might be necessary to digest genomic DNA with two different restriction enzymes. When dealing with a cDNA sequence, preference should be given to restriction enzymes whose target sites are close to the 5' end of the known cDNA.
2. Digest 50–100 µg of the DNA sample with the chosen restriction enzyme and separate the resulting restriction fragments, along with DNA size markers at the side, by preparative agarose gel electrophoresis. The choice of the agarose (low-melting-point vs normal agarose) will depend on the DNA extraction method (*see* step 6).
3. Cut the gel parallel to the direction of DNA migration and stain the part containing the DNA size markers with EtBr.
4. Realign the stained part of the gel with the unstained one on the UV transilluminator and determine the position of the wanted band in respect to DNA size markers.
5. Cut the agarose gel slice with a razor blade at a point corresponding to the position of the hybridization band. In order to avoid missing the wanted fragment, either cut a wide gel slice or a few slices around the estimated band point.
6. Extract DNA from the gel slice. The method depends on the type of agarose used. With normal agarose the extraction can be obtained by the use of the Qiaex kit (Qiagen, Chatsworth, CA) according to manufacturer's instructions. If using low-melting-point agarose, apply the standard procedure *(5)*.
7. Resuspend DNA in 10 µL of sterile distilled water.
8. Estimate DNA concentration by the agarose plate method *(5)*. Briefly, spot 1-µL aliquots of diluted sample and DNA standards on a 1% agarose slab gel containing 0.5 µg/mL ethidium bromide and estimate DNA concentration ion by means of an ultraviolet transilluminator.

## 3.2. Minilibrary Construction

1. DNA fragments eluted from the agarose gel slice are ligated to the cloning vector (e.g., pGEM-4Z, Promega) linearized with the restriction enzyme(s) used to digest genomic DNA, and dephosphorylated. The molar ratio of vector/insert may vary from 0.5–2. Use a final DNA concentration of 5–50 ng/μL. For an insert of 2 kb, use 100 ng of insert and 100 ng of vector DNA in a final ligation volume of 10 μL. Perform ligation overnight at 14–16°C *(5)*.

2. Add 1 μL of yeast tRNA (10 mg/mL), precipitate DNA with 0.1 vol of 3$M$ Na-acetate and 2.5 vol of ethanol *(5)* and resuspend the pellet in 10 μL of sterile distilled water (*see* Note 2).

3. Transform *E. coli* cells with ligation products to produce a minilibrary of size-selected DNA. This can be performed by high-voltage electroporation (e.g., Gene Pulser™, Bio-Rad). Prepare *E. coli* cells for electroporation as follows: Preinoculate bacteria from a single colony in 10 mL of LB medium and incubate overnight at 37°C rotating at 200 rpm. Transfer 4 mL of the preinoculum into 400 mL of LB medium and grow bacteria at 37°C rotating at 200 rpm. Collect cells as they reach 0.5–1 $A_{600}$. Collect and wash bacteria by centrifugation (1000$g$ at 4°C for 10 min) and progressively concentrate them resuspending the pellet in 400, 200, 20, 5, and 0.8 mL of cold (4°C) 10% sterile glycerol. Keep the cells on ice; either use them immediately or quickly freeze them in liquid nitrogen and then store them at –80°C up to several mo (*see* Note 1).

4. Add a 2-μL aliquot of ligation products to 30 μL of *E. coli* cells and electroporate under the following conditions: resistance 200 Ω, capacitance 25 mF, 12.5 kV/cm. Add 1 mL of SOC medium *(5)* and grow bacteria at 37°C, rotating 200 rpm for 1 h. Either use the minilibrary immediately or store it at –80°C after the addition of glycerol to a final concentration of 35%.

5. Take a 100-μL aliquot of the minilibrary and inoculate 100 mL of LB medium containing the proper antibiotic(s) (60 μg/mL of ampicillin in the case of pGEM-4Z). Grow bacteria at 37°C, rotating at 200 rpm until they reach 0.7–1 $A_{600}$.

6. Collect bacteria by centrifugation and extract total plasmid DNA using the Qiagen Plasmid Kit (Qiagen) according to manufacturer's instructions. Resuspend DNA in a final volume of 100 μL. Measure DNA concentration (*see* Section 3.1., step 8) and proceed to PCR.

## 3.3. Primer Design

1. The gene specific primer (P1 in Fig. 1) has to be synthesized on the basis of the available cDNA sequence; primer design should avoid selfcomplementarity and strong palindromic structure. The distance from the primer 3' end to the 5' end of the known cDNA sequence should be sufficient to identify the PCR product. A distance of about 100 nt is usually sufficient. Moreover, to allow cloning of the amplification product it may be useful to create a restriction site at the 5' end of the gene specific primer (*see* Section 3.5.).

2. The second primer (unspecific) corresponds in the case of pGEM-4Z to SP6 or T7 promoter sequences.

### 3.4. PCR

1. Assemble 10 μL PCR reactions containing 10–20 ng of plasmid DNA (*see* Note 2), 0.2 m*M* of each dNTP, 5–10 pmol of each primer, 1 μL of 10X *Taq* DNA polymerase buffer, and 0.3 U of *Taq* DNA polymerase. Overlay samples with a drop of light mineral oil.
2. Perform PCR using the following cycle conditions: 94°C for 3 min (initial denaturation); 92°C for 1 min (denaturation), 59°C for 1 min (annealing; *see* Note 3), and 72°C for 1 min (extension) for four cycles; 92°C for 1 min, 55°C for 1 min, 72°C for 1 min for 30 cycles; and 72°C for 10 min (final extension).

For preparative purposes, several reactions of 50-μL vol with the components proportionally increased can be used.

### 3.5. Cloning of PCR Products

If cloning of the PCR product is desired it is possible to create a restriction site on the gene-specific primer. For instance, in the case of the carrot *dhfr-ts (7,8)* C residues at positions 121 and 119 were substituted with G and A, respectively, to create an *Eco*RI site at the 5' end of the primer ($^5$'GCG$_{121}$AA$_{119}$TTCCATTAGTAGGATTTGCGA$^3$') whereas the second cloning site was present in the plasmid MCS.

Alternatively, cloning can be obtained by end-filling of amplification products with *Pol* IK (Klenow fragment), followed by blunt end ligation. Cloning can also be obtained by the use of the T-vector, which can be prepared by incubating the plasmid vector, linearized with a restriction enzyme that produces blunt-ends, with *Taq* DNA polymerase at 72°C for 2 h *(9)*. *See* Part II for specific protocols for cloning of PCR products.

## 4. Notes

1. If an electroporator is not available, the minilibrary can be obtained by adding recombinant plasmids to *E. coli* cells made competent by the calcium chloride method using standard procedures *(5)*. Also in this case resuspend transformed bacteria in 1 mL of growth medium.
2. When ligation products (obtained as described in Section 3.2., step 1) were used directly in PCR experiments using a *dhfr-ts* specific primer and either of the standard sequencing primers of the vector, no visible amplification products were observed although these were detected by Southern blot hybridization. A substantial increase in the amount of amplified products were observed, as judged by Southern blot hybridization, when PCR was performed on linearized plasmid DNA. However, also in this case, amplified bands were visible only occasionally in E + Br-stained agarose gels. A possible explanation of this fact is that the transformation step selects chimeric plasmids while eliminating linear ligation products of genomic DNA only.

3. Increase the specificity of the PCR by using a gene-specific primer with a $T_m$ value higher than that of Sp6 or T7, and step the PCR by using an annealing temperature for the first few cycles. This produces a linear increase in the amount of gene-specific products, thus, improving the efficiency of the single-sided PCR.

## References

1. Ochman, H., Gerber, A. S., and Hartl, D. L. (1988) Genetic applications of an inverse polymerase chain reaction. *Genetics* **120**, 621–623.
2. Does, M. P., Dekker, B. M. M., de Groot, M. J. A., and Offringa, R. (1991) A quick method to estimate the T-DNA copy number in transgenic plants at an early stage after transformation, using inverse PCR. *Plant Mol. Biol.* **17**, 151–153.
3. Hamill, J. D., Rounsley, S., Spencer, A., Todd, G., and Rhodes, M. J. C. (1991) The use of the polymerase chain reaction in plant transformation studies. *Plant Cell Rep.* **10**, 221–224.
4. Troutt, A. B., McHeyzer-Williams, M. G., Pulendran, B., and Nossal, G. J. V. (1992) Ligation-anchored PCR: a simple amplification technique with single-sided specificity. *Proc. Natl. Acad. Sci. USA* **89**, 9823–9825.
5. Sambrook, J., Fritsch, E. F., and Maniatis, T. (1989) *Molecular Cloning: A Laboratory Manual* (2nd ed.), Cold Spring Harbor Laboratory, Cold Spring Harbor, NY.
6. White, B. A. (ed.) (1993) *Methods in Molecular Biology, vol. 15: PCR Protocols: Current Methods and Applications.* Humana, Totowa, NJ.
7. Luo, M. Z., Piffanelli, P., Rastelli, L., and Cella, R. (1993) Molecular cloning and analysis of a cDNA coding for the bifunctional dihydrofolate reductase-thymidylate synthase of *Daucus carota. Plant Mol. Biol.* **22**, 427–435.
8. Luo, M. Z. and Cella, R. (1994) A reliable amplification technique with single-sided specificity for the isolation of 5′ gene regulating regions. *Gene* **140**, 59–62.
9. Marchuk, D., Drumm, M., Saulino, A., and Collins, F. S. (1991) Construction of T-vectors, a rapid and general system for direct cloning of unmodified PCR products. *Nucleic Acids Res.* **19**, 1154.

# 26

# An End-Trimming Method and Its Application to Amplify Adjacent cDNA and Genomic DNA Fragments by PCR

## Hiroyuki Iwahana and Mitsuo Itakura

## 1. Introduction

An end-trimming method was developed to amplify adjacent cDNA fragments by PCR *(1,2)*. In an attempt to clone a novel cDNA, the 5' and/or 3' end are often missing. A well-known PCR technique for cloning the missing sequence is a method of rapid amplification of cDNA ends (RACE) *(3)*. Although the missing 5' and 3' ends of cDNA are amplified using a sequence-matched primer and a dT-primer as in RACE, multiple bands usually appear. This makes it difficult to select the band that contains the desired sequence. We devised an end-trimming method to clone adjacent DNA fragments, which specifically extended from a sequence-matched primer, in the fixed direction in a plasmid vector.

The rationale of the method is shown schematically in Fig. 1. A sequence-matched primer and a 5'-ATCG-primer are synthesized so that they anneal to the known sequence. A 5'-ATCG-primer has an extra 5'-ATCG-3' sequence at the 5' end of a sequence-matched primer locating in the downstream. A dT-primer has a 17 nucleotide (nt) stretch of T and endonuclease recognition sites of *Xho*I, *Sal*I, and *Cla*I. The first round of PCR is performed with a sequence-matched primer and a dT-primer using synthesized cDNA as a template. The second round of PCR is the seminested PCR with a 5'-ATCG-primer and a dT-primer using the diluted first PCR product as a template. The final PCR product is end-trimmed by T4 DNA polymerase. T4 DNA polymerase has a 5'→3' polymerase and a 3'→5' exonuclease activity *(4)*. At the presence of dCTP, T4 DNA polymerase removes nucleotides from the 3' end of DNA fragment and stops removing nucleotides at the position of C. As a result, a 5'-ATC overhang is created at the 5'-ATCG-primer side of the PCR-amplified DNA fragment.

From: *Methods in Molecular Biology, Vol. 67: PCR Cloning Protocols: From Molecular Cloning to Genetic Engineering* Edited by: B. A. White  Humana Press Inc., Totowa, NJ

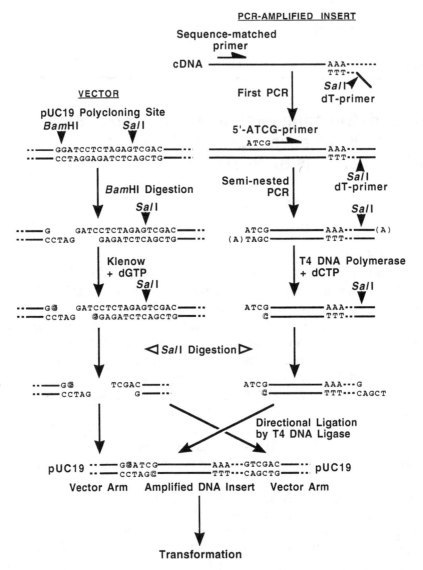

**Fig. 1.** Schematic presentation of the end-trimming method. The first PCR amplification was performed with a sequence-matched primer and a dT-primer using cDNA as a template. The seminested PCR was carried out with a 5'-ATCG-primer and a dT-primer using the 10-fold diluted first PCR product as a template. The PCR-amplified fragment was treated with T4 DNA polymerase in the presence of only dCTP and was digested with *Sal*I. pUC19 was digested with *Bam*HI and was filled with only dGTP using Klenow fragment of *E. coli* DNA polymerase I. pUC19 treated with Klenow fragment was digested with *Sal*I. After these procedures, the PCR-amplified fragment and pUC19 were ligated with T4 DNA ligase and transformation was performed.

The 5'-ATC overhang is complementary to a 5'-GAT overhang, which is created by *Bam*HI-digestion and filling with dGTP by Klenow fragment of *E. coli* DNA polymerase I *(4)*. Thus, the PCR product, which is end-trimmed by T4 DNA polymerase and digested by *Sal*I, can be cloned into a plasmid vector, that is *Bam*HI-digested and filled with dGTP and *Sal*I-digested, in the fixed direction (*see* Notes 1 and 2, and Table 1). Nucleotide sequencing is performed from the 5'-ATCG-primer side. If the newly cloned sequence is identical with the known sequence beyond the primer sequence, the desired DNA fragment is cloned.

The end-trimming method is useful because:

1. Fewer extraneous nucleotides are added to the ends of the primers, improving the specificity of the sequence-matched primer and decreasing primer cost;
2. Only one or zero restriction digests have to be carried out, thereby decreasing the chance that the end will be cut at a restriction site in the unknown portion, which makes the cDNA end unclonable or truncated in length; and
3. This procedure confines the direction of an inserted DNA fragment.

The method of PCR with end-trimming and cassette ligation (ETCL-PCR) was developed following the end-trimming method to clone the sequence adjacent to the known sequence by PCR *(5,6)*. The rationale of the method is shown schematically in Fig. 2. The DNA containing a desired sequence produces the 5' end overhang after digesting with endonucleases. The endonuclease-digested ends are end-trimmed by filling with a single nucleotide using Klenow fragment *(4)*. This procedure prevents the endonuclease-digested DNA fragments from ligating each other. The synthesized cassettes are ligated with end-trimmed DNA fragments. After that, nested PCRs are performed using two sets of a sequence-matched primer and a cassette annealing primer. As shown in Fig. 2, in the initial PCR, a primer targeted to a known sequence can work from the first cycle of PCR, whereas a complementary primer to a cassette can work only from the second cycle of PCR. Therefore, a DNA fragment, which elongates from a primer annealing to a known sequence, is the only specific fragment to be amplified. The PCR-amplified DNA fragment is cloned into a plasmid vector and the sequence is determined. As in an end-trimming method, if a part of newly cloned sequence, except for a primer sequence, is identical with the known sequence, DNA fragment adjacent to the known sequence is cloned.

The ETCL-PCR is useful because:

1. An end-trimmed DNA and a cassette do not selfligate (*see* Note 3).
2. A DNA fragment, which elongates from a primer annealing to a known sequence, is the only specific sequence to be amplified; and
3. Changing the 5' end sequence adaptor cassette (AC) oligomers makes it possible to apply ETCL-PCR to other restriction enzymes, which produce the 5' overhang, in addition to those stated in this protocol.

**Table 1**
**Combinations of Restriction Enzymes Applicable for an End-Trimming Method and 5'-Ends of PCR Primers Attached to Sequence-Matched Primers**

| Restriction enzymes[a] | Cloning site of a vector | | | | PCR products as an insert | |
|---|---|---|---|---|---|---|
| | Restriction sequences | Enzyme-digested ends | End-trimmed vectors[b] | | End-trimmed inserts[c] | 5'-end o PCR prime |
| *Bln*I | 5'-CCTAGG-3' / 3'-GGATCC-5' → | 5'-C / 3'-GGATC-5' → | 5'-CC / 3'-GGATC-5' | | | |
| *Nhe*I | 5'-GCTAGC-3' / 3'-CGATCG-5' → | 5'-G / 3'-CGATC-5' → | 5'-GC / 3'-CGATC-5' | | | |
| *Spe*I | 5'-ACTAGT-3' / 3'-TGATCA-5' → | 5'-A / 3'-TGATC-5' → | 5'-AC / 3'-TGATC-5' | + | 5'-TAGC—3' / G → | 5'-TAGC—3' |
| *Xba*I | 5'-TCTAGA-3' / 3'-AGATCT-5' → | 5'-T / 3'-AGATC-5' → | 5'-TC / 3'-AGATC-5' | | | |
| *Bam*HI | 5'-GGATCC-3' / 3'-CCTAGG-5' → | 5'-G / 3'-CCTAG-5' → | 5'-GG / 3'-CCTAG-5' | | | |
| *Bgl*II | 5'-AGATCT-3' / 3'-TCTAGA-5' → | 5'-A / 3'-TCTAG-5' → | 5'-AG / 3'-TCTAG-5' | + | 5'-ATCG—3' / C → | 5'-ATCG—3' |
| *Fba*I | 5'-TGATCA-3' / 3'-ACTAGT-5' → | 5'-T / 3'-ACTAG-5' → | 5'-TG / 3'-ACTAG-5' | | | |
| *Bsp*HI | 5'-TCATGA-3' / 3'-AGTACT-5' → | 5'-T / 3'-AGTAC-5' → | 5'-TC / 3'-AGTAC-5' | + | 5'-ATGC—3' / G → | 5'-ATGC—3' |
| *Nco*I | 5'-CCATGG-3' / 3'-GGTACC-5' → | 5'-C / 3'-GGTAC-5' → | 5'-CC / 3'-GGTAC-5' | | | |

| Enzyme | Recognition | | Digested | | Filled in[b] | | | | Final |
|---|---|---|---|---|---|---|---|---|---|
| *(top, partial — name cut off)* | 3'-CAGCTG-5' | → | 3'-CAGCT-5' | | 5'-GT<br>3'-CAGCT-5' | + | 5'-CGAT—3'<br>   **A** | → | 5'-CGAT—3' |
| *Xho*I | 5'-CTCGAG-3'<br>3'-GAGCTC-5' | → | 5'-C<br>3'-GAGCT-5' | | 5'-C**T**<br>3'-GAGCT-5' | + | 5'-CGAT—3'<br>   **A** | → | 5'-CGAT—3' |
| *Apa*LI | 5'-GTGCAC-3'<br>3'-CACGTG-5' | → | 5'-G<br>3'-CACGT-5' | | 5'-G**T**<br>3'-CACGT-5' | + | 5'-GCAT—3'<br>   **A** | → | 5'-GCAT—3' |
| *Hind*III | 5'-AAGCTT-3'<br>3'-TTCGAA-5' | → | 5'-A<br>3'-TTCGA-5' | | 5'-A**A**<br>3'-TTCGA-5' | + | 5'-GCTA—3'<br>   **T** | → | 5'-GCTA—3' |
| *Spl*I | 5'-CGTACG-3'<br>3'-GCATGC-5' | → | 5'-C<br>3'-GCATG-5' | | 5'-C**G**<br>3'-GCATG-5' | + | 5'-TACG—3'<br>   **C** | → | 5'-TACG—3' |

[a] Six-base restriction enzymes that create 5' overhangs with four different bases are recommended because end-trimming with a Klenow fragment of *E. coli* DNA polymerase I is possible with one dNTP.

[b] Restriction enzyme-digested ends are filled with one dNTP, which is shown by an outlined letter, by Klenow fragment.

[c] PCR products are treated by T4 DNA polymerase in the presence of one dNTP.

[d] PCR primer should have different bases next to the compatible bases. It is at this point that 3'-exonuclease activity of T4 DNA polymerase stops in the presence of one dNTP as shown by an outlined letter.

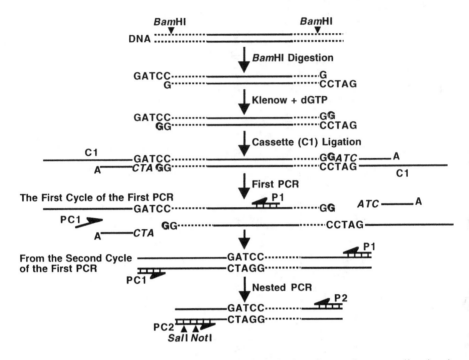

**Fig. 2.** Schematic presentation of PCR with end-trimming and cassette ligation in the case of *Bam*HI-digestion. DNA was digested with *Bam*HI and partially filled with dCTP using Klenow fragment of *E. coli* DNA polymerase I. C1 cassette (Fig. 3) was ligated to the *Bam*HI-digested and end-trimmed DNA. The initial PCR was performed using P1, an outer primer annealing to the known sequence, and PC1, an outer primer annealing to cassette. P1 can work as a primer from the first cycle of the initial PCR, but PC1 cannot work as a primer until the second cycle of the initial PCR, because the dephosphorylated 5' end of C1 is not ligated to the end-trimmed DNA. The 10-fold diluted initial PCR product was used as a template in the nested PCR. The nested PCR was carried out using P2, an inner primer annealing to the known sequence, and PC2, an inner primer annealing to the cassette. PC2 has two endonuclease recognition sites of *Sal*I and *Not*I. Solid lines, broken lines, and bold lines indicate known sequences, unknown sequences, and cassettes, respectively.

## 2. Materials

1. A system for synthesizing first- and second-strand cDNA, or a system for preparing genomic DNA.
2. Restriction enzymes: *Bam*HI, *Bgl*II, *Fba*I, *Mbo*I, *Bln*I, *Nhe*I, *Spe*I, *Xba*I, *Sal*I, *Xho*I, or *Not*I and 10X buffers as supplied by manufacturer.
3. T4 DNA polymerase and buffer (50 m$M$ Tris-HCl, pH 8.0, 5 m$M$ MgCl$_2$, 5 m$M$ dithiothreitol (DTT) and 50 µg/mL bovine serum albumin [BSA]).
4. Klenow fragment of *E. coli* DNA polymerase I.

5. Buffer K: 20 m$M$ Tris-HCl, pH 8.5, 10 m$M$ MgCl$_2$, 1 m$M$ DTT, 100 m$M$ KCl.
6. Oligomers: sequence-matched primers, 5'-ATCG-primers, a dT-primer, oligomers for constructing cassettes (Fig. 3), and cassette annealing primers (Fig. 3). *See* Note 4.
7. Deoxyribonucleoside triphosphates (dATP, dCTP, dGTP, and dTTP).
8. Reagents for PCR (*see* Chapter 1).
9. A system for purifying the PCR products. SUPREC™-O2 (Takara, Kyoto, Japan) is used in an end-trimming method. Slightly modified crush and soak method *(7)* is used in an end-trimming and cassette ligation method. This requires 0.5$M$ ammonium acetate, 1 m$M$ EDTA, pH 8.0, and 1% SDS.
10. Plasmid vectors: pUC19 and pBluescript II SK(+) (Stratagene, La Jolla, CA). *See* Note 5.
11. A system for DNA ligation. We use DNA Ligation Kit Ver. 1 (Takara).
12. A system for transforming competent *E. coli* cells. We used a Transformation Kit JM109 (Nippon Gene, Tokyo, Japan) preparing plasmids in a small scale *(8)*.
13. Reagents for polyacrylamide gel electrophoresis.
14. A system for determining a nucleotide sequence. We used the *Taq* Dye Primer Cycle Sequencing Core Kit (401112) in the 373A DNA sequencer (Perkin Elmer, Applied Biosystems Division, Foster City, CA).
15. A computer system for analyzing nucleotide sequence data. We use the SDC-GENETYX Version 8.0 program (SDC, Tokyo, Japan).
16. Distilled-autoclaved (DA) water.
17. 3$M$ sodium acetate, pH 5.2.
18. Ethanol, 99.5 and 70%.

## 3. Method

### 3.1. End-Trimming Method

#### 3.1.1. Primer Design

1. A sequence-matched primer and a 5'-ATCG-primer are designed according to the known sequence. A 5'-ATCG-primer has a 5'-ATCG-3' sequence at the 5' end of a known sequence and should be in the downstream of the sequence-matched primer.
2. The sequence of a dT-primer is 5'-GA<u>CTCGAGTCGACATCGA</u>TTTTTTTTTT-TTTTTT-3'. The underlined letters indicate the sequence to create restriction sites for *Xho*I, *Sal*I, and *Cla*I.

#### 3.1.2. cDNA Synthesis

1. Synthesize cDNA from 1 μg of poly(A)$^+$ RNA using a kit for cDNA synthesis according to manufacturer's instructions in a final volume of 100 μL.
2. Ethanol precipitate the synthesized cDNA. Add 10 μL of 3$M$ sodium acetate and 275 μL of 99.5% ethanol. Mix by vortexing and incubate at –20°C for at least 30 min. Centrifuge at 18,500$g$ at 4°C for 20 min. Discard the supernatant and wash the pellet with 70% ethanol and let pellet dry.
3. Dissolve the precipitated cDNA in 20 μL of DA-water and store at –20°C.

MC 5'HO-CCTCTTCGCTATTACGCCAGTCGACGCGGCCGCAAATC-OH 3'

+

AC1 5'HO-*ATC*GATTTGCGGCCA-OH 3'

AC2 5'HO-*TAG*GATTTGCGGCCA-OH 3'

AC3 5'HO-*CGA*GATTTGCGGCCA-OH 3'

↓

C1 5'HO-CCTCTTCGCTATTACGCCAGTCGACGCGGCCGCAAATC-OH 3'
3'HO-ACCGGCGTTTAG*CTA*-OH 5'
Cassette for *Bam* HI, *Bgl* II, *Fba* I and *Mbo* I

C2 5'HO-CCTCTTCGCTATTACGCCAGTCGACGCGGCCGCAAATC-OH 3'
3'HO-ACCGGCGTTTAG*GAT*-OH 5'
Cassette for *Bln* I, *Nhe* I, *Spe* I and *Xba* I

C3 5'HO-CCTCTTCGCTATTACGCCAGTCGACGCGGCCGCAAATC-OH 3'
3'HO-ACCGGCGTTTAG*AGC*-OH 5'
Cassette for *Sal* I and *Xho* I

P1 5'HO-CCTCTTCGCTATTACGCCAG-OH 3'

P2 5'HO-AATAAGTCGACGCGGCCGCAAATC-OH 3'

**Fig. 3.** Schematic presentation of the construction of cassettes. The main cassette oligomer (MC oligomer) was mixed with adaptor cassette oligomers (AC oligomers) AC1, AC2, and AC3, respectively, in an equal amount in molar value. The mixtures were heat-denatured and annealed by cooling on ice. We named cassettes consisting of AC1, AC2, and AC3 as C1, C2, and C3, respectively. C1 is a cassette for *Bam*HI, *Bgl*II-, *Fba*I-, and *Mbo*I-digested and end-trimmed DNAs. C2 is a cassette for *Bln*I, *Nhe*I-, *Spe*I-, and *Xba*I-digested and end-trimmed DNAs. C3 is a cassette for *Sal*I- and *Xho*I-digested and end-trimmed DNAs. PC1 is an outer primer annealing to cassettes and PC2 is an inner primer annealing to cassettes. The underlined letters indicate the sequence to create restriction enzyme recognition sites for *Sal*I and *Not*I.

### 3.1.3. PCR Amplification

1. Perform the first PCR with a sequence-matched primer and a dT-primer (use primers at a concentration of 1 $\mu M$ in all PCR reactions) and 1 $\mu$L of the synthetic cDNA as a template in a standard condition in a total volume of 10 $\mu$L. Dilute the first PCR product 10-fold with DA-water. Perform the second PCR using a 5'-ATCG-primer and a dT-primer, and the 10-fold dilution of the first PCR product as a template in a total volume of 100 $\mu$L.
2. Amplify by PCR using the following cycle profiles: 30–40 main cycles: 94°C for 1 min (denaturation), 50°C for 1 min (annealing), and 72°C for 3 min (extension); and 72°C for 7 min (final extension).
3. Inactivate *Taq* DNA polymerase in the second PCR product with phenol/chloroform extraction.
4. Remove primers, free deoxyribonucleoside triphosphate (dNTPs), and small DNA fragments using SUPREC-02 according to manufacturer's directions.

### 3.1.4. End Trimming of a PCR-Amplified DNA Fragment

In this reaction, the 3' to 5' exonuclease activity of T4 DNA polymerase is stopped at the first C from the 3' end and a 5'-ATC overhang is created at the 5'-ATCG-primer side of the PCR product. Because the 5' end of dT-primer is G, the 3' end of the dT-primer side is C. Thus, the dT-primer side is not end-trimmed.

1. End-trim the 3' end of the PCR-amplified DNA fragment with T4 DNA polymerase in a total volume of 50 $\mu$L in T4 DNA polymerase buffer, containing 50 m$M$ Tris-HCl, pH 8.0, 5 m$M$ MgCl$_2$, 5 m$M$ DTT, and 50 $\mu$g/mL BSA. Contents: A PCR-amplified DNA fragment (0.2–0.6 $\mu$g); 100 $\mu M$ dCTP; and 0.6 U T4 DNA polymerase; Incubation: 11°C for 20 min.
2. Inactivate T4 DNA polymerase by heat at 75°C for 10 min, then remove protein with phenol/chloroform extraction.

### 3.1.5. Preparation of a Vector

1. Digest pUC19 (0.1 $\mu$g) with 5 U of *Bam*HI in buffer K (20 m$M$ Tris-HCl, pH 8.5, 10 m$M$ MgCl$_2$, 1 m$M$ DTT, 100 m$M$ KCl) in a total volume of 20 $\mu$L for 2–3 h.
2. Fill the *Bam*HI-digested end with dGTP using Klenow fragment: 20 $\mu$L *Bam*HI-digested pUC19, 0.5 $\mu$L dGTP (10 m$M$), 0.25–0.5 $\mu$L Klenow (2–4 U/$\mu$L), and incubate at 30°C for 15 min.
3. Inactivate enzymes with heat at 75°C for 10 min, then remove protein with phenol/chloroform extraction. This vector is named pUC19 (*Bam*HI+G).

### 3.1.6. Sal I Digestion and Ligation

1. Combine the pUC19 *(Bam*HI+G) (in 20 $\mu$L) and the end-trimmed PCR product (in 50 $\mu$L) and perform ethanol precipitation by adding 7 $\mu$L of 3$M$ sodium acetate and 175 $\mu$L of 99.5% ethanol, and process as described in Section 3.1.2., step 2.

2. Digest DNAs with *Sal*I in buffer H (50 m*M* Tris-HCl, pH 7.5, 10 m*M* MgCl₂, 1 m*M* DTT, 100 m*M* NaCl) in a total volume of 30 μL at 37°C for 2–3 h. Ethanol precipitate by adding 30 μL of 5*M* ammonium acetate and 150 μL of 99.5% ethanol, and process as described in Section 3.1.2., step 2.

3. Ligate the end-trimmed and *Sal*I-digested PCR product (0.2–0.6 μg) with *Sal*I-digested pUC19 *(Bam*HI+G) (100 ng) using T4 DNA ligase and DNA Ligation Kit Version 1 in a total volume of 10 μL.

4. Transform 50 μL of competent JM109 cells (using Transformation Kit JM109) with 5 μL of recombinant pUC19, then prepare plasmid DNAs with a small-scale method.

5. Check the sizes of inserted fragments by double digestion with *Eco*RI and *Pst*I in buffer H, followed by electrophoresis on an 8% polyacrylamide gel of 138 (width) × 130 (height) × 2 mm (thickness) at 200 V for 90 min.

### 3.1.7. Nucleotide Sequencing and Computer-Aided Analysis

1. Perform DNA sequencing of the recombinant pUC19 using the *Taq* Dye Primer Cycle Sequencing Core Kit in the 373A DNA sequencer.

2. Analyze the nucleotide sequence identity between known and unknown sequences using the SDC-GENETYX Version 8.0 program.

## 3.2. PCR with End-Trimming and Cassette Ligation Method

### 3.2.1. Constructing Cassettes

1. Synthesize the oligomers shown in Fig. 3 *(see* Note 6).

2. Anneal an MC oligomer to AC1–AC3 oligomers. Mix 10 nmol of MC oligomer with 10 nmol each of AC1–AC3 in a total volume of 100 μL of DA-water.

3. Heat-denature these mixtures at 90°C for 3 min, then cool on ice for 5 min.

4. Resulting cassettes 1–3 (C1–C3) have the 5' overhang sequences of 5'-ATC-3', 5'-TAG-3', and 5'-CGA-3', respectively.

### 3.2.2. Endonuclease Digestion

1. Divide DNA into small aliquots for digestion with three groups of enzymes. Each aliquot should contain 0.5–1 μg DNA.

2. Digest DNA (0.5–1 μg) with 5–10 U of group 1 enzymes, *Bam*HI, *Bgl*II. *Fba*I, or *Mbo*I in a buffer recommended by the manufacturer in a total volume of 30 μL at 37°C for between 2 h and overnight. In the same way, digest DNA with 5–10 U of group 2 enzymes, *Bln*I, *Nhe*I, *Spe*I, or *Xba*I, with 5–10 U of group 3 enzymes, *Sal*I or *Xho*I *(see* Note 7).

### 3.2.3. End Trimming

With these procedures, endonuclease-digested ends of DNA in each group are end-trimmed with only G, C, or T, respectively.

1. Add 1 μL of 5 m*M* dGTP, dCTP, or dTTP to the endonuclease-digested DNA in 30 μL in groups 1, 2, and 3, respectively.

2. Add 2 U of Klenow fragment of *E. coli* DNA polymerase I to DNA in each group. Incubate these mixtures at 30°C for 15 min.
3. Inactivate Klenow fragment by heat at 75°C for 10 min.
4. Ethanol precipitate by adding 30 µL of 5*M* ammonium acetate and 150 µL of 99.5% ethanol, and process as described above.

### 3.2.4. Cassette Ligation

1. Add 1 µL of C1–C3 to ethanol precipitated DNA in groups 1–3, respectively.
2. Ligate each cassette with DNA as described in Section 3.1.6., step 3.
3. Perform ethanol precipitation in each group by adding 10 µL of 5*M* ammonium acetate and 50 µL of 99.5% ethanol, and process as described above. Dissolve the ethanol precipitated DNA in 10 µL of DA-water. This step is neccessary because the buffer for ligation reactions is inhibitory to PCR.

### 3.2.5. Initial PCR Amplification

1. Assemble a 10 µL PCR reaction mixture for each DNA. Each mixture contains 1 µL of each end-trimmed and cassette-ligated DNA as a template and 1 µ*M* of P1, an outer primer designed to anneal to the known sequence, and 1 µ*M* PC1, an outer primer annealing to cassettes.
2. Perform PCR using the following cycle profiles: 30 main cycles: 95°C for 1 min (denaturation), 55°C for 1 min (annealing), 72°C for 3 min (extension); and 72°C for 7 min (final extension).
3. Dilute PCR products 10-fold with DA-water, then use as templates in the nested PCRs.

### 3.2.6. Nested PCR Amplification

1. Assemble the second PCR mixture in a 10 µL final volume containing 1 µL of the diluted DNA from the first PCR as a template and 1 µ*M* of P2, an inner primer annealing to the known sequence, and 1 µ*M* of PC2, an inner primer annealing to cassettes.
2. Amplify by PCR using the same cycle profiles as the initial PCR.

### 3.2.7. Electrophoresis and Isolation of PCR Products

1. Electrophorese the PCR products on an 8% polyacrylamide gel as described in Section 3.1.6., step 5.
2. Stain DNA fragments with ethidium bromide and visualize them with UV transillumination.
3. Cut out the discrete bands from the polyacrylamide gel.
4. Extract DNA fragments by the "crush and soak" method with slight modification as described previously *(7)*. Briefly, crush the gel slice in a 1.5-mL tube and suspend it in about 2 vol of elution buffer containing 0.5*M* ammonium acetate, 1 m*M* EDTA, pH 8.0, and 1% SDS.
5. Incubate at 37°C for several hours with shaking and centrifuge at 18,500*g* for 2 min at 4°C. Transfer the supernatant to a fresh microfuge tube.

6. Resuspend the pellet in 0.5 vol of elution buffer and centrifuge as in step 5.
7. Combine the two supernatants and centrifuge once more. Transfer the supernatant to a fresh microfuge tube. Avoid transferring fragments of polyacrylamide. Add 2 vol of 99.5% ethanol and incubate at −20°C for at least 30 min.
8. Centrifuge at 18,500g at 4°C for 20 min. Discard supernatant and dissolve the pellet in 200 μL of DA water. Add 20 μL of 3*M* sodium acetate and 550 μL of 99.5% ethanol and precipitate as described in Section 3.1.2., step 2.
9. Quantitate the recovered DNA by spectrophotometric measurement at the wavelength of 260 nm.

## 3.2.8. Cloning of PCR-Amplified DNA Fragments

1. Mix the recovered DNA (0.01–0.02 μg) and pBluescript II SK(+) (0.1 μg).
2. Digest the mixed DNA with 5 U of *Not*I and an endonuclease, of which recognition site is introduced into P2 and located in the multicloning site of pBluescript, in the buffer recommended by the manufacturer at a total volume of 20 μL.
3. Perform ethanol precipitation for endonuclease-digested DNA and pBluescript by adding 20 μL of 5*M* ammonium acetate and 100 μL of 99.5% ethanol, and process as described in Section 3.1.2., step 2.
4. Ligate the DNA fragment with pBluescript as described in Section 3.1.6., step 3.
5. Transform the competent JM109 with recombinant pBluescript.
6. Prepare plasmid DNA by a small-scale method compatible with sequencing method.
7. Determine nucleotide sequence and analyze the nucleotide sequence identity between the known sequence and the part of a newly cloned DNA fragment (*see* Note 8).

## 4. Notes

1. Although we have used *Bam*HI to create an end-trimmed plasmid vector, pUC19 (*Bam*HI+G), our end-trimming method can be applied to restriction endonucleases other than *Bam*HI. Especially six-base restriction enzymes that create 5' overhangs with four different bases can be used because end-trimming with Klenow fragment is possible with one dNTP. We showed some examples of restriction endonucleases, recognition sequences, and sequences added to 5' ends of sequence-matched primers in Table 1. Plasmid vectors having these endonuclease recognition sites in their cloning sties can be used as a cloning vector for PCR-amplified DNA fragments. However, it is theoretically possible to use any restriction enzymes, even blunt-end or 3' overhang cutters. In these cases, T4 DNA polymerase is used to create 5' overhangs in the presence of one dNTP.
2. The combination of an end-trimmed vector and an end-trimmed insert makes the directional cloning possible (Fig. 1). The end-trimming method is applicable to both 5' and 3' sides of the PCR-amplified DNA fragments and makes it easy to clone them into plasmid vectors. For example, pBluescript is double-digested with *Bam*HI and *Not*I, then both ends of double-digested vector are partially filled with dGTP using Klenow fragment. The *Bam*HI- and *Not*I-digested ends have 5'-

GAT and 5'-GG overhangs, respectively. The DNA fragment, which is amplified with 5'-ATCG-primer and 5'-CCG-primer and end-trimmed by T4 DNA polymerase in the presence of only dCTP, is cloned in the fixed direction in this site.

3. In ETCL-PCR, the end-trimmed DNA does not ligate to itself, because restriction endonuclease-digested DNA is partially filled with only one kind of dNTP using Klenow fragment (Fig. 2). Moreover, each synthesized cassette also does not selfligate, because it has no phosphates at the 5' ends and no complementary ends to itself (Fig. 3). Because of these characteristics, the end-trimmed DNA is expected to ligate exclusively to a compatible cassette.

4. To make a compatible end to the end-trimmed vector, PCR primer should have different bases next to the compatible bases. It is at this point that the 3'-exonuclease activity of the T4 DNA polymerase stops in the presence of one dNTP.

5. Because PC2, a complementary primer to a cassette, has *Sal*I and *Not*I recognition sites (Fig. 3), the PCR-amplified DNA fragments can easily be cloned into a plasmid, such as pBluescript.

6. A mismatched A is introduced at the 3' end of an AC oligomer so that it inhibits AC oligomer from working as a PCR primer on a cassette (Figs. 2 and 3).

7. The average sizes of DNA fragments produced by 4- or 6-bp recognition restriction enzymes are theoretically estimated as $4^4 = 256$ bp or $6^4 = 4096$ bp, respectively. We used one 4-base cutter and nine different 6-base cutters. Therefore, it is estimated that there is one 4-base cutter recognition site in every 256 bp and there is one 6-base cutter recognition site in every $4096/9 = 455$ bp, on average. Thus, the probability of amplifying a DNA fragment using ETCL-PCR with these 10 different restriction endonucleases is very high.

8. Because the haploid genome sizes of human, rat, and mouse are about $3 \times 10^9$ bp, a sequence of 16 bp stretch ($4^{16} = 4.3 \times 10^9$) is unique in the genome of those species. Therefore, if the newly cloned sequence is longer than 16 bp, except for the part of the primer that is identical with the known sequence, the newly cloned DNA fragment is considered to be correct. However, it should be confirmed that a DNA fragment can be amplified with a primer annealing to the known sequence and a primer annealing to the newly cloned sequence.

## References

1. Iwahana, H., Yamaoka, T., Mizutani, M., Mizusawa, N., Ii, S., Yoshimoto, K., and Itakura, M. (1993) Molecular cloning of rat amidophosphoribosyltransferase. *J. Biol. Chem.* **268**, 7225–7237.

2. Iwahana, H., Mizusawa, N., Ii, S., Yoshimoto, K., and Itakura, M. (1994) An end-trimming method to amplify adjacent cDNA fragment by PCR. *BioTechniques* **16**, 94–98.

3. Frohman, M. A., Dush, M. K., and Martin, G. R. (1988) Rapid production of full-length cDNAs from rare transcripts: amplification using a single gene-specific oligonucleotide primer. *Proc. Natl. Acad. Sci. USA* **85**, 8998–9002.

4. Tabor, S., Struhl, K., Scharf, S. J., and Gelfand, D. H. (1987) DNA-dependent DNA polymerases, in *Current Protocols in Molecular Biology* (Ausubel, F. M.,

Brent, R., Kingston, R. E., Moore, D. D., Seidman, J. G., Smith, J. A., and Struhl, K., eds.), Wiley, New York, pp. 3.5.1–3.5.15.

5. Iwahana, H., Tsujisawa, T., Katashima, R., Yoshimoto, K., and Itakura, M. (1994) PCR with end trimming and cassette ligation: a rapid method to clone exon-intron boundaries and a 5'-upstream sequence of genomic DNA based on a cDNA sequence. *PCR Methods Applic.* **4**, 19–25.

6. Iwahana, H., Honda, S., Tsujisawa, T., Takahashi, Y., Adzuma, K., Katashima, R., Yamaoka, T., Moritani, M., Yoshimoto, K., and Itakura, M. (1995) Rat genomic structure of amidophosphoribosyltransferase, cDNA sequence of aminoimidazole ribonucleotide carboxylase, and cell cycle-dependent expression of these two physically linked genes. *Biochim. Biophys. Acta.* **1261**, 369–380.

7. Iwahana, H., Yoshimoto, K., Tsujisawa, T., and Itakura, M. (1994) T-cassette ligation: a method for direct sequencing and cloning of PCR-amplified DNA fragments. *PCR Methods Applic.* **3**, 219–224.

8. Sambrook, J., Fritsch, E. F., Maniatis, T. (1989) Small-scale preparations of plasmid DNA, in *Molecular Cloning: A Laboratory Manual* (2nd ed.), Cold Spring Harbor Laboratory, Cold Spring Harbor, NY, pp. 1.25–1.30.

# 27

## Anchoring a Defined Sequence to the 5' Ends of mRNAs

*The Bolt to Clone Rare Full-Length mRNAs*

**Jean Baptiste Dumas Milne Edwards, Olivier Valdenaire, and Jacques Mallet**

## 1. Introduction

### 1.1. Amplification of Low Amounts of Messenger RNAs

Among numerous applications, the PCR *(1,2)* provides a convenient means to clone 5' ends of rare messengers and to generate cDNA libraries from tissue available in amounts too low to be processed by conventional methods (e.g., screening of cDNA libraries). Basically, the amplification of cDNAs by the PCR requires the availability of the sequences of two stretches of the molecule to be amplified. A sequence can easily be imposed at the 5' end of the first strand cDNAs (corresponding to the 3' end of the mRNAs) by priming the reverse transcription with a specific primer (for cloning the 5' end of rare messenger). Several strategies have been devised to tag the 3' end of the ss-cDNAs (corresponding to the 5' end of the mRNAs). We *(3)* and others have described strategies based on the addition of a homopolymeric dG *(4,5)* or dA *(6,7)* tail using terminal deoxyribonucleotide transferase (TdT) ("anchor PCR" *[4]*). However, this strategy has important limitations. The TdT reaction is difficult to control and has a low efficiency. More importantly, however, the return primers containing a homopolymeric (dC or dT) tail generate nonspecific amplification, a phenomenon that prevents the isolation of low abundance mRNA species *(3)*, and/or interferes with the relative abundance of primary clones in the library. To circumvent these drawbacks two approaches have been devised. First, a strategy based on "a cRNA enrichment" procedure has been

From: *Methods in Molecular Biology, Vol. 67: PCR Cloning Protocols: From Molecular Cloning to Genetic Engineering* Edited by: B. A. White Humana Press Inc., Totowa, NJ

useful to eliminate nonspecific-PCR products and to allow detection and cloning of cDNAs of low abundance *(3)*. More recently, to avoid the nonspecific amplification caused by the annealing of the homopolymeric tail oligonucleotide, we have developed a novel anchoring strategy that is based on the ligation of an oligonucleotide to the 3' end of ss-cDNAs. This strategy is referred to as "SLIC" for single-strand ligation to ss-cDNA *(8)*.

The SLIC method has been used in identifying 5' ends of many messengers *(8–12)*. Most of the studies were done starting with the knolwedge of the sequence in nucleic acids. Recently, the SLIC strategy has also been used to isolate a large fragment of a messenger of which only a short stretch of the amino-acid sequence was known. This involves the use of degenerate oligonucleotides to perform primer extension experiments and PCR experiments after ligation. This is discussed in this chapter.

SLIC strategy can be used to generate cDNA library from a low amount of cells. The potential of this method to generate a cDNA library will be discussed in Chapter 32.

## *1.2. Single-Strand Ligation-Mediated Anchor PCR: The SLIC Strategy*

The SLIC strategy is illustrated in Fig. 1. The ss-cDNA is primed from a known sequence within the RNA (oligonucleotide PEX, Figs. 1 and 2). After RNA hydrolysis (Sections 2.6. and 3.5.) and removal of the primers (Sections 2.5. and 3.4.), a modified oligonucleotide, A5'_NV (Figs. 1 and 2), is ligated to the 3' end of the ss-cDNA in the presence of T4 RNA ligase (Sections 2.7. and 3.6.), that has also been shown to ligate short single-stranded DNA fragments *(13,14)*.

Three precautions are needed to target the ligation of the A5'_NV oligonucleotide specifically to the 3' end of the ss-cDNA and to avoid self ligation and/or circularization. First, the oligonucleotide that primes the ss-cDNA synthesis must not contain a phosphate at its 5' end. Second, the 3' end of the A5'_NV oligonucleotide must be blocked; this was performed by adding a dideoxyribonucleotide to the 3' end of the primer (*see* Sections 2.7.1. and 3.6.1.). Finally, the oligonucleotide used to prime the ss-cDNA synthesis must be removed prior to carrying out (Sections 2.5.1. and 3.4.) the ligation experiment since it can compete with the ss-cDNA for ligation to A5'_NV. As the strategy described in this chapter requires many oligonucleotides, location and sequences of the oligonucleotides used are given in Fig. 2.

## *1.3. Performing the SLIC Strategy with Degenerate Oligonucleotides*

To clone a cDNA out of a single peptide sequence or to look for new members of a gene family, the SLIC strategy may be followed using degenerate

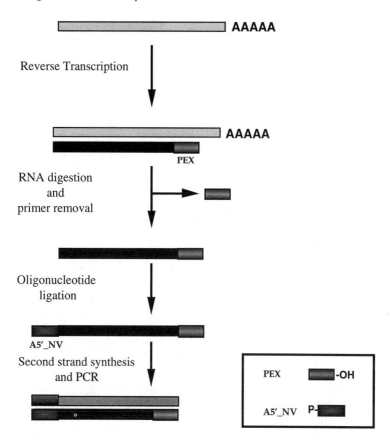

**Fig. 1.** Outlines of the single-strand ligation to cDNA strategy. After carrying out a reverse transcription primed with a specific primer PEX, the primer is removed to avoid its ligation to A5'_NV and the RNA are digested. The A5'_NV oligonucleotide is ligated to the 3' end of the ss-cDNA and PCR is carried out. To avoid the formation of concatemers, A5'_NV must have no hydroxyl group to its 3' end and a phosphate group to its 5' extremity. Primer PEX must have no phosphate group at its 5' end to avoid circularization or concatemerization of the ss-cDNA.

oligonucleotides. This may be especially useful when a single short peptide sequence of the protein (or a single conserved domain in the case of a protein family) is available. In our hands, the SLIC strategy allowed the isolation of cDNAs even when regular reverse transcription PCR experiments carried out with degenerate oligonucleotides failed to yield any result. Disposing of a single 30-residue peptide sequence of the rat oxydosqualene cyclase (Fig. 3), we have cloned a 577-bp cDNA fragment, which was used as a probe to isolate by a conventional screening a full-length cDNA encoding this protein.

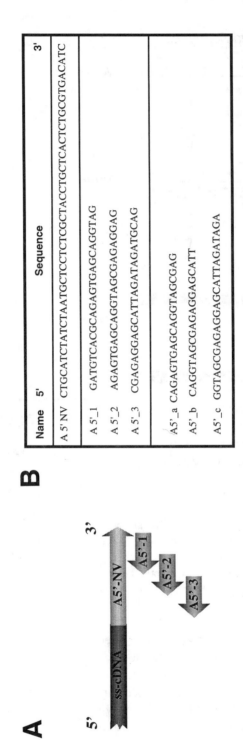

**Fig. 2.** Location (**A**) and sequences (**B**) of the primers used to isolate the 5' ends of mRNAs. (**A**): A5'_NV is ligated to the 3' end of the ss-cDNA. Three oligonucleotides A5'_1,2,3 are designed from the sequence of A5'_NV to be used in PCR experiments. Each primer is overlapping its neighbor by 10 bases. As the $T_m$ of the first set of oligonucleotides was relatively high (55–60°C), Philippe (10) has designed three other oligonucleotides, A5'_a, b, and c, which have a $T_m$ of 50°C.

The following is the content of table B:

| Name | 5' | Sequence | 3' |
|------|-----|----------|-----|
| A 5' NV | CTGCATCTATCTAATGTCCTCTCGCTACCTGCTCACTCTGCGTGACATC | | |
| A 5'_1 | GATGTCACGCAGAGTGAGCAGGTAG | | |
| A 5'_2 | AGAGTGAGCAGGTAGCGAGAGGAG | | |
| A 5'_3 | CGAGAGGAGCATTAGATAGATGCAG | | |
| A 5'_a | CAGAGTGAGCAGGTAGCGAG | | |
| A 5'_b | CAGGTAGCGAGAGGAGCATT | | |
| A 5'_c | GGTAGCGAGAGGAGCATTAGATAGA | | |

**Fig. 3.** Example of oligonucleotides used to perform the SLIC strategy with degenerated oligonucleotides to isolate the 5' end of rat oxydosqualene cyclase (OSC). Primer extension was performed with OSC1 using 1 μg of poly A+ RNA. Two nested PCR were performed with A5'_1, OSC2 and A5'_2, OSC3 primer pairs using three annealing temperatures: 58.5°C ($T_m$max), 54.5°C ($T_m$med) and 51.5°C ($T_m$min). Amplification products were Southern blotted and probed with OSC4. PCR perfomed at annealing temperature $T_m$med and $T_m$min revealed that strong positive signals corresponding to the rat OSC cDNA fragments of 577 base pairs.

## 1.4. Conclusion

Anchoring a defined sequence at the 3' end of the ss-cDNAs is a crucial issue. The nonspecific amplification associated with the use of homopolymeric tails prevents the isolation of low abundant mRNAs. In contrast, single-strand ligation-mediated anchor PCR (the SLIC strategy) yields single-stranded molecules with two defined ends containing specific sequences usable as targets for PCR primers. Thus, the problems encountered using the SLIC strategy are only those characteristic of classical PCR amplifications.

The addition of a known sequence to the 3' end of the ss-cDNA provides an anchor sequence to target the priming of the second strand synthesis. In contrast with other methods used to synthesize the second strand cDNA, it allows the synthesis of a complete cDNA molecule without cutting the end of the molecule. Thus, the SLIC strategy could be useful in the generation of regular cDNA libraries.

The isolation of 5' ends of messengers starting from the knolwedge of a peptide sequence can be difficult, especially if the known peptide corresponds to the N-terminus of the protein. To clone a cDNA fragment located downstream from the peptide sequence, we suggest performing a *Taq* polymerase extension with a specific biotinylated primer, using random primed cDNA as a template. The biotinylated single-stranded DNAs are then isolated with streptavidin beads, ligated to the protected A5'_NV oligonucleotide, and used as a template for nested PCR.

## 1.5. Further Applications of the SLIC Strategy

We can anticipate that the SLIC strategy, which is likely to replace tailing in all the strategies involving anchor PCR, will be most useful in cloning promoter sequences, identifying extremities of long clones and in the construction of subtracted libraries.

Molecular cloning of promoter sequences and extremities of long clones (*YAC* or genomic) can be performed according to a unique strategy. A primer extension experiment carried out starting from the known sequence (genomic matrix or polylinker sequence) using a 5'-biotinylated oligonucleotide would lead to a single-stranded DNA molecule tagged with biotin. Neosynthetized ss-DNA could then be collected through streptavidin chromatography and ligated to A5'_NV before being subjected to nested PCR amplification and cloned.

## 2. Materials

### 2.1. Oligonucleotides

#### 2.1.1. Choice of the Oligonucleotides

1. The sequence of the oligonucleotides A5'_NV and the related primers (A5'_1,2,3) have been arbitrarily chosen. When we started this work, we used oligonucle-

5'        ?     ←—————  **known sequence**  —————-▶  3'

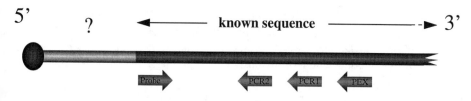

**Fig. 4.** Location and name of the oligonucleotides designed from the sequence of the messenger. At least four oligonucleotides must be used to isolate the 5' end of the messenger. PEX is used to prime the reverse transcription. PCR1 and PCR2 are designed to be used with primers A5'_1 and A5'_2 in PCR experiments. PROBE is designed to be used in Southern blot experiments as well as in PCR experiments to check the presence of the single-strand cDNA.

otides BM-3' and BM-5' *(8)* with restriction sites to facilitate the cloning of the PCR products. However, the cutting of these restriction sites was very inefficient in our hands. For these reasons, we omit the use of restriction sites in the sequences of our new primers (A5'_NV and A3'_NV). Since this first work, and as the investigations increased, some restriction sites (*Eco*RI, *Sfi*I) have been found to be well cut in PCR primers provided 3–5 bases are added between the restriction site and the 5' end of the primer. Thus, if necessary, it is possible to add sequences of restriction sites in the sequences of A5'_NV more details can be obtained in ref. *15,* but we prefer the use of modified PCR primers (e.g., A5'_2 or 3). These primers tailored with the desired restriction at the 5' end of the sequence provided in Fig. 2 can be used on an aliquot of the PCR prior to cloning of the vectors.

2. For all the oligonucleotides used, secondary structures and self complementarity should be avoided. Compatibility in $T_m$ and nonannealing of each pair of oligonucleotides (PCR1, A5'_1 and PCR2, A5'_2 or 3) should be checked.

3. Oligonucleotides related to A5'_NV are part of A5'_NV, overlapping each other by approx 10 bases. The oligonucleotides related to A5'_NV are anticomplementary to A5'_NV.

4. To clone the 5' ends of mRNAs, the sequence of the oligonucleotides is directly deduced by the messenger that has to be cloned. At least four oligonucleotides need to be used:

   a. An oligonucleotide for primer extension experiment (PEX in Figs. 1 and 2).

   b. Two (at least) for PCR experiments (PCR1 and PCR2 in Fig. 4). The sequences of these oligonucleotides as the sequence of the primer used in primer extension experiment is antisense. The compatibility of these oligonucleotides and their A5'_NV partners must be checked. In most of the SLIC experiments, the isolation of a 5' sequence can be performed with two sets of PCR primers (PCR1 and 2). When the messenger is very rare, a third set PCR primer (A5'_3 and PCR3) can be used to ampify the corresponding sequence to a yield which can be cloned.

c. One as probe for Southern blotting experiments. This oligonucleotide must be chosen close to the known 5' end of the messenger. It is preferable to choose an oligonucleotide sense. Thus, this oligonucleotide could be used with one of two oligonucleotides previously chosen to check the length of the primer extension product prior to carrying out the SLIC procedure (*see* Note 1a,b).

5. All the oligonucleotides except A_5' NV can be used as crude preparations, after ammonia deprotection and lyophilization. Since A5'_NV is a long oligonucleotide it must be purified (using polyacrylamide gels or HPLC) before use. Protocols are well described elsewhere *(16)*.

### 2.1.2. Functionalization of the Oligonucleotides

1. To avoid selfligation of the primer extension products, the oligonucleotide used to prime the synthesis of the cDNAs must not have a phosphate group at its 5' end. This is the common status of chemically synthesized oligonucleotides.
2. To allow the ligation of the 5' end of A5'_NV to the 3' end of the ss-cDNAs, its 5' end must have a phosphate group. This group can be added during the synthesis of the oligonucleotide or after synthesis by kination. To avoid selfligation of the oligonucleotide, its 3' end must not have a hydroxyl group. This can be obtained by addition of a dideoxyribonucleotide to the 3' end of the oligonucleotide using the activity of the terminal transferase enzyme (*see* Section 3.6.1.) or more conveniently during the synthesis of the primer by using modified phosphoramidites (ask your oligonucleotide supplier).

### 2.1.3. Choice of Degenerate Oligonucleotides

Primer extension and PCR oligonucleotide degeneracy should be minimized, especially with respect to the three or four bases at the 3' end of the primers. If these first 3' nucleotides have to include a fourfold base degeneracy position, it is advised to avoid the use of any inosine residue, but to keep the degeneracy. However, inosines may be used in other positions. An example is shown in Fig. 3.

### 2.2. Extraction of RNAs

All the aqueous solutions must be filtered through a 0.22-μm filter prior to storage. All these reagents can be conveniently replaced by the use of the RNAZOL'B (Biotecx Laboratory Inc.).

1. Starting material: pelleted cells stored at –80°C on collection or pieces of organs frozen in liquid nitrogen stored at –80°C. Ice.
2. Guanidium thiocyanate 5$M$ (Fluka Biochemica, #50990) stored at 4°C.
3. $N$ lauryl sarcosyl 5% (in pasteurized or ddH$_2$O) stored at room temperature.
4. 1$M$ sodium citrate, pH 7.0, stored at –20°C.
5. β-mercaptoethanol, stored at 4°C.
6. Lysis buffer: 4$M$ guanidinium thiocyanate, 25 m$M$ sodium citrate, pH 7.0, 0.5% $N$ lauryl sarcosyl, 10 m$M$ β-mercaptoethanol.
7. 1$M$ sodium acetate, pH 4.0, stored at –20°C.

8. Phenol saturated with water stored at 4°C.
9. Chloroform.
10. DT4O (Dextran T 40, Pharmacia, Piscataway, NJ, #17–0270–01). Prepare a stock solution at 5 g/L in ddH$_2$O and store at 4°C.
11. 100% ethanol aliquoted and stored at –20°C.
12. 70% ethanol stored at 4°C.
13. Tips, tubes, and all the material used in this manipulation must be very clean and at least sterilized. Wearing gloves during all the manipulation of RNAs will help to avoid contamination of the preparation by RNases. The same precautions should be followed in the synthesis of cDNAs. These precautions represent the lowest level of protection against RNase and it is advisable to carefully read ref. *17*.

## 2.3. Synthesis of the cDNAs

*See* step 13 of the previous section.

1. Water bath at 90°C and then at 42°C. Dry ice powder. Ice.
2. Acetylated bovine serum albumin (BSA).
3. RNasin (Promega Biotech, Madison, WI).
4. β-mercaptoethanol.
5. 1$M$ Tris-HCl, pH 8.3.
6. 1$M$ KCl.
7. 1$M$ MgCl$_2$.
8. dNTP solution (100 m$M$; Pharmacia or Boehringer Mannheim, Indianapolis, IN).
9. 20 m$M$ sodium pyrophosphate.
10. $^{32}$P dCTP (400 Ci/mmol; Amersham #PB 10165).
11. AMV reverse transcriptase (Stratagene [La Jolla, CA] or Promega Biotech).
12. All these reagents can be replaced by the use of the Superscript II kit (Stratagene).
13. The primer that has been selected for primer extension experiment (PEX) in solution in ddH$_2$O at 100 ng/μL.

## 2.4. Desalting Samples with Spin Columns

1. Trisacryl GF 0.5$M$ (IBF/LKB: #259112). Microfuge tubes 0.5, 1.5, and 2-mL needle syringe. A glass bead 3 mm in diameter (Ateliers Cloup [France], #00 065.03). A bench-top centrifuge.
2. Preparation of the resin: Suspend 1 vol of resin in 1 vol of TE 0.1 (1 m$M$ Tris-HCl, pH 7.5, 0.1 m$M$ EDTA). Decant the resin (this could be speeded up by mild centrifugation at 50$g$ for 7 min). Remove the supernatant and repeat these operations three times. Resuspend the resin in a small volume of TE 0.1 and store at 4°C up to 5 mo.

## 2.5. Removal of the Primers

To avoid ligation of the primers used in the synthesis of the cDNAs to A_NV, these primers need to be removed prior to carrying out ligation. The length of primer used to perform the primer extension is the critical factor in choosing the method of elimination. If the length is lower than 25 nt, PrimeErase Quick Push Columns (Stratagene) or ammonium acetate/2-propanol differential precipitation are efficient to quickly remove primers. In other cases AcA 34 (IBF/LKB) is recommended.

1. PrimeErase Quick Push Columns (Stratagene, #400705) are provided ready to use (material and buffers).
2. Ammonium acetate/2-propanol differential precipitation: 1 bench centrifuge for Eppendorf tubes, $4M$ ammonium acetate, 2-propanol, 75% ethanol.

## 2.6. RNA Hydrolysis and Precipitation of ss-cDNA

1. A water bath at 50°C.
2. $2N$ NaOH.
3. $2.2N$ acetic acid.
4. DT 40 *(see above)*.
5. 10 m$M$ LiCl.
6. Ethanol, 100 and 75%.

## 2.7. Single-Strand Ligation

### 2.7.1. Functionalization of the Ligated Oligonucleotide

1. Terminal transferase supplied with its buffer and 25 m$M$ CoCl$_2$ (Boehringer, #220582).
2. $^{32}$P ddATP 3000Ci/mmol. (Amersham [Arlington Heights, IL], #PB10233).
3. 8% Polyacrylamide gel *(14)*.
4. $^{32}$P-kinased A5'_NV *(see* Section 2.5.).
5. Spin-X cartridge (Costar, #8160).
6. Dry ice powder.

### 2.7.2. Ligation of the ss-cDNAs to the Modified Oligonucleotide

1. Purified ss-cDNAs.
2. RNA ligase buffer. 10X: 500 m$M$ Tris-HCl, pH 8.0, 100 m$M$ MgCl$_2$, 100 μg/L BSA, ATP (500 μ$M$). Sonicate (to disrupt possible DNA contaminant which could be amplified in further PCR experiments). Filter through 0.22 μm membrane. Store at −20°C.
3. 40% PEG 6000 (Appligène, Ilkirch, France) made in ddH$_2$O and stored at −20°C.
4. Modified oligonucleotide A5'_NV.
5. T4 RNA Ligase (New England Biolabs, Beverly, MA).

## 2.8. PCR Amplification

1. Thermostable DNA polymerase (*Taq* pol), buffer, and dNTP purchased from Boehringer have been successfully used, programmable thermal cycler machine mineral oil.
2. Primers.

## 2.9. Flying Southern and Hybridization Procedure

The protocols to prepare these solutions are well described in ref. *18*.

1. 0.4$N$ NaOH.
2. Filter paper sheets (at least 1-cm thickness) larger than the gel.
3. Hybond N+ (Amersham) or other positively charged membrane (the type of membrane is important since the fixation of DNAs is performed in alkaline conditions) cut at the size of the agarose minigel.
5. Four sheets of Whatman 3MM paper cut at the size of the gel.
6. A centrifuge allowing the centrifugation of microtiter plates and the appropriate gondola.
7. 50X Denhardt's solution.
8. 20X SSC.
9. 0.5$M$ sodium phosphate, pH 7.0.
10. 20% SDS.
11. 0.5$M$ EDTA.
12. Denatured Herring sperm DNA, 4 mg/mL.
13. The $^{32}$P-kinased oligonucleotide probe 3. Prepare by adding 100 ng of primer, 2.5 µL of γ-ATP (3000 Ci/mmol, Amersham, #PB10168), and 1 U of terminal deoxyribonucleotide transferase (Boehringer) in a final volume of 10 µL in 1X buffer (50 m$M$ Tris-HCl 50, pH 7.5, 10 m$M$ MgCl$_2$, 0.1 m$M$ spermidine, 0.1 m$M$ EDTA, 5 m$M$ DTT). Incubate at least 30 min at 37°C. Remove unincorporated nucleotides using Tris-acryl GF 0.5 spin columns. Typically half of the radioactivity is incorporated to the oligonucleotide.
14. An incubator at 42°C.

## 2.10. Cloning of the PCR Products

As cloning strategies, screening of the libraries and isolation of clones of interest are the basic knowledge of all molecular biologists. They are well described in classical literature *(18)* and are not detailed here. We limit our discussion to the generation of clonable PCR products.

The complexity of the PCR product mixture is the most important criterion in the choice of cloning strategy. If the PCR products are amplified 5' ends of messengers (PCR product exhibiting a weak complexity with the SLIC strategy), they can be cloned in blunt-ended vectors after polishing the 3' ends with T4 DNA polymerase after kinasing of the PCR products.

Many kits are now available to clone PCR products *(19)* and it is possible to use one of them (e.g., TA cloning kit [Invitrogen, La Jolla, CA], PCR Script [Stratagene; *see also* Chapter 9]). However, in our hands the methods described below always give good results with a high number of clones.

1. Kinasing of PCR products.
2. Blunt ending: Buffer for restriction enzymes A (Boehringer), dNTP 2 m$M$. T4

DNA, polymerase [Boehringer], a 16°C water bath, ice, and a spun column (Sections 2.4. and 3.3.).

3. Cloning vectors, *Eco*RV cut phosphorylated pBluescript II (Stratagene).

## 3. Methods

### 3.1. Extraction of Total RNA (see Note 2)

All manipulations must be carried out in a RNase-free environment (*see* Section 2.2., item 13, and ref. *17*). Pelleted cells, organs, or fragments of organs can be used as starting material. On collection they must be stored at −80°C as quickly as possible.

1. For $10^4$ cells, add 40 µL of lysis buffer and 0.5 µg of DT 40 to the sample.
2. If cells are used, vortex 10 min to shear DNA. If organs are used, a plastic micropotter is recommended to homogenize the tissue.
3. Sequentially add 2.5 µL (*see* Note 1) of $1M$ sodium acetate, pH 7.4, 50 µL of water saturated phenol, and 10 µL of chloroform. Mix after each addition.
4. Leave 10 min on ice.
5. Spin 10 min at 4°C.
6. Collect the aqueous phase in a clean tube.
7. Add 280 µL of 100% ethanol.
8. Precipitate (2 × 5 min) in liquid nitrogen.
9. Spin 12,000$g$ for 30 min at 4°C.
10. Wash one or two times with 70% ethanol.
11. Vacuum dry (no more than 3 min; longer drying times stick the RNAs to the tube).
12. Dissolve the pellets in 10–20 µL of $H_2O$. Store at −80°C.

### 3.2. Synthesis of the ss-cDNA (see Note 3)

#### 3.2.1. Primer Extension Experiment

1. In a final volume of 18 µL, dilute the RNAs and add 6 pmol of primer PEX.
2. Heat the tubes at 90°C for 5 min.
3. Cool slowly to the annealing temperature of the primer.
4. To a final volume of 50 µL, add 0.1 mg/mL BSA, 0.1 U/µL RNasin, 70 m$M$ β-mercaptoethanol, 10 m$M$ Tris-HCl, pH 8.3 (at 42°C), 8 m$M$ KCl, 1.6 m$M$ $MgCl_2$, 1 m$M$ dNTP, 4 m$M$ sodium pyrophosphate, and 15 U of AMV reverse transcriptase.
5. Incubate for 45 min at 42°C.
6. Store at −20°C until use (*see* Note 3).

#### 3.2.2. Priming cDNAs with Degenerate Oligonucleotides

Proceed as described in Section 3.2.1. with the following modifications.

1. Use 20 pmol of degenerate primer.
2. Estimate the annealing temperature using the sequence of the oligonucleotid that has the higher content in A,T.

### 3.3. Desalting Samples with a Spin Column

1. Pierce the 0.5-mL microfuge tube with a needle.
2. Put the glass bead in the bottom of the tube.
3. Place the 0.5-mL tube in the 2-mL microfuge tube.
4. Fill the 0.5-mL microfuge tube with resin.
5. Spin in a swing out centrifuge at 700$g$ for exactly 2 min.
6. Transfer the microcolumn in a 1.5-mL microfuge tube.
7. Load the microcolumn with the sample to be desalted (do not fill with volume >30 µL).
8. Spin in a swing out centrifuge at 700$g$ for exactly 2 min.
9. Remove the column and store the eluate until use.

### 3.4. Removal of the Primers

#### 3.4.1. PrimeErase Quick Push Columns

Follow the supplier instructions.

#### 3.4.2. Ammonium Acetate/2-Propanol Differential Precipitation (see *Note 4*)

1. Add 1 vol of ammonium acetate and 2 vol of 2-propanol. Mix.
2. Incubate at room temperature at least 10 min.
3. Spin at 8000–10,000$g$ for 10 min at room temperature.
4. Remove supernatant and add 5 vol of 75% ethanol. Resuspend the pellets.
5. Spin at 8000–10,000$g$ for 10 min at room temperature.
6. Dry the pellets.
7. Resuspend in ddH$_2$O.

### 3.5. Alkaline Hydrolysis and Precipitation of the RNAs

1. Add 0.15 vol of 2$N$ NaOH.
2. Incubate for 30 min at 50°C.
3. Add 0.15 vol of 2.2$N$ acetic acid.
4. Add 0.03 vol of 0.5 µg/µL DT40, 0.07 vol of 10 m$M$ LiCl, 2.8 vol of 100% ethanol and mix well.
5. Freeze/thaw twice for 5 min in liquid nitrogen.
6. Spin for 30 min at 12,000$g$ at 4°C.
7. Wash with 70% ethanol.
8. Dry the pellets under vacuum no longer than 3 min.
9. Dissolve the pellets in 5 µL of water and store the purified ss-cDNAs at –80°C.

### 3.6. Single-Strand Ligation

#### 3.6.1. Protection of the Oligonucleotide

The oligonucleotide used in the ligation has to be synthesized with a 5'-phosphate end and its 3' hydroxyl must be removed. One way to do this is to tail the 3' end of the oligonucleotide with a dideoxyribonucleotide.

1. Mix 500 ng of oligonucleotide in 25 µL containing 5 µL of 5X terminal transferase buffer, 1 µL of 25 m$M$ CoCl$_2$, 100 m$M$ of ddATP, and 2.5 µCi of (γ$^{32}$P) ddATP.
2. Add 25 U of terminal transferase.
3. Incubate for 1 h at 37°C.
4. Heat for 10 min at 75°C.
5. Load on a 8% polyacrylamide gel. In a neighboring lane load a sample of the nontailed but $^{32}$P-kinased oligonucleotide. Perform electrophoresis.
6. Excise the correct band (after autoradiography of the wet gel).
7. Transfer the band into SPIN-X cartridge (Costar #8160), add 200 µL of water.
8. Freeze in dry ice powder. Thaw.
9. Spin for 5 min at 10,000$g$.
10. Precipitate as described in Section 3.4.2.
11. Dissolve the pellets in 50 µL of ddH$_2$O (approx 5 ng/µL).

### 3.6.2. Ligation of the ss-cDNAs to the Modified Oligonucleotide (see Note 5)

1. Ligated sample. Mix: 1 µL of purified ss-cDNAs, 1 µL of 10X RNA ligase, 0.5 µL of the modified oligonucleotide, 6.25 µL of 40% PEG 6000, and 1 µL of T4 RNA ligase.
2. Incubate for 48 h at 22°C.
3. Remove 5 µL of each sample and store at –20°C until use.
4. Incubate the remaining mixture for another 48 h period.
5. Store the samples at –20°C until use.

### 3.7. PCR-Amplification (see Note 6)

Conditions for PCR amplification depend on the $T_m$ of the oligonucleotides used. The primers used in this study have been successfully used with annealing temperatures of 51 or 55°C. A signal corresponding to the amplification of the 5' end of the mRNA could be obtain after the first PCR experiment but it is more generally obtained after the second (nested PCR).

### 3.7.1. First PCR

1. In a 500-µL microfuge tube mix half of the ligation product: 10 µL of 10X PCR buffer, 0.2 µ$M$ dNTPs, primers for the first amplification (A5'_1/PCR1), ddH$_2$O to 100 µL, and 1 U of *Taq* Pol. Overlay with 100 µL of mineral oil.
2. Perform the following PCR cycles with a first denaturation at 93°C for 3 min, then: 35 cycles: 94°C for 30 s, 55°C for 30 s, and 72°C for 1 min. Cool down the reaction tubes to 4°C

### 3.7.2. Second PCR

1. Use 1 µL of the first PCR reaction in 50 µL final volume of a reaction mixture similar to the reaction mixture of the first PCR substituting primers A5'_2 and PCR2.
2. Perform PCR as described for the first PCR.

### 3.7.3. PCR Using Degenerated Oligonucleotides

Perform the PCR as described in Sections 3.7.1. and 3.7.2. with the following modifications:

1. Use a fivefold excess of the degenerate primers for both reactions (500 ng for a 20-residue oligonucleotide in 50 μL of reaction)
2. To increase the success of the manipulation, it is recommended to perform three different PCR experiments using three different annealing temperatures. Determine $T_m$ max ($T_m$ of the oligonucleotide having the higher GC content), $T_m$ min ($T_m$ of the oligonucleotide having the higher AT content), and $T_m$ med, which is the mean of the two previous temperatures. Perform the nested PCR at these three different sets of temperatures.

### 3.7.4. Analysis of the PCR Products

Analysis of the PCR products can be performed with 1% agarose electrophoresis as described in ref. *19*. It is recommended to check the specificity of the PCR with Southern blotting experiments. A fast and efficient protocol, named "Flying Southern" is given in Section 3.8.

### *3.8. Flying Southern and Hybridization Procedure*

This protocol is derived from ref. *20*. It allows quick and accurate analysis of the specificity of PCR products using agarose minigels (7.5 × 10.5 cm). This protocol requires the use of a centrifuge that accepts microtiter plates.

1. After UV-visualization of the nucleic acids, soak the agarose gel in 0.4$N$ NaOH.
2. Soak a Hybond N$^+$ membrane (Amersham) in water and then in 6X SSC.
3. Prepare in a plastic box: Filter paper sheets (1 cm total thickness), Hybond N$^+$ membrane, agarose gel (avoiding bubbles between the membrane and the gel), and four sheets of Whatman 3MM paper soaked in 0.4$N$ NaOH.
4. Spin in microtiter plate gondola at 1000$g$ for 20–30 min (agarose gels from 0.8–2%).
5. Remove the Whatman paper.
6. Mark the position of the wells on the membrane.
7. Remove the membrane from the gel and rinse twice for 5 min in 6X SSC.
8. Prehybridize in 1 mL (7.5 × 10.5 cm membrane) of buffer (1X Denhardt's, 6X SSC, 25 m$M$ sodium phosphate, pH 7.0, 25 m$M$ EDTA, 250 μg/mL denatured herring sperm, and 1% SDS). Load the buffer in a Petri dish, then put the membrane DNA face toward the bottom of the dish above the buffer. Avoid air bubbles. Close the Petri dish.
9. Incubate 20 min at 42°C.
10. Remove the prehybridization buffer and replace it by 300 μL of prehybridization buffer containing 2–3 $10^6$ cpm of $^{32}$P-kinased oligonucleotide. Place the membrane as described above.
11. Incubate 30 min at 42°C.
12. Wash from 6X–1X SSC with 1% SDS at 42°C. Check radioactivity with a bench monitor.

## 3.9. Cloning PCR Products (see *Note 7*)

1. After having removed primers (protocol 3.4.2.) for a final volume of 20 µL, add 2 µL of buffer A (Boehringer), 1 µL dNTP 2 m$M$, and 2 µL of T4 DNA polymerase (Boehringer). To limit the activity of the enzyme all the components must be kept at 4°C.
2. Incubate 15 min at 16°C. Warm to 75°C for 10 min to denature the enzyme.
3. Remove nucleotides and salts using a Tris-acryl GF 0.5 spin column.
4. Ligate to blunt-ended cloning vector following classical protocols *(17)* or supplier's instructions (*see* Part II).

## 4. Notes

1. With the increasing knowledge in 5' ends of messengers, it seems that alternative splicing occurs more frequently than expected in these regions. In these cases, the use of a PROBE oligonucleotide (Fig. 4) located far from the oligonucleotides used to prime the ss-cDNA and the following PCRs could lead to the isolation of only one or a few forms of messenger. In this case, it is useful to use at least two oligonucleotides to screen the PCR products obtained: One could be located close to the oligonucleotides used to prime PCR experiments and the other located at the end of the known sequences. Examples of the use of such screening methods have been described in refs. *8* and *10* to screen products of alternative promoters *(8)* or alternative splicing in 5' untranslated regions *(10)*.
2. We have successfully extracted total RNAs (34 µg) from punches of rat *raphe dorsalis*, multiplying by 5 all the volumes indicated.
3. To clone 5' ends of messengers, the labeling of the ss-cDNAs is not necessary provided the activity of the enzyme has been checked earlier. A more informative experiment in the cloning of 5' ends of mRNAs is to carry out PCR with the ss-cDNAs as matrix and two specific primers. One of the primers has to be chosen close to the known 5' ends of the cDNA clones (PROBE Fig. 4). After PCR, verify the specificity of the product by Southern blotting (Protocol 3.8.). This allows an accurate verification of the synthesis of an ss-cDNA long enough to pursue cloning experiments.
4. In this case, it is recommended to perform the alkaline hydrolysis prior to carrying out the fractionated precipitation. This allows the retention of single-strand cDNA and the removal of PEX in a single step.
5. During this manipulation it is recommended to prepare the following controls, which will be used in the PCR experiments following SLIC.
   a. Prepare a nonligated sample composed of the same mixture as described in Section 3.6.2., step 1 without the enzyme. This nonligated control will be used to determine if nonspecific amplification using the ss-cDNA as matrix has occurred.
   b. Prepare another control containing the mixture described in Section 3.6.2., step 1 minus ss-cDNA and the enzyme. This control will reveal the presence of contaminant in the SLIC mix when compared to the PCR control.

6. In all the PCR experiments it is usual to prepare a control sample with the PCR mixture minus DNA. This allows the detection of contaminant in the PCR mix.
7. Check that the PCR product is kinased. PCR primers are classically synthesized without phosphate group at their 5' end. It is recommended to perform a limited PCR (10 cycles) with kinased primer before blunt-ending or kinase the PCR products.

## References

1. Mullis, K. B. and Faloona, F. (1987) Specific synthesis of DNA in vitro via a polymerase-catalyzed chain reaction. *Methods Enzymol.* **155,** 335–350.
2. Saïki, R. K., Gelfand, D. H., Stoffel, S., Scharf, S. J., Higuchi, R., Horn, G. J., Mullis, K. B., and Erlich, H. A. (1988) Primer-directed enzymatic amplification of DNA with a thermostable DNA polymerase. *Science* **239,** 487–491.
3. Delort, J., Dumas, J. B., Darmon, M. C., and Mallet, J. (1989) An efficient strategy for cloning 5' extremities of rare transcripts permits isolation of multiple 5'-untranslated regions of rat tryptophan hydroxylase mRNA. *Nucleic Acids Res.* **17,** 6439–6448.
4. Loh, E. Y., Elliot, J. F., Cwisla, S., Lanier, L. L., and Davn, M. M. (1989) Polymerase chain reaction with single-sided specificity: analysis of T cell receptor β chain. *Science* **243,** 217–220.
5. Belyavsky, A., Vinogradova, T., and Rajewsky, K. (1989) PCR-based cDNA library construction: general cDNA libraries at the level of a few cells. *Nucleic Acids Res.* **17,** 2919–2932.
6. Frohman, M. A., Dush, M. K., and Martin, G. R. (1988) Rapid production of full-length cDNAs from rare transcripts: amplification using a single gene-specific oligonucleotide primer. *Proc. Natl. Acad. Sci. USA* **85,** 8998–9002.
7. Ohara, O., Dorit, R. L., and Gilbert, W. (1989) One-sided polymerase chain reaction: the amplification of cDNA. *Proc. Natl. Acad. Sci. USA* **86,** 5673–5677.
8. Dumas Milne Edwards, J. B., Delort, J., and Mallet, J. (1991) Oligodeoxyribonucleotide ligation to single-stranded cDNAs: a new tool for cloning 5' ends of mRNAs and for constructing cDNA libraries by in vitro amplification. *Nucleic Acids Res.* **19,** 5227–5232.
9. Bejanin, S., Habert, E., Berrard, S., Dumas Milne Edwards, J. B., Loeffler, J.-P., Mallet, J. (1992) Promoter element of the rat choline acetyl transferase gene allowing nerve growth factor inducibility in transfected primary cultured cells. *J. Neurochem.* **4,** 1580–1583.
10. Philippe, J. M., Renaud, F., Desset, S., Laurent, M., Mallet, J., Courtois, Y., and Dumas Milne Edwards, J. B. (1992) Cloning of two different untranslated exons of bovine acidic growth factor by the single-strand ligation to single-stranded cDNA methodology (SLIC). *Biochem. Biophys. Res. Commun.* **188,** 843–850.
11. Valdenaire, O., Vernier, P., Maus, M., Dumas Milne Edwards, J. B., and Mallet, J. (1994) Transcription of the rat D2 receptor gene from two promoters. *Eur. J. Biochem.* **220,** 577–584.

12. Martaseq, P., Amadro, J. M., Delfau-Larue, M.-H., Dumas Milne Edwards, J. B., Montagne, J. J., de Verneuil, H., Labbe, P., and Grandchamp, B. (1994) Molecular cloning, sequencing, and functional expression of a cDNA encoding human coprophyrinogen oxidase. *Proc. Natl. Acad. Sci. USA* **91,** 3024–3028.
13. Moseman-McCoy, M. I., and Gumport, R. I. (1980) T4 ribonucleic acid ligase join single-strand oligo(deoxyribonucleotides). *Biochemistry* **19,** 635–642.
14. Tessier, D. C., Brousseau, R., and Vernet, T. (1986) Ligation of single-stranded oligodeoxyribonucleotide by T4 RNA ligase. *Anal. Biochem.* **158,** 171–178.
15. Delidow, B. C. (1993) Molecular cloning of polymerase chain reaction fragments with cohesive ends, in *PCR Protocols: Current Methods and Applications* (White, B. A., ed.), Humana, Totowa, NJ, pp. 217–228.
16. Sambrook, J., Fritsch, E. F., and Maniatis, T. (1989) *Molecular Cloning: A Laboratory Manual* (2nd ed.), Cold Spring Harbor Laboratory, Cold Spring Harbor, NY, pp. 11.21–11.28.
17. Blumberg, D. D. (1987) Creating a ribonuclease free environment, in *Molecular Cloning Techniques. Methods in Enzymology,* vol. 152 (Berger, S. L. and Kimmel, A. R., eds.), Academic, New York, pp. 20–24.
18. Sambrook, J., Fritsch, E. F., and Maniatis, T., (1989) *Molecular Cloning: A Laboratory Manual* (2nd ed.), Cold Spring Harbor Laboratory, Cold Spring Harbor, NY. Composition of the solutions is respectively given in pp. B.15, B.13, B.21, B.13, B.11, B.15.
19. Sambrook, J., Fritsch, E. F., and Maniatis T. (1989) *Molecular Cloning: A Laboratory Manual* (2nd ed.), Cold Spring Harbor Laboratory, Cold Spring Harbor, NY, Chapters 1 and 2.
20. Wilkins, R. J. and Snell, R. G. (1987) Centrifugal transfer and sandwich hybridization permit 12-hour Southern blot analyses. *Nucleic Acids Res.* **15,** 7200.

# 28

## Rapid Directional Walk Within DNA Clones by Step-Out PCR

**Umadevi V. Wesley and Cedric S. Wesley**

### 1. Introduction

Genomic segments of many model experimental organisms are now available as segments within cosmid, P1, or YAC vectors *(1–3)*. Even cDNA portions of these segments are becoming available. Preliminary analyses with any of these clones would include sequencing and mapping (either on the genome or within a set of clones or subclones). Generally, such analysis would begin with sequences at the termini of inserts and proceed directionally into the internal sequences. PCR, with one specific primer and one nonspecific primer, is an ideal procedure to generate sequential and ordered fragments for use as sequencing templates or probes in nucleic acid hybridizations *(4)*. These fragments would extend from a specific position within the known sequences of either the vector arm or the insert (via the specific primer) into different positions within the adjacent unknown sequence (via the nonspecific primer). This "Step-Out PCR" procedure generates fragments in a matter of h and eliminates the need for time-consuming procedures, such as restriction enzyme digestion/analysis, ligations, and transformation into hosts *(5–8)*.

The procedure involves initial, random priming with a nonspecific primer by low temperature annealing, followed by DNA synthesis with a thermostable polymerase (Fig. 1). A specific primer is then added at a temperature above 60°C, and PCR initiated with the annealing step set at a temperature that is optimal for generation of reasonable yields and fewer products (4; *see* Note 1). The nonspecific primer would be made up of a specific tri- or hexa-nucleotide sequence (either the most frequent sequence in the genome or a restriction site) at the 3' end, followed by four nucleotides of complete degeneracy and an arbi-

From: *Methods in Molecular Biology, Vol. 67: PCR Cloning Protocols: From Molecular Cloning to Genetic Engineering* Edited by: B. A. White  Humana Press Inc., Totowa, NJ

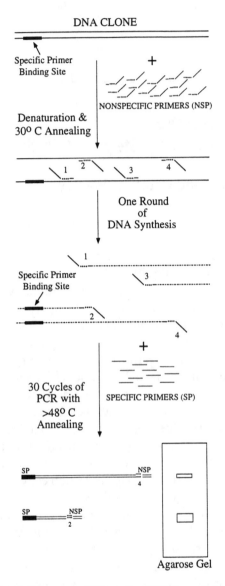

**Fig. 1.** Diagram of "Step-Out PCR" procedure.

trary array of nucleotides to facilitate annealing at higher temperatures (*see* Note 2). An example of such a primer is 5'-GTCAGTCAGTCAGANNNNGAG-3' *(4)*. The specific primer would be a vector derived primer (SP6, T7, T3, M13 forward, M13 reverse, and so forth, for phage, cosmid, or plasmid vectors; 1091 or1208 for YACs *[7]*) (*see* Note 3) or a primer based on the known sequence within the insert. This procedure was extensively employed for positional cloning of a gene *(9)*.

## 2. Materials

1. Thermal cycler.
2. Thin-walled 0.65-mL PCR tubes (Island Scientific Cat. # 510).
3. Thermostable polymerase (Perkin-Elmer, Norwalk, CT, Amplitaq Polymerase, 5 U/mL).
4. 10X PCR buffer (for Amplitaq polymerase: 15 m$M$ MgSO$_4$, 100 Tris-HCl, pH 8.3, at 25°C, 500 m$M$ KCl, 0.01 w/v gelatin).
5. 1.25 m$M$ dNTP.
6. PCR template: cosmid, P1, phage, plasmid, or YAC clone.
7. Specific primer, 10 ng/mL (one of T7, SP6, T3, m13 forward, m13 reverse, and so forth, primers).
8. Nonspecific primer, 25 ng/mL.
9. Reagent-quality water.
10. Mineral oil (Perkin-Elmer; *see* Note 4).
11. Heating block at 65–75°C.
12. Water bath at ~100°C (i.e., boiling) for denaturing templates.
13. Bucket of ice.
14. Agarose, SeaPlaque low melt agarose (FMC Bioproducts, Rockland, ME), and agarose electrophoresis equipment.
15. 6X loading dye *(10)*.
16. Ethidium bromide, 10 mg/mL in water.
17. New razor blades.
18. 5X TB (54 g Tris base, 27.5 g boric acid).
19. For fragment labeling:
    a. 10X hexanucleotide mix for random labeling of DNA (Boehringer Mannheim, Indianapolis, IN);
    b. 0.5 m$M$ dATP;
    c. 0.5 m$M$ dGTP;
    d. 0.5 m$M$ dCTP;
    e. 0.5 m$M$ dTTP;
    f. [$\alpha$-$^{32}$P] dCTP 10 mCi/mL or 0.4 m$M$ Biotin-14-dATP (BRL); and
    g. Klenow DNA polymerase or exo-Klenow DNA polymerase.
20. For fragment sequencing:
    a. ATP ($\gamma$ATP-$^{33}$P) (Dupont NEN, Boston, MA; cat. #NEG-302H);
    b. 10X kinase buffer: 700 m$M$ Tris-HCl, pH 7.6, 100 m$M$ MgCl$_2$, 5m$M$ dithiothreitol;
    c. T4 polynucleotide kinase 10 U/mL (New England Biolabs, Beverly, MA);
    d. AmpliCycle sequencing kit (Perkin/Elmer, Norwalk, CT, part no. N808-0175); and
    e. Sequencing equipment and chemicals *(10)*.

## 3. Methods

### 3.1. Generation of "Step-Out PCR" Fragments

1. Mix appropriate amount of DNA (1–10 ng for plasmid, lambda, or cosmid clones; 10–100 ng for P1 or YAC clones; a fraction of a plasmid or lambda or cosmid colony) and water in a thin-walled PCR tube such that the final volume is 24 µL (21 µL for YAC templates).
2. Boil the sample for 4 min and chill on ice for 3–4 min.
3. Quick spin in a microcentrifuge at room temperature and keep on ice.
4. For each tube of sample, mix 5 µL of 10X PCR buffer, 8 µL of 1.25 m$M$ dNTP solution, 3 µL of 25 ng/µL nonspecific primer (6 µL for YAC templates), and 0.25 µL of Amplitaq polymerase in a separate tube (*see* Note 5).
5. Mix 16.25 µL of the solution in step 4 with the DNA solution; layer the top of the solution with a drop or two of mineral oil.
6. Perform one cycle of 95°C (30 s), 30°C (4 min), and 75°C (4 min) with a 30–75°C ramp of 4 min. Either instruct the machine to hold the temperature at 75°C following completion of the cycle or let it terminate without connection to any other cycle (in which case the block will stay above 65°C for quite a while, enabling you to complete the procedure indicated in step 8).
7. During step 6, mix 4 µL of the 25 ng/µL specific primer, 2 µL of the 25 ng/µL nonspecific primer, and 4 µL of water in a separate tube. Set up as many such tubes as there are PCR reactions. Set a pipetman to volume >40 µL and wait for the completion of the PCR cycle in step 6.
8. At the end of step 6, pick up the 10 µL of the primer mix prepared in step 7, go through the oil and mix the primer solution with the PCR solution by way of quick up and down pipeting. Operating quickly, do the same to all the PCR reactions. Avoid letting the block temperature drop below the subsequent PCR annealing temperature (i.e., 52°C).
9. Perform PCR: 95°C (30 s), 52°C (1 min), 75°C (1 min) for 35 cycles.

### 3.2. Excision of "Step-Out PCR" Fragments

1. Mix 20 µL of the PCR reactions with 5 µL of 6X loading dye; electrophorese in a 1% low-melt agarose gel in 0.5X TB buffer until the bromophenol blue dye reaches the bottom of the gel. Stain the gel with ethidium bromide and destain with tap water.
2. Label an appropriate number of 1.5-mL microfuge tubes and arrange them in the order of gel loading. Keep a box of new razor blades ready.
3. Place the gel on Saran Wrap, visualize DNA with a hand held long-wave UV lamp, and excise the predominant or the desired sized band from each lane with a razor blade. Transfer the gel slice into the appropriate 1.5-mL tube labeled in step 2. Use a new blade for each fragment. Use one edge of the blade for trimming the gel away from the band of interest. Use the other unused edge for actual excision of the band of interest and transfer of the gel slice into the 1.5-mL tube.
4. Melt the gel slices by placing them in a 65–70°C water bath for about 5 min. The fragment is ready for use in reamplification if the yield of DNA in the band is low

(step 5; *see* Note 6), or for preparation of a probe for use in nucleic acid hybridizations (Section 3.3.) or for use as a template in DNA sequencing (Section 3.4., *see* Note 7).

5. Mix 1 μL of the melted gel solution of step 4 with 27.75 μL of sterile water, 5 μL of the PCR 10X buffer, 8 μL of the 1.25 m*M* dNTP solution, 4 μL of the 25 ng/μL specific primer, 4 μL of the 25 ng/μL nonspecific primer, and 0.25 μL of Amplitaq polymerase.
6. Layer the top with a drop or two of mineral oil and perform PCR as follows: 30 cycles: 95°C for 30 s, 52°C for 45 s, and 75°C for 1 min.
7. Repeat steps 1–4. There is no need to gel purify these templates because of the high yields of the template. The low quantities of unincorporated nucleotides and primers present in the solution will not interfere with the sequencing reaction. One microliter of the reamplified PCR solution could be directly taken into the sequencing step (Section 3.4., step 3).

## 3.3. Labeling of the "Step-Out PCR" Fragment

### 3.3.1. $^{32}P$ Isotope-Labeled Probe

1. Mix 3 μL of the melted gel solution with 6 μL of sterile water in a screwcap microfuge tube and boil for 7 min.
2. In a separate tube mix 2 μL of 10X hexanucleotide, 1 μL of 0.5 m*M* dATP, 1 μL of 0.5 m*M* dGTP, 1 μL of 0.5 m*M* dTTP, 5 μL of [a-$^{32}$P] dCTP, and 1 μL of Klenow DNA polymerase.
3. Immediately after boiling of the template, quick spin the tube and place it in a 37°C water bath.
4. After 3 min at 37°C, quickly mix the enzyme solution with the template solution and return to 37°C.
5. Incubate at 37°C for 1–2 h. The tube could be submerged in a scintillation vial filled with 37°C water, which in turn is submerged in a 37°C water bath, to prevent evaporation of fluid on top of the tube during incubation.
6. At the end of incubation, the solution could be used as a probe either directly or after passing through a Sephadex G-25, G-50 spin, or gravity column *(10)*. In the latter case, the labeling solution will have to be diluted with 80 μL of TE and stored at room temperature or 37°C until loading onto the column.

### 3.3.2. Biotin Labeled Probe for In Situ Hybridizations

1. Mix 6 μL of the melted gel solution with 19.5 μL of sterile water in a screwcap microfuge tube and boil for 7 min.
2. In a separate tube: mix 4 μL of 10X hexanucleotide, 2 μL of 0.5 m*M* dCTP, 2 μL of 0.5 m*M* dGTP, 2 μL of 0.5 m*M* dTTP, 2.5 μL of Biotin-14-dATP, and 2 μL of Exo-Klenow DNA polymerase.
3. Immediately after boiling of the template, quick spin the tube and place it in a 37°C water bath.
4. After 3 min at 37°C, quickly mix the enzyme solution with the template solution and return to 37°C.

5. Incubate at 37°C overnight. The tube should be submerged in a scintillation vial filled with 37°C water, which in turn should be submerged in a 37°C water bath, to prevent evaporation of fluid on top of the tube during the long incubation.

6. At the end of incubation, add 80 µL of TE to the labeling mix and pass it through a Sephadex G-25 or G-50 spin column *(10)*. Reduce the volume to about 10 µL, but never completely dry down. Use about 1–2 µL of this solution/slide of chromosome spreads or tissue sections.

## *3.4. Sequencing of the "Step-Out PCR" Fragment*

1. Mix 3.38 µL of water, 0.62 µL of 10X Kinase buffer, 1 µL of the 25 ng/mL primer, 1 µL of ATP (γATP-$^{33}$), and 0.25 µL of T4 polynucleotide kinase (10 U/µL) in an microfuge tube. Incubate at 37°C for 45 min, thereafter boil for 5 min, and store at –20°C until use. These labeled primers can be stored up to 2 wk before use.

2. Aliquot 4.5 µL of G, A, T, C termination mixes from the AmpliCycle Sequencing Kit to appropriately labeled thin-walled PCR tubes and store on ice.

3. Mix 1–2 µL of the gel solution (Section 3.2., step 4), or 1 µL of reamplification solution (Section 3.2., step 5), or an appropriate quantity of DNA extracted from the gel solution with enough water to yield 13 µL of template solution, boil for 2 min, quick spin in a centrifuge, and place it in a 65°C heat block (either water block or thermocycler held at that temperature).

4. Place the G, A, T, C termination tubes in the 65°C heat block.

5. Add 3 µL of the labeled primer and 4 µL of the 10X cycling mix from the AmpliCycle Sequencing Kit to the template solution, mix and immediately aliquot 4 µL of this solution to each of the four termination tubes. Work quickly to minimize the amount of time the tube spends out of the block.

6. Overlay the top of the solution with a drop of mineral oil and perform PCR as follows: 25 cycles: 95°C for 30 s, 52°C for 1 min, 75°C for 25 s.

7. Terminate the reactions by going through the oil and mixing 4 mL of the stop solution with each of the sequencing reactions.

8. Perform sequencing electrophoresis and autoradiography as described in ref. *10*.

9. The sequence derived from the nonspecific primer will be from the region farthest from the specific primer. This sequence can be used to synthesize the specific primer for the next round of amplification of adjacent fragments.

## 4. Notes

1. Experiments with various nonspecific primers of the indicated design suggest that their optimal annealing temperature generally lies between 48 and 52°C. This may be caused by the location of the degenerate sites within the primer *(11)*. Since PCR is a competitive process, the smaller of the several fragments that get initiated in the first round of amplification will come to predominate in the end.

2. The arbitrary array of nucleotides could be restriction sites or promoter sequences useful for further manipulations of the fragments. It should be kept in mind that a number of restriction sites have dyad symmetry and may form strong secondary structures that reduce the stringency of priming.

3. For balanced yields of PCR product, the specific primer should have an annealing temperature between 54 and 58°C.
4. Some brands of mineral oils inhibit PCR amplification. In my experience, the Perkin Elmer mineral oil, both within the tube and in the block, gives consistent and good results.
5. If a number of reactions are being set up, master mixes can be made with sufficient allowance for pipeting errors.
6. Yields can be assessed visually. The random priming method of labeling DNA and *Taq* polymerase sequencing do not require high levels of template. If the band appears fairly intense under long-wave UV light, it is unnecessry to reamplify the fragment. If you are not sure whether the yield is good, are not proficient in sequencing, or are sequencing a lot of templates, a reamplification may be prudent since more template will overcome several sequencing problems.
7. If desired, DNA from the gel slice could be extracted. Qiagen gel extraction kits, are quick and reliable. Take care not to take up any extraction system matrix into subsequent reactions.
8. Sequenase enzyme or kit can also be used to sequence the PCR fragment with the specific primer. However, *Taq* polymerase and high temperature sequencing is necessary for sequencing with the nonspecific primer since only a fraction of all the primers will perfectly match the template sequence. At high temperatures, only these hybrids will promote synthesis and therefore, yield unambiguous bands. In general, sequencing reactions with the nonspecific primer will require slightly longer exposure to the film during autoradiography.

# References

1. Gibson, T. J., Rosenthal, A., and Waterson, R. H. (1987) *Gene* **53**, 283–286.
2. Pierce, J. C., Saver, B., and Sternberg, N. (1992) *Proc. Natl. Acad. Sci. USA* **89**, 2056–2060.
3. Burke, D. T., Carle, G. F., and Olson, M. V. (1987) *Science* **236**, 806–812.
4. Wesley, C. S., Myers, M. P., and Young, M. W. (1994) *Nucleic Acids Res.* **22**, 538,539.
5. Traver, C. N., Sulston, J., Brenner, S., and Karn, J. (1989) *Proc. Natl. Acad. Sci. USA* **86**, 5898–5902.
6. Silverman, G. A., Ye, R. D., Pollack, K. M., Sadler, J. E., and Korsmeyer, S. J. (1989) *Proc. Natl. Acad. Sci. USA* **86**, 7485–7489.
7. Riley, J., Ogilvie, D., Finniear, R., Jenner, D., Powell, S., Anand, R., Smith, J. C., and Markham, A. F. (1990) *Nucleic Acids Res.* **18**, 2887–2890.
8. Hermanson, G. G., Hoekstra, M. F., McElligott, D. L., and Evans, G. A. (1991) *Nucleic Acids Res.* **18**, 4943–4948.
9. Myers, M. P., Wager-Smith, K., Wesley, C. S., Young, M. W., and Sehgal, A. (1995) *Science* **270**, 805–808.
10. Sambrook, J., Fritsch, E. F., Maniatis, T. (1989) *Molecular Cloning: A Laboratory Manual* (2nd ed.), Cold Spring Harbor Laboratory, Cold Spring Harbor, NY.
11. Wesley, C. S., Ben, M., Kreitman, M., Hagag, N., and Eanes, W. F. (1990) *Nucleic Acids Res.* **18**, 599–603.

# 29

## Inverse PCR

### An Efficient Approach to Cloning cDNA Ends

**Sheng-He Huang**

### 1. Introduction

Since the first report on cDNA cloning in 1972 *(1)*, this technology has developed into a powerful and universal tool in isolation, characterization, and analysis of both eukaryotic and prokaryotic genes. However, the conventional methods of cDNA cloning require much effort to generate a library that is packaged in phage or plasmid, and then surveyed in a large number of recombinant phages or plasmids. There are three major limitations in these methods. First, a substantial amount (at least 1 mg) of purified mRNA is needed as starting material to generate libraries of sufficient diversity *(2)*. Second, the intrinsic difficulty of multiple sequential enzymatic reactions required for cDNA cloning often leads to low yields and truncated clones *(3)*. Finally, screening of a library with hybridization technique is time-consuming. PCR technology can simplify and improve cDNA cloning. Using PCR with two gene-specific primers, a piece of known sequence cDNA can be specifically and efficiently amplified and isolated from very small numbers (<$10^4$) of cells *(4)*. However, it is often difficult to isolate full-length cDNA copies of mRNA on the basis of very limited sequence information. The unknown sequence flanking a small stretch of the known sequence of DNA cannot be amplified by the conventional PCR. Recently, anchored PCR *(5–7)* and inverse PCR *(8–10)* have been developed to resolve this problem. Anchored PCR techniques have the common point that DNA cloning goes from a small stretch of known DNA sequence to the flanking unknown sequence region with the aid of a gene-specific primer at one end and a universal primer at other end. Because of only one gene-specific primer in the anchored PCR it is easier to get a high level of nonspecific ampli-

From: *Methods in Molecular Biology, Vol. 67: PCR Cloning Protocols: From Molecular Cloning to Genetic Engineering* Edited by: B. A. White Humana Press Inc., Totowa, NJ

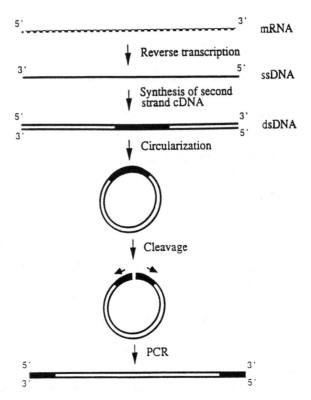

**Fig. 1.** Diagram of IPCR for cDNA cloning. The procedure consists of five steps: reverse transcription, synthesis of second-strand cDNA, circularization of double-strand cDNA, reopening the circle DNA, and amplification of reverse DNA fragment. The black and open bars represent the known and unknown sequence regions of double-stranded cDNA, respectively.

fication by PCR than with two gene-specific primers *(10,11)*. The major advantage of inverse PCR (IPCR) is to amplify the flanking unknown sequence by using two gene-specific primers.

At first, IPCR was successfully used in the amplification of genomic DNA segments that lie outside the boundaries of the known sequence *(8,9)*. I have developed a new procedure that extends this technique to the cloning of unknown cDNA sequence from total RNA *(10)*. Double-stranded cDNA is synthesized from RNA and ligated end-to-end (Fig. 1). Circularized cDNA is nicked by selected restriction enzyme or denatured by NaOH treatment *(12,13)*. The reopened or denatured circular cDNA is then amplified by two gene-specific primers. Recently, this technique has been efficiently used in cloning full-length cDNAs *(14–16)*. The following protocol was used to amplify cDNA ends for the human stress-related protein ERp72 *(10)* (Fig. 2).

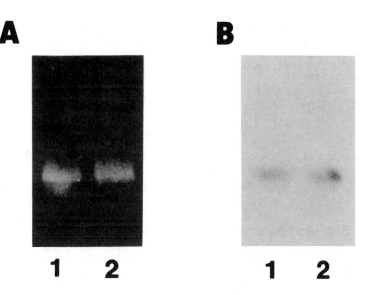

**Fig. 2.** Application of IPCR to amplifying the joining region (280 bp) from 5' (160 bp) and 3' (120 bp) sequences of human Erp72 cDNA. Amplified DNAs from CCRF/CEM cells sensitive (lane 1) and resistant (lane 2) to cytosine arabinoside stained by ethidium bromide **(A)** or hybridized with $^{32}$P-labeled ERp72 cDNA **(B)**. *See text* for the sequences of the primers and the parameters of IPCR.

## 2. Materials
### 2.1. First-Strand cDNA Synthesis

1. Total RNA prepared from human CCRF/CEM leukemic lymphoblast cells *(17,18)*.
2. dNTP mix (10 m$M$ of each dNTP).
3. Random primers (Boehringer Mannheim, Indianapolis, IN). Prepare in sterile water at 1 μg/μL. Store at –20°C.
4. RNasin (Promega, Madison, WI).
5. Actinomycin D (1 mg/mL). Actinomycin D is light sensitive and toxic. It should be stored in a foil-wrapped tube at –20°C.
6. MMLV reverse transcriptase.
7. 5X first-strand buffer: 0.25$M$ Tris-HCl, pH 8.3, 0.375$M$ KCl, 50 m$M$ MgCl$_2$, 50 m$M$ DTT, and 2.5 m$M$ spermidine. The solution is stable at –20°C for more than 6 mo.

### 2.2. Second-Strand Synthesis

1. 10X second-strand buffer: 400 m$M$ Tris-HCl, pH 7.6, 750 m$M$ KCl, 30 m$M$ MgCl$_2$, 100 m$M$ ammonium sulfate, 30 m$M$ DTT, and 0.5 mg/mL of bovine serum albumin (BSA). The solution is stable at –20°C for at least 6 mo.
2. NAD (1 m$M$).

3. RNase H (2 U/μL).
4. *Escherichia coli* DNA polymerase I (5 U/μL).
5. *E. coli* DNA ligase (1 U/μL).
6. Nuclease-free H₂O.
7. T4 DNA polymerase.
8. 200 m*M* EDTA, pH 8.0.
9. GeneClean (Bio 101, La Jolla, CA).
10. TE buffer: 10 m*M* Tris-HCl, pH 7.6, 1 m*M* EDTA. Sterile filter.
11. DNA standards. Prepare 1-mL aliquots of a purified DNA sample at 1, 2.5, 5, 10, and 20 μg/mL in TE buffer. Store at –20°C for up to 6 mo.
12. TE/ethidium bromide: 2 μg/mL of ethidium bromide in TE buffer. Store at 4°C for up to 6 mo in a dark container.

## 2.3. Circularization and Cleavage or Denaturation

1. 5X ligation buffer (supplied with T4 DNA ligase).
2. T4 DNA ligase (1 U/μL).
3. T4 RNA ligase (4 μg/μL).
4. Hexaminecobalt chloride (15 μ*M*).
5. Phenol:CHCl₃:isoamyl alcohol (25:24:1).
6. 3*M* sodium acetate, pH 7.0.
7. Absolute ethanol.
8. 70% Ethanol.

## 2.4. Inverse PCR

1. 10X PCR buffer: 100 m*M* Tris-HCl, pH 8.3, 500 m*M* KCl, 15 m*M* MgCl₂, 0.01% (w/v) gelatin.
2. 15 m*M* MgCl₂.
3. Deoxyoligonucleotides were synthesized on an Applied Biosystems (Foster City, CA) 380B DNA synthesizer and purified by OPEC column from the same company. The primer pairs were selected from the 5' and 3' sequence of the cDNA coding for human ERp72 stress-related protein (5'-primer: 5'-TTCCTCCTCCT-CCTCCTCTT-3'; 3'-primer: 5'-ATCTAAATGTCTAGT-3') *(10)*.
4. Light mineral oil.
5. *Taq* DNA polymerase.

## 3. Methods
## 3.1. First-Strand cDNA Synthesis (19)

1. Perform reverse transcription in a 25-μL reaction mixture, adding the following components: 5.0 μL 5X first-strand buffer, 2.5 μL dNTP mix, 2.5 μL random primers, 1.0 U RNasin, 1.25 μL actinomycin D, 250 U MMLV reverse transcriptase, 15–25 μg of total RNA (heat denature RNA at 65°C for 3 min prior to adding to reaction), and to a final vol of 25 μL nuclease-free H₂O.
2. Incubate at 37°C for 60 min.

## 3.2. Second-Strand Synthesis (20)

1. Add components to the first-strand tube on ice in the following order: 12.5 µL 10X second-strand buffer, 12.5 µL 1 m*M* NAD, 0.5 µL RNase H (2 U/µL), 5.75 µL *E. coli* DNA polymerase I (5 U/µL), 1.25 µL *E. coli* ligase (1 U/µL), and 67.5 µL of nuclease-free water.
2. Incubation at 14°C for 2 h.
3. Heat the reaction mix to 70°C for 10 min, spin for a few seconds, and then put in ice.
4. Add 4 U of T4 DNA polymerase and incubate at 37°C for 10 min to blunt the ends of double-stranded cDNA.
5. Stop the reaction with 12.5 µL of 0.2*M* EDTA and 200 µL of sterile H$_2$O.
6. Concentrate and purify the sample with GeneClean. Resuspend the DNA in 100–200 µL of sterile H$_2$O.
7. Estimate the DNA concentration by comparing the ethidium bromide fluorescent intensity of the sample with that of a series of DNA standards on a sheet of plastic wrap *(21)*. Dot 1–5 µL of sample onto plastic wrap on a UV transilluminator. Also dot with 5 µL of DNA standards. Add an equal vol of TE buffer containing 2 µg/mL of ethidium bromide, mix by repipeting up and down. Use proper UV shielding for exposed skin and eyes.

## 3.3. Circularization and Cleavage (see Notes 1–4)

1. Set up the circularization reaction mix containing the following components: 100 µL (100 ng DNA) of the purified sample, 25 µL of 5X ligation buffer, and 6 µL of T4 DNA ligase. Finally, add 2 µL of T4 RNA ligase or 15 µL of 15 µ*M* hexaminecobalt chloride (*see* Note 5).
2. Incubate at 18°C for 16 h.
3. Boil the ligated circular DNA for 2–3 min in distilled water or digest with an appropriate restriction enzyme to reopen circularized DNA.
4. Purify the DNA sample with GeneClean as described in Section 3.2., step 6 or extract with water-saturated phenol/CHCl$_3$ and then precipitate with ethanol *(20)*.

## 3.4. Inverse PCR (see Note 6)

1. Add 1/10 of the purified cDNA to 100 µL of amplification mix *(22)*: 10 µL 10X PCR buffer, 10 µL 15 m*M* MgCl$_2$, 10 µL dNTP mix (2.5 m*M* of each), 10 µL 5'-primer (10 pmol/mL), 10 µL 3'-primer (10 pmol/mL), 10 µL cDNA, 39.5 µL of nuclease-free H$_2$O, and 0.5 µL *Taq* DNA polymerase (2.5 µ/mL).
2. Cap and vortex the tubes. Spin briefly in a microfuge. Cover each reaction with a few drops of light mineral oil to prevent evaporation.
3. Put a drop of mineral oil into each well of the thermal cycler block that will hold a tube. Load the reaction tubes.
4. Amplify by PCR using the following cycle profile: 25 cycles: 94°C for 1 min (denaturation), 65°C for 2 min (annealing), and 72°C for 4 min (elongation).

## 4. Notes

1. For maximum efficiency of intramolecular ligation, low concentration of cDNA should be used in the ligation mix. High density of cDNA may enhance the level of heterogeneous ligation, which creates nonspecific amplification.
2. Cleavage or denaturation of circularized double-strand cDNA is important since circular double-strand DNA tends to form supercoil and is a poor template for PCR *(23)*. Circularized double-strand DNA is only good for amplification of a short DNA fragment.
3. The following can be considered to introduce nicks in circularized DNA.
   a. Boiling is a simple and common way. Because of the unusual secondary structure of some circular double-strand DNA, sometimes this method is not sufficient in nicking and denaturing circular double-strand DNA.
   b. A second method is selected restriction enzyme digestion. The ideal restriction site is located in the known sequence region of cDNA. In most cases, it is difficult to make the right choice of a restriction enzyme because the restriction pattern in the unidentified region of cDNA is unknown.
   c. If an appropriate enzyme is not available, EDTA-oligonucleotide-directed specific cleavage may be tried *(24,25)*. Oligonucleotide linked to EDTA-Fe at T can bind specifically to double-stranded DNA by triple-helix formation and produce double-stranded cleavage at the binding site.
4. Alkali denaturation has been successfully used to prepare plasmid DNA templates for PCR and DNA sequencing *(12,13,26)*. This method should be feasible in denaturing circularized double-strand cDNA.
5. Inclusion of T4 RNA ligase or hexaminecobalt chloride can enhance the efficiency of blunt-end ligation of double-strand DNA catalyzed by T4 DNA ligase *(27)*.
6. IPCR can be used to efficiently and rapidly amplify regions of unknown sequence flanking any identified segment of cDNA or genomic DNA. This technique does not need construction and screening of DNA libraries to obtain additional unidentified DNA sequence information. Some recombinant phage or plasmid may be unstable in bacteria and amplified libraries tend to lose them *(23)*. IPCR eliminates this problem.

## Acknowledgments

I would like to acknowledge John Holcenberg for his invaluable comments and generous support. I especially thank Chun-Hua Wu and Bing Cai for their technical assistance and Kai-Jin Wu for her art work.

## References

1. Verma, I. M., Temple, G. F., Fan, H., and Baltimore, D. (1972) In vitro synthesis of double-stranded DNA complimentary to rabbit reticulocyte 10S RNA. *Nature* **235**, 163–169.
2. Akowitz, A. and Mamuelidis, L. (1989) A novel cDNA/PCR strategy for efficient cloning of small amounts of undefined RNA. *Gene* **81**, 295–306.

3. Okayama, H., Kawaichi, M., Brownstein, M., Lee, F., Yokota, T., and Arai, K. (1987) High-efficiency cloning of full-length cDNA; construction and screening of cDNA expression libraries for mammalian cells. *Methods Enzymol.* **154**, 3–28.
4. Brenner, C. A., Tam, A. W., Nelson, P. A., Engleman, E. G., Suzuki, N., Fry, K. E., and Larrick, J. W. (1989) Message amplification phenotyping (MAPPing): a technique to simultaneously measure multiple mRNAs from small numbers of cells. *Biotechniques* **7**, 1096–1103.
5. Frohman, M. A. (1990) RACE: Rapid amplification of cDNA ends, in *PCR Protocols: A Guide to Methods and Applications.* (Innis, M. A., Gelfand, D. H., Sninsky, J. J., and White, T. J., eds.), Academic, San Diego, CA, pp. 28–38.
6. Shyamala, V. and Ames, G. F.-L. (1989) Genome walking by single-specific-primer polymerase chain reaction: SSP-PCR. *Gene* **84**, 1–8.
7. Huang, S.-H., Jong, A. Y., Yang, W., and Holcenberg, J. (1993) Amplification of gene ends from gene libraries by PCR with single-sided specificity. *Methods Mol. Biol.* **15**, 357–363.
8. Ochman, H., Gerber, A. S., and Hartl, D. L. (1988) Genetic applications of an inverse polymerase chain reaction. *Genetics* **120**, 621–625.
9. Triglia, T., Peterson, M. G., and Kemp, D. J. (1988) A procedure for in vitro amplification of DNA segments that lie outside the boundaries of known sequences. *Nucleic Acids Res.* **16**, 8186.
10. Huang, S.-H., Hu, Y. Y., Wu, C.-H., and Holcenberg, J. (1990) A simple method for direct cloning cDNA sequence that flanks a region of known sequence from total RNA by applying the inverse polymerase chain reaction. *Nucleic Acids Res.* **18**, 1922.
11. Delort, J., Dumas, J. B., Darmon, M. C., and Mallet, J. (1989) An efficient strategy for cloning 5' extremities of rare transcrips permits isolation of multiple 5'-untranslated regions of rat tryptophan hydroxylase mRNA. *Nucleic Acids Res.* **17**, 6439–6448.
12. Cusi, M. G., Cioe', L., and Rovera, G. (1992) PCR amplification of GC-rich templates containing palindromic sequences using initial alkali denaturation. *Biotechniques* **12**, 502–504.
13. Lau, E. C., Li, Z.-Q., and Slavkin, S. C. (1993) Preparation of denatured plasmid templates for PCR amplification. *Biotechniques* **14**, 378.
14. Green, I. R. and Sargan, D. R. (1991) Sequence of the cDNA encoding ovine tumor necrosis factor-a: problems with cloning by inverse PCR. *Gene* **109**, 203–210.
15. Zilberberg, N. and Gurevitz, M. (1993) Rapid Isolation of full-length cDNA clones by "Inverse PCR:" purification of a scorpion cDNA family encoding α-neuro-toxins. *Anal. Biochem.* **209**, 203–205.
16. Austin, C. A., Sng, J.-H., Patel, S., and Fisher, L. M. (1993) Novel HeLa topoisomerase II is the IIβ isoform: complete coding sequence and homology with other type II topoisomerases. *Biochim. Biophys. Acta* **1172**, 283–291.
17. Delidow, B. C., Lynch, J. P., Peluso, J. J., and White, B. A. (1993) Polymerase Chain Reaction: Basic Protocols. *Methods Mol. Biol.* **15**, 1–29.

18. Davis, L. G., Dibner, M. D., and Battey, J. F. (1986) *Basic Methods in Molecular Biology,* Elsevier, New York.
19. Kru, M. S. and Berger, S. L. (1987) First strand cDNA synthesis primed by oligo(dT). *Methods Enzymol.* **152,** 316–325.
20. Promega (1991) *Protocols and Applications* (2nd ed.), pp. 199–238.
21. Sambrook, J., Fritch, E. F., and Maniatis, T. (1989) *Molecular Cloning,* 2nd ed., Cold Spring Harbor Laboratory, Cold Spring Harbor, New York.
22. Saiki, R. K., Gelfand, D. H., Stoffel, S., Scharf, S. J., Higuchi, R., Horn, G. T., Mullis, K. B., and Erlich, H. A. (1988) Primer-directed enzymatic amplification of DNA with a thermostable DNA polymerase. *Science* **239,** 487–491.
23. Moon, I. S. and Krause, M. O. (1991) Common RNA polymerase I, II, and III upstream elements in mouse 7SK gene locus revealed by the inverse polymerase chain reaction. *DNA Cell Biol.* **10,** 23–32.
24. Strobel, S. A. and Dervan, P. B. (1990) Site-specific cleavage of a yeast chromosome by oligonucleotide-directed triple-helix formation. *Science* **249,** 73–75.
25. Dreyer, G. B. and Dervan, P. B. (1985) Sequence-specific cleavage of single-stranded DNA: oligodeoxynucleotide-EDTA.Fe(II). *Proc. Natl. Acad. Sci. USA* **82,** 968–972.
26. Zhang, H., Scholl, R., Browse, J., and Somerville, C. (1988) Double-strand DNA sequencing as a choice for DNA sequencing. *Nucleic Acids Res.* **16,** 1220.
27. Sugino, A., Goodman, H. M., Heynecker, H. L., Shine, J., Boyer, H. W., and Cozzarelli, N. R. (1977) Interaction of bacteriophage T4 RNA and DNA ligases in joining of duplex DNA at base-paired ends. *J. Biol. Chem.* **252,** 3987.

# 30

## Rapid Amplification of Gene Ends (RAGE) from Gene Libraries by Anchored PCR

**Sheng-He Huang and Ambrose Y. Jong**

## 1. Introduction

Isolation of a full-length gene on the basis of limited sequence information is often troublesome and challenging. Tremendous effort is needed to isolate a specific gene by screening cDNA or genomic libraries by oligonucleotide or nucleic acid probes. In those methods, basically, nucleic acid probes are used in a screening process to check whether or not a plaque or a colony contains the sequence of interest. There have been attempts to isolate specific DNA fragments using immobilized DNA, in which particular DNA fragments were enriched by hybrid selection and then the concentrated library was screened by a specific DNA probe *(1,2)*. Recently, PCR has been applied to the cloning of genes. Friedmann et al. *(3)* first used PCR to screen λgt11 library with two gene-specific primers. This protocol can be effectively used to isolate a particular DNA fragment between two specific primers or to generate nucleic acid probe from cDNA libraries. The unknown sequences flanking the fragment between the two specific primers cannot be amplified by this method.

Anchored PCR or single-specific-primer PCR *(4)* and inverse PCR *(5)* have been adapted to cloning of full-length cDNAs with the knowledge of a small stretch of sequence within the gene. Both methods start from mRNA and are good for cDNA cloning when a cDNA library is not available. Cloning of full-length cDNA is usually far more difficult than any other recombinant DNA work because the multiple sequential enzymatic reactions often result in low yield and truncated clones *(6)*. Shyamala and Ames *(7)* extended the use of anchored PCR to amplify unknown DNA sequences from genome as the basis of using a short stretch of known sequence for designing a gene-specific primer.

From: *Methods in Molecular Biology, Vol. 67: PCR Cloning Protocols: From Molecular Cloning to Genetic Engineering* Edited by: B. A. White  Humana Press Inc., Totowa, NJ

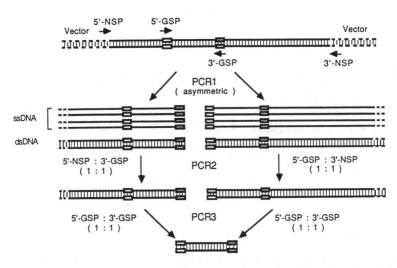

**Fig. 1.** The scheme for rapid amplification of gene ends (RAGE). 5'-NSP is λgt11 forward primer (5'-GACTCCTGGAGCCCG-3'). 3'-NSP: λgt11 reverse primer (5'-GGTAGCGACCGGCGC-3'). 5'-GSP: 5ASPR with *Eco*RI restriction site (5'-AGA-CTGAATTCGGTACCGGCGGTACTATCGCTTCC-3'. 3'-GSP: 3ASPB containing *Bam*HI site (5'-CTGATGGATCCTGGCAGTGGCTGGACGC-3').

We developed a much simpler method to isolate full-length cDNA and flanking genomic DNA from gene libraries by anchored PCR. The principle of this technique is schematically depicted in Fig. 1. Briefly, two ends of a gene were amplified by two nonspecific primers (NSP) complementary to the vector sequences flanking the polylinker region and two gene-specific primers (GSP) complementary to 5' and 3' parts of the known gene sequence.

We used the yeast gene coding for asparaginase II *(8)* as a model to verify this method. Two ends of the yeast asparaginase II gene *(8)* were amplified from λgt11 yeast genomic library by this method. The size of the full-length gene is 1.7 kb and there is 353 bp overlapping sequence between the two GSPs (5ASPR and 3ASPB). The 3' gene end from 5ASPR to the extreme 3' end of the gene and the 5' gene end from 3ASPB to the very beginning of 5' part of the gene are 1.15 and 0.9 kb, respectively. Since there may be two different orientations for each insert in the gene library, four reactions are performed for the amplication of two gene ends in the first round of PCR. Figure 2 shows that there was only one orientation for the asparaginase II gene clone and the expected sizes of the two amplified gene ends were obtained. The same size fragment was amplified from the gene library and the two purified gene ends by using the two GSPs as primers. The gel purified PCR DNA fragments were sequenced and the sequence data were consistent with the literature *(8)*.

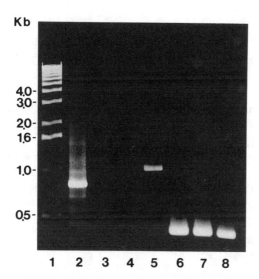

**Fig. 2.** Gel pattern of PCR products. Fragments amplified from the purified 5' (lane 6) and 3' (lane 7) gene ends and λgt11 yeast genomic library (lanes 2–5 and 8) were resolved in 1.5% agarose gel and stained with ethidium bromide. Lane 1 was 1 Kb ladder DNA marker (BRL). There were five pairs of primers used in PCR: 5'-NSP and 3'-GSP (lane 2); 3'-NSP and 3'-GSP (lane 3); 5'-NSP and 5'-GSP (lane 4); 3'-NSP and 5'-GSP (lane 5) and 5'-GSP and 3'-GSP (lanes 6–8).

Recently, we amplified and isolated 5'-cDNA fragment of 5-hydroxytryptamine 2 (5-HT2) receptor by this technique *(9)*. 5'-cDNA region of 5-HT2 receptor was successfully amplified from λSWAJ-2 mouse brain cDNA library with a GSP (5'-TTCTGCCTGAGACTAAAAAGGGTTAAGCCCTTATGATGGCA-3') and a NSP (*Xba* 1-T15 adaptor primer: 5'-GTCGACTCTTAGAT-3') *(9)*.

## 2. Materials

1. *Taq* DNA polymerase (Perkin/Elmer, Foster City, CA).
2. dNTP mix (10 m*M* of each; Perkin/Elmer).
3. Yeast genomic library (ClonTech, Palo Alto, CA). *See* Note 1.
4. lSWAJ-2 mouse brain cDNA library (ClonTech). *See* Note 1.
5. 5' and 3' λgt11 primers (ClonTech).
6. GeneClean™ kit (Bio101, La Jolla, CA).
7. Sequenase kits (United States Biochemical Corporation, Cleveland, OH)
8. 10X *Taq* DNA polymerase buffer: 67 m*M* Tris-HCl, pH 8.8, 6.7 m*M* MgCl$_2$, 170 µg/mL bovine serum albumin (BSA), 16.6 m*M* ammonium sulfate (New England Biolab, Beverly, MA) was prepared as described *(4; see* Note 2).
9. Oligodeoxyribonucleotides were synthesized on an Applied Biosystems 380B DNA synthesizer and purified by OPEC column from the same company.
10. Programmable thermal controller.

## 3. Methods

The following conditions were used to amplify the ends of the yeast asparaginase II gene from a λgt11 yeast genomic library.

### 3.1. PCR 1

1. Perform asymmetric PCR (*see* Note 2) in a 50-µL reaction mixture containing: 1-µL aliquot of λgt11 yeast genomic library (about $1 \times 10^7$ *pfu*), 2.5 pmol nonspecific primer (NSP), 50 pmol gene-specific primer (GSP) (*see* Note 3), 5 µL of dimethylsulfoxide, 5 µL of 10X *Taq* DNA polymerase buffer, 1.5 m*M* of each dNTP, and 2.5 U of *Taq* DNA polymerase. Before adding the polymerase, heat the PCR cocktail to 94°C for 3 min to disrupt the phage particles.
2. Amplify using the following parameters: 35 cycles: 94°C for 1 min, 48°C for 30 s, and 72°C for 8 min. The major product, single-stranded DNA, can be used for sequencing and Southern blotting analyses.

### 3.2. PCR 2

1. Dilute the first PCR product at 1:10 in $H_2O$.
2. Amplify 1-µL aliquots as described in PCR 1, with the exception of using equal amounts (50 pmol) of the NSP and GSP. The selected dsDNA fragments are available for cloning, which can be facilitated by incorporating restriction sites in the primers.

### 3.3. PCR 3

PCR 3 is used to test the products of PCR 2 since amplification of the two GSPs should produce the same fragment from a gene library and the two purified gene ends.

1. Assemble a PCR with 1 µL of 1:10 diluted PCR 2 product or 1-µL aliquot of a gene library in the third round of PCR using two GSPs (50 pmol each).
2. Amplify for 35 cycles at 94°C for 30 s, 48°C for 30 s, and 72°C for 4 min.
3. Purify DNA fragments from PCR 1, 2, and 3 by GeneClean (BIO 101) from agarose gel and sequence by the dideoxy chain termination method *(10)* with Sequenase (USB) with the aid of NP-40 *(11)*. (*See* Notes 4 and 5.)

## 4. Notes

1. The successful isolation of clones from cDNA or genomic libraries is dependent on the quality of the library. There is a big difference between the primary and the amplified libraries because different recombinant clones may grow at very different rates, resulting in unequal distribution of the recombinants in the amplified library *(16)*. It may be better to use the primary library for PCR amplification. If a short stretch of known DNA sequence (more than 100 bp) is available, it is easy to test the quality of the library by PCR amplification with two GSPs. In case the library is not good, an alternative strategy is to use inverse or anchored PCR to isolate the clone from genomic DNA or self-made cDNA.

2. The most obvious and common problem for PCR with single-sided specificity is nonspecific amplification of DNA fragments without significant homology with the gene of interest *(13,14)*. In this study we found two things that improve specific amplification. At first we used PCR buffer from Perkin Elmer Cetus and failed to amplify specific products for yeast asparaginase II gene. It appears that the *Taq* DNA polymerase buffer from New England Biolab is much better for PCR with single specific primer. Second, asymmetric PCR with a relatively large amount of GSP was performed in the first round of PCR in order to enhance the specificity of amplification.

3. This method does not appear to be suitable for DNA amplification with degenerate primers based on the highly conserved regions of a protein from other species or limited amino acid sequence data because degeneracy of primers can create more problems with nonspecific amplification. This limit may be reduced by incorporation of deoxyinosine into wobble positions of degenerate oligonucleotides *(12)*.

4. This method can be used for directional genome walking from known into unknown flanking regions of the chromosomal DNA with genomic libraries. PCR amplification of large DNA fragments (up to 35 kb) has been achieved by the combination of a high level of an exonuclease-free, N-terminal deletion mutant of *Taq* DNA polymerase, Klentaq1, with a very low concentration of a thermostable DNA polymerase exhibiting a 3'-exonuclease activity *(Pfu, Vent,* or *Deep, Vent) (15)*.

5. The two gene ends with overlapping sequence can be linked simply by two ways. First, ssDNAs of the two gene ends from PCR1 can be annealed and end filled by Klenow in presence of random primers. Second, when the sequence information is obtained from the two gene ends, a full-length cDNA or gene can be amplified by two specific primers that represent the sequences at the extreme 3' and 5'-ends of the gene or cDNA.

## Acknowledgments

We would like to acknowledge John Holcenberg for his generous support. We especially thank Chun-Hua Wu and Bing Cai for their technical assistance and Kai-Jin Wu for her art work. This work was supported by the T. J. Martell Foundation.

## References

1. Schott, H. and Bayer, E. (1979) Template chromatography. *Adv. Chrom.* **17**, 181–229.
2. Tsurui, H., Hara, E., Oda, K. Suyama, A., Nakada, S., and Wada, A. (1990) A rapid and efficient cloning method with a solid-phase DNA probe: application for cloning the 5'-flanking region of the gene encoding human fibronectin. *Gene* **88**, 233–239.
3. Friedmann, K. D., Rosen, N. L., Newman, P. J., and Montgomery, R. R. (1988) Enzymatic amplification of specific cDNA inserts from λgt11 libraries. *Nucleic Acids Res.* **16**, 8718.

4. Frohman, M. A., Dush, M. K., and Martin, G. R. (1988) Rapid production of full-length cDNAs from rare transcripts: amplification using a single gene-specific oligonucleotide primer. *Proc. Natl. Acad. Sci. USA* **85**, 998–9002.

5. Huang, S. H., Hu, Y. Y., Wu, C. H., and Holcenberg, J. (1990) A simple method for direct cloning cDNA sequence that flanks a region of known sequence from total RNA by applying the inverse polymerase chain reaction. *Nucleic Acids Res.* **18**, 922.

6. Okayama, H., Kawaichi, M., Brownstein, M., Lee, F., Yokota, T., and Arai, K. (1987) High-efficiency cloning of full-length cDNA: construction and screening of cDNA expression libraries for mammalian cells. *Methods Enzymol.* **154**, 3–28.

7. Shyamala, V. and Ames, G. F.-L. (1989) Genome walking by single-specific-primer polymerase chain reaction: SSP-PCR. *Gene* **84**, 1–8.

8. Kim, K. W., Kamerud, J. Q., Livingston, D. M., and Roon, R. J. (1988) Asparaginase II of *Saccharomyces cerevisiae:* characterization of the ASP3 gene. *J. Biol. Chem.* **263**, 11,948–11,953.

9. Huang, S.-H., Jong, A. Y., Yang, W., and Holcenberg, J. (1993) Amplification of gene ends from gene libraries by polymerase chain reaction with single-sided specificity. *Methods Mol. Biol.* **15**, 357–363.

10. Sanger, F., Nickler, S., and Coulson, A. R. (1977) DNA sequencing with chain-terminating inhibitors. *Proc. Natl. Acad. Sci. USA* **74**, 5463–5467.

11. Bachmann, B., Lucke, W., and Hunsmann, G. (1990) Improvement of PCR amplified DNA sequencing with the aid of detergents. *Nucleic Acids Res.* **18**, 1309.

12. Patil, R. V. and Dekker, E. E. (1990) PCR amplification of an *Escherichia coli* gene using mixed primers containing deoxyinosine at ambiguous positions in degenerate amino acid codons. *Nucleic Acids Res.* **18**, 3080.

13. Frohman, M. A. (1990) RACE: rapid amplification of cDNA ends, in *PCR Protocols: A Guide to Methods and Applications* (Innis, M. A., Gelfand, D. H., Sninsky, J. J., and White, T. J., eds.), Academic, San Diego, CA, pp. 28–38.

14. Loh, E. Y., Elliott, J. F., Cwirla, S., Lanier, L. L., and Davis, M. M. (1989) Polymerase chain reaction with single-sided specificity: analysis of T cell receptor d chain. *Science* **243**, 217–220.

15. Barnes, W. M. (1994) PCR amplification of up to 35-kb DNA with high fidelity and high yield from bacteriophage templates. *Proc. Natl. Acad. Sci. USA* **91**, 2216–2220.

16. Frischauf, A.-M. (1987) Construction and characterization of a genomic library in λ. *Methods Enzymol.* **152**, 190–199.

# 31

## Isolation of Coding Sequences from Yeast Artificial Chromosome (*Yac*)

*Clones by Exon Amplification*

**Fernando Gibson and Steve D. M. Brown**

### 1. Introduction

Exon amplification is a technique designed to address a central problem in mammalian molecular genetics—how to extract coding sequences from large tracts of genomic DNA. As shown in Fig. 1, the technique (also known as exon trapping) exploits the ability of the eukaryotic splicing machinery to detect splice sites flanking exon sequences in pre-mRNA molecules. The original exon trapping vector pSPL1 developed by Buckler et al. *(1)* and its subsequently improved derivative pSPL3 *(2)* allow segments of genomic DNA to be cloned into an HIV-tat intron that is flanked by the 5' and 3' splice sites and exons of the viral gene. Recombinant clones are transfected into COS-7 cells, which support high levels of transcription driven by the SV40 early promoter of the vector. During in vivo splicing, the 5' and 3' splice sites flanking an exon contained within the genomic insert are paired with the HIV-tat splice sites, with the result that the genomic exon is retained in the mature cytoplasmic poly A$^+$ RNA. Reverse transcription of the cytoplasmic RNA is followed by PCR using primers specific for the HIV-tat exons to amplify the "trapped exon." After a secondary (nested) PCR amplification, the PCR products are cloned into a suitable plasmid vector.

Initial experiments with pSPL1 demonstrated the effectiveness of this system in isolating exon sequences from individual λ (15–20 kb) and cosmid (40 kb) clones *(1)*. However, pSPL1 was subsequently modified in order to improve the efficiency and specificity of the technique *(2)*. The crucial modification incorporated into the second generation vector pSPL3 was the inclusion of recognition sites for the rare cutter enzyme *Bst*XI that allowed the elimina-

From: *Methods in Molecular Biology, Vol. 67: PCR Cloning Protocols: From Molecular Cloning to Genetic Engineering* Edited by: B. A. White Humana Press Inc., Totowa, NJ

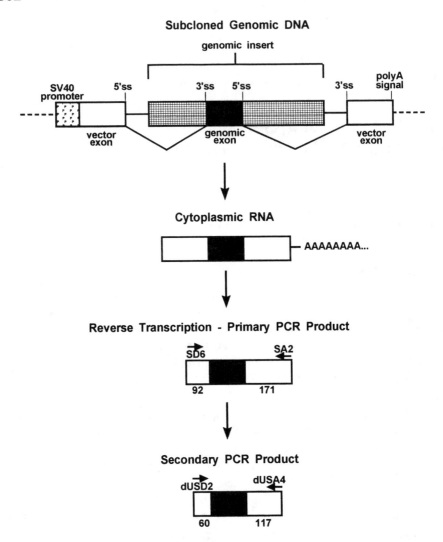

**Fig. 1.** An overview of exon amplification. ss, splice site. SD6 and SD2, primary PCR primers. dUSD2 and dUSA4, secondary PCR primers. SD6 and dUSD2 are the sense primers. Numbers below boxes indicate the number of bases from the first nucleotide of the primer to the 3' terminus of the vector exon. Note that the secondary PCR primers are nested (internal to the primary PCR primers). *See text* for details.

tion of two unwanted, but previously predominant RNA-PCR products. These result from:

1. Vector–vector splicing, which occurs when there is no exon in the subcloned genomic fragment (Fig. 2);

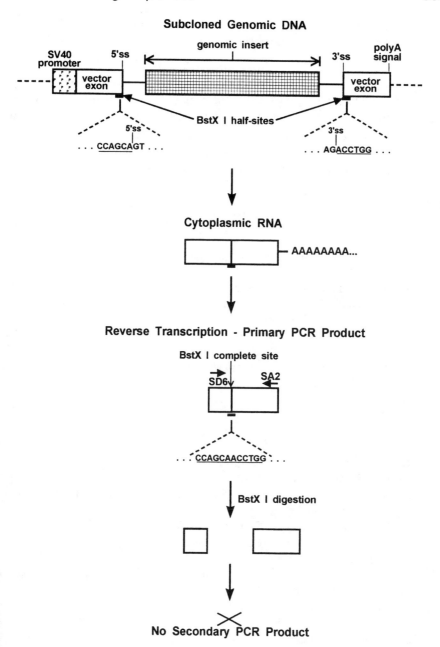

**Fig. 2.** Elimination of vector–vector splice products. *Bst*XI half-sites are located immediately adjacent to the vector 5' and 3' splice sites. When no exon is present in the cloned genomic fragment, vector–vector splicing generates a complete *Bst*XI site. Cleavage of the primary PCR product prevents secondary PCR amplification.

2. The presence of a 3' cryptic splice site in the genomic fragment (Fig. 3). The improved vector enabled the technique to be applied to much larger regions of genomic DNA up to 3 Mb *(2)*.

Exon amplification has developed into a highly efficient method that can be applied to large target regions of cloned genomic DNA including P1 clones (75–100 kb), bacterial artificial chromosomes (*Bacs*) (100–300 kb), cosmid pools (300–400 kb), and largest of all, *Yacs* (150–2000 kb). In this chapter, we describe a step-by-step procedure for applying the established technique of exon amplification to the direct isolation of exon sequences from *Yacs*.

## 2. Materials

1. A pulsed-field gel electrophoresis (PFGE) system.
2. A Gelase™ kit (Cambio).
3. Equipment and reagents for agarose gel electrophoresis.
4. Restriction endonucleases.
5. pSPL3 plasmid (Gibco-BRL, Gaithersburg, MD).
6. Calf intestinal alkaline phosphatase (CIP) (Boehringer Mannheim, Indianapolis, IN).
7. T4 DNA ligase and 10X T4 DNA ligation buffer (New England Biolabs, Beverly, MA).
8. *Escherichia coli* XL2-Blue ultracompetent cells and XL1-blue competent cells (Stratagene, La Jolla, CA).
9. SOC medium (Gibco-BRL).
10. 10 mg/mL Ampicillin.
11. LB medium and 2X YT medium (Gibco-BRL)
12. LB agar (Gibco-BRL)
13. TE buffer: 10 m$M$ Tris-HCl, pH 7.5, 1 m$M$ EDTA.
14. Tris-buffered phenol (Gibco-BRL).
15. Phenol:chloroform:isoamyl alcohol (PCI) (25:24:1) (Gibco-BRL).
16. Absolute ethanol and 70% ethanol.
17. COS-7 cells (American Type Culture Collection).
18. An electroporation system. We recommend the Gene Pulser (Bio-Rad).
19. Trypsin-EDTA: 0.25% trypsin, 1 m$M$ EDTA.
20. Supplemented Dulbecco's Modified Eagle Medium (S-DMEM) (Gibco-BRL).
21. Reagents and equipment for PCR.
22. Dulbecco's phosphate-buffered saline (D-PBS) (Gibco-BRL).
23. TKM buffer: 10 m$M$ Tris-HCl, pH 7.5, 10 m$M$ KCl, 1 m$M$ MgCl$_2$.
24. 10% Triton X-100 (BDH).
25. 5% SDS.
26. 5$M$ NaCl.
27. Diethyl pyrocarbonate (DEPC)-treated H$_2$O.
28. 5X reverse transcriptase (RT) buffer: 250 m$M$ Tris-HCl, pH 8.3, 375 m$M$ KCl, 15 m$M$ MgCl$_2$.

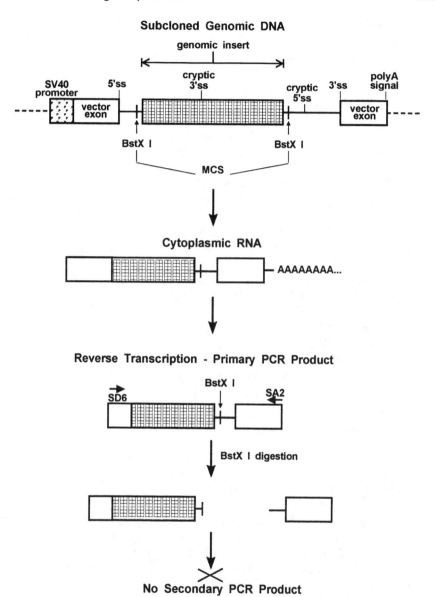

**Fig. 3.** Elimination of cryptic 3' splice site products by *Bst*XI digestion. Genomic fragments are cloned into the pSPL3 multiple cloning site (MCS), which is flanked by *Bst*XI sites. When a cryptic 3' splice site is present in the genomic fragment, it is joined to the vector 5' splice site, and cryptic 5' splice site in the vector intron is activated and joined to the vector 3' splice site. The resulting primary PCR product contains the 3' portion of the MCS, including a *Bst*XI site. *Bst*XI cleavage eliminates secondary PCR amplification as in Fig. 2.

29. 0.1*M* DTT.
30. 10 m*M* dNTPs.
31. SuperScript™ II reverse transcriptase (Gibco-BRL).
32. RNase H.
33. Uracil DNA glycosylase (UDG) (Gibco-BRL).
34. pAMP1 plasmid (Gibco-BRL).
35. Qiaex gel extraction kit (Qiagen).
36. VisiGel™ separation matrix (Stratagene).
37. Wizard™ miniprep and midiprep systems (Stratagene).
38. 6% Polyacrylamide gel mix.
39. A DNA sequencing gel electrophoresis system.
40. Sequenase™ Version 2.0 DNA Sequencing Kit (USB).
41. [$^{33}$P]dATP (>1000 Ci/mmol) (Amersham).
42. XAR-5 X-ray film (Kodak, Rochester, NY).
43. Primary PCR primers SD6 (5' TCT GAG TCA CCT GGA CAA CC 3') and SA2 (5' ATC TCA GTG GTA TTT GTC AGC 3').
44. Secondary PCR primers dUSD2 (5' CUA CUA CUA CUA GTG AAC TGC ACT GTG ACA AGC TGC 3') and dUSA4 (5' CAU CAU CAU CAU CAC CTG AGG AGT GAA TTG GTC G 3').
45. SP6 Promoter Primer (5 µg/mL) (Gibco-BRL).
46. Yeast tRNA (Gibco-BRL).
47. DNase-free RNase A (Sigma, St. Louis, MO).
48. 10X TBE buffer: 1*M* Tris, 0.9*M* boric acid, 10 m*M* EDTA.

## 3. Methods
### 3.1. Purificaton of Yac DNA

1. Embed yeast spheroplasts in agarose blocks using standard procedures *(3)* with two modifications: suspend spheroplasts in an equal vol of 1.6% agarose; use preparative scale (500 µL) block formers.
2. Load the agarose blocks into a single "gash" well traversing the width of the gel. Perform preparative PFGE through 1% low melting-point agarose to separate the *Yac* from the host yeast chromosomes. It is very important that there is a clear separation of the *Yac* from the host chromosomes to minimize contamination with yeast gene sequences (*see* Note 1).
3. Using UV light, locate and cut out the *Yac* band.
4. Cut the gel slice into small (~1 cm³) cubes and incubate in 10 vol of 1X GELase buffer for 1 h with gentle rotation. Replace with fresh buffer and incubate for a further 1 h.
5. Pour off buffer and incubate at 75°C for 15 min or until the agarose is completely molten.
6. Incubate at 40°C for 15 min.
7. Add 1 U of GELase enzyme/200 mg of molten agarose and incubate at 40°C for 2 h with occasional gentle mixing.

8. Add an equal vol of PCI and mix thoroughly by inversion.
9. Transfer the mixture to a suitable centrifuge tube and spin at 14,000$g$ for 5 min at 4°C.
10. Transfer the top (aqueous) layer to a fresh centrifuge tube without disturbing the interphase, then add 1/10 vol of 5$M$ NaCl, 10 μg carrier yeast tRNA, and 2 vol of ethanol. Mix and incubate at −70°C for 20 min.
11. Spin at 14,000$g$ for 1 h at 4°C.
12. Locate the pellet and carefully remove the supernatant. Wash the pellet with 5 mL of ice-cold 70% ethanol. Respin at 14K for 15 min.
13. Carefully remove as much of the supernatant as possible and dry the pellet at 50°C for 2 min.
14. Resuspend the *Yac* DNA pellet in 100 μL of TE buffer.
15. In a microcentrifuge tube, mix a 10-μL aliquot of *Yac* DNA with 1 μL of 10 mg/mL DNase-free RNase, and incubate at 37°C for 30 min. Spin briefly, then add 2.75 μL of 5X loading buffer. Load onto a 0.7% agarose gel with 50 ng of uncut DNA in an adjacent lane.
16. Assess the yield of *Yac* DNA obtained.
17. Repeat steps 2–16 until 1–2 μg of *Yac* DNA has been accumulated.
18. Concentrate the pooled *Yac* DNA to 50 ng/μL as follows. Add 0.5 vol of 7.5$M$ ammonium acetate and 2 vol of ethanol, mix, and incubate at −70°C for 20 min. Spin at 14,000$g$ for 30 min at 4°C. Discard the supernatant and wash pellet with ice-cold 70% ethanol. Dry pellet and resuspend in 20 μL TE buffer.

### 3.2. Subcloning Yac DNA into pSPL3

1. In a microcentrifuge tube, digest 2 μg of pSPL3 with 10 U of *Bam*HI for 2 h at 37°C.
2. Dephosphorylate with 16 U of CIP for 1 h at 37°C.
3. Add 1 vol of PCI, mix and spin at 14K for 3 min at room temperature.
4. Transfer the upper phase to a fresh tube and ethanol precipitate the dephosphorylated pSPL3 as in Section 3.1.1., step 8.
5. Digest 1 μg *Yac* DNA with 10 U of *Bam*HI, 10 U of *Bgl*II, and 0.5 mg/mL DNase-free RNase A for 5 h at 37°C.
6. Extract with 1 vol of PCI and ethanol precipitate as above.
7. Resuspend pellet in 20 μL TE buffer.
8. Set up a ligation as follows: 5 μL of pSPL3X *Bam*HI, 12 μL of *Yac* × *Bam*HI and *Bgl*II, 2 μL of 10X T4 ligase buffer, and 1 μL of T4 DNA ligase.
9. Prepare a control ligation without *Yac* DNA.
10. Incubate ligations at 16°C overnight.
11. Following the manufacturer's protocol, transform XL2-Blue ultracompetent cells with 2 μL of the ligation. Store the rest of the ligation at −20°C.
12. Plate out all of the transformed cells onto two 135-mm LB agar + 100 μg/mL of ampicillin plates (half the transformants on one plate, half on the other).
13. Incubate the plates at 37°C overnight.
14. The moment of truth! Examine the plates: if the subcloning procedure has worked well, there will be 10–20X more colonies on the pSPL3 + *Yac* plate than on the

pSPL3-only plate. This represents a low nonrecombinant background of 5–10%. Store the plates at 4°C.

15. Transform the rest of the pSPL3 + *Yac* ligation to obtain a library of recombinants equivalent to at least a fivefold coverage of the *Yac*. The number of recombinants required can be simply calculated as follows: Size of *Yac* (kb) × 5 (*see* Note 2).

### 3.3. Isolation and Analysis of pSPL3 + Yac Plasmid DNA

1. Apply 10 mL of LB medium onto each plate and disperse the colonies into the LB with a glass spreader.
2. Using a 10-mL pipet, transfer the cell suspension into a centrifuge tube and pellet the cells at 6000g for 5 min.
3. Prepare plasmid DNA using the Wizard midiprep system following the manufacturer's protocol.
4. Determine the concentration of the plasmid DNA by measuring its absorbance at 260 nm.

$$1 \ A_{260} \text{ unit of DNA} = 50 \ \mu g/mL \qquad (1)$$

5. Digest 3 μg of the pSPL3 + *Yac* plasmid library with 15 U of *Bst*XI for 2 h at 55°C to excise the inserts.
6. Examine the complexity and size distribution of the library by electrophoresing the digests through a 1% agarose gel. A good library will yield a strong smear of insert bands with an average size of ~1 kb.

### 3.4. Transfection of COS-7 Cells

1. Passage the COS-7 cells 24 h prior to transfection by placing $6 \times 10^6$ cells into 40 mL of S-DMEM in a 160-cm$^2$ tissue culture flask. Incubate the cells at 37°C in a 5% $CO_2$ incubator.
2. When the cells have reached 75–85% confluence, harvest them as follows.
3. Remove the S-DMEM and add 5 mL of trypsin-EDTA. Incubate at room temperature for 5 min.
4. Detach the cells by vigorously tapping the flask.
5. Suspend the cells in 5 mL S-DMEM and determine the cell density with a hemocytometer.
6. Pellet the cells by centrifuging at 300g for 8 min at 4°C.
7. Wash the cells twice in ice-cold D-PBS.
8. Mix $5 \times 10^6$ cells with 5 g of pSPL3 + *Yac* plasmid DNA in a final vol of 0.7 mL of ice-cold D-PBS.
9. Transfer the cell/DNA mixture to an ice-cold 0.4-cm electroporation cuvet. Leave on ice for 10 min.
10. Gently resuspend the cells by inversion, then electroporate at 1.2 kV, 25 F with a time constant of 0.5–0.6 (Bio-Rad Gene Pulser).
11. Place cells on ice for 10 min, then add to 10 mL of S-DMEM in a 100-mm well of a 6-well tissue culture dish.
12. Incubate for 48–72 h at 37°C in a 5% $CO_2$ incubator.

## 3.5. Isolation of Cytoplasmic RNA

Use RNase-free conditions throughout *(3)*.

1. Remove the S-DMEM.
2. Carefully add 10 mL of ice-cold D-PBS, then remove. Wash twice more with D-PBS.
3. Add 5 mL of trypsin-EDTA and incubate at room temperature for 5 min. Loosen cells by rapping the side of the dish.
4. Add 5 mL of S-DMEM to stop the trypsin digestion.
5. Transfer the cells to a 15-mL conical centrifuge tube and centrifuge at $300g$ for 8 min at 4°C.
6. Pour off the supernatant and remove as much residual solution as possible.
7. Resuspend cells in 200 µL ice-cold TKM buffer.
8. Add 10 µL of 10% Triton X-100. Mix gently by inversion and incubate on ice for 5 min.
9. Centrifuge at $450g$ for 5 min at 4°C to pellet the nuclei.
10. Transfer the supernatant to a microcentrifuge tube then add 15 µL of 5% SDS and 200 µL of Tris-buffered phenol. Mix well and centrifuge at $10,000–12,000g$ for 5 min at 4°C.
11. Transfer the upper phase to a fresh microcentrifuge tube. Add 200 µL of PCI and mix well. Centrifuge at $10,000–12,000g$ for 5 min at room temperature.
12. Transfer the upper phase to a fresh tube. Add 10 µL of $5M$ NaCl and 750 µL of ethanol. Mix and incubate at –70°C for 20 min.
13. Centrifuge at $10,000–12,000g$ for 45 min at 4°C.
14. Discard the supernatant and carefully wash the pellet with ice-cold 70% ethanol. Dry the pellet at 50°C for 2 min.
15. Resuspend in 40 µL of DEPC-treated water. Remove 4 µL for $A_{260}$ quantitation using a narrow-bore (400 µL) quartz cuvet.

$$1 \ A_{260} \text{ unit of RNA} = 40 \ \mu g/mL \quad (2)$$

16. Store RNA at –70°C.

## 3.6. Reverse Transcription

1. Add the following components to a 0.5-mL microcentrifuge tube: 5 µg RNA, 1 µL SA2 primer ($20M$), and DEPC-treated water to a final volume of 12 µL.
2. Incubate at 70°C for 5 min to denature the RNA, then chill 1 min on ice.
3. Briefly centrifuge, then add the following: 4 µL of 5X RT buffer, 2 µL of $0.1M$ DTT, 1 µL of 10 m$M$ dNTPs, and 1 µL of SuperScript reverse transcriptase.
4. Mix and briefly centrifuge. Incubate at 42°C for 1 h.
5. Incubate at 55°C for 5 min.
6. Add 1 µL of RNase H (1–4 U/µL), mix and incubate at 55°C for 10 min.
7. Briefly centrifuge and store at –20°C.

## 3.7. Primary PCR

1. Assemble a PCR as follows: 5 µL of 10X PCR buffer, 28 µL of water1 µL of 10 m$M$ dNTPs, 5 µL of SD6 primer (10 µ$M$), 5 µL of SA2 primer (10 µ$M$), 5 µL of RT reaction, and 1 µL of *Taq* polymerase (5 U/µL).

2. Mix and centrifuge briefly. Overlay with 50 μL of mineral oil.
3. Perform the PCR using the following cycle parameters: 95°C for 5 min (initial denaturation); 95°C for 1 min, 60°C for 1 min, and 72°C for 5 min for 6 cycles; soak file at 4°C.

### 3.8. BstXI Digestion of Primary PCR Products

1. Add 50 U of *Bst*XI to the primary PCR.
2. Mix and incubate at 55°C overnight.
3. Add a further 10 U of *Bst*XI and incubate for 2 h at 55°C.

### 3.9. Secondary PCR

1. Assemble a PCR as follows: 5 μL of 10X PCR buffer, 28 μL of water, 1 μL of 10 m*M* dNTPs, 5 μL of dUSD2 (10 μ*M*), dUSA4 (10 μ*M*), 5 μL of *Bst*XI-treated primary PCR product, and 1 μL of *Taq* polymerase (5 U/μL).
2. Mix and centrifuge briefly. Overlay with 50 μL of mineral oil.
3. Perform the PCR using the following cycle parameters: 95°C for 5 min (initial denaturation); 35 cycles: 95°C for 1 min, 60°C for 1 min, 72°C for 3 min; soak file at 4°C.

### 3.10. Cloning the Secondary PCR Products

1. Electrophorese 25 μL of the secondary PCR through a 2% low-melting point agarose gel. Use X174 × *Hae*III size markers. The size of the vector–vector splice product is 177 bp (Fig. 2). This may still be visible despite the exhaustive *Bst*XI digestion of the primary PCR product. Products larger than 177 bp, therefore, represent exon trap products—these will be visible as a smear of fragments 200–700 bp in size.
2. Cut out the gel slice containing the exon trap smear, then cut the slice into two pieces such that one piece contains the medium and largest sized fragments, whereas the other piece contains the smallest fragments. This counteracts the preferential cloning of the smallest PCR products.
4. Purify the DNA from the agarose pieces with the Qiaex Gel Extraction Kit, following the manufacturer's protocol.
5. Set up the following reaction (*see* Note 3): 10 μL of secondary PCR product, 2 μL of pAMP1 (25 ng/μL), 2 μL of 10X PCR buffer, 5 μL of water, and 1 μL of UDG (1 U/μL).
6. Mix and incubate at 37°C for 30 min.
7. Transform 5 μL of the UDG reaction into XL1-Blue competent cells following the manufacturer's protocol. Plate out 200 μL of the transformed cells onto LB agar + 100 μg/mL of ampicillin. Incubate plates at 37°C overnight.

### 3.11. Preliminary Analysis of Exon Trap Clones

1. Determine the number of clones that are to be analyzed (*see* Note 4).
2. Inoculate single colonies into 5 mL of 2X YT medium + 100 μg/mL of ampicillin. Incubate at 37°C overnight.

3. For each overnight culture, set up a PCR as follows: 5 µL of 10X PCR buffer, 31 µL of water, 1 µL of 10 m*M* dNTPs, 5 µL of dUSD2 (10 µ*M*), 5 µL of dUSA4 (10 µ*M*), 2 µL of overnight culture, and 1 µL of *Taq* polymerase (5 U/µL).

4. Mix and centrifuge briefly. Overlay with 50 µL of mineral oil.

5. Perform the PCR using the following cycle parameters: (initial denaturation) 95°C, 6 min; 35 cycles: 95°C, 1 min, 60°C, 1 min, and 72°C, 1 min; soak file at 4°C.

6. Electrophorese 5 µL of each reaction through a midi-size VisiGel following the manufacturer's protocol. Use X174 × *Hae*III markers.

7. Determine the number of unique clones by carefully comparing the sizes of the PCR products. Different clones tend to have different sizes.

8. Use 3 mL of the overnight culture to prepare plasmid DNA from each unique clone with the Wizard miniprep system, following the manufacturer's protocol.

### 3.12. Double-Stranded Sequencing of the Exon Trap Clones

1. To a 0.5-mL microcentrifuge tube, add 9 µL of plasmid DNA and 1 µL of 2*M* NaOH. Mix and incubate at 37°C for 20 min.

2. Centrifuge briefly, then add 2.5 µL of 3*M* sodium acetate, pH 5.2, and 37.5 µL of ethanol.

3. Incubate at –70°C for 20 min.

4. Centrifuge at 14,000*g* for 20 min at 4°C.

5. Discard the supernatant and wash the pellet with the ice-cold 70% ethanol. Dry the pellet at 50°C for 2 min.

6. Resuspend the pellet by adding the following: 4 µL of water, 2 µL of 5X Sequenase buffer, and 4 µL of SP6 promoter primer (5 µg/mL).

7. Using the Sequenase Version 2.0 DNA Sequencing Kit, perform the sequencing reactions as described in the manufacturer's protocol. Use [33P]dATP in the labeling reaction.

8. Use a premixed 6% polyacrylamide gel mix to pour a 30 × 40-cm sequencing gel by standard procedures *(3)*.

9. Prerun the gel for 1 h in 1X TBE buffer at 70 W.

10. Wash out the wells and load the gel with 2 µL of each sequencing reaction.

11. Run the gel at 70 W for 3–4 h until the top dye is 1–2 cm from the bottom of the gel.

12. Dry the gel and expose to X-ray film overnight.

13. Read the sequence.

### 3.13. Further Analysis of Exon Trap Sequences

1. Use a DNA sequence analysis software package (e.g., GCG, MacMolly) to confirm that each exon has an open reading frame.

2. Carry out a DNA and protein sequence database search with each exon using the BLAST program *(4)*.

3. Hybridize exons back to Southern blots of digested *Yac* DNA to confirm that the exons are derived from the *Yac*, and to position the exons on the *Yac* restriction map.

4. Hybridize exons to Multiple Tissue Northern Blots (Clontech) to detect the cognate transcript and analyze the tissue specificity of expression.

5. Design PCR primers from the exon sequences and prescreen cDNA libraries by PCR to rapidly identify appropriate libraries for conventional hybridization-based cDNA library screening.

6. Isolate and characterize the target cDNAs.

## 4. Notes

1. Using recombination-mediated chromosome fragmentation, a set of yeast host strains have recently been constructed that contain defined karyotypic alterations (L. Borbye, M. Johnston, and E. D. Green, personal communication). On PFGE, each strain exhibits a characteristic electrophoretic karyotype that is devoid of chromosomes over a defined size interval. This allows any *Yac* to be purified free of endogenous yeast chromosomes, regardless of its size. This approach should prove very useful when the target *Yac* comigrates with yeast chromosome(s).

2. The number of recombinants required for a fivefold coverage of the *Yac* can be calculated as follows: Size of *Yac* (kb)/2 kb [theoretical average size of DNA fragments produced by two 6-cutter restriction enzymes] × 2 [each fragment can insert in either orientation] × 5 (fivefold coverage).

3. The secondary PCR primers contain dUMP residues at their 5' ends. Treatment of the PCR products with UDG removes the uracils from the products' 5' termini. This disrupts base-pairing and generates 3' overhangs that simultaneously anneal to the complementary pAMP1 vector overhangs *(5)*. This procedure results in unidirectional cloning of the secondary PCR products such that the SP6 strand of the vector represents the coding (sense) strand of the exon trap clone.

4. The first thing to be borne in mind when embarking on the seemingly daunting task of analyzing the exon trap sequences is the high degree of redundancy among the clones. In our experience of exon trapping with *Yac* clones, one exon is isolated/~40 kb of *Yac* genomic DNA (6; Gibson and Brown, unpublished data). In addition, up to 50% of the total clones recovered consist of either artefactual intronic sequences (derived from vector or *Yac* DNA) or repetitive sequences. For a 300 kb *Yac*, this gives a total count of only 15 unique clones, in marked contrast to the 100–300 colonies typically obtained after transformation of the UDG reaction. Nevertheless, to minimize the possibility of missing an exon trap product, we recommend sequencing at least 3X the estimated number of unique clones. Thus, for a 300 kb *Yac*, at least 45 clones should be sequenced.

## References

1. Buckler, A. J., Chang, D. D., Graw, S. L., Brook, D. J., Haber, D. A., Sharp, P. A., and Housman, D. E. (1991) Exon amplification: a strategy to isolate mammalian genes based on RNA splicing. *Proc. Natl. Acad. Sci. USA* **88,** 4005–4009.

2. Church, D. M., Stotler, C. J., Rutter, J. L., Murrel, J. R., Trofatter, J. A., and Buckler, A. J. (1994) Isolation of genes from complex sources of mammalian genomic DNA using exon amplification. *Nature Genetics* **6,** 98–105.

3. Sambrook, J., Fritsch, E. F., and Maniatis, T. (1989) *Molecular Cloning: A Laboratory Manual,* 2nd ed., Cold Spring Harbor Laboratory, Cold Spring Harbor, NY.

4. Altschul, S. F., Gish, W., Miller, W., Myers, E., and Lippman, D. J. (1990) Basic local alignment search tool. *J. Mol. Biol.* **215,** 403–410.

5. Rashtchian, A., Buchman, G. W., Schuster, D. M., and Berninger, M. (1992) Uracil DNA glycosylase-mediated cloning of PCR-amplified DNA: application to genomic and cDNA cloning. *Anal. Biochem.* **206,** 91–97.

6. Gibson, F., Lehrach, H., Buckler, A. J., Brown, S. D. M., and North, M. A. (1994) Isolation of conserved sequences from yeast artificial chromosomes by exon amplification. *Biotechniques* **16,** 453–458.

# V

## LIBRARY CONSTRUCTION AND SCREENING

# 32

# cDNA Libraries from a Low Amount of Cells

**Philippe Ravassard, Christine Icard-Liepkalns, Jacques Mallet, and Jean Baptiste Dumas Milne Edwards**

## 1. Introduction
### 1.1. Application of the SLIC Strategy to cDNA Libraries: General Considerations

Conventional cDNA library construction often requires a minimum available amount of material (typically 1 or 2 μg of polyA$^+$ RNA). For complex organs, such as brain, or certain species, such as humans, as well as subsets of cell types, this condition is often difficult to fulfill. Amplification by PCR can be used to circumvent this limitation because it is a powerful method to obtain working quantities of low-abundance DNAs. To effectively apply this method, known sequences need to be attached to the ends of the single-stranded cDNA (ss-cDNA). One at the 5' end of the ss-cDNA is added during the priming of the synthesis, the other, at the 3' end, is covalently attached by ligation using the SLIC strategy as described elsewhere (Chapter 27). With known DNA sequences attached to both ends of the synthetized cDNA, minute quantities can be amplified with sequence-specific primers to provide sufficient material to successfully generate and screen cDNA libraries. The overall scheme is illustrated in Fig 1.

Obviously, the goal in constructing such a library is to maintain the representation of every mRNA in the total RNA population, even low-abundant ones. Since the reverse transcription and the single-strand ligation do not modify the overall proportion of each molecule, only the PCR step may introduce an important bias in cDNA representation. Thus, for cDNA library construction using this methodology, it is important to optimize the synthesis of the sequence-tagged cDNA, and to take steps to limit any amplification bias.

From: *Methods in Molecular Biology, Vol. 67: PCR Cloning Protocols: From Molecular Cloning to Genetic Engineering* Edited by: B. A. White Humana Press Inc., Totowa, NJ

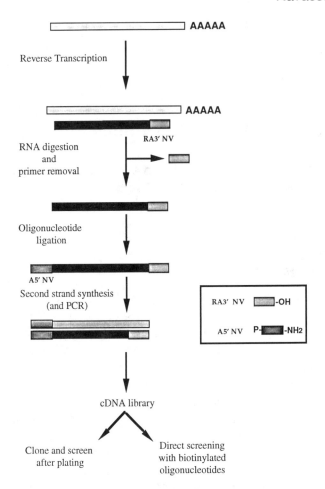

**Fig. 1.** Constructing cDNA libraries using the SLIC strategy. Reverse transcription is carried out with a random primer RA3' NV. RNAs are digested and primer is removed to avoid its ligation to the A5' NV oligonucleotide. A5' NV bears a phosphate group at its 5' extremity to allow ligation to the 3' end of the ss-cDNA. To avoid formation of concatemers A5' NV bears an amino group at its 3' end. Two rounds of nested PCR are performed to generate a ready-to-use cDNA library.

## 1.2. Simultaneous PCR Amplification of a Complex DNA Mixture Generates an Important Size Bias

The sequence-tagged cDNAs correspond to a large population of molecules with indentical ends. The only difference between these molecules is their relative size and sequence. Therefore, keeping the representativity of the cDNAs after PCR requires a constant amplification yield regardless of the size and sequence of the original cDNA population. This, unfortunately cannot be

| Name | 5'            Sequence            3' |
|---|---|
| A3'NV | ATCGTTGAGACTCGTACCAGCAGAGTCACGAGAGAGACTACACGGTACTGG(T)$_{14}$ |
| RA 3'NV | ATCGTTGAGACTCGTACCAGCAGAGTCACGAGAGAGACTACACGGTACTGG(N)$_9$ |
| A3'_1 | ATCGTTGAGACTCGTACCAGCAGAG |
| A3'_2 | TCGTACCAGCAGAGTCACGAGAGAG |
| A3'_3 | CACGAGAGAGACTACACGGTACTGG |

Fig. 2. Oligonucleotides used for 3' anchored PCR. The three oligonucleotides A3'_1,2,3 are designed from A3' NV and RA3' NV to be used in PCR experiments. Note that these three primers have to be used with A5'_1,2,3 (*see* Chapter 27).

achieved by PCR. In fact, coamplification of short and long molecules with the same primers always leads to a selective amplification of the shorter ones, even though the longer ones were originally more abundant *(1)*.

Attempts to enlarge the average size of the PCR products have been made with the help of long-range PCR procedures, such as *Pfu* dilutions *(2), Taq* extender (Stratagene, La Jolla, CA) and Expand PCR system (Boehringer, Indianapolis, IN). The results were not significantly different if compared to the classical PCR techniques (data not shown).

In conclusion, the size bias represents the major limitation in constructing PCR cDNA libraries. Thus, to avoid bias during the amplification step the average size of the cDNA library must be between 0.8 and 1 kbp, which is far below the usual average length of a ss-cDNA.

### 1.3. ss-cDNA Synthesis

The constraint caused by the size limitation of the ss-cDNA forbids the use of oligo-dT to prime the reverse transcription. If such a priming strategy is chosen, this will lead to a 3'-UTR-cDNA library. To obtain a cDNA library representative of the sequences of all messenger RNA, priming with random primer RA3' NV must be performed.

Experimentally, the ss-cDNA is synthesized with a random primer RA3' NV (Fig. 2) and a radiolabled nucleotide. The average size is determined by alkaline gel electrophoresis and autoradiography. The incubation time with the reverse transcriptase is calibrated in order to generate an ss-cDNA whose average length is between 0.8 and 1 kbp. With those conditions the representation of any mRNA will be optimal in the library.

## 1.4. Amplification of the cDNA Library
### Based on the SLIC Strategy

We have used this strategy to generate a cDNA library from newborn rat cervical superior ganglia (CSG). Total RNA was prepared from one single CSG, and polyA$^+$ RNA was prepared with oligo dT coated magnetic beads (Dynabeads mRNA purification kit). Half of the material, which corresponds to polyA$^+$ RNA of about 5000 cells, was used to synthesize ss-cDNA primed with RA3' NV. An incubation time of 30 minutes was optimum to generate an average size of 1 kbp. After removal of the primer, the ss-cDNA was ligated to A5' NV (note that after each step the different primers are removed). These ss-cDNA were amplified by two rounds of nested PCR. To increase the specificity of the PCR reaction, we have used the touchdown PCR protocol *(3)*. The primers used for the nested PCR were A5'_1 × A3'_1 and A5'_2 × A3'_2, respectively. One-twentieth of the reaction was cloned in a blunt-end vector, yielding $2 \times 10^5$ colonies. Analysis by direct PCR on colonies of 96 randomly chosen clones indicated that the average size of the library was about 900 bases. The striking result is that the size dispersion around the mean is extremely low when compared with a conventional library. Finally 5000 primary clones were screened with a TH oligonucleotide, yielding two positive TH clones (0.04%). Thus, no major distortion had been introduced in the abundance of TH clones, since TH represents 0.05% of CSG mRNA.

## 1.5. Direct Screening of the PCR Library
### with Biotinylated Oligonucleotides

One of the major difficulties in making a cDNA library is the cloning of the double-stranded cDNA (ds-cDNA). We tested a direct screening protocol of uncloned ds-cDNA. After the second nested PCR the amount of ds-cDNA was about 2–5 μg. We directly hybridized the denatured ds-cDNA with a biotinylated TH specific primer. After the hybridization reaction, probe-cDNA hybrids were separated from unhybridized DNA using streptavidin-coated magnetic beads (Dynabeads M-280). After various washing steps the captured cDNA was amplified using the third nested PCR primers (A5'3 and A3'3) directly onto the beads. The PCR product was cloned and more than 85% of them were TH clones as analyzed by partial sequencing. Biotinylated cDNA probes, instead of oligonucleotides, can also be used to screen a PCR library *(4)*.

## 1.6. Application of the SLIC Method to Subtractive Libraries

Subtraction cloning strategies could be modified and certainly improved taking advantage of the generation of cDNA molecules exhibiting two defined extremities. Basically, tracer cDNA synthesized on mRNA from source A is

| Name | 5' | Sequence | 3' |
|------|-----|----------|-----|
| B3'NV | | AGCATCAGCCAAGACCCAGCGAAGCACCCTCGAGCTACCGCG(T)$_{14}$ | |
| RB3'NV | | AGCATCAGCCAAGACCCAGCGAAGCACCCTCGAGCTACCGCG(N)$_9$ | |
| B3'_1 | | AGCATCAGCCAAGACCCAGCGAAGC | |
| B3'_2 | | AGCGAAGCACCCTCGAGCTACCGCG | |
| B5'NV | | CTGACGACGAATTCCTTAGCGATTGGTGAGGGCATTCGGGTC | |
| B5'_1 | | GACCCGAATGCCCTCACCAATCGCT | |
| B5'_2 | | CAATCGCTAAGGAATTCGTCGTCAG | |

**Fig. 3.** Alternative oligonucleotides used for subtractive cDNA library construction. For the B oligonucleotides the same modifications (5' phosphate and 3' amino) as the A primers have to be used.

hybridized to sequences of driver mRNA, isolated from a different but usually related source B. The tracer cDNAs that do not become hybridized with driver mRNA represent an enriched population of sequences expressed only in A cells. These are used for constructing an A-cell specific cDNA library.

The SLIC strategy provides DNA molecules with two defined ends. This offers the opportunity to work on cDNA from populations A and B with two different sets of SLIC primers, A (A5' NV and RA3' NV) and B (B5' NV and RB3' NV) (Fig. 3). During the PCR amplification of the tracer population B, a pair of biotinylated primers is used. Thus, this population can be captured and pure single-stranded molecules immobilized on magnetic beads. Then, after hybridization with the amplified A population, the unhybridized population corresponds to the A-cell specific sequences and can be used easily to generate substracted libraries or probes. This strategy gives for the first time the opportunity to realize such subtractive libraries with a very small amount of input material.

## 1.7. Conclusion

The SLIC method is a powerful and unique tool to synthesize cDNA and substractive libraries from a limited number of cells. It is unfortunately extremely difficult to generate full-length libraries. Nevertheless, cloning 5' or 3' ends of a cDNA is no longer a limiting step, since anchored PCR can be easily performed (*see* Chapter 27). In this case, the same ss-cDNA that was used to generate the library can also be used as a matrix to isolate both ends of the incomplete clone.

## 2. Materials

### 2.1. Oligonucleotides

1. PCR library primers: For the 5' end of the ss-cDNA, use RA3' NV and the related oligonucleotides (A3'_1,2,3). For the 3' end use A5'NV, A5'_1,2,3. All oligonucleotides must bear 5' and 3' hydroxyl group. Only A5'NV must have a phosphate group to its 5' end to allow ligation to the 3' end of the ss-cDNA. To avoid selfligation of A5' NV, its 3' end must be protected with an amino group. Those modifications are performed by any oligonucleotide suppliers. Apply the same rules when using B5' NV and RB3' NV. Note that only two related oligonucleotides are available for both ends.
2. Biotinylated screening primers: Ask your oligonucleotide supplier for 5' biotinylated primers with a seven carbon spacer. Do not use an 11 carbon spacer, because this will dramatically decrease the capture yield. If this primer is designed to hybridize in the middle of a molecule, add 4–6 random nucleotides at the 5' end to facilitate the interaction with the streptavidin magnetic beads.

### 2.2. RNA Extraction (see Note 1)

1. Starting material: Fresh pelleted cells, store at –80°C on collection. Fresh pieces of organs, or tissue frozen in liquid nitrogen and stored at –80°C.
2. PolytronR TP1200 (if organs are used).
3. RNAZol reagent (BIOPROBE).
4. DT40 (Pharmacia, [Piscataway, NJ]; #17-0270-01) Prepare a 5 mg/mL stock solution in ddH$_2$O. Aliquot and store at –20°C.
5. CHCl$_3$.
6. Isopropanol.
7. 100% Ethanol.
8. 70% Ethanol.
9. ddH$_2$O.
10. Dynabeads mRNA purification kit (Dynal, Lake Success, NY).

### 2.3. Synthesis of the ss-cDNA

1. Water bath at 70 and 42°C.
2. Dry-ice powder and ice.
3. 10X FSB buffer: 1$M$ Tris-HCl, pH 8.4 (at 42°C), 1.2$M$ KCl, 100 m$M$ MgCl$_2$. Aliquot and store at –20°C.
4. 100 m$M$ DTT.
5. Acetylated bovine serum albumin, 5 mg/mL (RNase-free; Life Technologies/BRL, Gaithersburg, MD).
6. RNasin, 36 U/µL (Promega Biotech, Madison, WI).
7. dNTPs, 10 m$M$ each (Use lithium-free dNTPs).

8. 20 m$M$ sodium pyrophosphate (PPi).
8. [α$^{32}$P] dATP, 3000 Ci/mmol (Amersham, Arlington Heights, IL).
9. AMV reverse transcriptase 10 U/μL (Promega Biotech, Madison, WI).
10. RA3' NV (50 ng/μL).

## 2.4. Synthesis Yield Determination

1. DE81 Whatman paper.
2. 0.5$M$ Na$_2$HPO$_4$.
3. 100% Ethanol.
4. Aqueous scintillation cocktail.

## 2.5. Removal of Primers

1. Prep-A-Gene DNA purification Kit (Bio-Rad [Richmond, CA]; #732-6010). The silica matrix used in this kit does not bind RNA or small DNA molecules (cut off around 100 nucleotides) under oxidizing conditions.
2. Water baths at 90 and 65°C.

## 2.6. Ligation of the ss-cDNA to the Modified Oligonucleotide (see Chapter 27)

Modified oligonucleotide: 5' phosphate and 3' NH$_2$ A5'NV (or B5'NV).

## 2.7. PCR Amplification

PCR reagents (*see* Chapter 1), including primers.

## 2.8. Direct Screening of the PCR Library with Biotinylated Oligonucleotides

1. Biotinylated primer (*see* Section 2.1.2.).
2. 10 mg/mL streptavidin Dynabeads (Dynabeads M-280 Streptavidin, DYNAL, Lake Success, NY).
3. Magnetic concentrator (Dynal MPC).
4. Rotating wheel.
5. 20X SSPE solution: 200 m$M$ NaH$_2$PO$_4$, pH 7.4, 3.6$M$ NaCl, 20 m$M$ EDTA *(5)*.
6. 20X SSC: 300 m$M$ sodium citrate, pH 7.0, 3$M$ NaCl *(5)*.
7. 50X Denhardt's solution: 1% each of BSA, Ficoll 400, and PVP *(5)*.
8. 20% SDS.
9. Sonicated salmon sperm DNA.
10. 10X PCR buffer (*see* Chapter 1).
11. 42°C incubator.

## 2.9. Blunt-End Cloning of PCR Products

1. Unpurified PCR products.
2. T4 DNA polymerase (4 U/μL, Amersham).
3. Agarose gel (choose agarose concentration according to the PCR product length).

4. QIAEX II purification kit (Qiagen, Inc., Chatsworth, CA).
5. 3 and 8 m*M* ATP.
6. T4 polynucleotide kinase 5 U/μL (Amersham).
7. 10X T4 polynucleotide kinase buffer (Amersham).
8. Dephosphorylated *Sma*I pUC19 vector (Appligene).
9. T4 DNA ligase, 4 U/μL.
10. Electrocompetent XLI blue cells.
11. *Escherichia coli* electroporation equipment.

### 2.10. Direct PCR on Colonies

1. Inoculating needles.
2. PCR reagent, including primers. With pUC vectors, use M13 universal sequencing primer and M13 reverse sequence primer.

## 3. Methods

### 3.1. RNA Extraction

All manipulations must be carried out in an RNase-free environment and with PCR anticontamination material. In our hands, the best extraction yield for low amount of material is obtained with the RNAZol kit. When working with tissues, use a polytronR TP1200 to homogenize in the RNAzol solution. Follow the supplier's instructions with two important modifications.

1. Add 1 μL of DT40 (5 mg/mL) in the RNAZol solution prior to the homogenization step. This will increase the extraction yield.
2. At the end, resuspend the pellet in 20 μL of ddH$_2$O and store at –80°C.

If the extraction of polyA$^+$ RNA is required, use the Dynabeads mRNA purification kit (Dynal); do not use the Dynabeads mRNA DIRECT kit. Elute from the magnetic beads with 20 μL of ddH$_2$O instead of elution buffer and store at –80°C.

### 3.2. Synthesis of the ss-cDNA (see Note 2)

The final volume for the reverse transcription is 50 μL.

### 3.2.1. AMV RTase Preincubation Mix

1. Add, on ice, in a sequential manner the following reagents: 1.5 μL H$_2$O, 3 μL 10X FSB, 2.5 μL 0.1*M* DTT, 3 μL 10 m*M* each dNTP (lithium free), 1 μL 5 mg/mL BSA, 5 μL [α$^{32}$P] dATP (100 U/μL) (3000 Ci/mmol), 1 μL RNasin (36 U/μL), 10 μL 20 m*M* PPi, and 1 μL AMV RTase.
2. Preincubate on ice for 30 min.

### 3.2.2. RNA Mix Preparation

Prepare this mix during the preincubation of the AMV reverse transcriptase.

1. Dilute the RNA in 17 μL of ddH$_2$O (*see* Note 3).
2. Add 1 μL of 50 ng/μL RA3'NV (or any anchored random primer).

3. Add 2 μL of 10X FSB.
4. Heat the tubes at 70°C for 15 min.
5. Spin and freeze in dry-ice powder.
6. Let it thaw on ice.

### 3.2.3. Reverse Transcription

1. Assemble the preincubation mix and the RNA mix and incubate at 42°C for 20 min to 1 h.
2. To stop the reaction, add 1 μL of 0.5*M* EDTA. To determine the optimal incubation time, perform five cDNA synthesis in 10 μL final reaction volume and stop the incubation every 10 min after 20 min initial reaction time. Load 5–10 μL on an alkaline agarose gel *(5)* and measure the average length. The optimal reaction time will be the one that gives an average size of 0.8–1 kbp.

### 3.3. Synthesis Yield Determination

1. Take 1 μL of the cDNA and dilute it to 10 μL in ddH$_2$O.
2. Spot 5 μL of the dilution on two pieces of DE81 Whatman filter.
3. Wash one filter only in 0.5*M* Na$_2$HPO$_4$ for 10 min.
4. Repeat step 3 two more times and dry this filter in 100% ethanol.
5. Dry both filters in air for 15 min.
6. Add 10 mL of aqueous scintillation cocktail to the washed and unwashed filters.
7. Count the $^{32}$P activity. The washed filter activity corresponds to the incorporated activity *(I)*, whereas the unwashed one corresponds to the total activity *(T)*.
8. The synthesized ss-cDNA mass *(M)* is given by the formula:

$$M = (I/T) \times (\text{total dATP mass during reaction}) \times 4 \qquad (1)$$

In the conditions used, M(ng) = *(I/T)* × 792. The overall yield should be between 25 and 30% of the starting mass of RNAs.

### 3.4. Removal of Primers

Use Prep-A-Gene DNA purification kit. Follow the supplier's instructions with the following important modifications.

1. To remove primer after the ss-cDNA synthesis, heat for 5 min at 90–95°C to denature the RNA:DNA heteroduplexes. Add 150 μL of binding buffer, mix, then add 5 μL of resuspended matrix. Mix well. Incubate for 10 min at room temperature.
2. To remove primer after ligation or PCR, do not perform this denaturation step.
3. For purification do not use a starting volume smaller than 50 μL.
4. Carefully remove all the wash buffer after the last wash. Traces of ethanol can be removed by drying the tubes for 3 min in a SpeedVac or equivalent rotary vacuum desiccator.
5. Elute with 5–10 μL of ddH$_2$O for 5 min at 65°C, then spin for 30 s and collect supernatant.
6. Alternative procedures can be followed as described in Chapter 27.

### 3.5. Ligation of the ss-cDNAs to the Modified Oligonucleotide (see Note 4)

1. Perform as described in Chapter 27.
2. Remove primer before the PCR amplification step.

### 3.6. PCR Amplification (see Notes 5 and 6)

The general conditions used for both PCR amplifications are: 50-μL reaction volume, hot start, and touchdown PCR.

1. First PCR: Use A5'_1 and A3'_1 and half of the purified ligation mixture.
2. Second PCR: Use A5'_2 and A3'_2 and one tenth of the purified first PCR.
3. For both PCRs:
   a. Use as final concentration 200 μ$M$ of dNTPs, 0.8 μ$M$ of each primer and 1.5 m$M$ of MgCl$_2$.
   b. Hot start: Add 0.5 μL of *Taq* DNA polymerase (5 U/μL) below the mineral oil when the reaction mixture reaches 80°C.
   c. Perform the following touch down PCR cycles: denaturation 93°C for 3 min; 2 cycles of 94°C, 30 s/70°C, 45 s/72°C, 1.5 min; 2 cycles of 94°C, 30 s/69°C, 45 s/72°C, 1.5 min; 2 cycles of 94°C, 30 s/68°C, 45 s/72°C, 1.5 min; 2 cycles of 94°C, 30 s/67°C, 45 s/72°C, 1.5 min; 2 cycles of 94°C, 30 s/66°C, 45 s/72°C, 1.5 min; 25 cycles of 94°C, 30 s/65°C, 45 s/72°C, 1.5 min; cool down to 4°C.
   d. Remove primer after each PCR.
   After the second nested PCR the amount of ds-cDNA is about 2–5 μg.

### 3.7. Direct Sceening of the PCR Library with Biotinylated Oligonucleotides

#### 3.7.1. Hybridization with the Biotinylated Oligonucleotide (see Note 7)

1. Use 500 ng of purified ds-cDNA. The volume should not exceed 8 μL.
2. Add 2 μL of the 100 ng/μL biotinylated oligonucleotide.
3. Adjust volume to 10 μL.
4. Heat denature for 5 min at 95°C.
5. Immediately add 100 μL of hybridization buffer: 5X SSPE, 5X Denhardt's solution, 1% SDS.
6. Incubate overnight at 42°C.

#### 3.7.2. Separation of Probe-cDNA Hybrids

After the hybridization reaction, probe-cDNA hybrids are separated from unhybridized DNA using Streptavidin-coated magnetic beads.

1. Prehybridize Dynabeads with salmon sperm DNA by washing 20 μL of 10 mg/mL Dynabeads twice with 50 μL of hybridization buffer containing 250 μg/mL salmon sperm DNA. Incubate for 2 h at room temperature on a rotating wheel.
2. Mix the Dynabeads with the probe-cDNA solution. Incubate 15–30 min at room temperature on a rotating wheel.

3. The hybrids captured by the beads are washed twice with 1X SSC, 1% SDS, then twice with 0.1X SSC, 1% SDS. Washes are performed at 42°C for 20 min each.
4. Wash twice with 1X PCR buffer, 5% SDS, for 5 min at room temperature.
5. Wash with 1X PCR buffer until SDS is completely removed. To do this, change the microtube after every wash.

### 3.7.3. PCR Amplification of the Captured cDNA

1. Transfer one fourth of the beads with the captured cDNA into a PCR tube. Make sure that all traces of SDS are removed.
2. Perform PCR amplification with A5'_3 and A3'_3. Use the same protocol as described in Section 3.6.

## 3.8. Blunt-End Cloning of PCR Products

For blunt-end cloning, the 3' overhanging extremities of the PCR product are removed with T4 DNA Polymerase (3'-5' exonuclease activity). Oligonucleotides usually have 5' hydroxyl ends. To allow ligation of the PCR product those extremities have to be phosphorylated by T4 Polynucleotide kinase (T4 PNK).

1. At the end of the amplification reaction, add to the PCR mixture 1 µL of T4 DNA polymerase (4 U/µL). Incubate for 20 min at 16°C. Do not allow the temperature to rise above 16°C.
2. Load the PCR product on a preparative agarose gel.
3. Cut the desired bands and purify the DNA with the QIAEX II purification kit. Follow the supplier's recommendations.
4. Elute DNA from the silica matrix with 10 µL of ddH$_2$O.
5. Add 1.5 µL of 10X T4 PNK buffer, 1 µL of 3 m*M* ATP, 1.5 µL of ddH$_2$O, and 1 µL of T4 PNK (5 U/µL).
6. Incubate at 37°C for 30 min.
7. Heat inactivate the enzyme at 75°C for 20 min.
8. The PCR product is ready for ligation. Use the same buffer as the phosphorylation reaction, a dephosphorylated blunt-end vector (e.g., pUC19 *Sma*I), and a final concentration of ATP of 0.8 m*M*.
9. Transform by electroporation and plate on the appropriate selection medium.

## 3.9. Direct PCR on Colonies (see Note 8)

1. Pick a colony with an inoculating needle.
2. Touch the bottom of a 0.5-mL microtube (or a well in a microtiter plate) with the needle.
3. Inoculate with the same needle, 3 mL of liquid bacterial growth medium in a 15-mL tube, and incubate at 37°C for 18 h.
4. Repeat steps 1–3 for all the colonies to be analyzed.
5. Prepare the PCR mixture on ice as follows. The volumes given are sufficient for one reaction: 2.5 µL 10X PCR buffer, 2 µL 2.5 m*M* dNTP, 2 µL 50 m*M* MgCl$_2$, 2 µL Primer 1 (50 ng/µL), 1 µL Primer 2 (50 ng/µL), 0.1 U *Taq* DNA polymerase, and ddH$_2$O to 25 µL.

6. Distribute the PCR mixture into every tube on ice and add 100 μL of mineral oil.
7. Place the tubes or the microtiter plate in the thermal cycler and run the following program: 3 min denaturation at 93°C; 35 cycles of 94°C for 30 s, 55°C for 45 s, and 72°C for 1 min/kb.
8. Analyze the PCR products on an agarose gel.
9. Prepare plasmid DNA of the positive clones from the cultures (prepared in step 3). Use this plasmid DNA for sequencing.

## 4. Notes

1. All the material used in this manipulation must be very clean and at least sterilized. Wear gloves throughout the manipulation to avoid RNase contamination. The same precautions should be followed in the synthesis of the ss-cDNA. These precautions represent the lower level of protection against RNase and it is advisable to read ref. *6* carefully. Since PCR has to be performed later on this material, use anticontamination tips and aliquot every solution.
2. During this manipulation, prepare controls that will be used in the ligation and PCR experiments. Prepare samples without AMV RTase and samples without RNA. This will lead to three different controls.
3. We have successfully used as little as 10 ng of polyA$^+$ RNA. Do not exceed 1 μg of total RNA.
4. Prepare nonligated samples composed of the same mixture as described in Section 3.5. without the T4 RNA ligase. Include each control of the cDNA synthesis.
5. After removal of the primers, perform PCR amplification. Do not forget to include a PCR control without DNA for both amplifications.
6. To analyze each PCR amplification, load 5 μL (one-tenth) of the PCR product on an agarose gel. After ethidium bromide staining, a signal could be observed after the first PCR but it is generally obtained after the second nested PCR.
7. We have used degenerated primers to screen the PCR library. The amount of primer and Dynabeads should be at least five times above the amount described in Section 3.7.
8. We have developed a direct PCR analysis to determine the sizes of inserted fragments and to rule out false positive recombinant clones.

## References

1. Boularand, S., Darmon, M. C., and Mallet, J. (1995) The human tryptophan hydroxylase gene: an unusual complexity in the 5' untranslated region. *J. Biol. Chem.* **270**, 3748–3756.
2. Barnes, W. M. (1994) PCR amplification of up to 35-kb DNA with high fidelity and high yield from λ bacteriophage templates. *Proc. Natl. Acad. Sci. USA* **91**, 2216–2220.
3. Don, R. H., Cox, P. T., Wainwright, B. J., Baker, K., and Mattick, J. S. (1991) "Touch down" PCR to circumvent spurious priming during gene amplification. *Nucleic Acids Res.* **19**, 4008.

4. Abe, K. (1992) Rapid isolation of desired sequences from lone linker PCR amplified cDNA mixtures: application to identification and recovery of expressed sequences in cloned genomic DNA. *Mammalian Genome* **2,** 252–259.
5. Sambrook, J., Fritsch, E. F., and Maniatis, T. (1989) *Molecular Cloning. A Laboratory Manual,* 2nd ed., Cold Spring Harbor Laboratory, Cold Spring Harbor, NY, SSPE, p. B.13, SSC, p. B.13, Denhart's solution, p. B.15, Alkaline gel electrophoresis, p. B.23.
6. Blumberg, D. D. (1987) Creating a ribonuclease free environment. *Methods Enzymol.* **152,** 20–24.

# 33

# Rapid and Nonradioactive Screening of Recombinant Libraries by PCR

**Michael W. King**

## 1. Introduction

The complexity of the genome of a particular organism or the relative abundance of a particular mRNA, within the cell type from which a cDNA library was constructed, affects the ability with which one can isolate a gene or cDNA clone of interest. With respect to genomic libraries, the number of clones needed to be screened to isolate a single-copy sequence is a function of the complexity of the genome and the average size of the cloned fragments in the library *(1)*. In the case of cDNA libraries, the frequency of a given clone of interest depends on the abundance of the messenger RNA. Highly abundant messages can represent 10% or more of total mRNA, whereas very rare messages can be as low as one in $10^6$. In addition, the representation of some sequences in a cDNA library, particularly the 5' ends of large mRNAs, will be less than expected because of the technical difficulties in converting the mRNA into full-length cDNA copies. In some cases, a particular sequence of interest can be depleted or lost at various steps of screening because of its inefficiency to be replicated relative to other clones in the library.

Prior to the advent of the PCR, the principal technique for screening bacteriophage lambda-based libraries involved plating on a lawn of *Escherichia coli,* transferring the phage particles contained in the resultant plaques to nitrocellulose filters and hybridizing the filters with radioactively labeled DNA or oligonucleotide probes. Unfortunately, this technique is all too often not sensitive enough for detecting low abundance mRNAs or may result in a high frequency of false positives. Recently, more highly sensitive methods for screening that utilize the PCR have been described *(2,3)*.

From: *Methods in Molecular Biology, Vol. 67: PCR Cloning Protocols: From Molecular Cloning to Genetic Engineering* Edited by: B. A. White  Humana Press Inc., Totowa, NJ

This chapter describes a PCR-based sib-selection method for the isolation of clones from recombinant DNA libraries prepared in λ-based vectors. The technique uses no radioisotopes and can be completed in as few as 7 d. The technique is amenable to the use of highly specific PCR primers as well as degenerate primers designed to isolate families of related clones. In this method a cDNA or genomic library is initially plated (at a lower density than for filter hybridization) and the phage from each plate is soaked from the plates in SM buffer *(4)* to generate the starting aliquots for PCR. Positive aliquots from the primary plating are identified and replated at lower and lower densities to generate subaliquots for secondary and tertiary screens until a positive clone is identified by PCR of phage soaked from a single plaque.

PCR-based library screening can be performed on both cDNA libraries and genomic libraries that are cloned in any variation of bacteriophage λ-based vectors. Screening can be performed from aliquots of unamplified libraries or amplified libraries. It is helpful, yet unnecessary, to know the titer (plaque forming units [pfu] per milliliter [pfu/mL]) of the library being screened. In general, good primary libraries have a titer of anywhere from 1,000,000–5,000,000 pfu in the entire initial packaging reaction (usually a 500-μL library). Libraries that have been amplified generally have a titer of $10^{10}$–$10^{11}$ pfu/mL. Unless it is known that the titer of a given library is uncharacteristically low, determining the precise titer of the starting aliquot is not necessary (*see* Section 3.5.).

## 2. Materials

1. Recombinant DNA library.
2. Oligonucleotide primers for PCR (*see* Section 3.1.).
3. Bacterial culture media, agar plates and top agar media.
4. SM Buffer *(4)*: 100 m$M$ NaCl, 8 m$M$ MgSO$_4$, 50 m$M$ Tris-HCl, pH 7.5.
5. Reagents for PCR (*see* Chapter 1).
6. Reagents for agarose gel electrophoresis (*see* Chapter 1).
7. CHCl$_3$.

## 3. Methods
### 3.1. Design of Primers

1. In designing primers for PCR one needs to consider several important factors, such as making the primers with a near 50% GC content, a high degree of specificity with respect to nucleotide sequences, and the absence of primer selfcomplementarity. Primers that are 18–21 nucleotides in length are optimal for this (and most) PCR-based techniques since they allow high annealing temperatures that result in greater specificity in the reaction. In most cases, where possible, the primers should be designed to be maximally useful in a PCR with 2 m$M$ MgCl$_2$ and an annealing temperature above 62°C.

**Table 1**
**Outline of Screening Protocol**

| Day | Activity | Identity |
|---|---|---|
| 1 | Start culture of host *E. coli* | |
| 2 | Plate 5–10 100-mm plates (approx 4000–5000 pfu/plate), grow 6–8 h, soak phage in SM overnight at 4°C | Primary screen |
| 3 | Process phage lysates, PCR screen | Primary screen |
| 3 | Start culture of host *E. coli* | |
| 4 | Plate five 100-mm plates, of 1 or all 1° positives (approx 500 pfu/plate), grow 6–8 h, soak phage in SM overnight at 4°C | Secondary screen |
| 5 | Process phage lysates, PCR screen | Secondary screen |
| 5 | Start culture of host *E. coli* | |
| 6 | Plate five 100-mm plates (approx 100 pfu/plate), grow 6–8 h, soak phage in SM overnight at 4°C, it is possible to pick single plaques at this stage | Tertiary screen |
| 7 | Process phage lysates, PCR screen | Tertiary screen |
| 7 | Start culture of host *E. coli* | |
| 8 | Plate one 100-mm plate (approx 50 pfu/plate), grow 6–8 h, pick single plaques into SM and elute overnight at 4°C | Final screen |
| 9 | Screen single plaque lysates by PCR | Final screen |

2. With the use of degenerate primer pairs, as for the isolation of families of related cDNAs or genes, it is extremely important that the 3'-nucleotide position of each primer set be nondegenerate in order to prevent an increase in nonspecific templating. In most cases, it is also optimal if the annealing temperature used with degenerate primers is not below 55°C, although examples are available where successful isolation of related cDNAs has been carried out using degenerate primer pairs with annealing temperatures of 52°C.

3. For increased specificity it is possible to add a nested primer that resides within the sequences to be amplified with the primary primer pair. This nested primer can then be used in a second PCR, in combination with one of the primary primers, to test positive PCRs for the presence of correct internal sequences.

## 3.2. Basics of Phage Growth

1. Each step in this PCR-based screening protocol begins with the overnight culture of the appropriate host *E. coli* (*see* Table 1). Grow the cells at 37°C with agitation in NZCYM media supplemented with 1 m*M* MgSO$_4$ and 0.3% maltose *(4)* to ensure optimal phage infection and growth. The minimum volume of the over-

night culture of cells depends on the number of plates that will be used. The standard volume of cells is 100 μL/100-mm plate.

2. The next morning, melt an aliquot of top agar and hold at 45°C. Prewarm the appropriate number of agar plates at 37°C for 15–20 min.

3. Dilute the library or plate lysate aliquots appropriately in SM buffer, and then add to an aliquot of the fresh overnight host *E. coli* that corresponds to the total volume needed for plating a 100 μL/100-mm plate. Adding the phage to the entire volume of cells instead of to individual aliquots ensures an equal distribution on each plate. Incubate the cell and phage at 37°C for 10 min to allow the infection cycle to initiate.

4. Separate the infected cells into individual 100-μL aliquots for plating. Carry out the plating by adding 3 mL of the melted 45°C top agar to each tube, pouring the solution onto individual plates and ensuring even spreading of the top agar prior to its resolidifying.

5. Invert the plates and incubate at 37°C for 6–8 h to allow plaques to form; *see* Sections 3.5.–3.8. for the screening protocol.

### 3.3. Preparation for Screening

1. It is critically important that prior to screening any library with this technique the library be tested for the presence of clones that contain sequences that will amplify with a given primer pair.

2. Test a 1-μL aliquot (undiluted) of the library to be screened using the standard PCR protocol described in Section 3.4. As described in Section 1., an amplified library will have approx $10^7$–$10^8$ phage in a 1-μL aliquot. Therefore, a 1-μL aliquot will contain 100–1000-fold more phage than is statistically necessary to screen to find a clone of a given sequence. If the primers are unable to amplify the correct fragment from this amount of phage, the library is either devoid of clones or the primer pair is not functional as expected.

### 3.4. Standard PCR

1. Perform all PCRs in a volume of 25 μL containing 1X *Taq* buffer, 2 mM $MgCl_2$ (or a concentration appropriate for a given primer pair), 30–35 pmol of each primer (200 ng; assuming primers of 18–21 nucleotides), and 200 μM dNTPs.

2. To ensure that the phage particles in the plate lysates are disrupted, "hot start" the PCRs. This is accomplished by an initial denaturation at 95°C for 10 min followed by holding the reaction at 80°C for 30–60 min.

3. Use the following cycle profiles for specific primers (*see* Notes for degenerate primers): 30 cycles: 95°C, 1–2 min (denaturation); $n$°C, 0.5–2 min (annealing temperature defined by primer sequences; *see* Note 1); and 72°C, $n$ min (extension time depends on product length; *see* Note 2).

### 3.5. Primary Screen

1. For the primary screen it is usually necessary to plate 10 aliquots of the library at a density in the range of 4000–5000 pfu/plate. This density will nearly lyse the entire bacterial lawn. If the titer of the library is unknown, most amplified librar-

ies can be plated using 1 μL of a $10^3$ dilution on the 10 plates (i.e., the equivalent of 0.1 μL/plate). In some cases it is possible to screen as few as five plates.

2. Incubate the plates upside down at 37°C. Stop the incubation when the plaques begin to merge with one another. This usually takes 6–8 h at 37°C.

3. Overlay the plate with 3.5 mL of SM buffer and let stand at 4°C overnight. It is possible to incubate the plates with SM buffer for 2 h at 37°C or room temperature for 4–5 h, however, the PCR results are sometimes smeared because of bacterial growth in the SM buffer. Also, the titer of the resultant lysates can be 100–1000-fold lower than the lysates prepared by 4°C overnight incubation. This latter fact is important to remember for subsequent screens.

4. Collect the SM buffer as a separate aliquot from each of the plates and remove agar and bacterial debris by centrifugation at 5000 rpm for 10 min in a JA17 rotor (Beckman high speed centrifuge) or an SS34 rotor (Sorval centrifuge).

5. Save 1 mL from each aliquot. Add $CHCl_3$ to 0.3% to prevent bacterial growth in these aliquots and allow for their longer term storage at 4°C.

6. Use a 1-μL aliquot for the PCR assay (*see* Section 3.4.).

7. Analyze a 10-μL aliquot of each PCR on an agarose gel (*see* Note 3) to determine which aliquot(s) have amplified the target of interest.

8. In most screens there should be at least one plate lysate exhibiting a positive signal by PCR.

9. Provided the library was tested for the presence of DNA that can be amplified with the primer pair being used (*see* Section 3.3.), it will be possible to find primary lysates that contain positive signals. It may be necessary to continue to screen more primary plates until a positive one is found (*see* Note 4).

## 3.6. Secondary Screen

1. It is not necessary to titer the primary plate lysates, although this can be done if accurate plaque numbers in the secondary screen are desired. However, based on the fact that the primary plates should have experienced near complete lysis of the bacterial lawn, the secondary screens (using five 100-mm plates) are plated using 1 μL of a $10^3$ dilution (i.e., 0.2 μL/plate) of each positive primary lysate. Plating at this density (approx 500 plaques/plate) is to ensure that the secondary lysates are dense enough to ensure enrichment of the clones of interest.

2. Allow plaques to grow as for the primary screen.

3. Prepare the phage lysates as for the primary screen using 3 mL of SM buffer.

4. Process the plate lysates and save a 1-mL aliquot of each with 0.3% $CHCl_3$ for long-term storage.

5. Screen a 1-μL aliquot of each in a 25-μL PCR as for the primary screen.

6. Analyze a 10-μL aliquot of each PCR by agarose gel electrophoresis.

## 3.7. Tertiary Screen

1. The tertiary screen is the last plate lysate screen. However, it is possible in some instances to proceed directly to the screening of single plaques from the secondary screen.

2. Plate five plates (100-mm plates) using 1 μL of a $10^4$ dilution of a positive secondary lysate (i.e., 0.2 μL/plate; *see* Note 5). In some cases it may be necessary to use a $10^5$ dilution, depending on the density of plaques in the secondary screen.
3. Allow plaques to grow as for the primary screen.
4. Prepare the phage lysates as for the primary screen using 3 mL of SM buffer.
5. Process the plate lysates and save a 1-mL aliquot of each with 0.3% $CHCl_3$ for long-term storage.
6. Screen a 1-μL aliquot of each in a 25-μL PCR as for the primary screen.
7. Analyze a 10-μL aliquot of each PCR by agarose gel electrophoresis. *See* Notes 6 and 7.

### *3.8. Single Plaque Screen*

1. Plate a single 100-mm plate from a tertiary (or a secondary) positive using 1 μL of $10^3$ or $10^4$ dilution.
2. The fold dilution used for the single plaque screening plate is less than for a tertiary screen and the amount of the dilution plated (1 μL) is more. This is because the density of plaques in the tertiary screen should have been in the range of >250/plate, such that the lysates from those plates will have a low titer.
3. Pick individual plaques into 100 μL of SM buffer using sterile glass Pasteur pipets to "scoop" the plaques out of the top agar.
4. Elute the phage particle from the plaque overnight at 4°C.
5. At this step it is best to screen a 3-μL aliquot in the standard 25-μL PCR. Volumes <3 μL can be used but have a tendency to give variable amplification; *see* Notes 8 and 9.

### 4. Notes

1. The annealing temperature ($T_m$) used during the PCR, for any given primer, is determined from the base composition of the primer. To calculate the annealing temperature use the following formula: $T_m = 2(A + T) + 4(G + C)$. As an example, for a 21-mer primer with 12 G + C and 9 A + T, the $T_m = 2(9) + 4(12) = 66°C$. For the majority of PCRs it is optimal to use an annealing temperature that is 2–4°C below the calculated $T_m$. However, increased specificity is obtained by annealing at the $T_m$.
2. The elongation time used in the PCR is determined by the length of the resultant product. Given the rate of the majority of DNA polymerases at approx 1000 bases/s, it would, in theory, be possible to use extremely short elongation times. However, in practice it has been observed that an elongation time of approx 1 min/1000 bases is optimal.
3. To analyze PCR products by agarose gel electrophoresis requires different percentages of agarose (dependent on product size) in order to obtain good resolution of products. In general use 1.5–2% gels for products of <500 bp, 1–1.5% for products of 500–1000 bp, and 0.7–1% for products >1000 bp.
4. Following the primary screen, each positive plate lysate will most probably represent a different type of cDNA or gene clone unless degenerate primers were utilized. Therefore, to maximize the possibility of obtaining full-length cDNA

clones or over-lapping genomic clones, each primary positive should be carried through to the secondary stage. However, to reduce the screening "load" one single primary positive at a time can be carried through to single plaque isolation.

5. When using specific primers, each positive secondary lysate represents the same single type of clone that was present in the primary lysate. Therefore, only one of the positive secondary lysates is carried through to the tertiary screen. Also, only a single tertiary positive is carried through to the single plaque screen.

6. When using degenerate primer pairs, the complexity of possible clone types in any given primary positive can be large. For this reason it is necessary to plate at least 10 plates for the secondary screens. Because of this complexity of clones in a primary positive, it is best to carry only one primary positive at a time through to single plaque isolations. In addition, since each of the secondary positives will likely represent different types of clones, each of them needs to be screened in the tertiary screen. The latter fact is also likely in many cases at the level of the tertiary (and beyond) screens.

7. The use of degenerate primers in this screening technique will require at least four and possibly as many as six rounds of plate lysate screening prior to the screening of single plaques.

8. The single plaque lysate is used to prepare phage DNA as well as a permanent stock of the clone by small-scale liquid lysis. Start an overnight culture of the appropriate host *E. coli* in NZCYM plus maltose and $MgSO_4$ (*see* Section 3.2.). The next morning add 25 μL of the 100 μL of SM (into which the positive plaque was eluted) to 100 μL of overnight cells and 100 μL NZCYM with maltose and $MgSO_4$. Incubate with agitation at 37°C for 10 min. Transfer this culture to 50 mL of NZCYM without maltose or $MgSO_4$. Incubate with agitation at 37°C until the cells in the culture begin to lyse. This takes approx 6 h and is visible as debris in the normally silky appearance of the growing *E. coli*. At this time add $CHCl_3$ to 0.5% and incubate an additional 10 min to accelerate the cell lysis as well as to prevent further growth of the cells. Centrifuge the cells and debris at 7000 rpm for 10 min. Save an aliquot of the supernatant as a stock of the clone, either at 4°C or by adding DMSO to 7% and storing at –80°C. To the remainder of the phage supernatant add RNase A and DNase I to 1.5 μg/mL and incubate at 37°C for 30 min. Precipitate the phage particles by addition of solid PEG 6000 to 10% (w/v) and solid NaCl to 0.5*M*. Place at 4°C overnight. Collect the precipitate by centrifugation at 9000 rpm for 25 min. Resuspend the precipitate in 0.5–1 mL of TE buffer *(4)*. Add Proteinase K to 150 μg/mL and incubate at 45°C for 45 min. Extract the released phage DNA with an equal volume of phenol then phenol/ $CHCl_3$ and again with $CHCl_3$. Precipitate the DNA by addition of 0.1 vol of 2*M* ammonium acetate, pH 5.0, and 2 vol of ethanol. The DNA should form a stringy precipitate immediately. It is best to remove the precipitating DNA by collecting it on a swirling glass rod. This reduces RNA and protein contamination that may affect restriction enzyme digestion. Centrifuge the precipitate at top speed in a microfuge, remove any supernatant, and resuspend the pellet in 50–100 μL of TER buffer *(4)*. Use 3–10 μL for restriction enzyme digestion.

9. The single plaque lysate is used for the rescue of phagemid DNA, containing the cDNA clone, from λZap (Stratagene, Inc., La Jolla, CA) or λZipLox (Life Technologies, Inc., Gaithersburg, MD) if the cDNA library was constructed with either of these vectors. The protocol for phagemid rescue from λZap is described. Start a culture of XL1–Blue overnight in NZCYM plus maltose and $MgSO_2$. Also start a culture of SOLR in LB media *(4)*. The next morning dilute 40 µL of the overnight XL1-Blue cells into 1 mL of NZCYM plus maltose and $MgSO_4$. Incubate with agitation at 37°C for 60 min. Transfer 200 µL of these cells to a new tube, add 25 µL of the 100 µL SM (into which the positive plaque was eluted) and 1 µL of ExAssist helper phage. Incubate with agitation at 37°C for 15 min. Add 3 mL of LB media *(4)* and incubate with agitation at 37°C for 2.5 h. Centrifuge the solution at 2000 rpm for 15 min. Transfer the supernatant to a new tube and heat at 70°C for 15 min. Centrifuge at 7000 rpm for 15 min. Save the supernatant in a sterile tube because this is a stock of the rescued single-strand phagemid. This solution can be stored at 4°C for up to 2 mo. To obtain colonies with the double-stranded phagemid, add 20–50 µL of the phagemid stock solution to 100 µL of the fresh overnight SOLR cells. Incubate with agitation at 37°C for 15 min. Spread 10–50 µL onto a single LB plus ampicillin plate and incubate overnight at 37°C. The double-stranded phagemid DNA can then be isolated from colonies by standard miniprep techniques *(4)*.

## References

1. Ausubel, F. M., Brent, R., Kingston, R. E., Moore, D. D., Smith, J. A., Seidman, J. G., and Struhl, K., eds. (1987) *Current Protocols in Molecular Biology,* John Wiley, New York.
2. Amaravadi, L. and King, M. W. (1994) A rapid and efficient, nonradioactive method for screening recombinant DNA libraries. *BioTechniques* **16,** 98–103.
3. Isola, N. R., Harn, H. J., and Cooper, D. L. (1991) Screening recombinant DNA libraries: a rapid and efficient method for isolating cDNA clones utilizing the PCR. *Biotechniques* **11,** 580–582.
4. Maniatis, T., Fritsch, E. F., and Sambrook, J., eds. (1992) *Molecular Cloning: A Laboratory Manual.* Cold Spring Harbor Laboratory, Cold Spring Harbor, NY, p. 443.

# 34

## Use of PCR for cDNA Library Screening

### Toru Takumi

### 1. Introduction

cDNA cloning from a library is now a routine laboratory practice. Hybridization screening with either radiolabeled or nonradiolabeled probes, which is laborious and time-consuming, has been commonly used. Application of the polymerase chain reaction (PCR) is surprisingly expanded *(1–3)*. Here a new application for cDNA library screening is reported: rapid cloning of full-length cDNAs by screening pools of cDNAs by PCR (RC-PCR) *(4)*. This PCR-based cDNA screening technique is applicable to both bacteria and phage libraries.

Providing that one has identified unique DNA sequences in a PCR clone or a partial-length cDNA clone, this RC-PCR screening technique is very simple and extremely rapid for isolation of a full-length cDNA from a library. This RC-PCR technique involves no radioisotopes and avoids labor intensive and expensive procedures, such as transferring bacterial colonies to filters and hybridizing them to radiolabeled DNA probes. RC-PCR enables us to isolate a single clone from the library in a period of a few days.

### 2. Materials

1. cDNA library (*see* Notes 1 and 2).
2. Reagents for PCR: 1X PCR buffer (10 m*M* Tris-HCl, pH 8.3, 50 m*M* KCl, 1.5 m*M* MgCl$_2$), dNTP mixture, *Taq* DNA polymerase, sterile water, sterile mineral oil.
3. Sequence-specific primers.
4. 96-well dishes.
5. LB medium.
6. Agarose gel for analysis of PCR products.

From: *Methods in Molecular Biology, Vol. 67: PCR Cloning Protocols: From Molecular Cloning to Genetic Engineering* Edited by: B. A. White  Humana Press Inc., Totowa, NJ

## 3. Methods

### 3.1. Primer Design

1. The sequence of primers should be specific for the target sequence. In the case of cloning members of a gene family, choose the region that is less conserved among the family.
2. The parameters for designing primers are the length (base pairs), the GC/AT ratio, and melting temperature *(5)*.
3. A 17–18-mer is thought to be sufficient for RC-PCR if a positive control gives a single specific band in an agarose gel. Longer primers are appropriate to amplify larger PCR products.
4. Amplification of 200–400 bp DNA is the most efficient *(5)*.

### 3.2. PCR Screening

1. Divide the cDNA library into ten fractions, each of which contains, for example, approx 100,000 clones. Spread each of the library over 150-mm LB agar plates including the appropriate antibiotics. After overnight growth, collect the colonies by a scraper and put them into microfuge tubes containing LB solution.
2. For each of the 10 fractions, assemble a 50-µL PCR in a 500-µL microfuge tube containing the following: 5.0 µL 10X PCR buffer, 200 µ*M* each dNTP mix, 0.15 µ*M* each primer, 2.5 µL of broth containing the colonies as the DNA template *(see* Note 3), sterile H$_2$O to 49.5 µL final volume, and 0.5 µL (2.5 U) *Taq* DNA polymerase.
3. After a 5-min denature at 94°C, the cycling parameters are the following: 25 cycles: 94°C for 45 s, 60 ~ 65°C for 1 min, and 72°C for 2 min. *See* Note 4.
4. Run an agarose gel to detect the positive fractions *(see* Chapter 1).
5. For the second screening, dilute the positive pool of cells in LB broth to a concentration of about 30,000 clones/mL and distribute 100 µL of this suspension (about 3000 clones) into each of the 96-wells of a microplate. Combine 10 µL from each well in a column (giving a total of 120 µL from each of the plate's 12 columns). Similarly, combine 10 µL from each well in a row (giving 80 µL from each of the eight rows). From each of the resulting 20 mixtures, directly use 2.5 µL as the DNA template of the next PCR.
6. Perform PCR as described in items 2 and 3 and analyze by gel electrophoresis.
7. Repeat screening and subdividing of the positive pools until a single clone is obtained. *See* Note 5.

## 4. Notes

1. As described, this protocol is applicable to both bacteria and phage libraries. In the case of screening of phage libraries, a high-phage titer is necessary for RC-PCR in the first screening. Once the titer of the positive pools drops to less than the order of 10$^3$ in a serial screening, it is recommended to amplify the phages on a small plate (10-cm) before proceeding. For other protocols of the screening of the phage libraries by PCR, *see* refs. *10–12*.

2. The quality of cDNA library determines whether or not it is possible to isolate a full-length cDNA or a partial one. Use of primers that correspond to the relatively upper stream region of the cDNA may help increase the chances of isolating a full clone.

3. Extraction of plasmid DNA is not necessary for PCR; intact bacterial colonies are sufficient for use as PCR templates *(8,9)*. LB solution containing bacterial colonies can be used directly as a template without any culture. In theory it is not necessary to incubate the medium if it has a positive clone. Incubation of the broth, however, may help to detect the positive signal, especially for the initial screening. It may depend on the number of colonies whether the medium should be incubated or not.

4. The PCR conditions should be optimized in each case by use of a positive control. Primer design, concentration of magnesium ($Mg^+$) ions, and annealing conditions, including temperature, are important parameters. In the case of phage libraries, note that phage dilution buffer already contains $Mg^+$ ions. In some cases, 10% of dimethylsulfoxide (DMSO) in the PCR buffer can help to synthesize a clean, specific PCR product. For a small fragment it may be necessary to amplify more than the 30 rounds *(11)*. Careful precautions to avoid contamination are essential because of the extremely high sensitivity of PCR *(13)*.

5. Here an example of this rapid cDNA cloning by PCR screening (RC-PCR) from a bacteria library is illustrated. In an attempt to clone gene family members of transforming growth factor-beta (TGF-β) receptors, a novel clone, termed clone B1, was isolated from GH3 rat pituitary tumor cells by PCR using degenerate oligonucleotide primers *(4,6)*. The method of RC-PCR was used to isolate a full-length B1 cDNA from the GH3 cell library *(6)*.

First, a bacterial library of GH3 cells was divided into 16 pools, each of which included approx 100,000 clones. As primers for PCR, two oligonucleotides that were expected to be highly specific for the B1 PCR product and that were derived from the sequence of the reverse transcription-polymerase chain reaction (RT-PCR) product were utilized: ATCGTGGTTCCGGGAGGCAGAGATC (25-mer, the sequence corresponding to nucleotides 726–750 of B1, *see* Fig. 1A in ref. *6*) as a 5'-primer and CTGATTTGGAGCAATGTCTATGGTG (25-mer, nucleotides 1095–1119 of B1, *see* Fig. 1A in ref. *6*) as a 3'-primer. These primers correspond to a region of the kinase domain that is poorly conserved among members of the TGF-β receptor family.

The screening by PCR was done as follows: After a 5 min denaturation step at 94°C, the cycling parameters were 94°C for 45 s, 65°C for 1 min, and 72°C for 2 min, for a total of 25 cycles. Three pools of clones generated a PCR product of the expected size. One pool generated a slightly larger PCR product; subsequently this fraction was shown to contain a B1 clone with an insertion of 34 nucleotides in the kinase domain.

For the second screening, this pool of cells was diluted in LB broth to a concentration of about 30,000 clones/mL and 100-μL aliquots of this suspension (about 3000 clones/aliquot) were distributed into 96-wells of a microplate. Ten

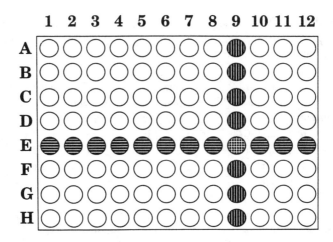

**Fig. 1.** Schematic representation of a 96-well microplate. The row and column were identified as described in the text.

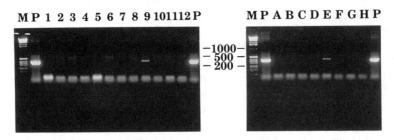

**Fig. 2.** Analysis of the products of second screening by gel electrophoresis. An aliquot of each PCR reaction, one for each column (1–12) and one for each row (A–H), was analyzed directly by electrophoresis through 2% agarose gel with ethidium bromide staining. In the reaction in lane P, the same PCR analysis was done on a plasmid containing the original B1 PCR product obtained as described in the text. Lane M contains a 1 kb DNA ladder (Life Technologies, Gaithersburg, MD).

microliters from each well in a column were combined (giving a total of 120 µL from each of the plate's 12 columns); similarly 10 µL from each well in a row were combined (giving 80 µL from each of the eight rows) (Fig. 1). From each of the resulting 20 mixtures, 2.5 µL were used directly in a PCR and analyzed by gel electrophoresis. Figure 2 shows the results of the second screening; the PCR products were analyzed directly by electrophoresis through 2% agarose gel and ethidium bromide staining. Two of the twenty pools, which correspond to column "9" and row "E," generated PCR products of the same size (394 bp) obtained from the original cloned B1 PCR product (lane P). With a total of only 20 PCR reactions (12 columns and eight rows) it was possible to identify a positive column and a positive row and thus a single positive pool (Fig. 1). The positive pool was again diluted, this time to about 300 clones/mL, and subdivided into 96-well plates (each well having 100 μL, or approx 30 clones). Again column and row aliquots were pooled and PCR performed using each of the resulting 20 pools. The one positive pool identified in this subsequent screening was finally streaked onto several LB plates and, after overnight growth, 96 individual colonies were inoculated into 100 μL of broth in each well of a microplate. As above, 20 PCR reactions were sufficient to identify a single positive clone. The cDNA isolated by this procedure contained 2276 bp and encodes a novel 505-amino-acid protein belonging to the family of type I receptor serine/threonine kinases *(6,7).*

## Acknowledgments

I thank Paul Burke for his comments. This work at the Whitehead Institute, Cambridge, MA, was supported in part by NIH grant HL 41484 to Harvey Lodish and a fellowship from the Human Frontier Science Program to the author. The support of a grant from the Kanae Foundation of Research for New Medicine to the author is also gratefully acknowledged.

## References

1. Innis, M. A., Gelfand, D. H., Sninsky, J. J., and White, T. J., eds. (1990) *PCR Protocols: A Guide to Methods and Applications.* Academic, San Diego, CA.
2. White, B. A., ed. (1993) *PCR Protocols: Current Methods and Applications,* in *Methods in Molecular Biology,* vol. 15. Humana, Totowa, NJ.
3. Mullis, K. B., Ferré, F., and Gibbs, R. A., eds. (1994) *The Polymerase Chain Reaction.* Birkhäuser, Boston, MA.
4. Takumi, T. and Lodish, H. F. (1994) Rapid cDNA cloning by PCR screening. *BioTechniques* **17,** 443,444.
5. Rychlik, W. (1993) Selection of primers for polymerase chain reaction, in *PCR Protocols: A Guide to Methods and Applications,* in *Methods in Molecular Biology,* vol. 15 (White, B. A., ed.), Humana, Totowa, NJ, pp. 31–40.
6. Takumi, T., Moustakas, A., Lin, H. Y., and Lodish, H. F. (1995) Molecular characterization of a type I serine-threonine kinase receptor for TGF-β and activin in the rat pituitary tumor cell line GH3. *Exp. Cell Res.* **216,** 208–214.
7. Moustakas, A., Takumi, T., Lin, H. Y., and Lodish, H. F. (1995) GH3 pituitary tumor cells contain heteromeric type I and type II receptor complexes for transforming growth factor β and activin-A. *J. Biol. Chem.* **270,** 765–769.
8. Güssow, D. and Clackson, T. (1989) Direct clone characterization from plaques and colonies by the polymerase chain reaction. *Nucleic Acids Res.* **17,** 4000.
9. Sandhu, G. S., Precup, J. W., and Kline, B. C. (1989) Rapid one-step characterization of recombinant vectors by direct analysis of transformed *Escherichia coli* colonies. *BioTechniques* **7,** 689,690.
10. Friedman, K. D., Rosen, N. L., Newman, P. J., and Montgomery, R. R. (1990) Screening of λgt11 libraries, in *PCR Protocols: A Guide to Methods and Applica-*

*tions* (Innis, M. A., Gelfand, D. H., Sninsky, J. J., and White T. J., eds.), Academic, San Diego, CA, pp. 253–258.

11. Yu, L. and Bloem, L. J. (1993) Use of polymerase chain reaction to screen phage libraries, in *PCR Protocols: A Guide to Methods and Applications,* in *Methods in Molecular Biology,* vol. 15 (White, B. A., ed.), Humana, Totowa, NJ, pp. 211–215.

12. Munroe, D. J., Loebbert, R., Bric, E., Whitton, T., Prawitt, D., Vu, D., Buckler, A., Winterpacht, A., Zabel, B., and Housman, D. E. (1995) Systematic screening of an arrayed cDNA library by PCR. *Proc. Natl. Acad. Sci. USA* **92,** 2209–2213.

13. Kwok, S. and Higuchi, R. (1989) Avoiding false positives with PCR. *Nature* **339,** 237,238.

# 35

## Generation and PCR Screening of Bacteriophage λ Sublibraries Enriched for Rare Clones (the "Sublibrary Method")

### Michael Lardelli

### 1. Introduction

The simple sublibrary method described in this chapter allows the detection and rapid isolation of rare clones from bacteriophage λ libraries. The method is based on the ability of PCR to detect clones present in a library at very low frequencies. Clones present at frequencies as low as one in 10,000,000, which would normally be impractical to isolate by conventional probe hybridization, can be rapidly isolated in this way *(1)*. The method can also be used to isolate clones that are initially undetectable by PCR in λ libraries.

The sublibrary method can be divided into three steps (*see* Fig. 1):

1. $10^6$ or $10^7$ clones from a λ library are divided into pools of 25,000 or 250,000 clones, respectively, which are then amplified to form sublibraries.
2. The sublibraries are screened for the presence of the desired clone using PCR.
3. The desired clone is isolated from a positive sublibrary by conventional techniques (i.e., hybridization of a labeled probe to blots from a sublibrary that has been plated out).

The sublibrary method is useful in many different cloning strategies. For example, when a novel DNA sequence has been identified by PCR it is often desirable to isolate a corresponding λ clone for sequence analysis since the thermostable polymerases used in PCR have considerably greater in vitro error rates than those of bacteriophage DNA polymerases operating within their bacterial hosts *(2)*. PCR-screening of λ sublibraries not only allows rapid isolation of rare bacteriophage clones but also ensures that an isolated clone contains the

From: *Methods in Molecular Biology, Vol. 67: PCR Cloning Protocols: From Molecular Cloning to Genetic Engineering* Edited by: B. A. White Humana Press Inc., Totowa, NJ

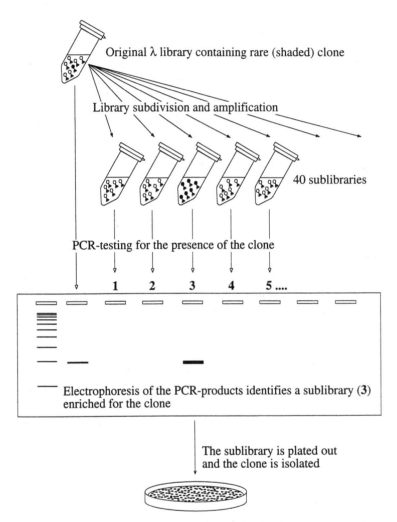

**Fig. 1.** The sublibrary method for detection and enrichment of rare clones in λ libraries.

entire region identified by the PCR. This avoids the necessity of isolating (by probe hybridization) multiple bacteriophage clones in order to be certain of covering the entire identified region.

Other advantages of the sublibrary method are the savings in time and expensive materials (nylon/nitrocellulose disks and radioactive isotopes) that it permits. It should also be possible to extend the method to the isolation of novel DNA sequences directly from λ libraries, for example, by use of degenerate PCR primers.

## 2. Materials

1. Dynazyme™ thermostable DNA polymerase and supplied buffer (from Finnzymes Oy [Espoo, Finland], Cat. No. F-501 L).
2. Double-distilled water.
3. Oligonucleotide primers for PCR that amplify the DNA region of interest.
4. 10 m$M$ dNTPs (2.5 m$M$ each of dATP, dCTP, dGTP, and dTTP).
5. Paraffin oil.
6. PCR machine and accessories (recommended: a PCR machine that accepts 96-well microassay plates with paraffin oil, e.g., a Techne [Cambridge, UK] Cyclogene Programmable Dri-Block™ cycler with Techne Hi-Temp 96™ microassay plates).
7. Micropipeters and pipet tips (e.g., Gilson, France) for liquid.
8. A DNA sample containing the sequence of interest (*see also* Note 2).
9. A cDNA or genomic DNA library in a λ vector.
10. Centrifuge(s) capable of accepting microtiter plates, 50-mL tubes and 1.5-mL microfuge tubes.
11. 6X gel loading buffer: 15% w/v Ficoll (Type 400; Pharmacia, Piscataway, NJ), 0.35% w/v Orange G (Sigma, St. Louis, MO), 60 m$M$ EDTA, pH 8.0.
12. Equipment and reagents for agarose gel electrophoresis.
13. DNA electrophoresis size markers (e.g., 1 Kb DNA Ladder; Life Technologies (Gibco-BRL [Gaithersburg, MD], Cat. No. 15615-016).
14. A sterile 100-mL conical flask for bacterial culture.
15. An *Escherichia coli* strain suitable for infection by, and growth of, bacteriophage λ, e.g., NM538 *(3)*.
16. NZYCM liquid medium, NZYCM agar, and top agar. Prepare according to ref. *4* by adding to 950 mL deionized water: 10 g of NZ amine, 5 g of NaCl, 5 g of bacto-yeast extract, 1 g of casamino acids, and 2 g of MgSO$_4$·7H$_2$O. Shake until solutes have dissolved. Adjust pH to 7.0 with 5$N$ NaOH. Adjust the volume to 1 L with deionized water. For NZYCM agar add 15 g bacto-agar/L. For NZYCM top agar add 7 g bacto-agar/L. Autoclave for 20 min at 15 lb/in.$^2$ on liquid cycle. Remelt the NZYCM agar and soft agar as required in a microwave oven on low power.
17. 20% maltose in distilled water (sterilized by filtration).
18. An oscillator incubator at 37°C.
19. A spectrophotometer and cuvets with a 1-cm light path to measure light absorbance at 600 nm.
20. A 50-mL centrifuge tube.
21. 10 m$M$ MgSO$_4$ sterilized by autoclaving.
22. SM solution: Prepare according to ref. *4* by combining 5.8 g of NaCl, 2 g of MgSO$_4$·7H$_2$O, 50 mL of 1$M$ Tris-HCl, pH 7.5, and 5 mL of 2% gelatin solution. Adjust volume to 1 L with double-distilled water. Sterilize by autoclaving.
23. 10-mL pipets.
24. Forty capped glass test tubes.
25. Chloroform.
26. 1.5-mL capped, polypropylene microfuge tubes.

27. 10-cm and 15-cm Diameter plastic Petri dishes (to be filled with NZYCM agar for plating out of bacteriophage).
28. An oven for incubation at 37°C.
29. 13-cm Nylon disks for bacteriophage blotting (e.g., NEN/Dupont, Boston, MA; cat. no. NEF-990A).
30. Reagents for bacteriophage blotting, radioactive probe hybridization, and posthybridization washing (e.g., *see* ref. *4*).
31. Needle (1–2-mm diameter).
32. DNA for probe manufacture (e.g., *see* Note 15).
33. Rediprime random primer labeling kit from Amersham (Little Chalfont, UK; cat. no. RPN1633) and ($\alpha^{32}$P)-dCTP.
34. Adhesive tape.
35. Thick card (the same size as the X-ray film used for autoradiography).
36. X-ray film suitable for autoradiography with $^{32}$P.
37. X-ray film developing solutions.

## 3. Methods

The sublibrary method described below is divided into five sections (3.1.–3.5.). First, the PCR for detection of the desired DNA region is tested. The PCR is then used to test λ libraries for the presence of clones containing the desired DNA region. A positive λ library is used to generate sublibraries. PCR is then used to test for sublibraries enriched for the desired clone. Finally, the desired clone is isolated from a positive sublibrary.

### 3.1. Testing the PCR Screen

1. Assemble a PCR premix and a polymerase premix as follows: (a) PCR premix: 2 μL of 10X Dynazyme™ buffer (*see* Note 1), 13 μL of water, 1 μL of primer A (20 μ*M*), 1 μL of primer B (20 μ*M*), 2 μL of dNTPs (10 m*M*). (b) Polymerase premix: 0.5 μL of 10x Dynazyme™ buffer, 4.0 μL of water, and 0.5 μL of Dynazyme (2 U/μL) (*see* Note 1).
2. Prepare a DNA sample for PCR:
   a. Place one drop of mineral oil in the well of a microassay plate (or PCR tube). Then pipet 1–2 μL of a suitable DNA sample (*see* Note 2) under the oil. Note that preparation of a negative (no DNA) control is essential (*see* Note 3).
   b. (Necessary only when the DNA sample is a bacteriophage library) Place the microassay plate (or PCR tube) in the PCR cycler and heat to 94°C for 2 min to denature bacteriophage capsids and inactivate any DNases.
3. Add the PCR mix and centifuge the microassay plate (or PCR tube) for a couple of seconds at 1000 rpm (approx 170*g*).
4. Place the microassay plate (or PCR tube) in the PCR cycler and heat to 94°C (e.g., "pause" the PCR cycler during the first denaturation step of the PCR cycling protocol given in step 5). Wait 30 s and then eject the polymerase premix into the well from just above the top of the oil (e.g., onto the wall of the plate well or

tube). The polymerase mix will fall through the oil and mix with the other PCR components.

5. Perform PCR cycling as follows: 35 cycles: denaturation at 94°C for 1 min, annealing at a suitable temperature for 1 min (*see* Note 4), temperature ramp of 0.5°C/s to 72°C, and elongation at 72°C for a suitable time (*see* Note 5). Then hold the PCRs at 4°C (or freeze them) until analysis.

6. Add 5 µL of 6X gel loading buffer to the top of the oil and centrifuge the plate or tube as in step 3.

7. Electrophorese 20 µL of the PCR on an agarose gel of suitable concentration and using suitable size markers (*see* Note 6). When loading the PCR onto the gel, draw the solution up and down in the pipet a couple of times to mix it thoroughly with the loading buffer. Avoid drawing oil into the tip.

8. If a specific PCR fragment of the expected size is revealed by the electrophoresis, then the PCR screening method probably works.

## 3.2. PCR-Testing of λ Libraries

1. Test the λ library for the presence of a clone containing the desired DNA sequence by performing PCR on 1–2 µL of undiluted λ library as described in steps 1–8 of Section 3.1. Always perform a negative (no DNA) control and a positive control using the DNA known (from Section 3.1.) to give a PCR signal (*see* Note 7).

2. If the PCR detects a desired clone in the λ library then proceed to the manufacture of sublibraries as described in the next Section (*see* Note 8).

## 3.3. Generation of λ Sublibraries

1. Grow a 50-mL culture of *E. coli* strain NM538 overnight at 37°C in NZYCM medium plus 0.2% maltose with vigorous shaking (*see* Note 9).

2. Measure the optical density (OD) of the culture relative to sterile NZYCM medium at 600 n*M* with a light path of 1-cm. The OD of an overnight culture of NM538 in NZYCM plus 0.2% maltose is usually between 3 and 5.

3. Centrifuge the culture in a 50-mL centrifuge tube at approx 6000*g* for 10 min. Pour off the supernatant and resuspend the cells in 10 mL of 10 m*M* MgSO$_4$. The cells can be stored for up to 2 d at 4°C (*see* Note 10).

4. Dilute the cells in NZYCM medium (prewarmed to 37°C) to a final volume of 40 mL so that the OD is calculated to equal 0.2.

5. Immediately infect the diluted culture with $10^6$ or $10^7$ plaque forming units (pfu) from the λ library. (Dilute the λ library as required in SM solution; *see* Note 11.)

6. Incubate the infected culture at 37°C for 10 min to allow bacteriophage absorption by the cells and then rapidly divide the culture into 40 × 1-mL aliquots in numbered, capped glass test tubes. Incubate the tubes at 37°C with vigorous shaking.

7. When cell lysis occurs (usually after 3–5 h of incubation, *see* Note 12) add three drops of chloroform to each tube and continue vigorous shaking for 5 min. The tubes now contain λ sublibraries. Store them (for a limited period; *see* Note 13) capped at 4°C. (Alternatively, transfer the contents to labeled, capped 1.5-mL polypropylene microfuge tubes; *see* Note 13.)

### 3.4. PCR Screening of the Sublibraries

Test the λ sublibraries for the presence of a clone containing the desired DNA sequence by performing PCR on 1–2 μL of each λ sublibrary as described in steps 1–8 of Section 3.1. A positive-control PCR using 1–2 μL of the original λ library and a negative (no DNA) control should also be performed for comparison with the sublibrary PCRs (*see* Note 14). An example of a successful sublibrary screening is shown in Fig. 2.

### 3.5. Isolation of the Desired Bacteriophage Clone from a Sublibrary

1. Determine the titer of a positive sublibrary by infecting 100-μL aliquots of NM538 at OD = 2.0 in 10 m$M$ MgSO$_4$ with 10-μL aliquots of $10^{-5}$, $10^{-7}$, and $10^{-9}$ dilutions of the positive sublibrary in SM solution. After absorption of the bacteriophage onto the bacteria for 10 min at 37°C, add 3 mL of NZYCM top agar plus 0.2% maltose (melt the NZYCM top agar, add the maltose, and then

---

**Fig. 2.** *(opposite page)* Illustration of the use of the sublibrary technique for isolating λ cDNA clones from transcripts of the mouse *Notch 3* gene (a member of the *Notch* family of genes encoding large transmembrane signal receptor proteins; ref. *5*). Sequencing of a genomic DNA clone revealed putative open reading frames with homology to members of the *Notch* gene family. Oligonucleotide primers corresponding to these sequences were tested in PCR on single stranded cDNA from 11.5-d-old mouse embryos. DNA fragments from the PCR were cloned and checked by sequencing for *Notch* homology. Primers amplifying *Notch 3* cDNA fragments were used to test a random-primed cDNA library from prepubescent testis mRNA (gift of Christer Höög). Two primer pairs, "pair A" (*see* GEL A) and "pair B" (*see* GEL B) produced signals from the library (*see* positive controls, "+") indicating the presence of clones encompassing the regions amplified by these primers. $10^6$ clones from the library were used to produce 40 sublibraries founded from 25,000 clones each. The sublibraries were then tested with the same primer pairs. Primer pair A detected three positive sublibraries, numbers 2, 15, and 31. Primer pair B detected *Notch 3* clones in sublibraries 23 and 36. This indicates that the frequency in the original library of clones spanning the region amplified by primer pair A is $3.3 \times 10^{-6}$ and, for primer pair B is $2 \times 10^{-6}$. *Notch 3* clones were isolated from these five sublibraries by probe hybridization. Note the stronger PCR signals in the positive sublibraries compared to the original library caused by clone enrichment (although the PCR reactions were not performed quantitatively). Note also that the positive signals from primer pair B could be detected in spite of strong background signals. "–" denotes negative controls. "M" denotes DNA size markers (Gibco-BRL/Life Technologies, Cat. No. 520-5615SA). The PCR tests on sublibraries 1, 20, 21, and 40 are marked as position references. This figure is modified from ref. *1* with the kind permission of Eaton Publishing, Natick, MA, USA.

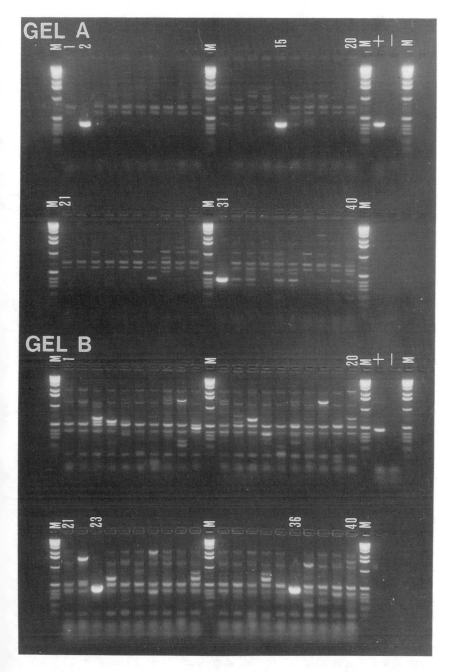

**Fig. 2.**

hold at 42°C), mix gently but rapidly, and pour onto a 10-cm Petri dish of NZYCM agar that has been slightly dried and prewarmed to 37°C. Rock the plate briefly to ensure even spreading of the top agar. When the top agar has set, invert the plate and incubate overnight at 37°C. Count the bacteriophage plaques produced and calculate the sublibrary titer using the following formula:

$$\text{Titer in pfu/mL} = (100 \times \text{No. of plaques})/\text{Sublibrary dilution} \qquad (1)$$

Proceed immediately to step 2 (*see* Note 13).

2. If the positive sublibrary was generated from 25,000 pfu (i.e., the initial 40-mL culture was infected with $10^6$ pfu) then plate out 100,000 pfu on two 15-cm diameter plates of NZYCM agar (i.e., 50,000 pfu/plate). If the positive sublibrary was generated from 250,000 pfu (i.e., the initial 40-mL culture was infected with $10^7$ pfu) then plate out $10^6$ pfu on twenty 15-cm plates or start a new round of sublibrary generation using $10^6$ pfu from the positive sublibrary (i.e., generate "subsublibraries").

   To plate out 50,000 pfu on a 15-cm plate, dilute an aliquot of the positive sublibrary to a concentration of $5 \times 10^6$ pfu/mL in SM solution and then add 10 μL of this to 300 μL of NM538 at OD = 2.0 in 10 m$M$ MgSO$_4$. Incubate 10 min at 37°C and then add 7 mL of NZYCM top agar plus 0.2% maltose (melted, then held at 42°C). Mix gently but rapidly and pour onto a 15-cm NZYCM agar plate slightly dried and prewarmed to 42°C. Ensure that the top agar is evenly spread. When the top agar has set, invert the plate and incubate at 37°C until the plaques are the size of small pin-pricks. Cool the plates to 4°C.

3. Blot the plated-out bacteriophage onto 13-cm nylon or nitrocellulose disks. Remember to stab the disks with a needle when they are lying on the plates so that the disks can be reoriented to the plate after hybridization and autoradiography. Hybridize the disks with a radiolabeled probe that detects the desired clone (*see* Note 15). Tape the washed and dried disks to card and tape the card to X-ray film for autoradiography. Remember to stab through the card and film with a pin so that the autoradiograph can be reoriented to the blots after development.

4. After autoradiography of the blots, pick out areas of the plates containing positive clones using a 1-mL micropipet with a cut off tip. Place these agar samples in individual 1.5-mL microfuge tubes containing 0.5 mL of SM solution and three drops of chloroform. Use a fresh pipet tip to break up the agar piece and invert the tube a number of times. Place tube overnight at 4°C. Repeat this procedure for an area of a plate lacking a positive signal (for use as a negative control in the following PCR).

5. Invert the tube again a couple of times and then centrifuge for 10 s at top speed in a microcentrifuge. Remove a 1-μL sample and perform PCR on this according to steps 1–7 of Section 3.1. A positive signal (properly controlled) indicates that the picked plate area contains a clone of the desired type (*see* Note 16).

6. Titrate the bacteriophage solution from the picked area, as in step 1, using undiluted, $10^{-2}$ and $10^{-4}$ dilutions of the solution. Plate out 1000 pfu on two 15-cm NZYCM plates (*see* Note 17) and allow plaques to grow to a diameter of 2 mm

before repeating steps 3 and 4. Positive clones should give a strong signal on the autoradiograph and do not need to be checked by PCR. If individual positive clones (plaques) cannot be picked, then pick an area containing a positive clone and repeat this step after plating out 100 pfu.

7. Amplify the isolated clone by infecting 10 mL of NM538 cells at OD = 0.2 in NZYCM medium (prepared as described in steps 1–4 of Section 3.3.) with 250,000 pfu followed by incubation at 37°C until lysis or by elution of plated-out bacteriophage in SM solution (*see* Note 18). Store at 4°C with added chloroform.

## 4. Notes

1. PCR conducted on samples from λ libraries can be problematic since media used for bacteriophage λ growth contain significant concentrations of $Mg^{2+}$ ions (that are required to prevent disassociation of the capsids). For example, standard NZYCM medium contains 8 m$M$ $MgSO_4$ *(4)*. The author has found that use of Dynazyme™ (a thermostable polymerase purified from *Thermus brockianus* by Finnzymes Oy) with the buffer supplied by the manufacturer allows PCR to be performed successfully without consideration for variations in $Mg^{2+}$ concentration in most cases. Other thermostable polymerases can be used in this procedure but these may be sensitive to small variations in $Mg^{2+}$ concentration. One solution to the problem is to adjust the $Mg^{2+}$ concentration in the PCR reaction buffer so that addition of an aliquot from a λ library gives the desired concentration. If the titer of the library is sufficiently high, it is also possible to make a 1 in 10 dilution of the library, thus reducing the $Mg^{2+}$ concentration while still providing enough bacteriophage for detection by PCR.

2. Before beginning the sublibrary method, the PCR must be tested for its ability to detect the desired sequence. The test substrate will depend on the type of library to be screened. Primers for screening cDNA libraries can be tested on single-stranded cDNA produced from whole cell mRNA or on a suitable dilution of a preexisting plasmid clone of the cDNA sequence (for example, a clone of a PCR fragment). When screening genomic λ libraries, the PCR primers should be tested on diluted genomic DNA. During PCR-testing of the λ library, 1–2 µL of the library is used in a reaction. If the library has a titer of $10^{10}$ pfu/mL ($10^7$ pfu/µL) and contains the desired clone at a frequency of between $10^{-6}$ and $10^{-7}$ (i.e., between 10 copies and 1 copy/µL), then this corresponds to a concentration of the desired clone in the library of between $1.7 \times 10^{-18}$ and $1.7 \times 10^{-19} M$, respectively, and in the PCR reaction of between $6.8 \times 10^{-20}$ and $6.8 \times 10^{-21} M$. Ideally, the PCR should detect the desired DNA sequence at this concentration. Higher titers in the original library allow the use of less efficient PCR. When ejecting samples under oil, check that the sample volume has actually been placed in the oil and has not clung to the outside of the micropipet tip.

3. The negative control is prepared and handled in exactly the same way as the other PCRs except that no DNA sample is added. To avoid crosscontamination with the other PCRs when testing a PCR screen, add PCR premix and polymerase premix to the negative control first.

4. The annealing temperature for the PCR reaction will depend on the primers used. For standard 18-mer PCR primers with a 50% GC content, try annealing first at 50°C. If this is not successful, then test annealing temperatures in the range 40–60°C.
5. The elongation time depends on the progressivity of the polymerase. The author usually assumes a progressivity of 1 kb/min and sets the elongation time at twice the time estimated for synthesis of the desired DNA fragment. Thus, elongation time in minutes = 2 × fragment length in kb.
6. Size markers should allow identification of the desired PCR product. DNA Ladder markers from Life Technologies (Gibco-BRL) allow identification of PCR fragments with a size of 200 bp–12 kb.
7. A 4-µL library sample may aid in the detection of very rare clones *(1)*, but in this case the reaction volume should be increased to 50–100 µL.
8. If PCR on a sample of the λ library does not produce an identifiable PCR product it may still be worth proceeding with the sublibrary procedure starting with $10^7$ clones. A suitable clone present in the original library at very low concentration may be sufficiently amplified in one of the sublibraries to be detected by the PCR.
9. The author has only used *E. coli* strain NM538 for generation of sublibraries. Other strains suitable for λ infection should also work but the initial OD of the culture that is infected for sublibrary generation may need to be adjusted slightly. Addition of maltose to the NZYCM medium induces synthesis of the maltose transporter protein to which bacteriophage λ binds when infecting a cell.
10. The sublibrary procedure as originally described *(1)* used maltose in both the overnight bacterial culture and after bacteriophage infection. However, the author has since found that higher sublibrary titers are obtained if the infected bacteria are cultured in medium lacking maltose. This may be partly because of the presence of fewer maltose receptors in the debris resulting from cell lysis that can absorb the bacteriophage. Dead cells in the initial cell culture will also needlessly absorb bacteriophage so only 1–2-d-old cultures should be used. NM538 cells resuspended in 10 m$M$ MgSO$_4$ and stored at 4°C can be used for up to a week for plating-out of bacteriophage.
11. In order to infect the bacteria with the correct amount of bacteriophage, the library should be titrated a short time before use using the method described in step 1 of Section 3.5.
12. If cell lysis fails to occur then the infection can be modified by decreasing the OD of the culture to be infected (e.g., from 0.2 to 0.1).
13. Unfortunately, despite their high titers, the sublibraries may be unstable. It is therefore desirable to proceed with PCR screening and bacteriophage isolation within a few days of their manufacture. Although PCR signals do not appear to be greatly affected by a decrease in a sublibrary's infectivity, the titer of a sublibrary should always be measured immediately before plating it out for clone isolation. Instability of sublibraries is not a great problem because generation of sublibraries is so simple and rapid that it is easily performed each time one wishes to isolate a new clone.

If the sublibraries are stored in microfuge tubes, then ensure that three drops of chloroform are always present in each tube. To remove some cell debris, the sublibraries may also be centrifuged (10 min at 16,000 rpm) and transferred to new microfuge tubes with chloroform.

14. To control for crosscontamination resulting from the pipeting technique when setting up the sublibrary PCRs, add PCR premix to the sublibrary samples first, then to the positive control, and finally to the negative control. Use the same order of addition for the polymerase mix.

15. Detailed descriptions of bacteriophage blotting and hybridization techniques are provided by reference 4. A suitable probe can be synthesized by excising the PCR product of a positive control from a low gelling temperature agarose gel after electrophoresis in Tris-acetate buffer (*see* ref. *4*) and then using this in a random-primed DNA synthesis reaction with the Amersham Rediprime kit (Cat. No. RPN 1633; *see* manufacturer's instructions).

16. This step ensures that clones detected by probe hybridization contain the entire region detected by the PCR. Clones containing partially overlapping regions of DNA and false positives resulting from artifactual spots on the autoradiograph are thus excluded.

17. To save time, the titration can be performed in duplicate on 15-cm plates. A pair of plates with approx 500 pfu each can then be used directly for blotting and clone isolation.

18. Detailed descriptions of all the techniques for handling bacteriophage clones can be found in ref. *4.*

## References

1. Lardelli, M. and Lendahl, U. (1994) Generating bacteriophage λ sublibraries enriched for rare clones. *Biotechniques* **16,** 420–422.

2. Mattila, P., Gelfand, D. H., Sninsky, J. J., and White, T. J. (1991) Fidelity of DNA synthesis by the *Thermococcus litoralis* DNA polymerase—an extremely heat stable enzyme with proofreading activity. *Nucleic Acids Res.* **19,** 4967–4973.

3. Frischauf, A.-M., Lehrach, H., Poustka, A., and Murray, N. (1983) Lambda replacement vectors carrying polylinker sequences. *J. Mol. Biol.* **170,** 827–842.

4. Sambrook, J., Frisch, E. F., and Maniatis, T., eds. (1989) *Molecular Cloning: A Laboratory Manual,* 2nd ed. Cold Spring Harbor Laboratory, Cold Spring Harbor, NY, pp. 2.2–2.125, 10.1–10.70, Appendices A.1, A.2, A.4, A.7, A.11.

5. Lardelli, M., Dahlstrand, J., and Lendahl, U. (1994) The novel *Notch* homolog mouse *Notch 3* lacks specific epidermal growth factor repeats and is expressed in proliferating neuroepithelium. *Mech. Dev.* **46,** 123–136.

# VI

# Differential and Subtractive Approach by cDNA Analysis and Cloning

# 36

# Normalization of cDNA Sequence Representation by Molecular Selection

**Thierry G. Coche**

## 1. Introduction

### 1.1. Rationale for Sequence Normalization

Consider the situation in which one would like to clone sequences that are differentially expressed between two developmental stages. An experiment could be set up in which cDNA from one developmental stage is taken as reference and cDNA from a later developmental stage is taken as target. The objective of the experiment is to isolate sequences that switch on or are upregulated during the developmental process. If the experiment is carried out and the differentially expressed sequences isolated by subtractive cloning are identified by sequencing, it will generally turn out that a high proportion of the sequences are trivially interesting, highly expressed sequences, such as globin, actin, or tubulin. This can be a frustrating experience and is not an artifact of subtractive cloning since it can be shown that the isolated sequences are effectively present at higher levels in the target mRNA. Rather, highly expressed sequences are preferentially cloned because the frequency with which a clone is retrieved by subtractive cloning is correlated to the fractional representation of the clone in the initial mRNA population. Consequently, if nothing is done to compensate for the heterogeneity in sequence representation before subtractive cloning, highly expressed sequences will represent an unacceptably high proportion of clones in the subtraction product. The objective of sequence normalization is to circumvent this unfavorable situation by reducing the differences in sequence representation between cDNA species and thereby provide an opportunity to investigate both high and low expression sequences. Normalization should be envisaged in all situations where the high fractional representation of certain

From: *Methods in Molecular Biology, Vol. 67: PCR Cloning Protocols: From Molecular Cloning to Genetic Engineering* Edited by: B. A. White  Humana Press Inc., Totowa, NJ

sequences will impede the access or exploitation of sequences expressed at lower levels (differential screening, catalog libraries, total probes, and so forth). As in subtractive cloning, sequence normalization relies exclusively on solution hybridization to discriminate between abundant and rare sequences. The kinetic basis of this discrimination is explained in Section 1.2. As in subtractive cloning (Chapter 37), a basic understanding of the mathematical principles behind sequence normalization should aid the bench scientist in deciding which conditions are most appropriate for his or her experiment. Normalization methods differ in the way normalized sequences are retrieved after the hybridization step. In general, this has been achieved by hydroxyapatite chromatography *(1–4)*. This approach has the reputation of being cumbersome and requires large amounts of starting material. The method described below relies on magnetic bead technology and molecular selection *(5)*. This method has the advantage of requiring smaller amounts of input material and implementation is quite straightforward.

## 1.2. Practical Mathematics of Normalization

Equation *(1)* in Chapter 37 describes the basic kinetic scheme for sequence hybridization between driver and tracer DNA molecules. Extrapolate this scheme to the situation where, instead of tracer and driver, the opposite strands of cDNA molecules are hybridized. The reaction scheme becomes:

$$C + C' \xrightarrow{\text{k}} CC'$$

and the differential equation describing this hybridization reaction is

$$d(CC')/(dt) = k\,(C)(C') \tag{1}$$

where $C$ and $C'$ represent the concentrations of the complementary strands of the cDNA molecules. In nonmathematical terms, the differential equation simply states that the rate of hybridization (rate of formation of double-stranded molecules) is proportional to the concentration of the reacting molecules. Consequently, if in a given sample two or more reactions are occurring simultaneously, the one with the highest concentration will proceed fastest. This is the rationale used in sequence normalization. A cDNA library is denatured and then reannealed to itself. As hybridization proceeds, abundant sequences will reanneal faster than the rare sequences. Thus, if nonhybridized sequences are isolated (or hybridized sequences are eliminated, as will be the case in this protocol) the resulting collection of molecules will present lower heterogeneity in sequence representation than the starting population. The question is: What experimental conditions should be set up and what degree of normalization can be expected? The answer to this question requires a rate equation that can be derived from Eq. *(1)*. Note that this is a special case in which $C$ is exactly equal to $C'$, so that the equation used in Chapter 37 is not applicable. Integration of Eq. *(1)* provides a rate equation that describes $C$ (or $C'$) as a function of time:

$$C(t) = C(0)/[1 + kC(0)t] \qquad (2)$$

where $C(0)$ is the initial DNA concentration. As was the case for excess hybridization kinetics, when multiple components are present in the reaction, the overall reaction is described by summing the individual components. The situation for a fast and slow component would be:

$$C(t) = [C(0)^F]/[1 + kC(0)^F t] + [C(0)^S]/[1 + kC(0)^S t] \qquad (3)$$

where $C(0)^F$ and $C(0)^S$ are the initial DNA concentrations in each component. Use of a spread-sheet program will allow us to visualize the results for a real-life example. We will use k = 0.28 µL/µg × h as in the preceding chapter and choose $C(0)$ = 1 µg/µL. We will also assume multiple kinetic components, but only visualize the contributions of a fast component (40% of total DNA) and a slow component (20% of total DNA). The curves generated using these values are shown in Fig. 1.

At $t = 0$, the abundance of the fast component is twice that of the slow component: (fast/slow) = (0.4/0.2) = 2. At $t = 50$ h, (fast/slow) is down to 1.2 and further incubation would not affect this value significantly. On the other hand, if the nonhybridized DNA at $t = 50$ is selectively amplified and submitted to a second cycle of denaturation/self-hybridization, normalization can be substantially improved. To calculate the effect of a second cycle of normalization, take the values at $t = 50$, multiply them by an amplification factor, and use these values for $t = 0$ of the second cycle of normalization. If the DNA concentration at the end of the first cycle is multiplied by five, (fast/slow) is reduced to 1.02 after a second 50 h normalization reaction. Using this approach it can also be shown that the higher the initial DNA concentration, the more dramatic the decrease in (fast/slow) will be.

## 1.3. Overview of Normalization by Molecular Selection

The steps in the normalization protocol are illustrated in Fig. 2. If normalization is performed in preparation for subtractive cloning, both target and reference sequences should be normalized. A solid phase cDNA library is prepared from total RNA. The free ends of the cDNA molecules are tagged with a linker molecule and submitted to a first cycle of thermal denaturation/self-hybridization. Reannealing of the most abundant molecules will generate double-stranded molecules, whereas the less abundant molecules will remain in single-stranded form. Hybridization is stopped by eliminating nonhybridized

*Coche*

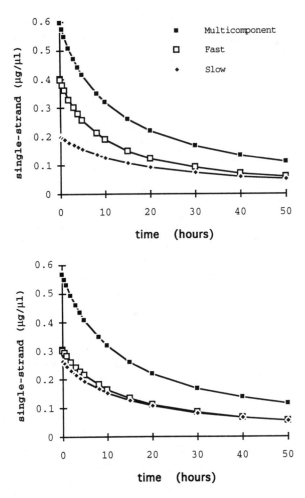

**Fig. 1.** Self-hybridization kinetics of a denatured cDNA library. Multicomponent curve is the sum of a fast (40%) and a slow (20%) component. The extent of normalization is directly visible as the reduction of the gap between the fast and slow component curves. **(Top)** First cycle of normalization with starting concentration of single-stranded DNA $D(0) = 1$ µg/µL. **(Bottom)** Second cycle of normalization after fivefold amplification of the DNA concentration at the end of the first cycle.

sequences in the supernatant by washing the beads. Molecular selection against double-stranded molecules is performed by restriction endonuclease digestion. As a result of DNA cleavage, the double-stranded molecules will lose their linker tag. Single-stranded DNA molecules are converted to double-strand by oligonucleotide-primed DNA synthesis using the appropriate tag primer and

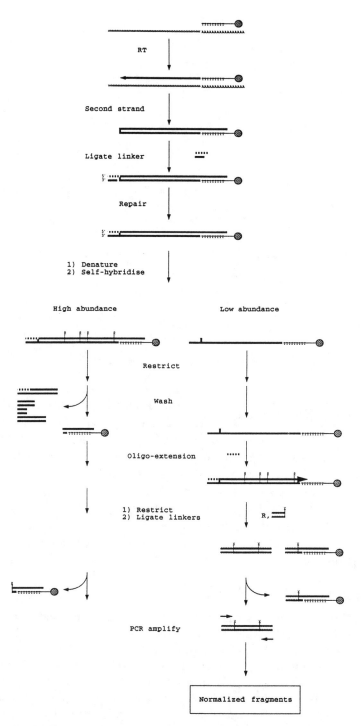

**Fig. 2.** Steps involved in the solid-phase normalization protocol.

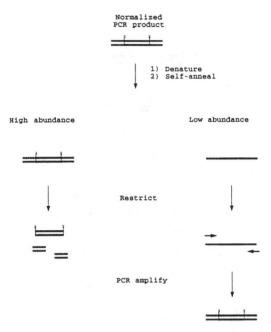

**Fig. 3.** Steps for "in solution" normalization.

T7 DNA polymerase. The newly synthesized DNA is then cleaved and linkers are added to both extremities of the resulting fragments. (Note: If the same enzyme is used for molecular selection and restriction before linker addition, no new fragments will be generated from the abundant sequences left on the beads since the sites in these sequences have already been cleaved.) The normalized fragments are amplified and purified.

Subsequent normalization cycles are performed "in solution" as opposed to "on beads" (Fig. 3). In these cycles, molecular selection against reannealed sequences is performed by cleaving off the linker sequences. Loss of linker sequences precludes further amplification.

## 2. Materials

1. DEPC-treated materials and reagents for preparation of total cellular RNA (Trisolv, Biotecx [Houston, TX] or equivalent).
2. Oligo-d(T)$_{25}$ magnetic beads with magnetic stand (Dynal, Lake Success, NY).
3. Reagents for reverse transcription and PCR (protocol describes Life Technologies reagents [Gaithersburg, MD]; adapt as necessary if other reagents are used).
4. Enzymes: T4 DNA ligase, T7 DNA polymerase, RNase H, *Rsa*I, *Alu*I, *Bam*HI, *Escherichia coli* DNA ligase, *E. coli* DNA polymerase I, T4 DNA polymerase, RNasin.

5. Solutions: TE, 20X SSC, 250 m*M* EDTA, 10 mg/mL tRNA, 0.1*M* ATP, 0.1*M* DTT, and 5*M* NaCl *(6)*.

6. 2X KGB: 200 m*M* potassium glutamate, 50 m*M* Tris-acetate, pH 7.5, 20 m*M* magnesium acetate, 100 μg/mL BSA, 1 m*M* β-mercaptoethanol.

7. 2X binding buffer: 20 m*M* Tris-HCl, pH 7.5, 1*M* LiCl, 2 m*M* EDTA.

8. 1X washing buffer: 10 m*M* Tris-HCl, pH 7.5, 0.15*M* LiCl, 1 m*M* EDTA.

9. 5X second-strand buffer: 94 m*M* Tris-HCl, pH 6.9, 453 m*M* KCl, 23 m*M* MgCl$_2$, 750 μ*M* βNAD$^+$, 50 m*M* (NH$_4$)$_2$SO$_4$.

10. 5X T7 DNA polymerase buffer: 615 m*M* Tris-HCl, pH 7.5, 307 m*M* MgCl$_2$, 385 m*M* NaCl.

11. BE buffer: 33 m*M* bis-Tris propane, pH 8.0, 0.3 m*M* EDTA.

12. Chloroform, isopropanol, ethanol.

13. "Lone linker," created by annealing of a 22- and a 23-mer oligonucleotide *(see* Chapter 37, Note 2).

14. DNA purification system (Prep-A-Gene affinity matrix, BioRad [Hercules, CA] or equivalent).

15. Agarose horizontal gel electrophoresis system.

## 3. Methods

### 3.1. Preparation of a Linker-Tagged Solid Phase cDNA Library

#### 3.1.1. RNA Extraction

There are now a number of reagents and methods available for RNA preparation. Below is a description of the Trisolv method (Biotecx), which I have successfully used to prepare embryonic brain RNA.

1. Put 2 mL of Trisolv reagent in a Dounce homogenizer. Keep on ice.

2. Add 200 mg of fresh or frozen tissue to the reagent and homogenize on ice. Work quickly to limit RNase activity.

3. Transfer the homogenized tissue to a Corex tube and incubate for 5 min at room temperature.

4. Add 400 μL chloroform, vortex and incubate for 3 min at room temperature.

5. Centrifuge for 15 min at room temperature at 12,000*g* in a SS34 rotor or equivalent. **From this step onward all materials and solutions in contact with the RNA must be RNase-free.**

6. Transfer the clear upper phase to a fresh tube.

7. Add 1 mL of isopropanol (RNA flocculates) and incubate for 10 min at room temperature.

8. Centrifuge for 5 min at room temperature at 12,000*g*.

9. Discard the supernatant and wash the pellet with 4 mL of cold 70% ethanol.

10. Centrifuge for 5 min at 4°C at 7500*g*.

11. Dry briefly under vacuum (SpeedVac). RNA pellet should remain humid. Overdrying will make resuspension of the pellet difficult.

12. Resuspend the RNA pellet in 500 μL of DEPC-treated water and keep on ice.

13. Measure OD at 260 nm using 10 μL of the RNA solution in a 500-μL cuvet (dilution = 50). Calculate RNA concentration: RNA conc. (μg/mL) = $OD_{260} \times 40 \times$ dilution. Yield for embryonic brain tissue is about 600 μg RNA for 200 mg of tissue.

## 3.1.2. Solid-Phase Reverse Transcription

If normalization is performed in preparation for subtractive cloning, follow the protocol exactly. If normalization is being applied to a single source of sequences, ignore instructions relative to the "Reference" sequences and adapt the protocol as necessary. Work under RNase-free conditions.

1. Label one 1.5-mL Eppendorf tube "Target" and a second "Reference." On ice, introduce 60 μg of the appropriate RNA into each tube and bring the volumes to 100 μL with DEPC-treated water. If RNA is too dilute, just add 1 vol of 2X binding buffer.
2. Heat RNA for 3 min at 70°C.
3. While RNA is heating, transfer 200 μL of magnetic bead solution to a fresh tube, and wash the beads twice with 200 μL of binding buffer. Finally, resuspend the beads in 200 μL of 2X binding buffer.
4. Transfer 100 μL of washed beads into each of the tubes containing the hot RNA.
5. Incubate at room temperature for 10 min.
6. Wash beads with attached RNA twice with 100 μL of 1X washing buffer and discard supernatants.
7. Wash beads twice with 50 μL of 1X reverse transcription (RT) buffer (supplied with the enzyme). Discard supernatants but do not let the beads dry out.
8. Mix in a separate tube: 22 μL of DEPC-treated water, 8 μL of 5X RT buffer, 4 μL of DTT 0.1M, 2 μL of 10 mM each dNTP, and 2 μL of RNasin.
9. Transfer 19 μL of the above solution to the "Target" and "Reference" tubes.
10. Preheat for 3 min at 37°C and add 1 μL Superscript II (200 U) in each tube.
11. Incubate for 15 min at 37°C and 45 min at 42°C. From here on, RNase-free conditions are not required.

## 3.1.3. Second-Strand Synthesis

1. On ice, add directly into each reverse transcription reaction: 92.9 μL of water, 32 μL of 5X second-strand buffer, 3 μL of 10 mM each dNTP, 6 μL of 0.1M DTT, 1.5 μL of *E. coli* DNA ligase (15 U), 4 μL of *E. coli* DNA polymerase I (40 U), and 0.56 μL of *E. coli* RNase H (1.4 U).
2. Incubate between 2 h and overnight at 16°C.
3. Add 2 μL (10 U) of T4 DNA polymerase and incubate for 5 min at 16°C.
4. Add 20 μL of 250 mM EDTA.

## 3.1.4. Ligation of "Lone" Linkers to the Free Ends of the cDNA Molecules

1. Prepare "lone" linkers by mixing 1 μL (100 pmol) of a 22-mer and 1 μL (100 pmol) of a complementary 23-mer in 2.5 μL of 2X KGB and 5.5 μL of water. Heat for 3 min at 80°C, and incubate for 10 min at room temperature.

2. Wash beads twice with 50 µL of 1X KGB.
3. Add to the beads 5 µL of 2X KGB, 11 µL of water, 1 µL of ATP 10 m*M*, 2 µL of annealed oligonucleotide solution (step 1), and 1 µL of T4 DNA ligase.
4. Incubate for 2 h at 37°C.
5. Wash beads twice with 1X *Taq* buffer.
6. Add to the beads: 2 µL of 50 m*M* MgCl$_2$, 5 µL of 10X reaction buffer, 2.5 µL of 1% W1 detergent, 1 µL of 10 m*M* each dNTP, 39.3 µL of water, and 0.2 µL of *Taq* polymerase (1 U).
7. Incubate for 10 min at 72°C.

## 3.2. First Cycle of Normalization

### 3.2.1. Self-Hybridization of the Thermally Denatured Library

1. Wash beads twice with 50 µL of 5X SSC and discard supernatant.
2. Add to the beads 0.5 µL of 10 m*M* EDTA, 0.1 µL of 10 mg/mL tRNA, 3.1 µL of water, and overlay with mineral oil (mix reagents separately and add appropriate volume).
3. Heat for 10 min at 94°C.
4. Add 1.3 µL of 20X SSC, and incubate between 24 and 48 h at 55°C.

### 3.2.2. Molecular Selection Against Reannealed Sequences by Endonuclease Restriction

1. Wash beads twice with 50 µL of 1X Life Sciences restriction buffer #1 (or equivalent) and discard supernatants.
2. Add to the beads 16 µL of water, 2 µL of 10X Life Sciences restriction buffer #1, 1 µL of *Rsa*I (10 U), and 1 µL of *Alu*I (10 U).
3. Incubate for 1 h at 37°C.
4. Wash beads twice with 50 µL of TE and resuspend in 50 µL of TE.
5. Heat beads for 3 min at 94°C.

### 3.2.3. Conversion of Single-Stranded Molecules to Double-Strand by Oligonucleotide-Primed DNA Synthesis

1. Wash beads twice with 50 µL of 1X T7 DNA polymerase buffer and discard supernatant.
2. Add to the beads 2 µL (40 pmol) of 23-mer oligonucleotide, 2 µL of 5X T7 polymerase buffer, and 6 µL of water.
3. Heat to 80°C, and transfer to 37°C for 5 min.
4. Add 2 µL of 5X T7 polymerase buffer, 1 µL of 0.1*M* DTT, 0.5 µL of 10 m*M* each dNTP, 4.5 µL of water, and 2 µL of DNA polymerase.
5. Incubate for 10 min at 37°C.

### 3.2.4. Restriction and "Lone" Linker-Tailing of Normalized Sequences

1. Prepare "lone" linkers by mixing 1 µL (100 pmol) of a 22-mer and 1 µL (100 pmol) of complementary 23-mer in 2.5 µL of 2X KGB and 5.5 µL of water. Heat for 3 min at 80°C, and incubate for 10 min at room temperature.

2. Wash beads twice with 50 μL of 1X KGB.
3. Add to the beads 5 μL of 2X KGB, 10 μL of water, 1 μL of Rsa I (10 U), 1 μL of 10 m*M* ATP, 2 μL of linkers, and 1 μL of T4 DNA ligase.
4. Incubate for 2 h at 37°C.

### 3.2.5. PCR Amplification of Normalized Fragments

1. Assemble PCR reaction with the following: 2 μL of ligation reaction (Section 3.2.4., step 4), 4 μL of 50 m*M* MgCl$_2$, 10 μL of 10X *Taq* buffer, 5 μL of 1% W1 detergent, 2 μL of 10 m*M* each dNTP, 1 μL (100 pmol) of amplification primer (23-mer), 75.6 μL of water, 0.4 μL of *Taq* polymerase (2 U), and overlay with mineral oil.
2. Precycle for 10 min at 72°C
3. Perform 30 cycles: 1 min at 94°C; 1 min at 60°C; and 2 min at 72°C.
4. Check 10 μL of the PCR reaction by electrophoresis on a 2% agarose gel. A smear with material of average size 300 bp should be apparent.
5. Purify the remaining PCR material using 15 μL of Prep-A-Gene affinity matrix to eliminate oligonucleotides, small-mol-wt material, and other reactants. Resuspend the purified material in 30 μL water. Check 1 μL by electrophoresis.

### 3.3. Subsequent Cycles of Normalization

### 3.3.1. Self-Hybridization of Normalized Fragments

1. Lyophilize between 5 and 10 μg of purified PCR product.
2. Resuspend in 4 μL of BE buffer and layer with mineral oil.
3. Heat for 3 min at 94°C.
4. Add 1 μL of 5*M* NaCl.
5. Incubate between 24 and 48 h at 55°C.

### 3.3.2. Molecular Selection Against Reannealed Fragments by Linker Cleavage

1. To the annealing reaction, add 33 μL of water, 5 μL of 10X Life Sciences restriction buffer #3, and 7 μL (70 U) of *Bam*HI.
2. Incubate for 1 h at 37°C.
3. Amplify restriction product and purify amplification product as in Section 3.2.5.

## 4. Notes

After normalization, target and reference fragments can be put through a subtraction protocol to isolate target-specific sequences. Since differences in expression levels have been eliminated by normalization, only target-specific sequences can be isolated by subtractive cloning. Although a number of subtraction protocols can, in principle, be used with the normalized fragments, two are in my opinion particularly well suited for this purpose.

The first is the method described by Klickstein *(7)*, which relies on the reconstitution of clonable ends to insert target-specific sequences into a plasmid. To

implement this method, the linkers at the extremities of target and reference molecules would be cleaved off to produce protruding extremities. Reference sequences would be blunt-ended by nuclease S1 treatment. An excess of reference fragments would then be hybridized to the target fragments and the product of the hybridization would be ligated to a compatible plasmid. Theoretically, only reannealed target molecules that have not been "soaked-up" by complementary reference sequences should produce recombinant plasmids.

The second method, which we have been using in the laboratory, is based on the RDA protocol described by Lisitsyn *(8)*. Although initially developed for cloning differences between genomes, the methodology can be applied to cDNA fragments with minor adaptations. The method relies on the ligation of 3'-recessed linkers to the ends of the target molecules and post-hybridization repair of the recessed ends to generate targets for PCR primers on the reannealed target molecules. The method can readily be applied to normalized fragments by replacing the linkers on the target molecules by appropriate RDA linkers.

## References

1. Patanjali, S. R., Parimoo, S., and Weissman, S. M. (1991) Construction of a uniform-abundance (normalized) cDNA library. *Proc. Natl. Acad. Sci. USA* **88**, 1943–1947.
2. Ko, M. S. (1990) An "equalized cDNA library" by the reassociation of short double-stranded cDNAs. *Nucleic Acids Res.* **18**, 5705–5711.
3. Soares, M. B., Bonaldo, M. F., Jelene, P., Su, L., Lawton, L., and Efstratiadis, A. (1994) Construction and characterization of a normalized cDNA library. *Proc. Natl. Acad. Sci. USA* **91**, 9228–9232.
4. Takahashi, N. and Ko, M. S. (1994) Toward a whole cDNA catalog: construction of an equalized cDNA library from mouse embryos. *Genomics* **23**, 202–210.
5. Coche, T. and Dewez, M. (1994) Reducing bias in cDNA sequence representation by molecular selection. *Nucleic Acids Res.* **22**, 4545,4546.
6. Sambrook, J., Fritsch, E. F., and Maniatis, T. (1989) *Molecular Cloning: A Laboratory Manual,* 2nd ed., Cold Spring Harbor Laboratory, Cold Spring Harbor, NY.
7. Klickstein, L. B. (1993) Production of a subtracted cDNA library, in *Current Protocols in Molecular Biology* (Ausubel, F. M., Brent, R., Kingston, R. E., Moore, D. D., Seidman, J. D., Smith, J. A., and Struhl, K., eds.), Wiley, New York, pp. 5.8.9–5.8.15.
8. Lisitsyn, N., Lisitsyn, N., and Wigler, M. (1993) Cloning the difference between two complex genomes. *Science* **259**, 946–951.

# 37

## Subtractive cDNA Cloning
## Using Magnetic Beads and PCR

Thierry G. Coche

## 1. Introduction

### 1.1. General Strategy
### for Cloning Differentially Expressed Sequences

Subtractive cloning is a method that facilitates the isolation of nucleotide sequences present in a test sample but absent or present at much lower levels, in a reference sample. A variety of situations lend themselves to this methodology and it has been used, for example, to characterize changes in levels of gene expression during development *(1–3)*, in pathological situations *(4–6)*, in inductive events *(7,8)*, and for the isolation of tissue-specific genes *(1,9–11)*. Current methodologies also allow detection of pathogens and integrated retroviruses present at exceedingly low levels in the cell *(12)*.

Historically, differences in levels of gene expression were initially investigated by differential screening strategies *(2,11)*. This was performed by preparing cDNA libraries from test and reference tissues and screening the test library in duplicate with labeled cDNA from the test and reference libraries. Construction of the cDNA libraries required large amounts of biological material, which limited the field of application of this methodology. Furthermore, statistical considerations restrict this method to the analysis of sequences expressed at relatively high levels. For example, if a 1% risk level of missing a clone during a screening procedure is accepted, then it can be calculated that 50,000 clones must be screened in order to find a differentially expressed sequence that is present at a level of 0.1% in the initial mRNA population. This frequency is not exceptional and many interesting genes, including regulatory genes, are expressed at much lower levels. For these reasons, differential screening is now mostly used for downstream processing of sequences isolated using newer methodologies.

From: *Methods in Molecular Biology, Vol. 67: PCR Cloning Protocols: From Molecular Cloning to Genetic Engineering* Edited by: B. A. White Humana Press Inc., Totowa, NJ

Recent advances in cloning technology and the advent of gene amplification by PCR in particular have allowed modern differential cloning strategies to evolve. The general scheme of these methods is to combine target sequences with an excess of reference sequences and to use solution hybridization to soak up sequences that are common to both sources. The excess of reference sequences "drives" the reaction. Target sequences that have not hybridized to reference sequences are then physically isolated and cloned. The protocols based on this scheme differ in the way target sequences are retrieved after hybridization. Enrichment of target molecules has been achieved using hydroxyapatite chromatography *(1,3,13)*, phenol extraction of biotin-streptavidin-labeled material *(4,9,14)*, selection by plasmid cloning *(15)*, selective labeling after chemical crosslinking *(16)*, differential amplification *(17)*, and the use of magnetic beads *(18–21)*.

In this chapter, I will describe a method that relies on magnetic bead technology. As with all subtractive cloning strategies, one of the crucial steps in this method is setting up correct experimental conditions for solution hybridization. Careful consideration of hybridization kinetics will help determine *a priori* whether the proposed experiment can be expected to succeed. Consequently, I have included a very laboratory-oriented paragraph on hybridization kinetics that I hope will aid the bench scientist in making decisions on how to adapt the protocol to his or her particular needs.

### 1.2. Practical Mathematics of Subtractive Hybridization

The kinetic scheme for solution hybridization between target sequences and reference (driver) sequences is:

$$D + T \xrightarrow{k} DT \tag{1}$$

where $D$ and $T$ are the concentrations of single-stranded driver and target sequences, $DT$ is the concentration of hybrid driver-target sequences and k is the rate constant for the reaction. The rate of formation of $DT$ is proportional to the product of the concentrations of the single-stranded molecules:

$$[d(DT)]/dt = k\,(D)\,(T) \tag{2}$$

This scheme is applicable to situations such as hybridization of excess driver mRNA to small amounts of target cDNA (re-annealing of cDNA can then be ignored) or hybridization of driver cDNA molecules of a given polarity to complementary target cDNA of the opposite polarity. The latter situation prevails in the protocol I will describe. The equation describing the time-course of single-stranded target molecules during hybridization can be derived *(22)* and adapted for direct use in the laboratory:

$$T(t) = T(0) - \{D(0)T(0)\,[1-e^{[T(0)-D(0)]kt}]\}/\{D(0) - T(0)\,e^{[T(0)-D(0)]kt}\} \qquad (3)$$

*T(t)* is the concentration ($\mu g/\mu L$) of single-stranded target molecules at time $t$ (hours), $T(0)$ and $D(0)$ are, respectively, the initial concentrations ($\mu g/\mu L$) of single-stranded target and driver molecules introduced into the reaction. In each case the single-strand DNA concentration is calculated using half the total amount of DNA in the reaction. k ($\mu L/\mu g \times h$) is the rate constant for the reaction. The value of k is given by:

$$k = (R/C) \cdot (3600/324.1) \qquad (4)$$

in which 3600 is the number of seconds in 1 h; 324.1 is the average molecular weight of a nucleotide, $C$ is the complexity of the collection of sequences (number of nucleotides of unique sequence), and $R$(L · mol/s) is the complexity-independent rate constant of the reaction. $R$ is, within certain limits, proportional to salt concentration, temperature, and square root of sequence length. A typical value for $R$ at 180 m$M$ Na, $T_m$ 25°C, for sequences of 500 nucleotides is $10^6$ (L · mol/s) *(22)*. Note that Eq. *(2)* is only valid in cases where $D(0) \neq T(0)$. The special case $D(0) = T(0)$ is considered in Chapter 36.

The practical implications of Eq. *(2)* can be visualized through a real-life example. In the case of a typical somatic cell, at any given time about 20,000 distinct mRNA molecules of average size 2000 nucleotides are expressed. For now, we will assume that all mRNA molecules are equally abundant. The sequence complexity of the mRNA isolated from such a cell (and the cDNA synthesized from this mRNA) is $C = 20{,}000$ (unique sequences) × 2000 (nucleotides) = $4 \times 10^7$ nucleotides. Taking $R = 10^6$, we may calculate k = ($10^6/4 \times 10^7$) × (3600/324.1) = 0.28 $\mu L/\mu g \times h$. Let us take $D(0) = 1$ $\mu g/\mu L$ and $T(0) =$ 0.05 $\mu g/\mu L$. The usual way to look at the data is to plot *T(t)* as a function of $R_{ot}$ values on a semilogarithmic scale. With the advent of microcomputers, we urge the reader to introduce all the variables and equations into a spreadsheet program and plot *T(t)* as a function of $t$ directly using Eq. 3. Once this is set up, all the variables can be changed at will to evaluate the influence of each variable on the rate of hybridization. The bold curve in Fig. 1 (■) illustrates the time-course of unhybridized target molecules as a function of time for our example.

When $D(0) = 1$ $\mu g/\mu L$, the reaction will terminate after about 15 h. When $D(0) \gg T(0)$, hybridization proceeds very rapidly. This situation is termed "pseudo-first order kinetics" because *D(t)* remains virtually constant over time and the rate of hybridization is dependent on *T(t)* exclusively (*see* Eq. *[1]*) with D constant. If $D(0)$ is not significantly higher than $T(0)$, termination may be delayed substantially. Using this approach, it is straightforward to see that increased sequence complexity, reduced salt concentration, reduced temperature, and reduced sequence length will delay termination of the annealing reaction.

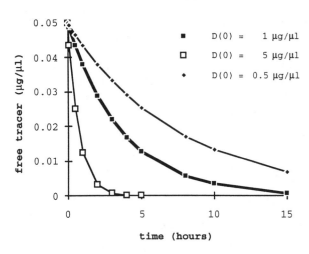

**Fig. 1.** Effect of target:driver ratio on hybridization kinetics of target sequences. $T(0) = 0.05$ μg/μL.

In summary, the message in Eq. *(1)* is that it is desirable to work with the highest achievable absolute concentrations of reacting sequences. In practice this is obtained by reducing the reaction volume to a minimum. The message in Eq. *(2)* is that one should set up conditions with high driver:target ratios and work at high temperature and high ionic strength. There is a catch here: As temperature is increased, thermal degradation of cDNA is accelerated, which can lead to accumulation of short cDNA fragments that in turn reduces the rate of hybridization. One must compromise by keeping both temperature and hybridization time within reasonable limits. Thermal degradation can be retarded by good buffering conditions.

In our real-life example, we made the simplifying assumption that all messengers were equally abundant. This does not affect the practical conclusions reached above, but will affect the absolute rate of hybridization. Because valid incubation times must be calculated for successful subtractive hybridization, we must take into consideration that sequences are not equally abundant. In our typical somatic cell there are at least two populations of molecules: an abundant (40%), highly repetitive population (low complexity) in which 10 mRNA species may be present at 5000 copies or more per cell and a less abundant (20%), lowly repetitive (high complexity) population in which 15,000 mRNA species may be present at an average number of 10/cell. The low complexity population will be characterized by a higher hybridization rate constant than the high complexity population. The presence of these kinetically heterogeneous sequence populations will lead to multicomponent hybridization profiles. From a mathematical point of view, each kinetic component is described

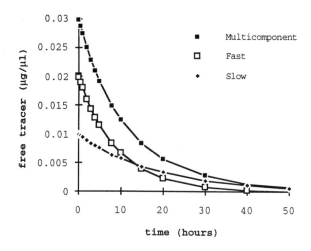

**Fig. 2.** Multicomponent hybridization kinetics. The multicomponent curve is the sum of a fast (40%) and a slow (20%) component.

by Eq. *(2)* with its own set of variables. The overall hybridization profile is obtained by summing the individual kinetic components. For the two components considered above, rate equation is:

$$T(t) = T(0)^F - \{C^F D(0)T(0)^F [1-e^{[T(0)^F - C^F D(0)]kt}]\}/\{C^F D(0) - T(0)^F e^{[T(0)^F - C^F D(0)]kt}\} +$$

$$T(0)^S - \{C^S D(0)T(0)^S [1-e^{[T(0)^S - C^S D(0)]kt}]\}/\{C^S D(0) - T(0)^S e^{[T(0)^S - C^S D(0)]kt}\} \quad (5)$$

where $T(0)^F$ and $T(0)^S$ are the initial concentrations of tracer in the fast and slow components. Note that correction factors multiply the values of $D(0)$. These corrective factors must be introduced to account for the fact that the driver is not kinetically homogeneous. Take the slow component for example: Only a fraction of the total mRNA is complementary to the cDNA in the slow component. Consequently, only a fraction of the total cDNA will "drive" the slow hybridization component. Thus, the total amount of driver must be multiplied by the fractional representation of the slow component in the driver. The correction factor for the slow component is given by:

$$C^s = T(0)^S/[T(0)^S + T(0)^F] \quad (6)$$

Introducing $D(0) = 1$ μg/mL, $T(0)^F = 0.4$ μg/mL, $T(0)^S = 0.2$ μg/mL, and $k = 0.28$ μL/μg × h into the above equations will produce the curves illustrated in Fig. 2.

Notice that the fast component will approach completion after about 40 h, whereas the slow component takes more than 50 h to approach completion. In summary, if the cDNA collection is heterogeneous, as is most often the case, incubation should be prolonged sufficiently to allow hybridization of the slowest component. It is important to realize that failure to prolong hybridization sufficiently will enrich the subtraction product in slow component sequences that are not necessarily differentially expressed. With the kinetic tools presented here and *a priori* information on the target sequence and its environment, it should be possible to outline an experimental strategy that will increase the odds for success of the subtractive cloning strategy.

### 1.3. Overview of Solid-Phase Subtractive Cloning

An overview of the solid-phase subtractive cloning strategy is provided in Fig. 3. Solid-Phase cDNA libraries from reference and target RNA are prepared. Target cDNA fragments are obtained by random priming of the target cDNA library and restriction of the reaction product. Linkers are ligated to the ends of the restriction fragments to allow PCR-amplification. After amplification, the target fragments are hybridized back onto the target cDNA library to restore the correct stoichiometry in target sequences and eliminate fragments that have the same polarity as the reference cDNA library. The high concentration in PCR product "drives" the hybridization reaction and ensures that sequences of low abundance will hybridize back onto the target library. Subtractive hybridization of the target fragments to the reference cDNA library is then carried out. Since the target fragments have linkers at their extremities, unhybridized target fragments in the supernatant can be amplified directly by PCR.

Subsequent cycles are performed according to a similar scheme (Fig. 4). Common sequences are eliminated from the target library by molecular selection. This is achieved by annealing the common sequences retained on the reference beads to the target library and restricting any double-stranded molecules by a cocktail of frequently cutting enzymes. The target library is thereby enriched in differentially expressed sequences. The subtracted fragments obtained in the previous cycle are annealed to the enriched target library and then reannealed to the reference library.

### 2. Materials

1. DEPC-treated materials and reagents for preparation of total cellular RNA (Trisolv, Biotecx [Houston, TX] or equivalent).
2. Oligo-d(T)$_{25}$ magnetic beads with magnetic stand (Dynal, Lake Success, NY).
3. Reagents for reverse transcription, random-priming, and PCR (the protocol uses Life Technologies [Gibco-BRL, Gaithersburg, MD] reagents; adapt as necessary if other reagents are used).
4. Enzymes: T4 DNA ligase, RNase H, Klenow, *Rsa*I, *Alu*I, and corresponding buffers as supplied with enzymes.
5. TE: 10 m$M$ Tris-HCl, pH 7.5, 1 m$M$ EDTA.
6. 1$N$ NaOH.

7. 20X SSC: 3$M$ NaCl, 0.3$M$ sodium citrate, pH 7.0.
8. 250 m$M$ EDTA.
9. tRNA (10 mg/mL solution).
10. ATP (0.1$M$ solution).
11. Dithiothreitol (DTT; 0.1$M$; *see* ref. *23*).
12. 2X KGB: 200 m$M$ potassium glutamate; 50 m$M$ Tris-acetate, pH 7.5; 20 m$M$ magnesium acetate; 100 µg/mL bovine serum albumin (BSA); 1 m$M$ β-mercaptoethanol.
13. 2X binding buffer: 20 m$M$ Tris-HCl, pH 7.5, 1$M$ LiCl, 2 m$M$ EDTA.
14. 1X washing buffer: 10 m$M$ Tris-HCl, pH 7.5, 0.15$M$ LiCl, 1 m$M$ EDTA.
15. Chloroform, isopropanol, ethanol.
16. "Lone linker," created by annealing of a 22- and a 23-mer oligonucleotide.
17. DNA purification system (Prep-A-Gene affinity matrix [BioRad, Hercules, CA] or equivalent).
18. Agarose horizontal gel electrophoresis system.

## 3. Methods

### 3.1. Preparation of Solid-Phase Target and Driver cDNA Libraries

### 3.1.1. RNA Extraction

A protocol for RNA extraction using the Trisolv (Biotecx) reagent is described in Chapter 36.

### 3.1.2. Solid-Phase Reverse Transcription

The following steps are carried out with the reference and target RNA simultaneously. Work under RNase-free conditions. For the first cycle of subtraction, one tube of target library and four tubes of reference library are prepared. If more than one cycle of subtraction will be performed prepare four extra reference tubes for each supplementary cycle of subtraction and adapt the protocol as necessary.

1. Label five 1.5-mL Eppendorf tubes: one "Target" and four "Reference." Introduce 60 µg of the appropriate RNA into each tube and bring the volumes to 100 µL with DEPC-treated water on ice (*see* Note 1). If RNA is too dilute, just add 1 vol of 2X binding buffer.
2. Heat RNA for 3 min at 70°C.
3. While RNA is heating, transfer 500 µL of magnetic bead solution to a fresh tube and wash the beads twice with 500 µL of binding buffer. Resuspend the beads in 500 µL of 2X binding buffer.
4. Transfer 100 µL of washed beads into each of the tubes containing the hot RNA.
5. Incubate for 10 min at room temperature.
6. Wash beads with attached RNA twice with 100 µL of 1X washing buffer and discard supernatants.
7. Wash beads twice with 50 µL of 1X reverse transcription (RT) buffer (supplied with the enzyme). Discard supernatants but do not let the beads dry out.

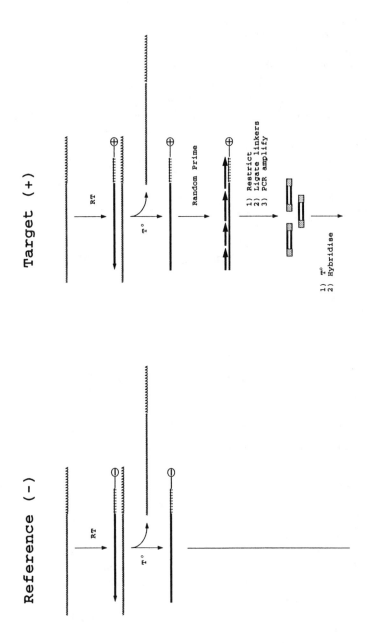

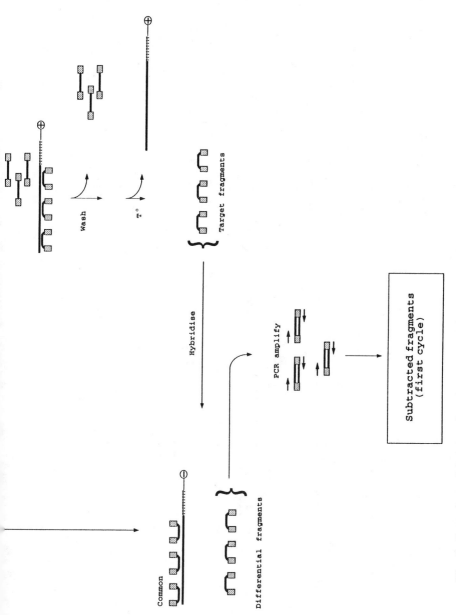

**Fig. 3.** Overview of the steps involved in the first cycle of the subtractive cloning protocol.

379

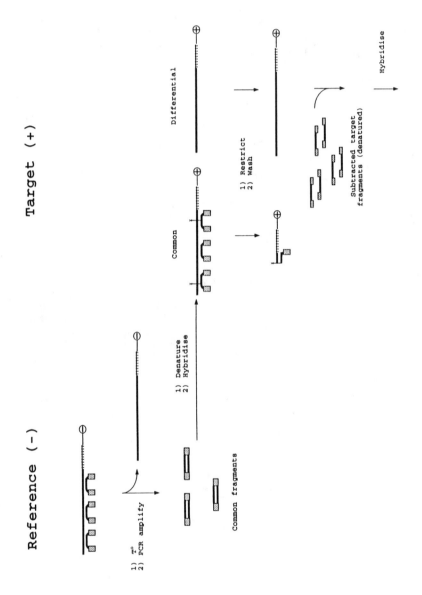

Reference (−)          Target (+)

1) T°
2) PCR amplify

1) Denature
2) Hybridise

Common fragments

Common          Differential

1) Restrict
2) Wash

Subtracted target
fragments (denatured)

Hybridise

380

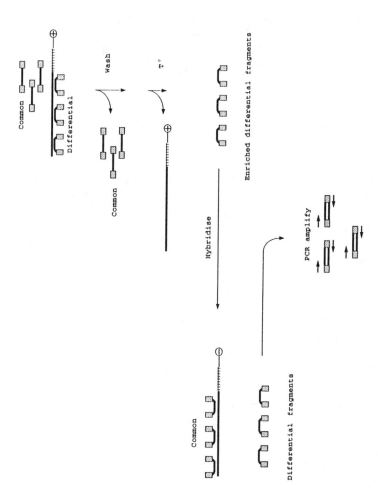

**Fig. 4.** Steps involved in subsequent cycles of subtraction.

8. In a separate tube, mix 55 µL of DEPC-treated water, 20 µL of 5X RT buffer, 10 µL of 0.1$M$ DTT, 5 µL of 10 m$M$ each dNTP, and 5 µL of RNasin.
9. Transfer 19 µL of the above solution to the "Target" and "Reference" tubes.
10. Preheat for 3 min at 37°C, and add 1 µL of Superscript II (200 U) in each tube.
11. Incubate for 15 min at 37°C, and 45 min at 42°C. The protocol may be interrupted here if necessary by adding 2 µL of 250 m$M$ EDTA and storing the tubes at –20°C. From here on, RNase-free conditions are not required.
12. Discard supernatants and add 50 µL of TE. Heat for 3 min at 95°C, and discard the supernatants that contain the mRNA. Repeat twice. Resuspend the beads in 20 µL of TE. Store the "Reference" tubes at –20°C, and keep the "Target" tube on ice.

## 3.2. Preparation of Target Sequences for Subtraction

### 3.2.1. Random Priming of Target Library

1. Resuspend the "Target" beads and transfer 5 µL to a newly labeled "Target 2" tube. Put the "Target" tube away at –20°C. Random prime the Target 2 sequences using a Random Prime Kit (example is using a Life Sciences kit): Add 45 µL of water, 4 µL of dATP solution, 4 µL of dGTP solution, 4 µL of dCTP solution, 4 µL of dTTP solution, 30 µL of Random Prime Mix, 2 µL of RNaseH (5 U), and 2 µL of Klenow (6 U).
2. Incubate for 1 h at room temperature.
3. Wash the beads twice with 100 µL of TE.

### 3.2.2. Restriction and "Lone" Linker-Tailing of Target Sequences

1. Prepare "lone" linkers by mixing 1 µL (100 pmol) of a 22-mer oligonucleotide and 1 µL (100 pmol) complementary 23-mer oligonucleotide in 2.5 µL 2X KGB and 5.5 µL water. Heat for 3 min at 80°C, and incubate for 10 min at room temperature (*see* Note 2).
2. Wash the "Target 2" beads twice with 50 µL of 1X KGB and discard the supernatants.
3. Add to the beads 5 µL of 2X KGB, 10 µL of water, 1 µL of *Rsa*I(10U), 1 µL of 10 m$M$ ATP, 2 µL of annealed oligonucleotide solution, and 1 µL of T4 DNA ligase (*see* Note 3).
4. Incubate for 2 h at 37°C.

### 3.2.3. PCR-Amplification of Target Sequences

1. Combine 2 µL of ligation reaction (from Section 3.2.2., step 4), 4 µL of 50 m$M$ MgCl$_2$, 10 µL of 10X *Taq* buffer (BRL), 5 µL of 1% W1 detergent, 2 µL of 10 m$M$ each dNTP, 1 µL (100 pmol) of amplification primer (23-mer), 75.6 µL of water, and 0.4 µL of *Taq* polymerase (2 U), and overlay with mineral oil.
2. Precycle for 10 min at 72°C, (*see* Note 4).
3. Perform 30 cycles of PCR: 1 min, 94°C; 1 min, 60°C; 2 min, 72°C.
4. Check 10 µL of the PCR reaction by electrophoresis on a 2% agarose gel. A smear with material of average size 300 bp should be apparent (*see* Note 5).

5. Purify the remaining PCR material using 15 μL of Prep-A-Gene affinity matrix to eliminate oligonucleotides, small-mol-wt material, and other reactants. Resuspend the purified material in 30 μL of TE or water.

### 3.2.4. Hybridization of Amplified Target Sequences to Target Library

1. Denature purified PCR product for 5 min at 95°C.
2. Thaw and resuspend beads in "Target" tube (Section 3.2.1., step 1). Transfer 5 μL to a fresh tube labeled "Target 3." Put "Target" tube back at –20°C.
3. Discard supernatant from "Target 3" and add denatured PCR product, 5.2 μL of 20X SSC, and 2 μL of 10 m$M$ EDTA.
4. Incubate for 24 h at 55°C.

## 3.3. First Cycle of Subtraction

### 3.3.1. Prepare Target Beads for Subtractive Hybridization

1. Transfer supernatant from "Target 3" beads (Section 3.2.4., step 4) to a fresh tube and store at –20°C.
2. To the "Target 3" beads, add 6.2 μL of water, 1 μL of 10 m$M$ EDTA, and 0.2 μL of tRNA (10 mg/mL). Store on ice.

### 3.3.2. Prepare Reference Beads for Subtractive Hybridization

1. Thaw "Reference" beads (Section 3.1.2., step 12) and pool the contents of the four tubes in a single tube. Capture the beads, discard the supernatant, and resuspend the beads in 85 μL of water and 15 μL of 1$N$ NaOH.
2. Incubate for 10 min at room temperature.
3. Add 50 μL of 20X SSC, mix, and discard supernatant.
4. Resuspend beads in 100 μL of 5X SSC and transfer to a fresh, 0.5-mL tube labeled "Sub."

### 3.3.3. Hybridization of Target Sequences to Driver Library

1. Discard supernatant from reference beads in the "Sub" tube (Section 3.3.2., step 4).
2. Melt "Target 3" sequences (Section 3.3.1., step 2), 3 min, 94°C. Quickspin.
3. Capture the beads and transfer the "Target 3" sequences in the supernatant onto the reference beads in the "Sub" tube. The target and reference sequences are together and ready for subtractive hybridization (*see* Note 6).
4. Overlay with 40 μL of mineral oil and heat for 1 min at 94°C.
5. Add 2.6 μL of 20X SSC and incubate for 24–48 h at 55°C. Resuspend the beads occasionally.

### 3.3.4. PCR-Amplification of Subtracted Target Sequences

1. Remove the tube from 55°C, capture the beads and transfer the hot supernatant with the subtracted target sequences to a fresh tube. Resuspend the "Sub" beads in 50 μL of TE if subsequent cycles of subtraction are performed. The mineral oil

can be extracted by water-saturated ether if necessary. Water (between 10 and 40 µL) can be added to the hybridization reaction to increase volume and avoid picking up beads with the supernatant. If water is added, adapt the amount used in the PCR reaction below.

2. Amplify 2 µL of the subtracted target sequences as described in Section 3.2.3. Do not precycle 10 min. Use "hot-start" PCR: Add *Taq* after the denaturation step of first cycle.

3. Check the result of amplification on a 2% agarose gel. Affinity purify as in Section 3.2.3., step 5.

## 3.4. Subsequent Cycles of Subtraction

### 3.4.1. Molecular Selection Against Target Sequences

1. Wash the "Sub" beads (Section 3.3.4., step 1) twice with 50 µL of TE. Resuspend in 50 µL of water and heat for 3 min at 94°C, to melt off annealed sequences. These sequences are common to the target and reference libraries.

2. Capture beads and PCR-amplify 2 µL of the supernatant as in Section 3.2.3. Check the amplification product on the gel. Affinity purify as before.

3. Hybridize amplified common sequences back onto the target library as in Section 3.2.4.

4. Wash beads twice with 50 µL of 1X Life Technologies (BRL) restriction buffer 1 (or equivalent) and discard supernatants.

5. Add to the beads 16 µL of water, 2 µL of 10X Life Technologies (BRL) restriction buffer 1, 1 µL of *Rsa*I (10 U), and 1 µL of *Alu*I (10 U).

6. Incubate for 1 h at 37°C.

7. Wash beads twice with 50 µL of TE and resuspend in 50 µL of TE.

### 3.4.2. Hybridization of Subtracted Sequences to Selected Target Library

Purified target sequences (Section 3.3.4., step 3) are hybridized back onto the selected target library (Section 3.4.1., step 7) as in Section 3.2.4.

### 3.4.3. Subtractive Hybridization and PCR Amplification

Go through the steps in Section 3.3. as previously, using the target sequences annealed to the selected target library (Section 3.4.2.). It is preferable to use a fresh reference library for each cycle of subtraction (*see* introductory note of Section 3.1.2.). If this is not possible the same reference library may be reused although thermal degradation of the library may lead to reduced subtraction efficiency.

## 4. Notes

1. About 1% of total RNA is polyA$^+$ mRNA, so that the 60 µg of total RNA contain about 600 ng of mRNA. The mRNA binding capacity of the Dynal Oligo(T)$_{25}$ beads is about 1 µg of mRNA for 100 µL of bead solution. The amount of beads should thus be sufficient to accommodate all the mRNA in the starting material.

Be careful not to reduce ionic strength when the mRNA is hybridized to the poly-dT of the magnetic beads because this would lead to dissociation of the hybrid and loss of the mRNA.

2. The following nonphosphorylated oligonucleotides were used to create the "lone" linker:

   a. 23-mer 5' ATCCTTCCCAGCCTG<u>GGATCC</u>CT 3'

   b. 22-mer 3' AGGAAGGGTCGGAC<u>CCTAGG</u>GA 5'

   The 23-mer oligonucleotide is also used to prime PCR-amplification of the target sequences. Note the one base 5' extension that is produced by annealing the two oligonucleotides. This extension and the absence of phosphate groups at the 5' ends of the oligonucleotides ensure that ligation of the linker to the cDNA molecules occurs in a single orientation and prevents formation of linker concatemers. High linker concentration "drives" the blunt-end ligation reaction, which can be completed within 2 h at 37°C. Also note the presence of the *Bam*HI restriction site (underlined) in the linker, which can be used to clone the subtraction products into a plasmid. Other oligonucleotides with identical characteristics can probably be substituted for the above oligonucleotides. If other oligonucleotides are used, check that they are suitable for use in PCR (widely available computer programs).

3. *Rsa*I cuts eukaryotic DNA on average about every 300 bp so that a significant proportion of cDNA molecules will yield at least one PCR-amplifiable fragment. Nevertheless, if differences in target and reference libraries are limited to only one or a few mRNA species, it may be wise to repeat the experiment with a different restriction enzyme (e.g., *Alu*I) to increase the odds of generating an amplifiable fragment. Both restriction enzymes and ligase are active in 0.5X KGB, so cleavage and ligation can be performed simultaneously.

4. The 10 min precycling step is crucial to repair the 3' ends of the cDNA molecules and synthesize the sequence to which the amplification primer hybridizes. If "hot-start" PCR is used, sequences must reanneal before *Taq* polymerase can fill in the 3' ends. Since the concentration of sequences in the reaction at this time is very low, very few sequences will have time to reanneal. The overall yield of the PCR reaction will be insufficient and lowly expressed sequences may be lost.

5. During agarose gel electrophoresis, use Orange G dye in the loading buffer instead of bromphenol blue. The former dye will not interfere with visualization of the PCR products.

6. The final concentrations of reference and target sequences during the first cycle of subtraction are, respectively, 0.24 and 0.015 µg/µL. The ratio of reference:target sequences is about 16:1. Sequences exclusively present in the target library and sequences upregulated by at least a factor of 16 in the target library should in theory remain unpaired during subtractive hybridization if the reaction is allowed to approach completion. Nonspecific hybridization and/or crosshybridization with related sequences may affect the actual results.

7. One advantage of this method is that it allows multiple cycles of subtraction to be performed. The approach adopted here is to use the sequences that have annealed

to the reference library to eliminate common sequences from the target library by molecular selection. This is achieved by hybridizing the common sequences to the target library and restricting the double-stranded molecules with frequent cutting enzymes. By doing this, the target library is enriched in differential sequences. The amount of common sequences left in the target library can be decreased by increasing the number of enzymes used in this reaction. The subtracted sequences from the previous cycle of subtraction are then annealed back onto this selected target library. Common sequences that fail to find a partner are eliminated. Consequently, the fractional representation of differential sequences increases. The cycle is completed by reannealing the target fragments retained on the target library back onto the reference library.

8. When subtractive hybridization cycles have been completed, the last PCR product is cloned into a plasmid. Differential screening of miniprep material in dotblot format can then be used to isolate differentially expressed sequences. Target and reference probes can conveniently be obtained by random-priming the target and reference libraries. In experiments carried out in this laboratory, about 30% of the clones examined by differential screening were differentially expressed.

## References

1. Timblin, C., Battey, J., and Kuehl, M. W. (1990) Application of PCR technology to subtractive cDNA cloning: identification of genes expressed specifically in murine plasmacytome cells. *Nucleic Acids Res.* **18,** 1587–1593.
2. Sargent, T. D. and Dawid, I. B. (1983) Differential gene expression in the gastrula of *Xenopus laevis. Science* **222,** 135–139.
3. Mathiopoulos, C. and Sonenshein, A. L. (1989) Identification of *Bacillus subtilis* genes expressed early during sporulation. *Mol. Microbiol.* **3,** 1071–1081.
4. Bowes, C., Danciger, M., Kozak, C. A., and Farber, D. B. (1989) Isolation of a candidate cDNA for the gene causing retinal degeneration in the *rd* mouse. *Proc. Natl. Acad. Sci. USA* **86,** 9722–9726.
5. Reynet, C. and Kahn, R. C. (1993) Rad: a member of the *Ras* family overexpressed in muscle of type II diabetic humans. *Science* **262,** 1441–1444.
6. Cole, S. P. C., Bhardwaj, G., Gerlach, J. H., Machie, J. E., Grant, C. E., Almquist, K. C., Stewart, A. J., Kurz, E. U., Duncan, A. M. V., and Deeley, R. G. (1992) Overexpression of a transporter gene in a multidrug-resistant human lung cancer cell line. *Science* **258,** 1650–1654.
7. Nedivi, E., Hevroni, D., Naot, D., Israeli, D., and Citri, Y. (1993) Numerous candidate plasticity-related genes revealed by differential cloning. *Nature* **363,** 718–721.
8. Parfett, C. L., Hofbauer, R., Brudzynski, K., Edwards, D. R., and Denhardt, D. T. (1989) Differential screening of a cDNA library with probes amplified in a heterologous host: isolation of murine GRP78 (BiP) and other serum-regulated low abundance mRNAs. *Gene* **82,** 291–303.
9. Swaroop, A., Xu, J., Agarwal, N., and Weissman, S. M. (1991) A simple and efficient cDNA library substraction procedure: isolation of human retina-specific cDNA clones. *Nucleic Acids Res.* **19,** 1954.

10. Hedrick, S. M., Cohen, D. I., Nielsen, E. A., and Davis, M. M. (1984) Isolation of cDNA clones encoding T cell-specific membrane-associated proteins. *Nature* **308**, 149–153.
11. Tedder, T. F, Streuli, M., Schlossman, S. F., and Saito, H. (1988) Isolation and structure of a cDNA encoding the B1 (CD20) cell-surface antigen of human B lymphocytes. *Proc. Natl. Acad. Sci. USA* **85**, 208–212.
12. Chang, Y., Cesarman, E., Pessin, M. S., Lee, F., Culpepper, J., Knowles, D. M., and Moore, P. S. (1994) Identification of herpesvirus-like DNA sequences in AIDS-associated Kaposi's sarcoma. *Science* **266**, 1865–1869.
13. Stürzl, M. and Roth, W. K. (1990) PCR-synthesized single-stranded DNA: a useful tool for 'hyb' and 'HAP' standardization for construction of subtraction libraries. *Trends Genet.* **6**, 106.
14. Lebeau, M.-C., Alvarez-Bolado, G., Wahli, W., and Catsicas, S. (1991) PCR driven DNA-DNA competitive hybridization: a new method for sensitive differential cloning. *Nucleic Acids Res.* **19**, 4778.
15. Klickstein, L. B. (1989) Production of a subtracted cDNA library, in *Current Protocols in Molecular Biology* (Ausubel, F. M., Brent, R., Kingston, R. E., Moore, D. D., Seidman, J. D., Smith, J. A., and Struhl, K., eds.), John Wiley, New York, pp. 5.8.9–5.8.15.
16. Hampson, I. N., Pope, L., Cowling, G. J., and Dexter, T. M. (1992) Chemical crosslinking subtraction (CCLS): a new method for the generation of subtractive hybridization probes. *Nucleic Acids Res.* **20**, 2899.
17. Lisitsyn, N., Lisitsyn, N., and Wigler, M. (1993) Cloning the difference between two complex genomes. *Science* **259**, 946–951.
18. Rodriguez, I. R. and Chader, G. J (1992) A novel method for the isolation of tissue-specific genes. *Nucleic Acids Res.* **20**, 3528.
19. Schraml, P., Shipman, R., Stulz, P., and Ludwig, C. U. (1993) cDNA subtraction library construction using a magnet-assisted subtraction technique. *Trends Genet.* **9**, 70,71.
20. Coche, T., Dewez, M., and Beckers, M.-C. (1994) Generation of an unlimited supply of a subtracted probe using magnetic beads and PCR. *Nucleic Acids Res.* **22**, 1322,1323.
21. Bröker, M. (1993) PCR-based constraction of subtractive cDNA library using magnetic beads. *BioTechniques* **4**, 610,611.
22. Britten, R. J. and Davidson, E. H. (1985) Hybridization strategy, in *Nucleic Acid Hybridization: A Practical Approach* (Hames, B. D. and Higgins, S. J., eds.), IRL, Oxford, UK, pp. 3–15.
23. Sambrook, J., Fritsch, E. F., and Maniatis, T., eds. (1989) *Molecular Cloning. A Laboratory Manual,* 2nd ed. Cold Spring Harbor Laboratory, Cold Spring Harbor, NY.

# 38

## Generation of a PCR-Renewable Source of Subtractive cDNA

### W. Michael Kuehl and James Battey

### 1. Introduction

Differential (+/-) first-strand cDNA screening methods identify clones corresponding to mRNAs that are expressed at a higher level in one of a pair of phenotypically different cells. This approach is limited by the fact that screening of libraries with labeled first-strand cDNAs synthesized from unfractionated mRNA can detect clones containing sequences representing approx 0.1% or more of the complexity of mRNA (i.e., mRNAs present at greater than about 200 copies/cell since a typical mammalian cell line contains approx 250,000 mRNAs).

To enhance the sensitivity of this approach, a number of laboratories *(1–4)* developed methodology for preparing subtractive cDNA probes: (1) First-strand cDNA from one cell type is hybridized to an excess of mRNA from a closely related cell type; (2) the small fraction of single-stranded cDNA is separated from the bulk of double-stranded mRNA-cDNA hybrid; and (3) small fragments of nonhybridizing cDNA are removed from the subtractive cDNA. Depending on the extent of relatedness of the mRNAs from the parental and subtractive partner cell lines, cDNA sequences unique to the parental cell line might be enriched as much as 50-fold in the subtractive cDNA. As a result of this enrichment, the subtractive cDNA probe could be used to identify clones containing differentially expressed sequences comprising approx 0.002% or more of the complexity of cellular mRNAs (i.e., greater than approx 5 copies of mRNA per cell). The subtractive cDNA could also be used to prepare subtractive cDNA libraries, so that many fewer clones need to be screened.

There are increasing numbers of important genes that have been identified using subtractive cDNA technology, but the method requires large amounts of

From: *Methods in Molecular Biology, Vol. 67: PCR Cloning Protocols: From Molecular Cloning to Genetic Engineering* Edited by: B. A. White  Humana Press Inc., Totowa, NJ

mRNA, is technically demanding, and results in very low yields of short cDNA fragments. A number of novel approaches, many of which involve preparation of cDNA libraries from each subtractive partner, appear to avoid some of these problems *(5–11)*. In this chapter, we describe a method that incorporates polymerase chain reaction (PCR) technology, which partially solves the need for large amounts of mRNA but also permits more flexibility in using the PCR renewable subtracted cDNA as a probe or for cloning *(11)*. A similar method has been used successfully for genomic DNA subtractions *(12,19)*.

The following is an outline of the PCR/subtractive cDNA method presented:

1. Synthesize first-strand cDNA from parental cell mRNA;
2. Hybridize cDNA to excess of mRNA from subtractive partner;
3. Remove cDNA:mRNA duplex from negatively selected single-stranded cDNA by hydroxyapatite (HAP) chromatography;
4. Hybridize single-stranded cDNA to mRNA from parent or related cell;
5. Remove single-stranded cDNA from positively selected cDNA:mRNA duplex by HAP chromatography;
6. Synthesize second-strand cDNA from mRNA:cDNA template with RNase H and DNA polymerase I;
7. "Polish" and kinase ends of double-stranded cDNA with T4 DNA polymerase and T4 kinase;
8. Add amplification adaptors with T4 ligase;
9. Remove excess adaptors and small cDNAs by Sepharose 4B chromatography;
10. PCR amplify subtractive cDNA;
11. PCR chase reaction to ensure subtractive cDNA is mostly homoduplexes;
12. Size-fractionate amplified, subtractive cDNA on Sepharose 4B.

## 2. Materials

### 2.1. First-Strand cDNA Synthesis

1. 5 µg of parental mRNA, selected twice on oligo(dT) cellulose (*see* Notes 1–3).
2. MMLV reverse transcriptase (RT), 200 U/µL.
3. 5X RT buffer: $0.25 M$ Tris-HCl, pH 8.3, $0.375 M$ KCl, 15 m$M$ MgCl$_2$, 50 m$M$ DTT.
4. Random hexamer at 0.25 µg/µL.
5. Adapter-dT$_{17}$ primer (e.g., 5'-GGACTCGAGGTATCGATGCTTTTTTTTTTTT-TTTTT(G,A,C)-3', which has *Xho*I and *Cla*I restriction sites) at 0.1 µg/µL.
6. Mixture of 10 m$M$ of each dNTP (i.e. dGTP, dATP, dCTP, dTTP), pH 7.0.
7. [α-$^{32}$P]dCTP, 3000 C$_i$/mmol.
8. RNasin, 30 U/µL (Promega, Madison, WI).
9. Bovine serum albumin (BSA), nuclease free, 10 mg/mL.
10. $0.5 M$ EDTA, autoclaved.
11. $2 M$ Tris-HCl, pH 7.4, autoclaved.
12. $2 N$ HCl.
13. $10 N$ NaOH.

14. Phenol.
15. CHCl$_3$.
16. TE buffer: 10 m$M$ Tris-HCl, pH 7.5, 0.1 m$M$ EDTA.
17. Sephadex G-50 column or spin column equilibrated with TE.
18. 100% Trichloroacetic acid.
19. 3$M$ Sodium acetate, pH 6.0, autoclaved.
20. Ethanol.
21. Siliconized 1.5-mL microfuge tubes (e.g., PGC Scientifics, Gaithersburg, MD, #505-201).

## 2.2. Hybridization of mRNA to cDNA

1. 50 μg of subtractive partner mRNA (selected twice on oligo [dT] cellulose).
2. 50 μg of poly(A)- subtractive partner RNA for carrier, and also providing rRNA for subtraction.
3. 5X Hybridization buffer: 3$M$ NaCl, 100 m$M$ Tris-HCl, pH 7.7, 10 m$M$ EDTA (autoclaved).
4. 1% SDS.
5. Mineral oil.
6. Siliconized 0.5-mL microfuge tube (e.g., PGC Scientifics #505-195).

## 2.3. Separation of cDNA and cDNA:mRNA by Hydroxyapatite Chromatography

1. Hydroxyapatite powder (Bio-Rad HTP, Richmond, CA).
2. 0.5$M$ Na phosphate buffer (PB), pH 6.8. Prepare by mixing equal volumes of 0.5$M$ monosodium phosphate and 0.5$M$ disodium phosphate.
3. 5$M$ NaCl.
4. Appropriate dilutions of PB in 150 m$M$ NaCl (*see* Section 3.3.2. and Note 4).
5. Chromatography system that can be used at 60°C (*see* Section 3.3.2. and Note 5).
6. Labeled test DNAs: single-stranded = first-strand cDNA (Section 3.1.); double-stranded = φX174 *Hae*III fragments labeled with Klenow enzyme and [α-$^{32}$P]dCTP to high specific activity.
7. 6$N$ HCl.
8. Nensorb™ 20 cartridge (DuPont-NEN, Boston, MA). Since the capacity is approx 20 μg of nucleic acid, add contents of additional cartridges to process larger samples.
9. 50% Ethanol.
10. Siliconized 1.5-mL microfuge tubes.

## 2.4. Positive Selection and Generation of Double-Stranded cDNA

1. 10 μg of parental (or related) mRNA (twice selected on oligo[dT] cellulose).
2. 5X Hybridization buffer, 1% SDS, mineral oil.
3. Reagents for HAP chromatography (as in Section 2.3.).
4. 4X Second-strand buffer: 80 m$M$ Tris-HCl, pH 7.4, 20 m$M$ MgCl$_2$, 40 m$M$ ammonium sulfate, 400 m$M$ KCl, nuclease-free BSA at 0.2 μg/μL.
5. dNTP mixture, each at concentration of 2.5 m$M$.

6. [α-$^{32}$P]dCTP (3000 C$_i$/mmol).
7. RNase H, 2 U/μL.
8. DNA polymerase I, 5 U/μL.
9. 0.5$M$ EDTA, pH 8.0.
10. Phenol.
11. CHCl$_3$.
12. TE buffer: 10 m$M$ Tris-HCl, pH 7.4, 0.1 m$M$ EDTA.
13. Sephadex G-50 column or spin column equilibrated with TE.
14. 20 μg of DNA-free RNA carrier, e.g., poly(A)-RNA.
15. 3$M$ Sodium acetate, pH 6.0.
16. Ethanol.
17. Siliconized 1.5-mL microfuge tubes.

## 2.5. PCR Amplification of Subtractive cDNA

1. Thermal cycler, e.g., Cetus/Perkin-Elmer (Norwalk, CT).
2. 10X T4 polymerase buffer: 700 m$M$ Tris-HCl, pH 7.7, 100 m$M$ MgCl$_2$, 50 m$M$ DTT.
3. dNTPs at 5 m$M$ and 10 m$M$ each.
4. 10 m$M$ ATP.
5. T4 DNA polymerase, 5 U/μL.
6. T4 polynucleotide kinase, 5 U/μL.
7. 0.5$M$ EDTA, pH 8.0.
8. Phenol.
9. CHCl$_3$.
10. 40 μg of DNA-free carrier RNA, e.g., poly(A)-RNA.
11. 3$M$ Sodium acetate, pH 6.0.
12. Ethanol.
13. Amplification adaptor at 500 ng/μL, e.g., mix 6 μg of 5'-AGCTAGAATTCGGT-ACCGTCGACC-3' with 5 μg of 5' phosphorylated-GGTCGACGGTACCGAA-TTCT-3'; dilute to 20 μL total volume with 2 μL 10X T4 polymerase buffer and H$_2$O, heat to 95°C for 5 min, and allow to cool gradually to room temperature. Stock solution can be stored at –20°C.
14. 5X T4 ligase buffer (e.g., BRL, Gaithersburg, MD).
15. T4 ligase (e.g., BRL).
16. TE buffer: 10 m$M$ Tris-HCl, pH 7.4, 1 m$M$ EDTA.
17. Sepharose CL-4B.
18. Sepharose 4B buffer: 0.1$M$ NaCl, 20 m$M$ Tris-HCl, pH 7.4, 1 m$M$ EDTA.
19. 10X PCR amplification buffer: 100 m$M$ Tris-HCl, pH 8.3, 500 m$M$ KCl, 15 m$M$ MgCl$_2$, 0.01% gelatin.
20. Amplification primer, e.g. 5'- AGCTAGAATTCGGTACCGTCGACC-3', 1 μg/μL.
21. *Taq* polymerase, 5 U/μL.
22. Mineral oil.
23. 1.5% Agarose gel and buffer (*see* Chapter 1).
24. φX174 *Hae*III markers (e.g., BRL).
25. Acrylamide gel components (*see* Chapter 1).

## 3. Methods
### 3.1. First-Strand cDNA Synthesis

The choice of subtractive partners is the single most important consideration in using this approach. Subtractive cDNA methods work optimally for qualitative differences in mRNA expression, but are less certain of achieving enrichment of cDNA sequences corresponding to mRNAs that are expressed at a quantitatively higher level (this may be possible by using only a minimal excess of the subtractive partner mRNA for the subtractive hybridization). In principle, the highest degree of enrichment of unique parental cDNAs occurs when the subtractive partner differs only minimally from the parent. However, this highly enriched pool of subtractive cDNAs is likely to include a significant fraction of cDNAs representing very rare (<1 copy mRNA/cell) and clone-specific differences. To minimize the presence of these two categories, positive selection can be done to a relatively low Rot (*see* Notes 6 and 7) using mRNA from a cell line that shares critical properties with the parental cell.

1. Place 5 µg of parental mRNA in 16 µL of $H_2O$ in a 1.5-mL microfuge tube (*see* Notes 1 and 2).
2. Incubate at 70°C for 3 min and then place on ice.
3. Add in the following order: 10 µL of 5X RT buffer, 2 µL of adaptor-$dT_{17}$ or random hexamer primer (*see* Note 8), 5 µL of solution containing each dNTP at 10 m$M$, 1 µL of RNasin, 30 U/µL, 1 µL of nuclease-free BSA, 1 µg/µL, 10 µL (100 µCi) of [$\alpha$-$^{32}$P]dCTP (3000 $C_i$/mmol), and 5 µL of MMLV reverse transcriptase, 200 U/µL.
4. Microfuge for 5 s. Mix gently and incubate at 37°C for 60 min.
5. Add 3 µL of 0.5$M$ EDTA and 2 µL of 10$N$ NaOH. Mix.
6. Incubate at 70°C for 30 min. Cool to room temperature.
7. Add 35 µL of 2$M$ Tris-HCl, pH 7.4, and mix. Add 10 µL of 2$N$ HCl and mix.
8. Extract with 100 µL of phenol:CHCl$_3$, and recover aqueous phase. Remove duplicate 1-µL aliquots and dilute each to 20 µL. Use a 2-µL aliquot from each dilution to determine total cpm. The remaining 36 µL can be analyzed on a denaturing acrylamide gel or an alkaline agarose gel to determine the size distribution of the first-strand cDNAs.
9. Remove labeled cDNA from unincorporated dNTPs by Sephadex G-50 chromatography (or by using a Sephadex G-50 spin column). Remove an aliquot from the excluded fraction to determine incorporated cpm.
10. The amount of first-strand cDNA in micrograms = (incorporated cpm/total cpm) × 66; this should be 0.5–2 µg (i.e., 10–40% yield). The specific activity of the first-strand cDNA is about 3000 cpm/ng (i.e., 1000–2000 Cerenkov cpm/ng or 1 $^{32}$P/18,000 nucleotides). The material can be monitored by Cerenkov counts in all subsequent procedures.
11. Precipitate the excluded fractions by adding 0.1 vol of 3$M$ Na-acetate and 2.5 vol of ethanol.

12. After 30 min on ice, microfuge the sample for 10 min at 4°C.
13. Discard supernatant. Add 1 mL ethanol, microfuge for 2 min at 4°C, and remove supernatant. Dry pellet in a vacuum centrifuge.

### 3.2. Hybridization of mRNA to cDNA (see Notes 6, 9, and 10)

1. Add 5 μg of poly(A)- and 50 μg of poly(A) + RNA from the subtractive partner (i.e., a 10-fold excess of mRNA compared to the parental mRNA used for first-strand cDNA synthesis) to the cDNA. Transfer sample to a 500-μL microfuge tube and dry in a vacuum centrifuge.
2. Dissolve sample in 6 μL of $H_2O$. Then add 2 μL of 5X hybridization buffer plus 2 μL of 1% SDS.
3. Mix, microfuge for 5 s, and overlay with 100 μL of mineral oil.
4. Boil sample for 5 min, then incubate at 70°C to allow the reaction to achieve a Rot of 3000. At a concentration of 5 μg/μL of mRNA, this will require 20 h (*see* Note 6).

### 3.3. Separation of cDNA and cDNA:mRNA by Hydroxyapatite Chromatography

Single- and double-stranded polynucleotides can be conveniently separated from one another by chromatography on hydroxyapatite (HAP) columns *(1,16)*. Empirically it has been found that single-stranded and double-stranded nucleic acids are differentially bound to HAP in solutions containing low concentrations of phosphate. The best separations occur at 60°C, with single-stranded polynucleotides generally eluting quantitatively at approx 120 m$M$, pH 6.8, sodium PB, and double-stranded polynucleotides eluting quantitatively at 300–400 m$M$ PB. Since the elution conditions may differ somewhat for different batches of HAP, it is recommended that the optimum PB concentrations for eluting single- and double-stranded polynucleotides be determined for each lot of HAP (*see* the following section).

### 3.3.1. Preparation of HAP

1. Mix 10 g of hydroxyapatite powder (Bio-Rad HTP) with 50 mL of 50 m$M$ PB, 150 m$M$ NaCl. Allow to settle for 10 min and remove fines.
2. Add 50 mL of 50 m$M$ PB, 150 m$M$ NaCl. Mix and heat 15 min in boiling water.
3. Mix and allow to settle for 10 min and remove fines. Note packed volume, which should be about 25 mL. Add 2 vol (approx 50 mL) of 50 m$M$ PB, 150 m$M$ NaCl to give a 3:1 slurry, which can be stored for several months at 4°C. The capacity is about 200 μg of double-stranded polynucleotide per milliliter of packed HAP.

### 3.3.2. HAP Fractionation of Single-Stranded cDNA from RNA-cDNA Duplexes (see Notes 4 and 5)

1. At the end of the hybridization, remove most of the mineral oil and dilute the aqueous phase with 250 μL of the 60°C HAP starting buffer. Transfer the sample to a 10-mL polypropylene tube, which contains about 0.5 mL of packed HAP and

is maintained at 60°C. Rinse the hybridization tube with another 250 μL of the HAP starting buffer (the hybridization tube can be Cerenkov-counted to ensure quantitative transfer of the cDNA).

2. Mix the sample/HAP slurry and allow it to settle for 5 min. Remix the slurry and allow it to settle for 5 min.
3. Mix the material a third time, pipet it into a column, and allow it to settle for 5 min.
4. Collect the flow-through in a microfuge tube.
5. Collect six 1-mL washes with the HAP starting buffer in six microfuge tubes.
6. Elute the column with six 1-mL aliquots of 450 m$M$ PB, 150 m$M$ NaCl into six microfuge tubes.
7. Determine which fractions contain single-stranded (flow-through and wash fractions) and double-stranded (450 m$M$ PB) cDNA from Cerenkov counts. It is also possible to dissolve the HAP with 1 mL of 6$N$ HCl to determine the portion of material that is not eluted; this is usually <5% of the total material applied to the column.
8. Pool the appropriate single-stranded fractions (e.g., following a subtractive hybridization step) and reapply to a second HAP column. It is not necessary to rechromatograph the double-stranded material following the positive selection hybridization step.
9. After addition of 10 μg of carrier RNA, quantitatively desalt and concentrate the appropriate single- and/or double-stranded fraction(s) by Nensorb 20 chromatography (according to the manufacturer's instructions), using 50% ethanol as an eluant. The sample is then dried by vacuum centrifugation (*see* Notes 11 and 12).

### 3.4. Positive Selection and Generation of Double-Stranded cDNA

The subtracted cDNA is positively selected by hybridization to a twofold excess of the parental mRNA or to mRNA derived from a cell that shares desired properties with the parental cell. Even if the positive selection is performed with the parental mRNA, it accomplishes several goals: (1) It selects against cDNA that has the same sense as mRNA (we find this to be a significant problem for random-primed first-strand cDNA but not for oligo(dT) primed cDNA); (2) it may provide some selection against contaminant DNAs; and (3) it provides a cDNA:mRNA duplex that serves as a good template for second-strand synthesis.

### 3.4.1. Positive Selection of First-Strand cDNA

1. Add 10 μg of poly(A)$^+$ RNA from the parental line to the tube containing the subtracted cDNA. Dry the sample by vacuum centrifugation and then dissolve it in 3 μL of H$_2$O.
2. Add 1 μL of 5X hybridization buffer and 1 μL of SDS.
3. Mix, microfuge for 5 s, and overlay with 100 μL of mineral oil.
4. Place tube in boiling water bath for 5 min and incubate at 70°C to a Rot of 500 (*see* Note 7).

5. Isolate mRNA-cDNA duplex by HAP chromatography as described in Section 3.3. Add 10 µg of carrier RNA.
6. A rough estimate of the maximum enrichment of sequences in the subtractive cDNA = (100/% ss cDNA from first subtractive HAP column) × (% ds cDNA from positive selection HAP column).
7. Desalt and concentrate sample by Nensorb chromatography.

### 3.4.2. Synthesis of Double-Stranded cDNA

1. Determine Cerenkov cpm of mRNA-cDNA duplex resuspended in 12 µL of $H_2O$.
2. Add the following: 10 µL of 4X second strand buffer, 5 µL of mixture containing each dNTP at 2.5 m$M$, 10 µL (100 µ$C_i$) of [α-$^{32}$P]dCTP (3000 $C_i$/mmol), 1 µL RNase H (2 U/µL), and 2 µL DNA polymerase I (5U/µL).
3. Microfuge for 5 s, mix gently, and incubate at 12°C for 2 h and then at 22°C for 1 h.
4. Add 1 µL of 0.5$M$ EDTA and 60 µL of $H_2O$. Mix and extract with 100 µL of phenol:CHCl$_3$ (1:1).
5. Remove 1 µL × 2 for total cpm. Separate the unincorporated nucleotides by G-50 Sephadex chromatography or a spin column. After addition of 10 µg of poly(A)-RNA as carrier and 0.1 vol of 3$M$ sodium acetate, the excluded fractions are precipitated with ethanol, washed with 70% ethanol, and dried by vacuum centrifugation.
6. Resuspend the sample in 50 µL of TE buffer. The amount (ng) of second-strand synthesis = (total Cerenkov cpm in excluded fraction − Cerenkov cpm in mRNA-cDNA duplex) × 16,500/total Cerenkov cpm in labeling reaction. The yield of second-strand should be comparable to the amount of first-strand cDNA present in the second-strand reaction.
7. The specific activity of second strand cDNA is approx 4000–8000 Cerenkov cpm/ng or about 1 $^{32}$P/4000 nucleotides.

### 3.5. PCR Amplification of Subtractive cDNA

The double-stranded subtractive cDNA is treated to ensure blunt ends that contain 5' phosphates. A large excess of amplification adaptors (containing *Sal*I, *Kpn*I, and *Eco*RI restriction sites) are then ligated to the subtractive cDNA. After removal of the excess amplification adaptors and fragments of cDNA less than several hundred base pairs by Sepharose 4B chromatography, the subtractive cDNA is amplified by PCR. It appears that extensive PCR amplification of a heterogeneous population of subtractive cDNAs results in some terminal "sterile" cycles in which denaturation is not followed by quantitative annealing of the amplification oligonucleotide and subsequent polymerization. As a consequence, much of the amplified cDNA is at least partially single-stranded, perhaps owing to hairpin structures and heteroduplexes resulting from interactions of the terminal amplification sequences. A PCR chase reaction has been designed to permit denaturation of heteroduplexes and hairpin structures under conditions that will not denature homoduplexes, so that the partially single-stranded structures can be converted to homoduplexes. The PCR-chased subtractive cDNA is then size-fractioned on Sepharose 4B, resulting in a PCR-renewable source of subtractive cDNA that can be used to generate a probe or an insert for cloning in an appropriate vector.

### 3.5.1. Polishing and Kinasing the Ends of the Subtractive cDNA

1. To 50 μL of double-stranded cDNA, add the following: 7 μL of 10X T4 polymerase buffer, 5 μL mixture of all four dNTPs at 5 m$M$ each, 5 μL of 10 m$M$ ATP, 2 μL of T4 DNA polymerase, and 2 μL of T4 polynucleotide kinase.
2. Microfuge for 5 s, mix gently, and incubate at 37°C for 30 min.
3. Add 2 μL of 0.5$M$ EDTA. Extract with 70 μL phenol:CHCl$_3$ (1:1) and 70 μL CHCl$_3$.
4. Add 10 μg of poly(A)-RNA, 0.1 vol of 3$M$ Na-acetate, and 2.5 vol of ethanol. Allow the sample to precipitate for 15 min on ice. Then collect precipitate by centrifugation at maximum speed in a microcentrifuge for 10 min, wash with ethanol, and dry by vacuum centrifugation. Resuspend sample in 5 μL of H$_2$O.

### 3.5.2. Amplification Adaptor Ligation to Double-Stranded cDNA

1. To 5 μL of double-stranded cDNA, add: 1 μL of amplification adaptor at 500 ng/μL (*see* Note 13), 1 μL of H$_2$O, 2 μL of 5X T4 ligase buffer, and 1 μL of T4 ligase.
2. Incubate at 14°C overnight.
3. Add 90 μL of TE buffer, mix, and extract with 100 μL of phenol-CHCl$_3$ (1:1).
4. Chromatograph sample (*see* Note 14) on 4-mL column (i.e., 5-mL pipet) of Sepharose CL-4B, eluting with 200-μL aliquots. The eluted fractions can be analyzed by Cerenkov counting. The appropriate samples are pooled and precipitated by addition of 10 μg of poly(A)-RNA and 2.5 vol of ethanol. Resuspend in 30 μL of TE.

### 3.5.3. PCR Amplification of Subtractive cDNA

1. Transfer 5 μL of subtractive cDNA into the tube to be used for PCR amplification.
2. Add the following: 10 μL of 10X PCR amplification buffer, 2 μL of a mix containing each dNTP at 10 m$M$, 1 μL of the amplification oligonucleotide (1 μg/μL), 81 μL of H$_2$O, and 1 μL of *Taq* polymerase.
3. Mix gently and overlay with 100 μL of mineral oil.
4. Amplify by PCR using the following cycle profile: 30 main cycles: 94°C, 1 min (denaturation), 50°C, 1 min (annealing), and 72°C, 1 min (extension). Remove 20-μL aliquots after 20, 25, and 30 cycles of amplification.
5. Analyze the three aliquots on a 1.5% agarose gel (*see* Chapter 1). Run φX174 *Hae*III size markers and 0.2 μg of the amplification oligonucleotide in separate lanes as controls. After the gel is stained with ethidium bromide, a heterogeneous smear of material with sizes up to 500 bp or more should be visible in one or more lanes. The presence of cDNA fragments with a unique size suggests the possibility of contaminants (*see* Note 15c).

## 3.5.4. PCR Chase of Amplified Subtractive cDNA

1. To 30 µL of amplified subtractive cDNA, add: 27 µL of 10X PCR buffer, 6 µL of a mix of dNTPs at 10 m$M$ each, 1 µL of amplification oligonucleotide (1 µg/µL), 235 µL of H$_2$O, and 1 µL of *Taq* polymerase.
2. Mix gently and amplify by PCR using the following cycle profiles: 2 cycles: 94°C, 1 min (denaturation), 50°C, 1 min (annealing), and 72°C, 2 min (extension); and 4 chase cycles: 50°C, 1 min (annealing), 72°C, 2 min (extension).
3. Extract the chased material with 300 µL of phenol:CHCl$_3$. After addition of 10 µg of carrier RNA and 30 µL of 3$M$ Na-acetate, precipitate the aqueous phase with 750 µL of ethanol. Collect the precipitate by centrifugation, wash with 70% ethanol, and dry by vacuum centrifugation. Dissolve the pellet in 50 µL of TE buffer.

## 3.5.5. Size Fractionation of PCR-Chased Subtractive cDNA (see Note 16)

1. Size-fractionate the PCR-chased cDNA by chromatography on a 4-mL Sepharose CL-4B column in a 5-mL pipet, collecting 200-µL fractions. Starting with fraction 6, subject 50-µL aliquots from each fraction to electrophoresis on a native 8% polyacrylamide gel (*see* Chapter 1), using φX174 *Hae*III fragments as size markers. Fractions containing essentially no material smaller than 350 bp are pooled, precipitated with ethanol after addition of 10 µg of carrier RNA, dried, and resuspended in 50 µL of TE buffer.
2. Estimate the approximate concentration of the size-fractionated, amplified subtractive cDNA by electrophoresis of a 5-µL aliquot on a native 8% acrylamide gel by comparing the ethidium bromide staining of the heterogeneous subtractive cDNA with φX174 *Hae*III standards run on an adjacent lane. This stock of double-stranded subtractive cDNA can be used to prepare a radiolabeled probe or to prepare an insert suitable for cloning. A portion of it can also be PCR amplified and chased to renew the stock as desired. *See* Notes 15 and 17–20.

## 4. Notes

1. Source of RNA. It is essential that trace contamination of the RNA (or any reagent used in the protocol for that matter) with DNA from any source be avoided since the contaminants can be amplified together with the very small quantities of cDNA that remain after subtraction, positive selection, and so forth. If possible, it is best to use cytoplasmic RNA for first-strand cDNA synthesis since this not only minimizes DNA contamination but also eliminates rare nuclear RNA species that may be enriched by subtraction. DNA can also be removed by one or more of the following procedures: precipitation of single-stranded polynucleotides (mainly RNA) with 2$M$ LiCl, centrifugation of RNA through 5.7$M$ CsCl, or treatment with DNase. Isolation of poly(A) + RNA by selection on oligo(dT) cellulose also removes trace DNA contaminants. We prefer to use two cycles of oligo(dT)-cellulose selection to isolate mRNA for first-strand cDNA synthesis, since one cycle of selection generates RNA that is substantially (approx 50%) contaminated with poly(A)- RNA, i.e., mostly rRNA. If RNA from only one cycle

of selection is used for hybridization, a correction for the actual amount of mRNA should be made in estimating Rot values (*see* Note 6).

2. Amount of RNA. The amount of starting material can be decreased at least several fold although a minimum hybridization volume of about 4 μL and adequate concentrations of RNA required to reach a high Rot are limiting. Also, contaminant DNA is a more significant possibility as the amounts of RNA are decreased.

3. DNA contamination. This is a major problem to be avoided. Major sources include carrier RNA as well as the poly(A)$^+$ RNAs used for hybridization.

4. For a new lot of HAP, test binding/elution properties of single- and double-stranded DNA labeled to high specific activity. This can be done by applying each sample in 50 m*M* PB, 150 m*M* NaCl and eluting with duplicate 1-mL aliquots of 80, 100, 120, 140, 160, 180, 200, and 450 m*M* PB, 150 m*M* NaCl. The lowest concentration of PB that quantitatively elutes single-stranded DNA and little or no double-stranded DNA is then used to equilibrate the HAP slurry, to dilute samples for application to the column, and for washes to recover single-stranded cDNA. Some 10–15% of first-strand cDNAs (even if actinomycin D is present during synthesis) elute with double-stranded DNA; the single-stranded fraction elutes quantitatively in the single-stranded fraction on a second HAP column.

5. The columns used for HAP chromatography must be maintained at 60°C. Although a jacketed column can be used, we prefer a simple homemade system that uses disposable 3-mL plastic syringes as columns, essentially as described elsewhere *(16)*.

6. The kinetics of hybridization of cDNA to an excess of mRNA *(13–15)* is described by the pseudo-first-order rate equation:

$$C_t/C_o = e^{-k \times \text{Rot}} \tag{1}$$

where $C_o$ = the initial concentration of single-stranded cDNA, $C_t$ = the amount of single-stranded cDNA remaining at time $t$, $k$ = the rate constant in units L/(mol nucleotide)/s, and Rot = the concentration of mRNA in mol of nucleotide/L multiplied by the time *(t)* in seconds. The rate constant, $k$, is determined from the following equation:

$$k = 10^6/C \tag{2}$$

where $C$ = the complexity of the mRNA in nucleotides. The constant $k$ is based on hybridization at standard conditions, i.e., $T_m - 25°C$, cation concentration = $0.18M$, and average size of single-stranded mRNA = 500 nucleotides. Since the complexity of mRNA in a typical mammalian cell line is approx $5 \times 10^8$ nucleotides (ca. 20 pg RNA or 0.3 pg mRNA/cell), $k = 2 \times 10^{-3}$. One can calculate that a cDNA derived from an mRNA present at 1 copy/cell is 50% double-stranded at a Rot of 345, 90% double-stranded at a Rot of 1150, 99% double-stranded at a Rot of 2300, and 99.9% double-stranded at a Rot of 3450. In contrast, cDNAs derived from mRNAs present at 10 and 0.1 copies/cell are 90% double-stranded at Rot of 115 and 11,500, respectively. Thus, during hybridization of cDNA to excess mRNA, a Rot of 3000 should be sufficient to ensure that all cDNAs derived from mRNAs present at least once per cell are present 99% or more as a double-stranded cDNA-mRNA duplex. For standard hybridization conditions, a 1 µg/µL concentration of mRNA gives a Rot of 11/h. For the hybridization conditions described below (i.e., cation concentration = 0.60$M$), mRNA at a concentration of 1 µg/µL gives a Rot (corrected to standard conditions) of 30/h *(15)*.

7. Rot for positive selection. By hybridizing only to a Rot of 500, 60% or more of sequences present ≥1 time/cell form a cDNA:mRNA duplex, whereas only 10% and 1% of sequences represented 0.1 or 0.01 time/cell, respectively, form a cDNA:mRNA duplex.

8. Choice of primer for cDNA synthesis. First-strand cDNA can be synthesized from the parental mRNA using either an adaptor-dT$_{17}$ primer (e.g. 5'-GGAC-TCGAGGTATCGATGCTTTTTTTTTTTTTTTTT[G,A,C]-3') or a random hexamer primer. Use of the adaptor-dT$_{17}$ primer results in the following advantages:

   a. First-strand cDNA is distinguished from contaminant DNAs (it should be noted that many subtractive cDNAs will lack adaptor sequences as a result of fragmentation during the procedure);

   b. An adaptor probe can be used to detect clones that contain adaptor sequences;

   c. cDNA fragments containing the adaptor-dT$_{17}$ sequences can be enriched relative to other sequences by oligo-dT (or oligo dA) chromatography or by a PCR procedure (*see* Note 18);

   d. Single-stranded cDNA with the same sense as mRNA can be isolated on oligo-dT cellulose);

   e. Detection of cDNA clones that contain adaptor sequences and are derived from the same mRNA (i.e., sibs) is efficient;

   f. Enriching for subtractive cDNA inserts that contain adaptor-dT sequences decreases its complexity, thus enhancing the sensitivity of this material as a probe to identify clones in the subtractive library.

   The random hexamer-primed cDNA has none of these advantages but does have three other potential advantages: (1) More mRNA sequences are represented so that inadvertent loss of representation of some mRNAs in the subtractive probe/library is minimized; (2) positive selection using mRNA from other species is more likely to be successful; (3) the subtractive cDNA library can be screened more reliably by conventional methods (e.g., antibodies), since clones in the library would not be biased toward the 3' ends of mRNA.

9. Correction of Rot to standard hybridization conditions. By definition, $k$ varies as a function of hybridization conditions, whereas Rot is independent of these conditions. For convenience, however, we use a fixed $k$ and correct Rot to standard conditions. Thus, for a given complexity of RNA, the fraction of cDNA in hybrid is calculated directly from the Rot corrected to standard hybridization conditions.

10. Hybridization conditions. To minimize fragmentation of cDNA, the hybridizations can be done at 52°C in 40% formamide, although the kinetics of hybridization are not as well worked out under these conditions *(9,13)*.

11. Desalting HAP fractions. The fractions can be concentrated by repeated n-butanol extractions, followed by G50 Sephadex chromatography to remove phosphate, and alcohol precipitation. We prefer to use Nensorb to optimize recoveries.

12. Alternative to HAP chromatography. Use of biotinylated driver mRNA or DNA has been used by others *(9,12)*. A recent report using oligo(dA) cellulose to remove single-stranded oligo(dT)-primed cDNA from cDNA:mRNA duplexes appears to provide another alternative *(10)*. Each alternative is alleged to cause less fragmentation of cDNA than HAP chromatography.

13. PCR adaptor. Other potential amplification adapters may work but should be tested with a model substrate (e.g., PBR322 cut with *Sau*3A) to verify that all steps, including the blunting and kinasing reactions, work.

14. An identical Sepharose 4B column can be calibrated with blue dextran (void volume), phenol red or labeled dNTP (included volume), and φX174 *Hae*III markers. The first third of the samples between the void volume and the included volume contain the larger cDNAs, with amplification adapters ligated at each end, but should be devoid of unligated adapters and adapters that have self-ligated.

15. Characterization of amplified subtractive cDNA. Ideally, one would like to know the extent of enrichment of subtractive sequences, the number of independently amplified subtractive cDNA sequences, and whether or not a substantial fraction of contaminant DNAs have been amplified together with the cDNA.

    a. Enrichment. A rough calculation of the maximum enrichment of subtractive sequences is given earlier. In the two cases tested (*11* and J. Battey, unpublished), this calculation was in agreement with a direct comparison of the frequency of clones in libraries prepared from subtractive cDNA and parental first-strand cDNA.

    b. Number of independently amplified subtractive cDNA sequences. Although it is difficult to accurately assess this parameter, a minimum estimate can be made based on the number of amplification cycles required to generate 1 μg of subtractive probe, i.e. 20, 25, and 30 PCR amplification cycles would correspond to amplification of a minimum of $1.8 \times 10^6$, $5.8 \times 10^4$, and 1800 independent sequences having an average size of 500 bp.

    c. Contaminant DNA. If the first-strand cDNA was synthesized with an adaptor-$dT_{17}$ primer and had an average size of 2000 bp at the time of synthesis and 400 bp after amplification, approx 20% of the amplified molecules of cDNA should contain an adaptor sequence. This can be confirmed by cloning the insert into a convenient vector and probing with an adaptor probe. If only a small fraction of clones with inserts contain adaptor sequences, it is likely that contaminant DNAs have been coamplified with the subtractive cDNA. It is possible to enrich the amplified subtractive cDNA for molecules containing adaptor sequences (*see* Note 18).

16. Size of subtractive cDNA. Subtractive cDNA prepared by the aforementioned classical protocol results in a substantial reduction in the size of first-strand cDNA, independent of any effect of PCR amplification. The reasons for this are poorly understood although thermal degradation during hybridization and some degradation during HAP chromatography may contribute to this result. The size selection of cDNA prior to amplification minimizes PCR selection of very short sequences, and the size selection after amplification further ensures that very short sequences are eliminated.

17. Yields of subtractive cDNA. If there is a theoretical 50-fold enrichment of subtractive cDNAs (Section 3.4.1.), the actual yield of positively selected first-strand cDNA in cDNA:mRNA duplex is several fold lower (e.g., 0.5% instead of 2.0%), because of sequential small losses from adsorption of cDNA to tubes, incomplete precipitation, incomplete recovery of material from HAP and Nensorb chromatography, and so forth. Although the total amount of cDNA should increase during the second-strand cDNA synthesis step, there may be an additional 90% loss of cDNA (i.e., fragments smaller than approx 200 bp) during the Sepharose 4B chromatography step following the ligation of amplification adapters. Thus, there is likely to be as little as 1–5 ng of cDNA available for the PCR reaction.

18. Enrichment of subtractive cDNAs containing adaptor-dT sequences. In addition to oligo(dA)-cellulose selection of subtractive cDNAs containing oligo(dT) sequences, it is possible to use PCR to enrich for subtractive cDNAs containing adapter sequences (cf ref. *17*): Add dG to the 3' ends of the amplified subtractive cDNA using terminal deoxynucleotidyl transferase. The dG-tailed subtractive cDNA can then be amplified with three oligonucleotides: (1) the adapter portion of the adaptor dT primer (the adaptor-dT$_{17}$ primer *per se* does not seem to work as well for this PCR reaction); (2) a second adapter sequence, including a stretch of 12 dC residues at its 3' end, in limiting amounts; and (3) the second adaptor sequence (without the 12 dC residues) at a higher level.

19. Screening libraries by hybridization with the subtractive cDNA probe. As noted in Section 1., a subtractive cDNA probe that is enriched 50-fold can detect clones containing sequences representing approx 0.002% of the mRNA sequences in a cell. Clones containing large inserts are more effectively detected than clones containing smaller inserts. Since subtractive cDNA libraries generally contain smaller inserts (e.g., 400 bp average) than conventional cDNA libraries (e.g., >2 kb), a subtractive probe will detect rare mRNAs more efficiently in a conventional cDNA library than in a subtractive cDNA library.

20. PCR-renewable cDNA to replace mRNA for hybridization. One method uses PCR-renewable cDNA for both the parental cDNA and the biotinylated subtractive partner cDNA *(18)*. This approach, which should decrease the need for large amounts of mRNA, is conceptually similar to the method described in this chapter. In principle, it could be used to further select the subtractive cDNA prepared by the method described in this chapter.

## References

1. Alt, F. W., Kellems, R. E., Bertino, J. R., and Schimke, R. T. (1978) Selective multiplication of dihydrofolate reductase genes in methotrexate-resistant variant of cultured murine cells. *J. Biol. Chem.* **163,** 1357–1370.
2. Timberlake, W. E. (1980) Developmental gene regulation in Aspergillus nidulans. *Dev. Biol.* **78,** 497–510.
3. Sargent, T. D. and Dawid, I. B. (1983) Differential gene expression in the gastrula of Xenopus laevis. *Science* **222,** 135–139.

4. Hedrick, S. M., Cohen, D. I., Nielsen, E. A., and Davis, M. M. (1984) Isolation of cDNA clones encoding T cell-specific membrane-associated proteins. *Nature* **308,** 149–153.
5. Klickstein, L. B. (1987) Production of a subtractive cDNA library, in *Current Protocols in Molecular Biology,* vol. 1 (Ausubel, F. M., Brent, R., Kingston, R. E., Moore, D. E., Seidman, J. G., Smith, J. A., and Struhl, K., eds.), Wiley, New York, pp. 5.8.6.–5.8.13.
6. Duguid, J. R., Rohwer, R. G., and Seed, B. (1988) Isolation of cDNAs of scrapie-modulated RNAs by subtractive hybridization of a cDNA library. *Proc. Natl. Acad. Sci. USA* **85,** 5738–5742.
7. Travis, G. H. and Sutcliffe, J. G. (1988) Phenol emulsion-enhanced DNA-driven subtractive cDNA cloning: isolation of low-abundance monkey cortex-specific mRNAs. *Proc. Natl. Acad. Sci. USA* **85,** 1696–1700.
8. Palazzolo, M. J. and Meyerowitz, E. M. (1987) A family of lambda phage cDNA cloning vectors, lambda SWAJ, allowing the amplification of RNA sequences. *Gene* **52,** 197–206.
9. Rubenstein, J. L. R., Brice, A. E. J., Ciaranello, R. D., Denney, D., Porteus, M. H., and Usdin, T. B. (1990) Subtractive hybridization system using single-stranded phagemids with directional inserts. *Nucleic Acids Res.* **18,** 4833–4842.
10. Batra, S. K., Metzgar, R. S., and Hollingsworth, M. A. (1991) A simple, effective method for the construction of subtracted cDNA libraries. *GATA* **8,** 129–133.
11. Timblin, C., Battey, J., and Kuehl, W. M. (1990) Application for PCR technology to subtractive cDNA cloning: identification of genes expressed specifically in murine playmacytoma cells. *Nucleic Acids Res.* **18,** 1587–1593.
12. Wieland, I., Bolger, G., Asouline, G., and Wigler, M. (1990) A method for difference cloning: Gene amplification following subtractive hybridization. *Proc. Natl. Acad. Sci. USA* **87,** 2720–2724.
13. Britten, R. J., Graham, D. E., and Neufeld, B. R. (1974) Analysis of repeating DNA sequences by reassociation. *Methods Enzymol.* **29,** 363–418.
14. Galau, G. A., Britten, R. J., and Davidson, E. H. (1977) Studies on nucleic acid reassociation kinetics: rate of hybridization of excess RNA with DNA, compared to the rate of DNA renaturation. *Proc. Natl. Acad. Sci. USA* **74,** 1020–1023.
15. Van Ness, J. and Hahn, W. E. (1982) Physical parameters affecting the rate and completion of RNA driven hybridization of DNA: new measurements relevant to quantitation based on kinetics. *Nucleic Acids Res.* **10,** 8061–8077.
16. Sambrook, J., Fritsch, E. F., and Maniatis, T. (1989) Separation of single-stranded and double-stranded DNA by hydroxyapatite chromatography, in *Molecular Cloning: A Laboratory Manual,* vol. 3. Cold Spring Harbor Laboratory, Cold Spring Harbor, NY, pp. E.30–E.33.
17. Frohman, M. A., Dush, M. K., and Martin, G. R. (1988) Rapid production of full-length cDNAs from rare transcripts: amplification using a single gene-specific oligonucleotide primer. *Proc. Natl. Acad. Sci. USA* **85,** 8998–9002.
18. Wang, Z. and Brown, D. D. (1991) A gene expression screen. *Proc. Natl. Acad. Sci. USA* **88,** 11,505–11,509.
19. Lisitsyn, N., Lisitsyn, N., and Wigler, M. (1993) Cloning the differences between two complex genomes. *Science* **259,** 946–951.

# 39

# The Use of PCR
# for Differential Screening of cDNA Libraries

**Mark G. Thomas, Sarah A. Hesse, Yvonne J. Foss,
and Farzin Farzaneh**

## 1. Introduction

Differential screening *(1)* is probably the most direct approach for the identification of new genes whose expression is associated with a change in physiological conditions. Traditionally, the approach involved the probing of duplicate plaque lifts of cDNA libraries with different radiolabeled cDNA populations. Following this, second- and often third-round screens of putative clones representing differentially expressed genes were necessary in order to be certain that the isolated clones were clonal and truly differentially expressed. In recent years, other approaches have been developed, including the production of subtracted cDNA libraries *(2–4)*, differential display *(5,6)*, and RNA fingerprinting *(7,8)*. Although these methods can greatly increase the speed with which new differentially expressed sequences are identified, they can introduce a bias in the type of sequences that are isolated. For this reason, many researchers prefer the laborious but reliable approach of cDNA library screening by plaque hybridization.

Although this method is sensitive and suffers very few "experimental weak links," it can be very costly and time-consuming for the following reasons:

1. Small changes in gene expression patterns or changes in the expression of low-transcript-number genes require the screening of a large number of clones (frequently in excess of $10^5$) in order to isolate the desired cDNAs.
2. cDNA libraries must be plated out at relatively low densities (2–6 plaques/cm$^2$) so that on the final autoradiographs individual plaques can be distinguished from one another.
3. Despite low plating densities, crosscontamination of plaques with phage from neighboring plaques is very common. This probably occurs because carrying out

From: *Methods in Molecular Biology, Vol. 67: PCR Cloning Protocols: From Molecular Cloning
to Genetic Engineering* Edited by: B. A. White  Humana Press Inc., Totowa, NJ

plaque lifts aids the dispersal of phage across the plate. In addition, there is usually a lag period of up to a week between carrying out the plaque lifts and the identification of putative differentially expressed clones, while the hybridizations and autoradiography are carried out. During this time, further crosscontamination of plaques on the stored plates can take place. For this reason, it is necessary to carry out a second-round screen using a separate agar plate for each clone.

4. False positives (or more correctly, false differences) can be common in the first-round screen. Since each first-round isolate requires at least a second-round screen involving a separate agar plate and two hybridization membranes, this can necessitate a large number of second-round screens.

Here we describe a simple but effective polymerase chain reaction (PCR)-based procedure *(9)* for second-round differential screening of phage *(10)*. This method allows a large number of phage isolates to be screened for the presence of differentially expressed sequences even if those isolates are contaminated with phage from neighboring plaques.

This procedure involves PCR-amplification of cDNA inserts using primers complementary to vector sequences flanking the cloning site *(11,12)*, followed by electrophoresis on duplicate agarose gels and Southern blotting *(13)*. The duplicate Southern blots can then be probed with cDNA synthesized from mRNA populations isolated under different conditions and the autoradiographs compared for differential expression of the amplified sequences (Fig. 1). Each cDNA clone present in a mixed phage isolate will give rise to an independent PCR product of discrete size that will be fractionated and resolved by the gel electrophoresis step. In addition, because mixed phage isolates are tolerated in this method, the initial screening can be carried out using higher than normal plaque densities.

Following the screening and identification of the putative clones it is desirable to obtain sequence information about the differentially expressed cDNAs. Although it is possible to satisfactorily sequence many cDNA fragments as a double-strand template by either manual or automated sequencing procedures, the isolation of single-strand templates can make the task much easier, resulting in better sequencing of longer fragments, including those the composition of which (e.g., high G/C content) makes their double-strand sequencing difficult. The amplification of the insert cDNA sequences using a biotinylated version of one of the flanking primers will allow the subsequent isolation of single-strand templates using streptavidin-coated magnetic beads *(14)*. A fraction of the biotinylated PCR product can be used for the screening purposes, although the remainder is kept for the later isolation of the single-strand templates. The ability to carry out rapid, direct sequencing circumvents the need to crosscheck different clones for homology by hybridization. In addition, solid-phase support sequencing allows the high stringency washing of the template-bound sequencing primer *(15)* in order to avoid the common problem of mispriming (Fig. 2).

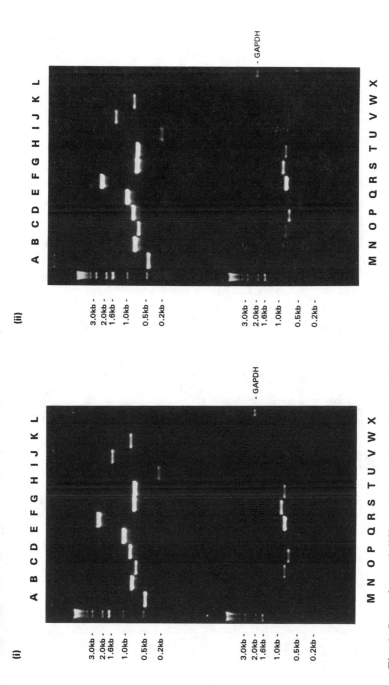

**Fig. 1.** Second-round differential screening of 24 putative differentially expressed cDNA clones. The cDNA inserts were amplified from selected clones using the λgt11 flanking primers and electrophoresed on duplicate 1% agarose gels (i and ii) as described in the text. (*continued*)

407

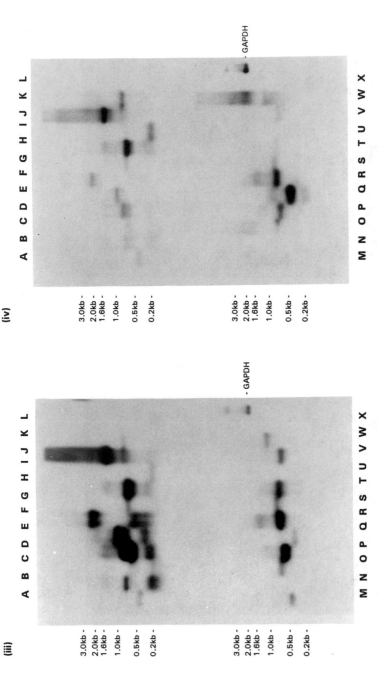

**Fig. 1.** (*continued*) Glyceraldehyde-3-phosphate dehydrogenase (GAPDH) cDNA was used as a control for Southern transfer and the blots were hybridized with [32]P-labeled cDNA probes according to the manufacturer's instructions for GeneScreen Plus. The cDNA probes were synthesized from poly (A)[+] RNA isolated from 12-wk-old fetal (iii) and adult heart (iv) The blots were washed in 0.2X SSC at 65°C, and autoradiographed for 16 h.

408

I  A  I  B  I  C  I  D  I  E  I

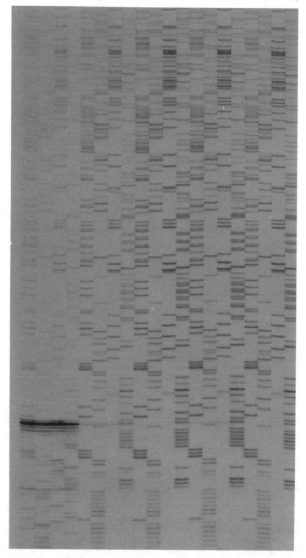

**Fig. 2.** Sequencing gel illustrating the effects of a sequencing primer stringency wash following primer annealing. The five sets of sequencing reactions each contain identical template DNA and sequencing primer combinations but differ in their treatment after sequencing primer annealing. In all cases, sequencing primer was annealed to Dynabead-bound template DNA as described in the text. Following this, sequencing reactions were carried out on sample 1 with no further processing. Samples B–E were washed as described in the text with the following variations: Samples B and C were washed once and twice, respectively, at 45°C, and samples D and E washed once and twice, respectively, at 55°C. Following washing, samples were sequenced as described in the text. Sequencing template was part of the mitochondrial d-loop region between positions 15,749 and 418 of the published sequence of Anderson et al. *(17)*.

One final problem associated with sequencing PCR-amplified cDNA inserts from cDNA libraries is the difficulty in reading through poly (A) sequences that are present at the 3' end of most cDNA sequences (*see* Fig. 3, lanes A, F, and G and Fig. 4A). This is probably because of random insertions and deletions caused by polymerase slippage occurring at high frequency within homopolymeric tracts. We also describe a method for overcoming this problem by the use of anchored oligo dT sequencing primers, which are suitable for both manual and automated sequencing of DNA *(16)*.

## 2. Materials

### 2.1. Second-Round Differential Screening

1. Phage isolates containing putative differentially expressed cDNAs.
2. Phage or SM buffer (100 m*M* NaCl, 8 m*M* MgSO₄, 5 m*M* Tris-HCl, pH 7.5, and 0.01% gelatin).
3. Oligonucleotide primers complimentary to vector-flanking sequences. The following is a list of primers that have been used successfully to amplify inserts from three popular cDNA cloning vectors.

   a. λgt10 primers:

   > 5' Biotin-CTTTTGAGCAAGTTCAGCCTGGTTAAG 3'
   > 5' GAGGTGGCTTATGAGTATTTCTTCCAGGGTA 3'

   b. λgt11 primers:

   > 5' Biotin-TCAACAGCAACTGATGGAAACCAG 3'
   > 5' TTGACACCAGACCAACTGGTAATG 3'

   c. λZap/Bluescript primers:

   > 5' ATGACCATGATTACGCCAAGCT 3'
   > 5' AACGACGGCCAGTGAATTGTA 3'

---

**Fig. 3.** *(opposite page)* Sequencing gel of samples amplified from cDNA clones selected from a λgt10 human fetal heart library. The cDNA inserts were amplified using the λgt10 flanking primers. Lanes A–E show sequences obtained from a λgt10 clone of the human ribosomal protein L18 cDNA using different strands of the cDNA and different primers. Lanes A–C are the biotinylated strand primed with 5 ng of unbiotinylated PCR primer (A), or either 5 ng (B) or 20 ng (C) of the 3' degenerate oligo dT primer mix [⁵'dT₍₁₈₎(A/G/C)(A/G/C/T)³']. Lanes D and E are the unbiotinylated (NaOH wash) strand of the same clone, both primed with the biotinylated PCR primer using dGTP (D) and dITP (E) sequencing reactions. Lanes F and G show another example of this phenonoma using the unbiotinylated strand of insert cDNA from an uncharacterized λgt10 clone. Both were primed with the biotinylated PCR primer using dGTP (F) and dITP (G) sequencing reactions.

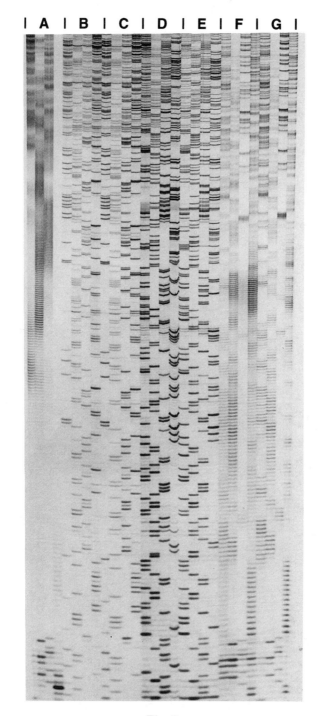

Fig. 3.

411

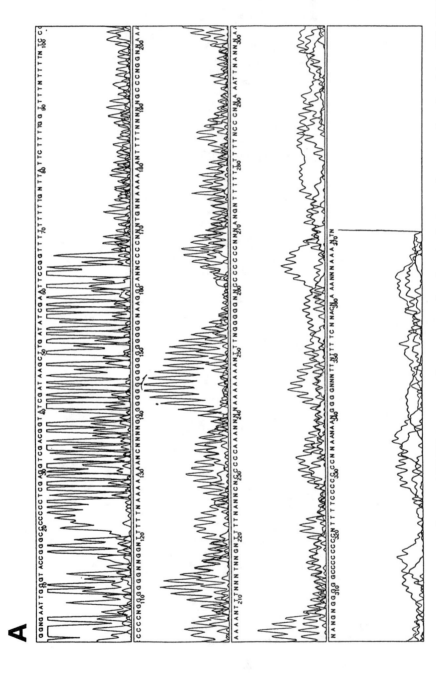

**Fig. 4.** Sequence output from an ABI 373A automated sequencer using dye terminator cycle sequencing chemistry (18). DNA template was obtained by PCR-amplification of the insert fragment in an uncharacterized clone from a λZAP11 mouse cDNA library using flanking primers. **(A)** Sequence obtained when one of the flanking primers was used as a sequencing primer. (continued)

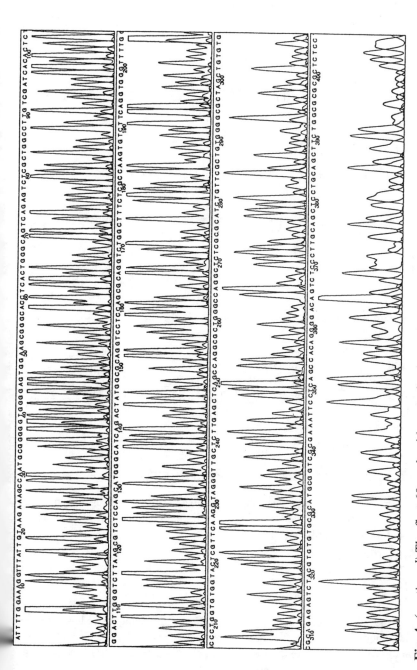

**Fig. 4.** (*continued*) The first 60 nucleotides are clearly resolved with strong intensity peaks but deletions and/or insertions in the homopolymeric region have resulted in a multiple frameshift completely obscuring the correct sequence beyond this region. (**B**) was obtained using the same template as for A but employing the degenerate oligo dT primer mix [5'dT$_{(25)}$(A/G/C) (G/A/C/T)$^{3'}$] as a sequencing primer.

413

4. A 10X PCR reaction buffer (400 m$M$ KCl, 100 m$M$ Tris-HCl, pH 8.8, and 1% Triton X-100) designed to be compatible with the high magnesium concentration of phage (SM) buffer.
5. Reagents for agarose gel electrophoresis.
6. Gel-loading buffer: 30% glycerol, 0.05% bromophenol blue, 0.05% xylene cyanol blue.
7. Hybridization membrane and reagents for Southern transfer, hybridization, and washing.

## 2.2. Sequencing cDNA

### 2.2.1. Direct Sequencing of cDNA Inserts Using Dynabeads and Employing a Sequencing Primer Stringency Wash

1. Dynabeads and a magnetic particle concentrator (MPC), Dynal (Lake Success, NY).
2. TWS 1 solution: 0.17% Triton X-100, 100 m$M$ NaCl, 10 m$M$ Tris-HCl, pH 7.5, 1 m$M$ EDTA.
3. TWS 2 solution: 0.17% Triton X-100, 1.5$M$ NaCl, 10 m$M$ Tris-HCl, pH 7.5, 1 m$M$ EDTA.
4. 8$M$ ammonium acetate and 3$M$ sodium acetate, pH 5.2.
5. Isopropanol.
6  70 and 100% ethanol.
7. 1$M$ NaOH, 4 m$M$ EDTA (fresh).
8. TE buffer: 10 m$M$ Tris-HCl, pH 8.0, 1 m$M$ EDTA.
9. 1X Sequenase® reaction buffer: 40 m$M$ Tris-HCl, pH 7.5, 50 m$M$ NaCl, 20 m$M$ MgCl$_2$.
10. Sequencing primer (5 ng/μL).
11. Reagents for sequencing.

### 2.2.2. Sequencing Using Anchored Oligo dT Primers

1. A 3' degenerate oligo dT primer mix [$^{5'}$dT$_{(18)}$(A/G/C)(A/G/C/T)$^{3'}$]. This mix can be made by synthesizing the following four oligonucleotides, which should then be mixed in an equimolar ratio:

    $^{5'}$dT$_{(18)}$(G/A/C)A$^{3'}$
    $^{5'}$dT$_{(18)}$(G/A/C)G$^{3'}$
    $^{5'}$dT$_{(18)}$(G/A/C)C$^{3'}$
    $^{5'}$dT$_{(18)}$(G/A/C)T$^{3'}$

2. For cycle sequencing used in the automated DNA sequencing procedure, primers with a higher melting temperature work more efficiently. This is because of the higher temperature in the extension phase of cycle sequencing (60 as opposed to 37°C). Therefore, the recommended primer is [$^{5'}$dT$_{(25)}$(A/G/C) (G/A/C/T)$^{3'}$], which can be made as described above.

## 3. Methods

### 3.1. Second-Round Differential Screening

1. Following first-round screening, core out plaques of interest using a sterile Pasteur pipet and place in 1 mL of sterile SM buffer. Add a drop of chloroform and store at 4°C.
2. For each phage isolate, assemble a 50-μL PCR reaction containing 200 μ*M* deoxynucleotides, 5 μL of the 10X PCR reaction buffer (*see* Section 2.), 0.15 μg of each vector-flanking primer, 10 μL of phage suspension in SM buffer, and 1 U of *Taq* DNA polymerase.
3. Amplify cDNA inserts using following cycle parameters: 35 main cycles: 95°C, for 3 min (initial denaturation); 93°C, for 1 min (denaturation), $T_a$°C, for 1 min (annealing), and 72°C, for 2 min (extension); and 72°C, for 10 min (final extension). The annealing temperature ($T_a$) depends on the primer pair used. For the primers given above, the following annealing temperatures are recommended

   λgt10 primers     $T_a$ = 59°C
   λgt11 primers     $T_a$ = 56°C
   λZap primers     $T_a$ = 57°C
4. Pour duplicate 1% agarose/TBE gels containing 5 μg/mL of ethidium bromide (it is important to ensure that both gels are similar in terms of thickness so that the Southern transfer of DNA is equal for both gels).
5. Extract each PCR reaction with 200 μL of chloroform in order to remove the mineral oil overlay (this is important; otherwise, the sample will float up to the top of the agarose gel when loaded and result in uneven loading of samples between gels). Mix 15 μL of each PCR sample with 5 μL of gel-loading buffer.
6. Load each gel with 7 μL of each sample, taking care not to lose any sample outside the wells. Constitutively expressed sequences may also be loaded on the gels as Southern transfer and hybridization controls. These may be in the form of whole plasmids, insert fragments, or PCR products of known constitutively expressed gene sequences. Run both gels in parallel as normal, then photograph both gels to ensure that they are similar (Fig. 1, i and ii).
7. Southern blot both gels using the procedure recommended by the hybridization membrane manufacturer (it should be possible to lay both gels side-by-side and place a single piece of hybridization membrane across both, which can be cut after Southern blotting).
8. Prehybridize duplicate Southern blots in separate containers, then add the radiolabeled cDNA probes (*see* Note 1) and hybridize for the recommended time (at least 16 h).
9. Wash blots in separate containers (*see* Note 2), expose to X-ray film, and compare autoradiographs (Fig. 1, iii and iv).

### 3.2. Sequencing cDNA

1. Precipitate the remainder of the PCR products by the addition of 1/3 vol of 8*M* ammonium acetate and an equal volume of isopropanol. Incubate at room temperature for 10 min, then spin in a microfuge for 12 min (*see* Note 3). Remove

supernatant and wash DNA pellets with 500 µL of 70% ethanol. Allow pellets to dry briefly.

2. Add 15 µL of TE buffer to DNA pellets and redissolve DNA by heating to 65°C, for 5 min.

3. Wash Dynabeads (10 µL for each sample) three times with TWS 1 solution and resuspend in TWS 2 solution (35 µL for each sample).

4. Add 35 µL of washed Dynabeads to each DNA sample and incubate with vigorous shaking for 30 min.

5. Separate Dynabeads using MPC and remove liquid (this may be saved because it contains unbound DNA).

6. Wash Dynabeads once with TWS 1 solution.

7. Add 32 µL of TE and 8 µL of 1*M* NaOH, 4 m*M* EDTA to Dynabead/DNA samples, vortex, and leave for 5 min.

8. Separate Dynabeads using MPC and remove the liquid. This contains the unbiotinylated strand of the sample DNA, which can be saved for the subsequent sequencing of the opposite strand.

9. Repeat steps 7 and 8. Pool the two NaOH washes and precipitate the unbiotinylated DNA by the addition of 10 µL of 3*M* sodium acetate, pH 5.2, and 2.2 vol of 100% ethanol (this should be stored at –20°C, until required, *see* Note 4).

10. Wash the Dynabead/DNA complex once with TWS 1 solution and once with TE.

11. To Dynabead/DNA complex add 2 µL of 5X Sequenase reaction buffer and 10 ng of sequencing primer (for choice of sequencing primer *see* Section 3.3.). Make up to a total volume of 10 µL.

12. Heat samples to 65°C, and allow to cool to 30°C, over a period of 15–30 min. While the samples are cooling, place the 1X Sequenase reaction buffer in a 50°C waterbath.

13. Place samples in the 50°C waterbath and add 100 µL of the prewarmed 1X Sequenase reaction buffer. Vortex samples briefly, then concentrate the Dynabead/DNA/primer complex using the MPC and remove the liquid.

14. Repeat step 13. Give the samples a final brief spin in a microfuge in order to collect any remaining liquid. Place samples in MPC and remove remaining liquid.

15. Add 10 µL of 1X Sequenase reaction buffer to each sample and sequence using Sequenase protocol.

## 3.3. Choice of Sequencing Primer

1. For the aforementioned PCR primer pairs, using the unbiotinylated primer as a sequencing primer with the Dynabead bound single-stranded DNA template will give good results. The unbiotinylated single-stranded DNA template obtained by precipitating the NaOH wash (*see* Note 3) may also be sequenced using the biotinylated PCR primer as a sequencing primer. In this case, a stringency wash of the sequencing primer is not possible and the sequencing ladder obtained may occasionally be poor.

2. If a sequencing ladder similar to that in Fig. 2 (lanes A, F, and G) is observed for some samples, or in the case of automated sequencing, an output similar to that in

Fig. 4A is obtained, it is probably caused by the presence of a poly (A) tail in the cDNA sequence. In such cases, an anchored oligo dT primer may be employed. Anchored oligo dT primers should be used at higher concentrations than normal (20 ng/reaction for manual sequencing using Dynabeads and a 25 ng/µL final concentration for automated cycle sequencing). When using the shorter oligo dT primer [$^5$dT$_{(18)}$(A/G/C)(A/G/C/T)$^3$] for manual sequencing using Dynabeads, stringency washes (*see* Section 3.2., step 13) should be carried out at lower temperatures (30°C).

## 4. Notes

1. It is acceptable to use probe recovered from initial first-round screen of the cDNA library since the signals expected from the Southern blots are high. We have successfully used 6-wk-old $^{32}$P labeled cDNA probe. In such cases, recovered probe should be stored at –20°C, and be heated to 95°C, for 10 min prior to being added to the blots.
2. We have found that even when blots are washed at very high stringency (0.1X standard saline citrate [SSC]/0.1% sodium dodecyl sulfate [SDS] at 70°C), strong signals remain. Because of the strong signals, autoradiographic exposure times are short. For this reason, it is practical to carry out washes with increasing stringency, exposing the blots to film between washes.
3. The precipitation of the PCR products with 1/3 vol 8$M$ ammonium acetate and an equal volume of isopropanol for 10 min will remove a large fraction of the primers remaining in the PCR reaction. This is an advantage, since free biotinylated primers will compete with PCR product for the streptavidin sites, thus necessitating the use of large volumes of Dynabeads. Incubations should not be carried out for more than 10 min since this may result in the precipitation of primers. An alternative is to use one of the commercially available PCR product purification cartridges.
4. The unbiotinylated strand of the DNA template should be prepared for sequencing as follows:
   a. Centrifuge sample from Section 3.2., step 9 in a microfuge for 15 min.
   b. Remove supernatant and wash pellet twice with 300 µL of 70% ethanol.
   c. Remove 70% ethanol and air-dry sample for 20 min on a bench or 10 min in a water bath.
   d. Redissolve DNA in 6 µL of TE buffer and heat sample to 65°C, for 10 min.
   e. Add Sequenase reaction buffer and 5 ng of sequencing primer and carry out primer annealing and sequencing using the Sequenase protocol.
5. Following the binding of DNA to the Dynabeads, it is important to make certain that the Dynabeads are periodically resuspended during all stages of template preparation and sequencing. If they appear to clump at any stage it is important to remedy this by thorough vortexing.

## References

1. Huynh, T. V., Young, R. A., and Davis, R. W. (1988) Constructing and screening cDNA libraries in λgt10 and λgt11. *DNA Cloning: A Practical Approach,* vol I (Glover, D. M., ed.), IRL, Oxford, UK.

2. Zeng, J., Gorski, R. A., and Hamer, D. (1994) Differential cDNA cloning by enzymatic degrading subtraction (EDS). *Nucleic Acids Res.* **22**, 4381–4385.

3. Wieland, I., Bolger, G., Asouline, G., and Wigler, M. (1990) A method for difference cloning: gene amplification following subtractive hybridization. *Proc. Natl. Acad. Sci. USA* **87**, 2720–2724.

4. Lemarie, P., Garrett, N., Kato, K., and Gurdon, J. B. (1993) Construction of subtracted cDNA libraries enriched for cDNAs for genes expressed in the mesoderm of early *Xenopus gastrulae*. *C.R. Acad. Sci. Paris, Sciences de la vie/Life Sciences* **316**, 938–944.

5. Liang, P. and Pardee, A. B. (1992) Differential display of eukaryotic mRNA by means of the polymerase chain reaction. *Science* **257**, 967–971.

6. Liang, P., Averboukh, L., and Pardee, A. B. (1993) Distribution and cloning of eukaryotic mRNAs by means of differential display: refinements and optimization. *Nucleic Acids Res.* **21**, 3269–3275.

7. Welsh, J., Chad, K., Dalal, S. S., Cheng, R., Ralph, D., and McClelland, M. (1992) Arbitrary primed PCR fingerprinting of RNA. *Nucleic Acids Res.* **20**, 4965–4970.

8. McClelland, M., Ralph, D., Cheng, R., and Welsh, J. (1994) Interactions among regulators of RNA abundance characterized using RNA fingerprinting by arbitrary primed PCR. *Nucleic Acids Res.* **22**, 4419–4431.

9. Mullis, K. B. and Faloona, F. (1987) Specific synthesis of DNA in vitro via a polymerase-catalyzed chain-reaction. *Meth. Enzymol.* **155**, 335.

10. Thomas, M. G., Hesse, S. A., Al-Mahdawi, S., Bui, T. D., Moniz, C. F., and Farzaneh, F. (1994) A procedure for second-round differential screening of cDNA libraries. *Biotechniques* **16**, 229–232.

11. Rosenberg, H. F., Corrette, S. E., Tenen, D. G., and Ackerman, S. J. (1991) Rapid cDNA library screening using the Polymerase chain reaction. *BioTechniques* **10**, 53–54.

12. Sang, J. and Thompson, N. J. (1994) An efficient procedure for obtaining long cDNA clones from phage library screening. *BioTechniques* **17**, 447–451.

13. Southern, E. M. (1975). Detection of specific sequences among DNA fragments separated by gel electrophoresis. *J. Mol. Biol.* **98**, 503–508.

14. Hultman, T., Stahl, S., Hornes, E., and Uhlen, M. (1989) Direct solid-phase sequencing of genomic and plasmid DNA using magnetic beads as solid support. *Nucleic Acids Res.* **17**, 4937–4946.

15. Thomas, M. G., Miller, K. W. P., Cook, Jr, C. E., and Hagelberg, E. (1994) A method for avoiding mispriming when sequencing with Dynabeads. *Nucleic Acids Res.* **22**, 3243,3244.

16. Thomas, M. G., Hesse, S. A., McKie, A. T., and Farzaneh, F. (1993) Sequencing of cDNA using anchored oligo dT primers. *Nucleic Acids Res.* **21**, 3915,3916.

17. Anderson, S., Bankier, A. T., Barrell, B. G., de Bruijn, M. H., Coulson, A. R., Drouin, J., Eperon, I. C., Nierlich, D. P., Roe, B. A., Sanger, F., Schreier, P. H., Smith, A. J., Staden, R., and Young, I. G. (1981) Sequence and organization of the human mitochondrial genome. *Nature* **290**, 457–465.

18. McBride, L. J., Koepf, S. M., Gibbs, R. A., Salser, W., Mayrand, P. E., Hunkapiller, M. W., and Kronick, M. N. (1989) Automated DNA sequencing methods involving polymerase chain reaction. *Clin. Chem.* **35**, 2196–2201.

# 40

# Identification and Cloning
# of Differentially Expressed Genes by DDRT-PCR

**Mikkel Rohde, Rene Hummel, Niels Pallisgaard, Tino Podstufka, Heidi Riedel, Henrik Leffers, and Michael Strauss**

## 1. Introduction

The invention of polymerase chain reaction (PCR) and its application to amplification of reverse transcribed cDNA copies of mRNA has opened new possibilities for the development of techniques for identification of changes in gene expression patterns. Two very similar variants of the same strategy were published in 1992 that were named mRNA differential display *(1)* and RNA arbitrary primed PCR *(2)*. In contrast to the standard RT-PCR technique, where individual mRNAs are detected in a qualitative and sometimes also quantitative manner using gene-specific primers, in these new techniques a set of short primers with arbitrary sequences is used to generate amplified cDNA fragments from a large number of mRNAs used before for DNA fingerprinting *(3)*. Basically, the identification of every individual mRNA species expressed should be possible provided a sufficiently large number of primer pairs is used *(4)*. The method of differential display (DDRT-PCR) has already been described in detail in a number of reviews *(5–10)*. Therefore, we give only a very brief description of the principle. First, RNA is prepared from the different cell types of interest. Second, RNA is reverse transcribed in 12 *(1,4)*, 4 *(11)*, or 3 *(12)* fractions using anchored oligo-dT primers (e.g., $T_{11}VV$ in Fig. 1) to generate single-stranded cDNA. Third, every cDNA fraction is amplified in a number of independent PCR reactions using various 10-mer upstream primers. Fourth, products of individual PCR reactions are separated on standard sequencing gels *(1,11)* or on nondenaturing polyacrylamide gels *(4,6,10)* and visualized, normally by autoradiography. Fifth, fragments differing in their

From: *Methods in Molecular Biology, Vol. 67: PCR Cloning Protocols: From Molecular Cloning to Genetic Engineering* Edited by: B. A. White  Humana Press Inc., Totowa, NJ

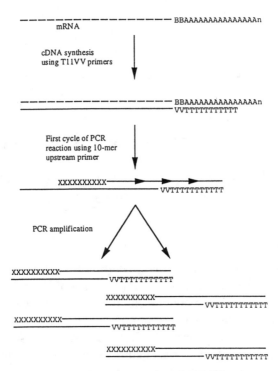

**Fig. 1.** Scheme of DDRT-PCR.

intensities between the parallel samples from different cells are cut out of the gel, reamplified, cloned into a suitable vector, and sequenced. Sixth, differential expression of the genes corresponding to the identified sequence tags is confirmed by an independent analysis. We have previously recommended nuclear run-on analysis (6,10). However, this method is not easy to apply to small amounts of cells and tissues. Therefore, we have recently optimized conditions for RNase protection analysis that we will describe here. Finally, if cloning of full-length cDNAs is the ultimate goal, simultaneous isolation of a larger number of cDNA clones from a suitable library can be carried out using a biotinylated fragment mix as probes and streptavidin-coated magnetic beads to purify specifically hybridizing cDNA clones. The cloning step is not an essential part of DDRT-PCR and it is usually omitted if only diagnosis of gene expression patterns is required.

## 2. Materials
### 2.1. Single-Strand cDNA Synthesis

1. DNA-free cytoplasmic RNA (see Note 1).

2. Single-strand cDNA synthesis kit (Life Technologies [Bethesda, MD], Cat. 8053SA), including 5X reverse transcription buffer, $0.1M$ dithiothreitol (DTT), and SuperScript reverse transcriptase, RNaseH-minus (*see* Note 2).
3. 100 $\mu M$ dNTP stock solution (e.g., Pharmacia, Piscataway, NJ).
4. Rnasin (40 U/$\mu$L; Promega, Madison, WI).
5. 12 anchored primers ($T_{11}$VN; *see* ref. *1* and Note 3).

## 2.2. DDRT-PCR

1. $T_{11}$VN primers (*see* Section 2.1.).
2. 24 arbitrary 10-mer primers (*see* ref. *4*).
3. $^{33}$P-dATP, 10.0 mCi/mL (Dupont/NEN, Boston, MA).
4. *Taq* polymerase and 10X PCR buffer (500 m$M$ KCl, 100 m$M$ Tris-HCl, pH 9.0, at 25°C, 0.1% gelatin, 1% Triton X-100).
5. 25 m$M$ MgCl$_2$.
6. 100 $\mu M$ dNTP mix (e.g., Pharmacia).

## 2.3. Electrophoretic Separation of Fragments

1. Acrylamide.
2. Bis-acrylamide.
3. Tris base.
4. Borate.
5. $0.5M$ EDTA, pH 8.0.
6. TEMED.
7. 10% Ammonium persulfate.
8. Loading solution: 40% Ficoll 400, 0.05% xylene cyanol FF, 0.05% bromophenol blue.

## 2.4. Fragment Amplification, Cloning, and Sequencing

1. TE: 10 m$M$ Tris-HCl, pH 7.5; 1 m$M$ EDTA.
2. Reagents for PCR (*see* Section 2.2.).
3. Reagents for agarose gel electrophoresis.
4. TA cloning kit (Invitrogen, San Diego, CA) and reagents for bacterial culture and plasmid minipreparations.
5. DNA sequencing kit (e.g., Promega, USB).

## 2.5. RNase Protection Analysis

1. 10X T7 transcription buffer: 400 m$M$ Tris-HCl, pH 8.0, 60 m$M$ MgCl$_2$, 50 m$M$ DTT, 10 m$M$ spermidine.
2. rUTP labeling mix: 2.5 m$M$ each of ATP, GTP, CTP, 0.2 m$M$ UTP, 10 m$M$ Tris-HCl, pH 7.5, 0.1 m$M$ EDTA.
3. $\alpha[^{32}P]$-UTP (800 Ci/m$M$).
4. RNasin (*see* Section 2.1.).
5. T7 RNA polymerase (40 U/$\mu$L).
6. DNase, RNase-free (Boehringer Mannheim, Indianapolis, IN).

7. TE (*see* Section 2.4.).
8. Phenol and chloroform.
9. 3*M* sodium acetate, pH 6.0.
10. Absolute ethanol.
11. Formamide loading buffer. Prepare by adding 10 mL of deionized formamide, 5 mg of xylene cyanol, and 5 mg bromophenol blue to 200 mL of 0.5*M* EDTA, pH 8.0.
12. Diethylpyrocarbonate (DEPC)-treated water.
13. Deionized formamide.
14. Extraction buffer: 10 m*M* Tris HCl, pH 7.5, 20 m*M* EDTA, pH 8.0, 0.5% SDS, 50 μg/mL of proteinase K.
15. 10X hybridization buffer: 40 m*M* PIPES, pH 6.4, 10 m*M* EDTA, 4*M* NaCl.
16. Digestion buffer: 100 m*M* Tris-HCl, pH 7.5, 50 m*M* EDTA, 2*M* sodium acetate.
17. RNaseOne (Promega).
18. RNase stop mix. Prepare fresh by mixing 20 μL of 10% *N*-lauryl sarcosine, 10 μL of 10 mg/mL proteinase K, and 5 μL of 1 μg/mL tRNA.
19. Reagents for 8% polyacrylamide:urea sequencing gel.
20. 16-Biotinylated-dUTP.
21. Denhardt's solution.
22. Streptavidin-coated magnetic beads and magnetic stand (e.g., Dynal, Lake Success, NY).
23. Formamide (deionized).
24. Xylene cyanol, bromophenol blue.

### 2.6. Isolation of Multiple cDNAs (Differential Libraries)

1. Biotinylation dNTP mix. Prepare a modified dNTP stock solution (*see* Section 2.1.) with 33% substitution of dTTP with 16-Biotin-dUTP *(13,14)*.
2. Reagents for PCR (*see* Section 2.2.).
3. Reagents for preparation of DNA from a total plasmid or λ cDNA library.
4. cDNA hybridization buffer: 750 m*M* NaCl, 50 m*M* Na$_2$HPO$_4$/NaH$_2$PO$_4$, pH 7.2, 2.5 m*M* EDTA, 5X Denhardt's solution, 0.05% SDS, 50% formamide.
5. Streptavidin-coated magnetic beads and magnetic stand (e.g., Dynal).
6. 10 m*M* Tris-HCl, pH 7.5, 2*M* NaCl, 0.5 m*M* EDTA.
7. 20X SSC: 3*M* NaCl, 0.3*M* sodium acetate, pH 7.0. Prepare 1, 0.5, and 0.2X SCC solutions by dilution of 20X SCC with DEPC water.
8. Reagents for bacterial transformation and plasmid DNA isolation.

### 3. Methods
### 3.1. Single-Strand cDNA Synthesis

In previous protocols we have given a detailed description of our method for isolation of starting RNA *(6,10)*. We have recommended the use of cytoplasmic RNA to avoid additional complexity originating from nuclear precursor

RNA. However, this is, in fact, a minor problem. Since a variety of commercial kits are available now to prepare clean total RNA, we recommend using one of these kits. The only important requirement is the complete absence of DNA, which is normally accomplished by treatment with RNase-free DNase. Some suppliers of RNA-purification kits guarantee DNA-free quality.

1. Set up 24 individual cDNA synthesis reactions for two cell types (2 cell types × 12 different primers) in a volume sufficient to generate cDNA for 24 PCR reactions each. To each of 24 reaction tubes add 3.0 μL of 25 μM $T_{11}VN$ primer. If you have 12 different downstream primers, set up 12 tubes/cell line; 1–12 for cell line A and 13–24 for cell line B.
2. Add 3.0 μL (0.2–1.0 μg) of total RNA from cell A to each of tubes 1–12 and of RNA from cell B to each of tubes 13–24. Add 7.5 μL of sterile water to all 24 tubes, incubate at 70°C, for 5 min, and place at 42°C, for 1 min.
3. Prepare a master mix by sequentially pipeting into a separate Eppendorf tube 150 μL of 5X reverse transcription buffer, 75 μL of 0.1M DTT, 150 μL of a 100 μM dNTP mix, 19 μL of 40 U/μL of RNasin, and 19 μL of sterile DEPC-treated water. Add 16.5 μL of the cDNA master mix to all tubes.
4. Add 1.5 μL (300 U) of SuperScript RNaseH-minus reverse transcriptase and incubate at 40°C, for 1 h, heat to 95°C, for 1 min, place on ice immediately, and store at 70°C, for further use.

## 3.2. DDRT-PCR

1. Prepare a master mix for 96 reactions (allowing for the analysis of two cDNA subfractions from two different cell types with two $dT_{11}VN$ downstream primers and 24 upstream primers) by sequentially pipeting into an Eppendorf tube: 234.0 μL 10X PCR buffer, 140.4 μL 25 mM $MgCl_2$, 11.7 μL [$^{33}$P]-dATP, 46.8 μL 100 μM dNTP mix, 22.0 μL *Taq* polymerase (5 U/μL), and 949.1 μL sterile water (DEPC treated).
2. From this master mix, transfer 324 μL to each of four tubes (A–D) followed by 54 μL (25 μM) of downstream primer (primer 1 in A and B, primer 2 in C and D) and 27 μL of cDNA (from cell type 1 in tubes A and C, from cell type 2 in tubes B and D), resulting in a total volume of 405 μL in each of the four tubes.
3. Prepare in a microtiter plate or in tubes arranged in a microtiter format (Perkin Elmer, Norwalk, CT) reaction mixtures by adding 15 μL of the premixed components from each of the four tubes to 24 wells or tubes (4 × 24). Add 5 μL (2 μM) of the 24 different upstream primers, one of the 24 different upstream primers to each well or tube.
4. Spin in a centrifuge that holds microtiter plates to mix the upstream primers with the premixed components and run PCR reactions in a suitable thermocycler (e.g., GeneAmp PCR system 9600, Perkin Elmer) equipped with a heated lid. Run the following cycles: 95°C, 2 min (initial denaturation); 40 cycles: 94°C, 30 s, 40°C, 1 min (*see* Note 4), 72°C, 1 min (*see* Note 5); and 72°C, 5 min (final elongation).

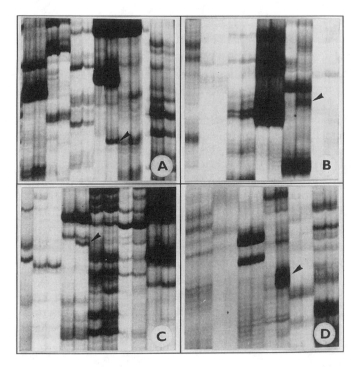

**Fig. 2.** Electrophoretic pattern of fragments generated by DDRT-PCR. Four regions of gels are shown where differences between two cell types were identified. The bands marked by arrows represent genes that are regulated by the retinoblastoma protein *(16)*. The left lane of every pair of lanes shows fragments from a RB-deficient tumor cell line (BT549) and the right one represents those from the same cell line transfected with the Rb gene.

### 3.3. Electrophoretic Separation of Fragments

1. Prepare stock solutions for nondenaturing gels. First, make 20X TBE by dissolving 215 g of Tris base, 110 g of borate, and 20 mL 0.5$M$ EDTA in water and make up to 1 L with water. Second, make a 6% gel solution by mixing 14.5 mL of 40% stock containing acrylamide and bis-acrylamide at a ratio of 19:1, 5.0 mL of 20X TBE, and 80.5 mL of water. Pass the solution through a 0.2 μm filter. Use 50 mL of the 6% gel solution for a 35 × 40 × 0.2-cm gel, add 40 μL TEMED and 200 μL 10% ammonium persulfate, and pour a sequencing-type gel.
2. Take half of the DDRT-PCR incubation mixture (10 μL); add 2 μL loading solution. Load 3 μL of each reaction with parallel samples from different cell lines side-by-side onto a prerun 6% polyacrylamide gel without urea and run in TBE buffer at 50 W. Finish the run when bromophenol blue runs out of the gel. Dry the gel on 3MM filter paper on a gel dryer or expose directly to X-ray film overnight. A typical result is given in Fig. 2.

## 3.4. Fragment Amplification, Cloning, and Sequencing

1. Align autoradiogram on the gel, mark the bands differing between the patterns from control cells and test cells, cut them out from the filter paper, and transfer to Eppendorf tubes. Add 50 μL of TE, shake and boil for 10 min, spin briefly at high speed, and transfer supernatant to a new tube.
2. To 26.5 μL of eluted DNA, add 5 μL of 10X PCR buffer (as in Section 2.2.), 3 μL of 25 m$M$ MgCl$_2$, 5 μL of a 500 μ$M$ dNTP mix, 5 μL of 2 μ$M$ downstream primer as in the original PCR, 5 μL of 2 μ$M$ upstream primer as in the original PCR, and 0.5 μL of *Taq* polymerase (5 U/μL).
3. Carry out amplification with the following cycle protocol: 94°C for 30 s, 45°C for 1 min, 72°C for 1 min for 30–40 cycles, and 72°C for 5 min.
4. Run 10 μL of the reaction mixture on a 2% agarose gel to check the size of the fragment.
5. Clone fragments according to standard procedures. We recommend using the TA cloning kit (Invitrogen, San Diego, CA) with 1–3 μL of the reamplification solution according to the instructions of the supplier (*see* Note 6).
6. Pick six white colonies from each cloned fragment, do plasmid minipreparations, and cut plasmids with *Eco*RI to check for inserts by agarose gel electrophoresis.
7. Carry out DNA sequence analysis using standard procedures, e.g., using the fmol sequencing kit (Promega) or the cycle sequencing kit (USB). *See* Note 7.

## 3.5. RNase Protection Analysis

*See* Note 8.

1. To generate a labeled RNA probe, mix in an Eppendorf tube: 1 μL 10X T7 transcription buffer, 1 μL 10X rUTP labeling mix, 2.5 μL (25 μCi) [$^{32}$P]-UTP, 0.75 μL RNasin, 1–4 μL DDRT-PCR DNA fragment (20–40 ng), 0.5 μL T7-RNA polymerase (20 U), and DEPC-treated H$_2$O to 10 μL total volume
2. Incubate at 37°C, for 1 h.
3. Add 1 μL of DNaseI and incubate at 37°C for 15 min.
4. Add 80 μL of TE buffer and extract once with phenol/chloroform. Add 10 μL of 3$M$ Na-acetate, pH 6.0, and 250 μL of ethanol, put in the cold for precipitation, spin, wash, and dry.
5. Dissolve pellet in 6 μL formamide loading buffer, denature at 95°C, for 3 min, place on ice briefly, load onto a small 6% sequencing-type gel, and run at 13 W for about 1 h.
6. Cut out the labeled band, extract overnight in 200 μL of extraction buffer, transfer solution to a new tube, re-extract gel piece with 100 μL of extraction buffer, combine supernatants, extract once with phenol/chloroform, and add 30 μL of 3$M$ Na-acetate, pH 6.0, and 2.5 vol of ethanol. After precipitation wash, dry, and dissolve in 20 μL of DEPC H$_2$O.
7. Prepare tubes with 1 μL (0.2–1 μg) of each of the cellular RNAs to be tested or 1 μL (0.8 μg) of tRNA (control sample). Add 500 cps (as determined by minimonitor, usually 1–2 μL) of the labeled RNA fragment in a final volume of 3 μL, 24 μL of deionized formamide, and 3 μL of 10X hybridization buffer and

mix. Incubate at 85°C, for 10 min, transfer tubes quickly to hybridization temperature (37–45°C) and incubate for 14–24 h.

8. Place tubes on ice, add 450 µL of ice-cold RNase digest mixture prepared from 40 µL of digestion buffer, 409 µL of water, and 1 µL of RNaseONE, and incubate at 15°C, for 30 min. Add 50 µL of RNase stop mix and incubate at 37°C, for 30 min. Extract once with 500 µL phenol/chloroform, add 8 µL of 3$M$ Na-acetate and 1 mL ethanol, precipitate, wash, and dry.

9. Dissolve pellet in 7 µL of formamide loading buffer (*see* Secton 2.3.). For reference purposes, mix 0.2 µL of the labeled fragment with 9.2 µL of formamide loading buffer. Heat samples to 95°C, for 3 min and load 3.5 µL onto a 8% small sequencing-type gel (0.4 mm); run at 11 W for 45–90 min. Expose to X-ray film.

### 3.6. Isolation of Multiple cDNAs (Differential Libraries)

*See* Note 9.

1. To generate biotinylated fragments as probes for isolation of the corresponding cDNAs from a library, carry out PCR as in Section 3.5. with a modified nucleotide stock solution with 33% substitution of dTTP by 16-Bio-dUTP *(13,14)*.

2. Prepare DNA from a total plasmid or λ cDNA library using standard protocols.

3. Dissolve 0.1 µg of biotinylated fragment(s) and 40 µg of library DNA in 50 µL of cDNA hybridization buffer. Incubate at 90°C, for 10 min and afterwards at 43°C, for 18 h.

4. Add 300 µL of 2$M$ NaCl, 10 m$M$ Tris-HCl, pH 7.5, 0.5 m$M$ EDTA containing streptavidin-coated magnetic beads (60 µg/0.1 µg probe) and incubate at 20°C for 1 h to allow for binding of the hybrid molecules to the beads.

5. Wash the beads on magnetic stand by incubating them sequentially in 1 mL of each of 1, 0.5, and 0.2X SSC at 65°C for 15 min in each step.

6. Heat the washed beads to 70°C, for 10 min, put on the magnetic stand, remove supernatant and use for transformation of *Escherichia coli*.

7. Use the total pool of transformants and isolate plasmid DNA. Repeat steps 3–6 once or twice to get sufficient enrichment of specific clones. An outline of the procedure is shown in Fig. 3.

### 4. Notes

1. RNA preparations must be free of DNA, which is normally guaranteed by incubation of the preparation with RNase-free DNase. Some suppliers of purification kits claim their kits to result in DNA-free RNA. We do not recommend to use purified polyA RNA because of low stability and lower reproducibility of the results.

2. The reverse transcription protocol also works successfully with AMV reverse transcriptase.

3. The protocol for DDRT-PCR given here is based on the original procedure involving 12 $T_{11}$VN primers *(1,4)*. It can easily be modified for use with either four $T_n$MN *(11)* or three $T_n$V *(12)* primers. Although the latter primer would be preferable if the primary goal is the detection of a limited number of genes and differences between two cell types, the original 12 primers should be used if a

Selection procedure

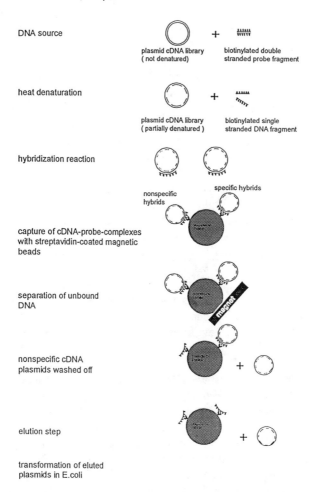

**Fig. 3.** Flow chart of clone isolation from cDNA libraries using biotinylated probe.

largely complete overview of all expressed genes is required. According to our experience, a two-nucleotide anchor provides a higher selectivity than a one-nucleotide anchor. Sequencing of more than 50 cloned fragments clearly showed the presence of two specific nucleotides in front of the polyA tract. Some 20% of the fragments originate by priming with the upstream primer from both sides. Those fragments are normally derived from coding parts of a mRNA, whereas fragments generated with the help of both primers are generally derived from the 3'-untranslated region. We have the impression that generation of only three cDNA pools using the $T_{11}V$ (or $T_{14}V$) primer results in too high a complexity of the individual pool, which does not allow a display of all possible fragments that can be generated together with one particular upstream primer. Some loss of information is likely to occur because of overloading of lanes with bands even if a larger number of upstream primers (80–100) is used to approach a total of 300 PCR reactions, which was regarded to be representative *(4,6)*. This variant is not only more time-consuming but also twice as expensive than the original version. Alternatively, there are various applications of the method where a complete overview of the expressed genes is not the primary goal *(8)*.

4. The original protocol used 40 or 42°C, for annealing throughout the cycles *(1,4)*. Since we and others have observed that most fragments originate from precise annealing in the first six or seven nucleotides of the 10-mer primer *(4–6,8)*, there was good reason to assume that the annealing temperature could be raised after the initial cycles to ensure exclusive amplification of fragments with perfectly matching 10-mer primers. Indeed, raising the annealing temperature up to 45°C, after three initial cycles for the remaining 35 cycles results in much cleaner band patterns without background and without significant loss of bands. However, this stringent protocol does not work with all PCR machines because it is very much dependent on the precision of temperature adjustment. It also does not work satisfactorily with the three $T_{11}VT$ primers (and sometimes with the $T_{11}AA$ primer). Therefore, we are now using only the nine $T_{11}VV$ primers.

5. Under standard conditions, only the lower abundant class of RNA is probably amplified in the linear range. If the number of cycles is reduced to 30, only very few bands are detectable. One would have to run at least three sets of analysis (30, 35, and 40 cycles) or three different concentrations of cDNA to compensate for this problem.

6. To analyze the sequence of DNA isolated from a band it is important to assure that one is dealing with just one DNA fragment. This is normally guaranteed by cloning and sequencing of individual clones. We have recently established an alternative procedure that circumvents cloning and allows carrying out sequencing reactions directly. It is based on the use of three $T_{14}V$ primers in cDNA synthesis and DDRT-PCR and subsequent reamplification of DNA from bands in four individual PCR reactions using four different $T7-T_{14}VN$ downstream primers where V is the same nucleotide as in the DDRT-PCR reaction. This procedure is based on the assumption that the chance for two fragments in the same band having the same nucleotide next to the anchoring nucleotide is only 25%. Normally, only one of the four downstream primers generates a strong fragment together with the upstream primer if 45°C are used for annealing. The resulting fragment can be sequenced directly using a T7 promoter primer. The T7 promoter can also be used to generate RNA probes for RNase protection *(see below)*. The sequence of the primer contains 23 nucleotides of the T7 promoter (TAATACGACTCACTATAGGGCCC) followed by 14X T and *VN* where *V* can be A, G, or C and *N* can be any of the four nucleotides. In the PCR reaction we prefer to use the Stoffel fragment of *Taq* polymerase because *Taq* polymerase does not care about precise base pairing at the 3' nucleotide.

7. DNA fragments that have been reamplified from the DDRT-PCR gels with the $T7-T_{14}VN$ primers and purified on a low-melting-agarose gel can be sequenced

directly without cloning using a primer that is complementary to the T7 sequence. Sequencing can be performed manually using [$^{35}$S]- or [$^{33}$P]-dATP and the AmpliCycle sequencing kit (Perkin-Elmer). Alternatively, sequencing can be performed on automated DNA sequenators (ABI 373/377, Applied Biosystems Inc.; ALF/ALF Express, Pharmacia Biotech). For the ABI machines we recommend using the T7 primer and the Dye Terminator Cycle Sequencing kit (Applied Biosystems Inc.). For sequencing on ALF machines we recommend using T7-fluorescein (ALF) or T7-CY5 (ALF Express) labeled primers and the cycle sequencing protocol described by Zimmermann et al. *(15)*.

8. The experience of many users of this method, including ourselves, shows that only 50% of the fragments identified by DDRT-PCR could be used to obtain signals on Northern blots. This could partly be caused by the limited length of some fragments (<300 bp) but also by the low abundance of the transcript. As mentioned before, the standard protocol for DDRT-PCR preferentially identifies lower abundant mRNAs as being differentially expressed. Therefore, it is advisable to use another method for verification of differential expression. We have previously shown that nuclear run-on transcript hybridization could identify >90% of the fragments corresponding to regulated genes *(6,10)*. This method has the additional advantage of requiring only one labeling reaction for every cell type and can analyze expression of a large number of genes at the same time. However, this method requires some skills and a larger amount of cells as starting material. The method of RNase protection, as described here, is a very sensitive technique that can be used alternatively. Of the three methods, it is the one that requires the lowest amount of RNA. However, as with Northern blotting, labeling of every individual fragment is required.

9. The final step in DDRT-PCR analysis is cloning of the full-length cDNAs corresponding to the fragments. There are three major options to do this. First, fragments can be used as labeled probes to screen a suitable cDNA library. This traditional method is only advisable if one is dealing with one or very few candidate genes. Second, the fragment that is normally derived from the very 3'-end of the mRNA can be used to specifically isolate the upstream part of the cDNA by a 5'RACE technique. Several kits are available for this purpose. The Marathon kit (Clonetech, Palo Alto, CA) is most appropriate for this purpose. This method is more rapid than library screening but is also difficult to use for larger numbers of genes. Third, individual fragments or, preferentially, a mixture of fragments, can be used to generate a minilibrary (which is, in essence, a differential library). The protocol given in Section 3.7. (Fig. 3) has been established with the aim of isolating 10–20 cDNA clones from a suitable library at the same time. It was originally developed for isolation of single clones *(14)*. CD69 clones were enriched 200-fold in a single step from a library that contained the specific cDNA with an abundance of 0.0001. The method can be applied to several genes at the same time. It is currently used in our laboratories to generate minilibraries of genes induced by the retinoblastoma gene product or by liver regeneration. The resulting libraries can be used for functional studies in transfection assays.

## References

1. Liang, P. and Pardee, A. B. (1992) Differential display of eukaryotic messenger RNA by means of the polymerase chain reaction. *Science* **257**, 967–971.
2. Welsh, J., Chada, K., Dalal, S. S., Cheng, R., Ralph, D., and McClelland, M. (1992) Arbitrarily primed PCR fingerprinting of RNA. *Nucleic Acids Res.* **20**, 4965–4970.
3. Welsh, J., and McClelland, M. (1990) Fingerprinting genomes using PCR with arbitrary primers. *Nucleic Acids Res.* **18**, 7213–7218.
4. Bauer, D., Muller, H., Reich, J., Riedel, H., Ahrenkiel, V., Warthoe, P., and Strauss, M. (1993) Identification of differentially expressed mRNA species by an improved display technique (DDRT-PCR). *Nucleic Acids Res.* **21**, 4272–4280.
5. Liang, P., Averboukh, L., and Pardee, A. B. (1994) Method of differential display. *Methods Mol. Genet.* **5**, 3–16.
6. Bauer, D., Warthoe, P., Rohde, M., and Strauss, M. (1994) Detection and differential display of expressed genes by DDRT-PCR. *PCR Methods Appl.* **4**, S97–S108.
7. McClelland, M. and Welsh, J. (1994) RNA fingerprinting by arbitrarily primed PCR. *PCR Methods Appl.* **4**, S66–S81.
8. Liang, P., Bauer, D., Averboukh, L., Warthoe, P., Rohrwild, M., Muller, H., Strauss, M., and Pardee, A. B. (1995) Analysis of altered gene expression by differential display. *Methods Enzymol.* **254**, 304–321.
9. Liang, P. and Pardee, A. B. (1995) Recent advances in differential display. *Curr. Opinion Immunol.* **7**, 274–280.
10. Warthoe, P., Bauer, D., Rohde, M., and Strauss, M. (1995) Detection and identification of expressed genes by differential display, in *PCR Primer* (Dieffenbach, C. W. and Dveksler, G. S., eds.), Cold Spring Harbor Laboratory, Cold Spring Harbor, NY, pp. 421–438.
11. Liang, P., Averboukh, L., and Pardee, A. B. (1993) Distribution and cloning of eukaryotic mRNAs by means of differential display: refinements and optimization. *Nucleic Acids Res.* **21**, 3269–3275.
12. Liang, P., Zhu, W., Zhang, X., Guo, Z., O'Connell, R. P., Averboukh, L., Wang, F., and Pardee, A. B. (1994) Differential display using one-base anchord oligo-dT primers. *Nucleic Acids Res.* **22**, 5763,5764.
13. Lo, Y.-H. D., Methal, W. Z., and Fleming, K. A. (1988) Rapid production of vector-free biotinylated probes using the polymerase chain reaction. *Nucleic Acids Res.* **16**, 8719.
14. Podstufka, T. (1994) Method for isolation of cDNA clones. Master thesis, Humboldt University Berlin.
15. Zimmermann, J., Wiemann, S., Voss, H., Schwager, C., and Ansorge, W. (1994) Improved fluorescent cycle sequencing protocol allows reading nearly 1000 bases. *BioTechniques* **17**, 302–307.
16. Rohde, M., Warthoe, P., Gjetting, T., Lukas, J., Bartek, J., and Strauss, M. (1996) The retinoblastoma protein modulates expression of genes coding for diverse classes of proteins including components of the extracellular matrix. *Oncogene,* in press.

# VII

## CLONING MEMBERS OF GENE FAMILIES

# 41

## Cloning Gene Family Members Using PCR with Degenerate Oligonucleotide Primers

Gregory M. Preston

## 1. Introduction

### 1.1. What Are Gene Families?

As more and more genes are cloned and sequenced, it is apparent that nearly all genes are related to other genes. Similar genes are grouped into families, such as the collagen and globin gene families. There are also gene superfamilies. Gene superfamilies are composed of genes that have areas of high homology and areas of high divergence. Examples of gene superfamilies include the oncogenes, homeotic genes, and myosin genes. In most cases the different members of a gene family carry out related functions. A detailed protocol for the cloning by degenerate oligonucleotide polymerase chain reaction (PCR) of members of the *Aquaporin* family of membrane water channels *(1,2)* will be discussed here.

### 1.2. Advantages of PCR Cloning of Gene Family Members

There are several considerations that must be taken into account when determining the advantages of using PCR to identify members of a gene family over conventional cloning methods of screening a library with a related cDNA, a degenerate primer, or an antibody. It is recommended that after a clone is obtained by PCR one uses this template to isolate the corresponding cDNA from a library, because mutations can often be introduced in PCR cloning. Alternatively, sequencing two or more PCR clones from independent reactions will also meet this objective. The following is a list of some of the advantages of cloning gene family members by PCR.

From: *Methods in Molecular Biology, Vol. 67: PCR Cloning Protocols*
Edited by: B. A. White  Humana Press Inc., Totowa, NJ

1. Either one or two degenerate primers can be used in PCR cloning. When only one of the primers is degenerate, the other primer must be homologous to sequences in the phage or bacteriophage cloning vector *(3,4)* or to a synthetic linker sequence, as with RACE PCR (*see* Chapter 24). The advantage to using only one degenerate primer is that the resulting clones contain all of the genetic sequence downstream from the primer (be it 5' or 3' sequence). The disadvantage to this anchor PCR approach is that one of the primers is recognized by every gene in the starting material, resulting in single-strand amplification of all sequences. This is particularly notable when attempting to clone genes that are not abundant in the starting material. This disadvantage can often be ameliorated in part by using a nested amplification approach with two degenerate primers to preferentially amplify desired sequences.
2. It is possible to carry out a PCR reaction on first-strand cDNAs made from a small amount of RNA, and in theory, from a single cell. Several single-stranded "minilibraries" can be rapidly prepared and analyzed by PCR from a number of tissues at different stages of development, or cell cultures under different hormonal conditions. Therefore, PCR cloning can potentially provide information about the timing of expression of an extremely rare gene family member, or messenger RNA splicing variants, that may not be present in a recombinant library.
3. Finally, the time and expense required to clone a gene should be considered. Relative to conventional cloning methods, PCR cloning can be more rapid, less expensive, and in some cases, the only feasible cloning strategy. It takes at least 4 d to screen 300,000 plaques from a λgt10 library. With PCR, an entire library containing $10^8$ independent recombinants (~5.4 ng DNA) can be screened in one reaction. Again, to ensure authenticity of your PCR clones, you should either use the initial PCR clone to isolate a cDNA clone from a library, or sequence at least two clones from independent PCR reactions.

## 1.3. Degenerate Oligonucleotide Theory and Codon Usage

Because the genetic code is degenerate, primers targeted to particular amino acid sequences must also be degenerate to encode the possible permutations in that sequence. Thus, a primer to a six-amino-acid sequence that has 64 possible permutations can potentially recognize 64 different nucleotide sequences, one of which is to the target gene. If two such primers are used in a PCR reaction, then there are 64 × 64 or 4096 possible permutations. The target DNA will be recognized by a small fraction (1/64) of both primers, and the amplification product from that gene will increase exponentially. However, some of the other 4095 possible permutations may recognize other gene products. This disadvantage can be ameliorated by performing nested amplifications and by using "guessmer" primers. A guessmer primer is made by considering the preferential codon usage exhibited by many species and tissues (*see* Section 3.1.). For instance, the four codons for alanine begin with GC. In the third position of this codon, G is rarely used in humans (~10.3% of the time) or rats (~8.0%), but often used in *Escherichia coli* (~35%) *(5)*. This characteristic of codon usage may be advantageously used when designing degenerate oligonucleotide primers.

## 1.4. Strategy for Cloning Aquaporin Gene Family Members

In a related methods chapter *(3)* I described the cloning by degenerate primer PCR of *Aquaporin*-1 (formerly CHIP28) from a human fetal liver λgt11 cDNA library starting with the first 35 amino acids from the *N*-terminus of the purified protein. A full-length cDNA was subsequently isolated from an adult human bone marrow cDNA library (4), and following expression in *Xenopus* shown to encode a water selective channel *(6)*. We now know that the *Aquaporin* family of molecular water channels includes genes expressed in diverse species, including bacteria, yeast, plants, insects, amphibians, and mammals *(1,2,7)*. We have recently used degenerate oligonucleotide primers designed to highly conserved amino acids between the different members of the *Aquaporin* family to clone novel *Aquaporin* gene family cDNAs from rat brain (*AQP*4) and salivary gland (*AQP*5) libraries *(8,9)*. In Section 3., I will describe the creation of a new set of degenerate primers that we are currently using to clone, by degenerate primer PCR, *Aquaporin* homologs from a number of different tissues and species. Section 3. has been broken up into three parts.

1. Section 3.1. describes the designing of the degenerate primers.
2. Section 3.2. describes the PCR-amplification with degenerate primers.
3. Section 3.3. describes the subcloning and DNA sequencing of the specific PCR-amplified products.

## 2. Materials

### 2.1. Design of Degenerate Oligonucleotide Primers

No special materials are required here, except the amino acid sequence to which the degenerate primers are going to be designed and a codon usage table *(5)*. If the degenerate primers are going to be designed according to a family of related amino acid sequences, these sequences should be aligned using a multiple sequence alignment program. A degenerate nucleotide alphabet (Table 1) provides a single letter designation for any combination of nucleotides. Some investigators have successfully employed mixed primers containing inosine where degeneracy was maximal, assuming inosine is neutral with respect to base pairing, to amplify rare cDNAs by PCR *(10,11)*.

### 2.2. PCR Amplification with Degenerate Primers

For all buffers and reagents, distilled deionized water should be used. All buffers and reagents for PCR should be made up in distilled deionized 0.2-μ filtered water that has been autoclaved (PCR-water) using sterile tubes and

**Table 1**
**The Degenerate Nucleotide Alphabet**

| Letter | Specification |
|--------|---------------|
| A | Adenosine |
| C | Cytidine |
| G | Guanosine |
| T | Thymidine |
| R | puRine (A or G) |
| Y | pYrimidine (C or T) |
| K | Keto (G or T) |
| M | aMino (A or C) |
| S | Strong (G or C) |
| W | Weak (A or T) |
| B | Not A (G, C, or T) |
| D | Not C (A, G, or T) |
| H | Not G (A, C, or T) |
| V | Not T (A, C, or G) |
| N | aNy (A, G, C, or T) |
| I | Inosine[a] |

[a]Although inosine is not a true nucleotide, it is included in this degenerate nucleotide list since many researchers have employed inosine-containing oligonucleotide primers in cloning gene family members.

aerosol blocking pipet tips to prevent DNA contamination (*see* Note 1). All plastic supplies (microfuge tubes, pipet tips, and so on) should be sterilized by autoclaving or purchased sterile.

1. 10X PCR reaction buffer: 100 m$M$ Tris-HCl, pH 8.3, at 25°C, 500 m$M$ KCl, 15 m$M$ MgCl$_2$, 0.1% w/v gelatin. Incubate at 50°C to melt the gelatin, filter sterilize, and store at –20°C (*see* Note 2).
2. dNTP stock solution (1.25 m$M$ dATP, dGTP, dCTP, dTTP) made by diluting commercially available deoxynucleotides with PCR-water.
3. Thermostable DNA polymerase, such as Amplitaq DNA Polymerase (Perkin Elmer Cetus, Norwalk, CT) supplied at 5 U/μL.
4. Mineral oil.
5. A programmable thermal cycler machine, available from a number of manufacturers, including Perkin Elmer Cetus, MJ Research, and Stratagene.
6. Degenerate oligonucleotide primers should be purified by reverse-phase high performance liquid chromatography (HPLC) or elution from acrylamide gels, dried down, resuspended at 20 pmol/μL in PCR-water, and stored at –20°C, preferably in aliquots.

7. The DNA template can be almost any DNA sample, including a single-stranded cDNA from a reverse transcription reaction, DNA from a phage library, and genomic DNA. The DNA is heat denatured at 99°C for 10 min and stored at 4 or –20°C.
8. Chloroform (*see* Note 3).
9. Tris-saturated phenol (*see* Note 3), prepared using ultra pure redistilled crystalline phenol as recommended by the supplier (Gibco-BRL [product #5509], Gaithersburg, MD). Use polypropylene or glass tubes for preparation and storage.
10. PC9 (*see* Note 3): Mix equal volumes of buffer-saturated phenol, pH >7.2, and chloroform, extract twice with an equal volume of 100 m$M$ Tris-HCl, pH 9.0, separate phases by centrifugation at room temperature for 5 min at 2000$g$, and store at 4 to –20°C for up to 1 mo.
11. 7.5$M$ AmAc for precipitation of DNA. Ammonium acetate is preferred over sodium acetate because nucleotides and primers generally do not precipitate with it. Dissolve in water, filter through 0.2-μm membrane, and store at room temperature.
12. 100% ethanol, stored at –20°C.
13. 70% ethanol, stored at –20°C.
14. *TE*: 10 m$M$ Tris, 0.2 m$M$ EDTA, pH 8.0. Dissolve in water, filter through 0.2-μm membrane, and store at room temperature.
15. 50X TAE: 242 g Tris-HCl base, 57.1 mL acetic acid, 18.6 g $Na_2(H_2O)_2$EDTA. Dissolve in water, adjust volume to 1 L, and filter through 0.2-μm membrane. Store at room temperature.
16. *Hae*III digested ϕX174 DNA markers. Other DNA molecular weight markers can be used depending on availability and the size of the expected PCR-amplified products.
17. 6X gel loading buffer (GLOB): 0.25% bromophenol blue, 0.25% xylene cyanol FF, 1 m$M$ EDTA, 30% glycerol in water. Store up to 4 mo at 4°C.
18. Agarose gel electrophoresis apparatus and electrophoresis grade agarose. For the optimal resolution of DNA products <500 bp in length, NuSieve GTG agarose (FMC BioProducts) is recommended.
19. Ethidium bromide (*see* Note 3). 10 mg/mL stock of ethidium bromide (EtBr) prepared in water and stored at 4°C in a brown or foil wrapped bottle. Use at 0.5–2.0 μg/mL in water for staining nucleic acids in agarose or acrylamide gels.
20. For the elution of specific PCR-amplified DNA products from agarose gels, several methods are available, including electroelution and electrophoresis onto DEAE-cellulose membranes *(12,13)*. Several commercially available kits will also accomplish this task. I have had some success with GeneClean II (Bio 101, La Jolla, CA) for PCR products >500 bp in length, and with QIAEX (Qiagen, Chatsworth, CA) for products from 50–5000 bp. If you do not know the approximate size of the PCR-amplified products and wish to clone all of them, the QIAquick-spin PCR purification kit is recommended (Qiagen) since this will remove all nucleotides and primers before attempting to clone. This kit is also recommended for purification of PCR products for secondary PCR-amplification reactions.

## 2.3. Cloning and DNA Sequencing of PCR-Amplified Products

1. From Section 2.2., items 8–14 and 20.
2. pBluescript II phagemid vector (Stratagene). A number of comparable bacterial expression vectors are available from several companies.
3. Restriction enzymes: *Eco*RV (for blunt-end ligation).
4. Calf intestinal alkaline phosphatase (CIP) (New England Biolabs, Beverly, MA).
5. Klenow fragment of *E. coli* DNA polymerase I (sequencing grade preferred) and 10 m*M* dNTP solution (dilute PCR or sequencing grade dNTPs).
6. T4 DNA ligase (1 or 5 U/µL) and 5X T4 DNA ligase buffer (Gibco-BRL).
7. Competent DH5α bacteria. Can be prepared *(12,13)* or purchased. Other bacterial strains can be substituted.
8. Ampicillin: 50 mg/mL stock in water, 0.2-µ filtered, stored in aliquots at –20°C (*see* Note 4).
9. LB media: 10 g bacto-tryptone, 5 g bacto-yeast extract, and 10 g NaCl dissolved in 1 L water. Adjust pH to 7.0. Sterilize by autoclaving for 20 min on liquid cycle.
10. LB-Amp plates: Add 15 g bacto-agar to 1000 mL LB media prior to autoclaving for 20 min on the liquid cycle. Gently swirl the media on removing it from the autoclave to distribute the melted agar. **Be careful:** The fluid may be superheated and may boil over when swirled. Place the media in a 50°C water bath to cool. Add 1 mL of ampicillin, swirl to distribute, and pour 25–35 mL/90-mm plate. Carefully flame the surface of the media with a Bunsen burner to remove air bubbles before the agar hardens. Store inverted overnight at room temperature, then wrapped at 4°C for up to 6 mo.
11. IPTG: Dissolve 1 g isopropylthiogalactoside in 4 mL water, filter through 0.2-µm membrane, and store in aliquots at –20°C.
12. X-Gal: Dissolve 100 mg 5-bromo-4-chloro-3-indolyl-β-ᴅ-galactopyranoside in 5 mL dimethylformamide and stored at –20°C in a foil wrapped tube (light sensitive).
13. Plasmid DNA isolation equipment and supplies *(12,13)* or plasmid DNA isolation kits, available from many manufacturers.
14. Double-stranded DNA sequencing equipment and supplies *(12,13)*, or access to a DNA sequencing core facility.

## 3. Methods

## 3.1. Design of Degenerate Oligonucleotide Primers

1. The first step in designing a degenerate primer is to select a conserved amino acid sequence, then determine the potential nucleotide sequence (or the complement of this sequence for a downstream primer), considering all possible permutations. If the amino acid sequence is relatively long, you can potentially design two or more degenerate primers. If only one is made, make it to sequences with a high (50–65%) GC content, because these primers can be annealed under more stringent conditions (e.g., higher temperatures). Figure 1 shows an alignment of the amino acid sequences for several members of the *Aquaporin* gene family in the two most highly conserved regions. Also shown is the consensus amino acid

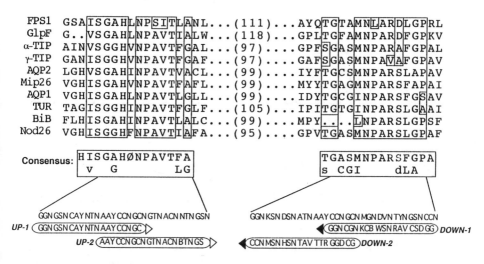

**Fig. 1.** Design of degenerate primers to amplify *Aquaporin* gene family members. **(Top)** The amino acid sequences of ten MIP family proteins, including the *S. cerevisiae* FPS1 *(23)*, *E. coli* GlpF *(24)*, α- and γ-tonoplast intrinsic proteins (TIP) of *Arabidopsis thaliana (25)*, the vasopressin-responsive water channel of rat renal collecting tubules *(AQP2) (26)*, the major intrinsic protein (MIP) of bovine lens fiber membranes *(27)*, human *Aquaporin*-1 *(4)*, turgor responsive gene (TUR) 7a from *Pisum stivum (28)*, the *Drosophila* neurogenic *big brain* protein *(29)*, and the *Rhyzodium* root Nodulin-26 peribacteroid membrane protein *(30)* were aligned by the PILEUP program of progressive alignments *(31)* using a gap weight of 3.0 and a gap length of 0.1 running on a VAX computer system. The two most highly conserved regions are shown, separated by the number of intervening amino acids. The most highly conserved amino acids are enclosed. **(Middle)** Below the aligned sequences, the consensus amino acid sequences are shown. **(Bottom)** From part of the consensus amino acid sequences, the degenerate nucleotide sequences were determined (using the degenerate nucleotide alphabet from Table 1) followed by the sequences for the degenerate oligonucleotide primers.

sequence, the degenerate nucleotide sequence, and the sequence of the primers we are currently using to isolate *Aquaporin* gene family members. Interestingly, not only are these two regions highly conserved among the different members of this gene family, but they are also highly related to each other, with the conserved motif being (T/S)GxxxNPAxx(F/L)G, that has been speculated to have resulted from an ancient internal duplication in a primordial bacterial organism, since this repeat has persisted in *Aquaporin* homologs from bacteria through plants and mammals *(1,6,14)*. These two regions are functionally related, both contributing to the formation of the water pore in *Aquaporin*-1 *(15)*.

2. The next step is to determine the number of permutations in the nucleotide sequence. There are 192 permutations ([2 × 4] × 3 × 4 × 2) in the sequence 5'-YTN-ATH-

GGN-GAR-3' which encodes the hypothetical amino acid sequence *Leu-Ile-Gly-Glu*. We can reduce the degeneracy by making educated guesses in the nucleotide sequence, i.e., by making a guessmer. The 3'-end of a primer should contain all possible permutations in the amino acid sequence, since *Taq* DNA polymerase will not extend a prime with a mismatch at the extending (3') end. If the above primer was to a human gene, a potential guessmer would be 5'-CTB-ATY-GGN-GAR-3', which only contains 64 permutations. This guessmer is proposed by taking into account the preferential codon usage for leucine and isoleucine in humans *(5)*.

3. The degeneracy of a primer can be reduced further by incorporating inosine residues in the place of *N*. The advantages of using inosine-containing primers is that they have a reduced number of permutations, and the inosine reportedly base pairs equally well with all four nucleotides, creating a single bond in all cases *(10)*. The disadvantage is that inosines reduce the annealing temperature of the primer. I have not employed inosine-containing primers in these studies.

4. It is often convenient to incorporate restriction endonuclease sites at the 5'-ends of a primer to facilitate cloning into plasmid vectors *(4,8,9)*. Different restriction sites can be added to the 5'-ends of different primers so the products can be cloned directionally. However, not all restriction enzymes can recognize cognate sites at the ends of a double-stranded DNA molecule equally well. This difficulty can often be reduced by adding a two to four nucleotide 5'-overhang before the beginning of the restriction enzyme site (*see* Note 5). Some of the best restriction enzymes sites to use are *Eco*RI, *Bam*HI, and *Xba*I. Catalogs from New England Biolabs have a list of the ability of different restriction enzymes to recognize short base-pair sequences. A potential pitfall of this approach would be the occurrence of the same restriction site within the amplified product as used on the end of one of the primers. Therefore, only part of the amplified product would be cloned.

5. The final consideration you should make is the identity of the 3' most nucleotide. The nucleotide on the 3'-end of a primer should preferably be G or C, and not be N, I, or T. The reason for this is that thymidine (and supposedly inosine) can nonspecifically prime on any sequence. Guanosines and cytidine are preferred since they form three H-bonds at the end of the primer, a degree stronger than an A:T base pair.

## 3.2. PCR Amplification and DNA Purification

The template for these reactions can be the DNA in a phage library or the first-strand cDNA from a reverse transcription reaction on RNA. A phage library with a titer of $5 \times 10^9$ pfu/mL would contain, in a 5-µL aliquot, $2.5 \times 10^7$ pfu (~1.5 ng of DNA). Prior to PCR-amplification, the DNA is heat denatured at 99°C for 10 min.

### 3.2.1. PCR Reaction (see Notes 1 and 6)

In all cases, the DNA template should also be PCR-amplified with the individual degenerate primers to determine if any of the bands amplified are derived from one of the degenerate primer pools. A DNA-free control is required to assess if there is contaminating DNA in any of the other reagents.

1. Pipet into 0.5-mL microcentrifuge tubes in the following order: 58.5 µL PCR-water that has been autoclaved; 10 µL 10X PCR reaction buffer (*see* Note 2); 16 µL 1.25 m*M* dNTP stock solution; 5.0 µL primer up-1; 5.0 µL primer down-1; and 5.0 µL heat-denatured library or cDNA (1–100 ng). If several reactions are being set up concurrently, a master reaction mix can be made up consisting of all the reagents used in all of the reactions, such as the PCR-water, reaction buffer, and dNTPs.
2. Briefly vortex each sample and spin for 10 s in a microfuge. Overlay each sample with 2–3 drops of mineral oil.
3. Amplify by hot-start PCR using the following cycle parameters. Pause the thermocycler in step 4-cycle 1, and add 0.5 µL Amplitaq DNA polymerase to each tube. 95°C, 5 min (initial denaturation); 94°C, 60 s (denaturation); 50°C, 90 s (annealing; *see* Note 7); 72°C, 60 s (extension); cycle 29 times to step 2; 72°C, 4 min; and 10°C hold.

### 3.2.2. DNA Isolation and Gel Electrophoresis Analysis

1. Remove the reaction tubes from the thermal cycler and add 200 µL chloroform. Spin for 10 s in a microfuge to separate the oil-chloroform layer from the aqueous layer. Carefully transfer the aqueous layer to a clean microfuge tube.
2. Remove the AmpliTaq DNA polymerase by extracting the aqueous phase twice with 100 µL PC9 (*see* Note 3). Spin for 2 min in a microfuge to separate the lower organic layer from the upper aqueous layer and transfer the aqueous layer to clean microfuge tube. This step is essential before digesting the DNA with restriction enzymes for directional cloning (*see* Section 3.3.) since the polymerase can precipitate, and in the presence of nucleotides, fill in recessed 3' termini on DNA.
3. AmAc-EtOH precipitation: To a 100 µL DNA sample add 50 µL 7.5*M* AmAc (50% vol). Vortex briefly to mix. Precipitate the DNA with 350 µL 100% ethanol (2–2.5 vol). Vortex the samples for 15 s and ice for 15 min. Spin down the DNA at 12,000*g* for 15 min at 4°C in a microfuge. Decant the aqueous waste. Add 250 µL 70% ethanol. Vortex briefly and spin another 5 min at 4°C. Decant the ethanol and allow the pellets to dry inverted at room temperature, or dry in a Speed-Vac for 2–10 min.
4. Resuspend in 20 µL PCR-water.
5. The next step is to resolve an aliquot (2–10-µL) of the PCR fragments by gel electrophoresis. Small DNA products (<300 bp) can be resolved at high resolution on 5–10% polyacrylamide gels *(12,13)*. Moderate-sized PCR products (150–1000 bp) should be resolved on 2–4% NuSieve agarose gels (in 1X TAE buffer). Larger PCR products (>500 bp) can be resolved on 0.8–2% agarose gels (1X TAE buffer).
6. After the bromophenol blue dye has reached the end of the gel, soak the gel for 5–30 min in about 10 vol of water containing 1 µg/mL EtBr (*see* Note 3). Then view and photograph the gel under UV light. As shown in Fig. 1, there is little

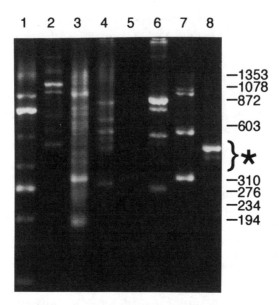

**Fig. 2.** Gel electrophoresis analysis of PCR-amplified DNA. DNA isolated from a human kidney cDNA library in bacteriophage λgt10 was amplified with degenerate primers up-1 (lanes 1, 5, and 6), up-2 (lanes 2, 7, and 8), down-1 (lanes 3, 5, and 7), and down-2 (lanes 4, 6, and 8). Reactions containing $5 \times 10^6$ *pfu* of heat-denatured phage DNA, 100 pmol of degenerate primers, and 1.5 m$M$ MgCl$_2$ in a 100 μL volume were subject to 40 cycles of PCR-amplification under the following parameters: 94°C for 60 s, 48°C for 90 s, and 72°C for 60 s. Following chloroform extraction and ethanol precipitation, the DNA was resuspended in 20 μL of water, and 5 μL was electrophoresed into a 4% NuSieve agarose gel in 1X TAE. The gel was stained with ethidium bromide and photographed. The relative mobility of *Hae*III digested ϕX174 DNA markers is shown on the right. The bracket shows the size range of known members of this gene family from the primers employed.

variability in the distance between the NPA motifs with the known members of the *Aquaporin* gene family. PCR-amplification of the known *Aquaporin*s cDNAs using the internal degenerate primers would generate products from 345–415 bp. A typical result is shown in Fig. 2.

### 3.2.3. Secondary PCR Amplifications and DNA Purification

Based on the results from gel electrophoresis of the PCR-amplified DNA products, a decision must be made on what to do next. The options are the following.

1. Amplify by PCR from the initial DNA sample under different conditions.
2. Amplify by PCR from a different DNA sample under the same conditions. (Different MgCl$_2$ concentration, annealing temperature, or primers, *see* Notes 2, 6, and 7.)

3. Gel purify a band(s) of DNA from the gel for cloning or to reamplify by PCR.
4. Purify all PCR-amplified DNA fragments for cloning or to reamplify by PCR.
5. Reamplify by PCR with the same or an internal pair of degenerate primers.

Options 1 and 2 are self explanatory. If you want to gel purify a particular band or group of bands from an agarose gel, a number of procedures and kits are available (*see* Section 2.2.). If you plan on immediately cloning a PCR band(s), you may want to run the rest of the initial PCR reaction on another gel to increase the recovery of DNA. It is also possible to recover specific DNA fragments from an acrylamide gel *(3,12,13)*. To purify all PCR-amplified DNA fragments from the remaining sample, a number of methods are available, including the QIAquick-spin PCR purification kit, which can be used instead of steps 1–3 in Section 3.2.2. (Qiagen). Finally, aliquots of DNA purified from a gel or from the initial PCR reaction (1–10%) can be reamplified by PCR with either the same or an internal pair of degenerate oligonucleotide primers (*see* Note 1).

When attempting to identify a gene family homolog from a tissue that is known to express a homolog(s), a number of tricks can be tried to enrich the final PCR sample for new homologs. Since the degenerate oligonucleotide primers are designed from the sequence of the known gene family members, these primers will likely be biased for those homologs. *Aquaporin*-1 is abundant in the capillaries around the salivary glands and throughout the body, but absent in the salivary gland *(16)*. To identify a salivary homology of the *Aquaporin* gene family, we used a rat salivary gland cDNA library, that also contained *Aquaporin*-1 cDNAs, presumably from the surrounding capillaries. We first amplified the cDNA library with an external set of degenerate primers, digested the PCR-amplified DNAs with the restriction enzyme *Pst*I (which cuts between the NPA motifs of rat *AQP*1), and reamplified with an internal pair of primers. We again digested with *Pst*I to digest the rat *AQP*1 DNAs, then cloned and sequenced the DNA fragments between 350 and 450 bp *(9)*. This strategy would not work if the resulting cDNA (*AQP*5) also contained a *Pst*I site. By trying different restriction enzymes that cut DNA infrequently (6–8 bp-recognition sites), a number of new homologs will preferentially be identified. Alternatively, after cloning the DNA products into bacterial expression vectors, bacterial colony lift hybridization can be used to identify colonies containing inserts for known gene family members *(3,12,13)*.

## 3.3. Cloning and DNA Sequencing of PCR-Amplified Products

### 3.3.1. Preparation of Vector for Ligation

1. For blunt-end ligations, digest 1 µg pBluescript II KS phagemid DNA (Stratagene) with 10 U *Eco*RV in a 50 µL vol. Incubate at 37°C for 2 h. For

cohesive-end ligations, similarly digest the vector with the appropriate restriction enzyme(s).

2. For both blunt-end ligations and cohesive-end ligations where the vector has been digested with only one restriction enzyme, it is necessary to remove the 5'-phosphate from the vector to inhibit the vector from self ligating. This is accomplished by treating the vector with CIP according to the manufacturer's recommendations. Note that 1 μg of a 3 kbp linear DNA molecule contains 1 pmol of 5'-overhangs (*Bam*HI), blunt-ends (*Eco*RV), or 3'-overhangs (*Pst*I), depending on the enzyme that digested it. Afterward add EDTA to 5 m*M* and heat-kill the enzyme at 65°C for 1 h. Adjust the volume to 50–100 μL with *TE* and extract once with Tris-saturated phenol, twice with PC9, and twice with chloroform. Back extract each organic layer with 50 μL *TE* and pool with the final sample. AmAc-EtOH precipitate (*see* Section 3.2.2.) and resuspend in 10 μL water.

3. If the insert is going to be directionally cloned into the vector, just extract once with 50 μL PC9, AmAc-EtOH precipitate (*see* Section 3.2.2.), and resuspend in 10 μL water.

### *3.3.2. Preparation of Inserts for Ligation*

AmpliTaq and other thermostable DNA polymerases often fail to completely fill in the ends of the double-stranded DNA products, thus leaving recessed 3' termini that can be filled in with the Klenow fragment of *E. coli* DNA polymerase I. This should be done whether or not the DNA is going to be digested with restriction enzymes added to the ends of the primers for directional cloning (*see* Section 3.1.).

1. AmAc-EtOH precipitate the DNA (*see* Section 3.2.2.) and resuspend in 15 μL water.

2. Add 2 μL 10X restriction enzyme reaction buffer. Klenow DNA polymerase works well in most restriction enzyme digestion buffers (10X REact 2 or 3 from Gibco-BRL). If the DNA is going to be subsequently digested with a restriction enzyme(s), use the buffer for that enzyme.

3. Add 2 μL 10 m*M* dNTP solution. Then add Klenow DNA polymerase (1 U/μg DNA) and incubate at room temperature for 15 min.

4. Heat-inactivate the enzyme at 75°C for 10 min. If the DNA is going to be directly used in ligation reactions it is not necessary to purify the DNA from the unincorporated dNTPs, since they will not inhibit T4 DNA ligase. To concentrate the DNA sample, proceed with step 6.

5. PCR products containing restriction sites on their ends should now be digested with the restriction enzymes. Incubate in the appropriate buffer, using 20 U of enzyme/μg of DNA and incubating for 2–4 h at the proper temperature.

6. Extract the DNA once or twice with PC9 and precipitate with AmAc-EtOH as described above (*see* Section 3.2.2.). Resuspend the final pellet in 5–10 μL water.

### 3.3.3. DNA Ligation and Bacterial Transformation

1. At this point it is often advantageous to run a small aliquot of the different DNA fragments on a gel to assess their approximate concentrations and purity. Ideally you want at least a 2:1 molar ratio of insert to vector in the ligation reactions. If necessary return to the above procedures to isolate more DNA for the ligation reaction.
2. Set up the ligation reactions with the vector and insert similar to the following:
   a. Reaction 1: 1 μL vector (10 ng; vector control);
   b. Reaction 2: 1 μL vector + 1 μL insert (~10 ng insert);
   c. Reaction 3: 1 μL vector + 4 μL insert.
   Then add 2 μL of 5X T4 DNA ligase buffer (Gibco-BRL) and water to 9.5 μL. If the buffer is more than 4-mo-old, the ATP may be depleted. Therefore, add fresh ATP to a final concentration of 1 m$M$.
3. For cohesive-end ligations add 0.5 μL of T4 DNA ligase (1 U/μL), gently mix, spin 5 s in a microfuge, and incubate at 15°C for 10–20 h. For blunt-end ligations add 1 μL of T4 DNA ligase (5 U/μL), gently mix, spin 5 s in a microfuge, and incubate at 25°C (or room temperature) for 1–12 h. Stop the reaction by heating at 75°C for 10 min and store the samples at –20°C.
4. Set up a bacterial transformation with competent DH5α bacteria or a comparable strain of bacteria. Be sure to include a positive control (10 ng undigested vector DNA) and a negative control (water). To 1.5-mL microfuge tubes, add half of the ligation mix (5 μL) or 5 μL of control DNA or water and 50 μL of competent bacteria (thawed slowly on ice); incubate on ice for 30 min. Heat-shock at 42°C for 2 min. Return to ice for 1 min. Add 200 μL of LB media containing 10% glycerol. Mix gently and allow bacteria to recover and express the ampicillin resistance gene by incubating at 37°C for 1 h.
5. Prewarm LB-Amp plates at 37°C for 45 min. About 30 min before plating the bacteria on the plates, add 40 μL of X-Gal and 4 μL IPTG and quickly spread over the entire surface of the plate using a sterile glass spreader. Spread 20–200 μL of the transformation reactions on these plates. Allow the inoculum to absorb into the agar and incubate the plates inverted at 37°C for 12–24 h (*see* Note 4). Afterward, placing the plates at 4°C for 2–4 h will help enhance the blue color development.

### 3.3.4. Plasmid DNA Minipreps and DNA Sequencing

1. Colonies that contain active β-galactosidase will appear blue, whereas those containing a disrupted *LacZ* gene will be white. Set up minicultures by inoculating individual white colonies into 2 mL of LB media containing ampicillin. After growing at 37°C overnight, isolate the plasmid DNA. Resuspend the DNA in 20–50 μL water or *TE*.
2. Digest 5–20 μL of the DNA with the appropriate restriction enzymes and analyze by agarose gel electrophoresis (*see* Section 3.2.3.).
3. Perform double-stranded DNA sequencing on recombinants containing inserts in the expected size range.

## 4. Notes

1. All PCR reactions should be set up in sterile laminar flow hoods using pipet tips containing filters (aerosol-resistant tips) to prevent the contamination of samples, primers, nucleotides, and reaction buffers by DNA. If the PCR reaction is going to be reamplified by PCR, all possible intervening steps should also be performed in a sterile hood with the same precautions to prevent DNA contamination. These precautions should also be extended to all extractions and reactions on the nucleic acid (RNA or DNA) through the last PCR reaction. Likewise, all primers, nucleotides, and reaction buffers for PCR should be made up and aliquoted using similar precautions. All buffers for PCR should be made with great care using sterile disposable plastic or baked glass, and restricted for use with aerosol-resistant pipet tips.

2. Standard PCR reaction buffers contain 15 m$M$ MgCl$_2$ (1.5 m$M$ final concentration). In many cases, changes in the MgCl$_2$ concentration will have significant consequences on the amplification of specific bands. In PCR-amplifying the four exons of the $AQP$1 gene, MgCl$_2$ concentrations between 0.7 and 1.0 m$M$ gave the best results *(17,18)*, however MgCl$_2$ concentrations between 0.5 and 5.0 m$M$ have been reported.

3. Organic solvents and ethidium bromide are hazardous materials. Always handle with tremendous caution, wearing gloves and eye protection. Contact your hazardous waste department for proper disposal procedures in your area.

4. Ampicillin-resistant bacteria secrete β-lactamase into the media, which rapidly inactivates the antibiotic in regions surrounding the growing bacterial colony. Thus, when bacteria are growing at a high density or for long periods (>16 h), ampicillin-sensitive satellite colonies will appear around the primary colonies (which are white in blue-white selections). This problem can be ameliorated (but not eliminated) by substitution of carbenicillin for ampicillin on agar plates.

5. When designing primers with restriction enzyme sites and 5'-overhangs, note that the 5'-overhang should not contain sequences complementary to the sequence just 3' of the restriction site, because this would facilitate the production of primer-dimers. Consider the primer 5'-ggg.<u>agatct</u>.CCCAGCTAGCTAGCT-3', which has a *Xba*I site proceeded by a 5'-ggg and followed by a CCC-3'. These 12 nucleotides on the 5'-end are palindromic, and can therefore easily dimerize with another like primer. A better 5'-overhang would be 5'-cac.

6. When cloning a gene from a recombinant library by PCR, remember that not all genes are created equally. Genes with high G:C contents have proven more difficult to clone than most. Several researchers have made contributions in a search for factors to enhance the specificity of PCR reactions. Nonionic detergents, such a Nonident P-40, can be incorporated in rapid sample preparations for PCR analysis without significantly affecting *Taq* polymerase activity *(19)*. In some cases, such detergents are absolutely required in order to reproducibly detect a specific product *(20)* presumably because of inter- and intrastrand secondary structure. More recently, tetramethylammonium chloride has been shown to enhance the specificity of PCR reactions by reducing nonspecific priming events *(21)*. Commercially available PCR enhancers are also available.

7. A critical parameter when attempting to clone by PCR is the selection of a primer annealing temperature. This is especially true when using degenerate primers. The primer melting temperature $(T_m)$ is calculated by adding 2° for A:T base pairs, 3° for G:C base pairs; 2° for N:N base pairs, and 1° for I:N base pairs. Most PCR chapters suggest you calculate the $T_m$ and set the primer annealing temperature to 5–10°C below the lowest $T_m$. Distantly related gene superfamily members have been cloned using this rationale *(22)*. However, I have found that higher annealing temperatures are helpful in reducing nonspecific priming, which can significantly affect reactions containing degenerate primers.

## Acknowledgments

I thank my colleagues, especially Peter Agre and William B. Guggino, for their support and helpful discussions. This work was supported in part by NIH grants HL33991 and HL48268 to Peter Agre.

## References

1. Reizer, J., Reizer, A., and Saier, M. H., Jr. (1993) The MIP family of integral membrane channel proteins: sequence comparisons, evolutionary relationships, reconstructed pathways of evolution, and proposed functional differentiation of the two repeated halves of the proteins. *Crit. Rev. Biochem. Mol. Biol.* **28,** 235–257.
2. Knepper, M. A. (1994) The *aquaporin* family of molecular water channels. *Proc. Natl. Acad. Sci. USA* **91,** 6255–6258.
3. Preston, G. M. (1993) Use of degenerate oligonucleotide primers and the polymerase chain reaction to clone gene family members, in *Methods in Molecular Biology,* vol. 15: *PCR Protocols: Current Methods and Applications.* (White, B. A., ed.) Humana, Totowa, NJ, pp. 317–337.
4. Preston, G. M. and Agre, P. (1991) Isolation of the cDNA for erythrocyte integral membrane protein of 28 kilodaltons: member of an ancient channel family. *Proc. Natl. Acad. Sci. USA* **88,** 11,110–11,114.
5. Wada, K.-N., Aota, S.-I., Tsuchiya, R., Ishibashi, F., Gojobori, T., and Ikemura, T. (1990) Codon usage tabulated from the GenBank genetic sequence data. *Nucleic Acids Res.* **18,** 2367–2411.
6. Preston, G. M., Carroll, T. P., Guggino, W. B., and Agre, P. (1992) Appearance of water channels in *Xenopus* oocytes expressing red cell CHIP28 protein. *Science* **256,** 385–387.
7. Chrispeels, M. J. and Agre, P. (1994) *Aquaporins*: water channel proteins of plant and animal cells. *TIBS* **19,** 421–425.
8. Jung, J. S., Bhat, B. V., Preston, G. M., Guggino, W. B., Baraban, J. M., and Agre P. (1994) Molecular characterization of an *aquaporin* cDNA from brain: candidate osmoreceptor and regulator of water balance. *Proc. Natl. Acad. Sci. USA* **91,** 13,052–13,056.
9. Raina, S., Preston, G. M., Guggino, W. B., and Agre, P. (1995) Molecular cloning and characterization of an *aquaporin* cDNA from salivary, lacrimal and respiratory tissues. *J. Biol. Chem.* **270,** 1908–1912.

10. Knoth, K., Roberds, S., Poteet, C., and Tamkun, M. (1988) Highly degenerate, inosine-containing primers specifically amplify rare cDNA using the polymerase chain reaction. *Nucleic Acids Res.* **16**, 10,932.

11. Chérif-Zahar, B., Bloy, C., Kim, C. L. V., Blanchard, D., Bailly, P., Hermand, P., Salmon, C., Cartron, J.-P., and Colin, Y. (1990) Molecular cloning and protein structure of a human blood group Rh polypeptide. *Proc. Natl. Acad. Sci. USA* **87**, 6243–6247.

12. Sambrook, J., Fritsch, E. F., and Maniatis, T., eds. (1989) *Molecular Cloning: A Laboratory Manual.* Cold Spring Harbor Laboratory, Cold Spring Harbor, NY.

13. Ausubel, F. M., Brent, R., Kingston, R. E., Moore, D. D., Seidman, J. G., Smith, J. A., and Struhl, K., eds. (1994) *Current Protocols in Molecular Biology.* Greene Publishing/Wiley-Interscience, New York.

14. Wistow, G. J., Pisano, M. M., and Chepelinsky, A. B. (1991) Tandem sequence repeats in transmembrane channel proteins. *TIBS* **16**, 170,171.

15. Jung, J. S., Preston, G. M., Smith, B. L., Guggino, W. B., and Agre, P. (1994) Molecular structure of the water channel through *aquaporin* CHIP: the hourglass model. *J. Biol. Chem.* **269**, 14,648–14,654.

16. Nielsen, S., Smith, B. L., Christensen, E. I., and Agre, P. (1993) Distribution of the *aquaporin* CHIP in secretory and resorptive epithelia and capillary endothelia. *Proc. Natl. Acad. Sci. USA* **90**, 7275–7279.

17. Smith, B. L., Preston, G. M., Spring, F. A., Anstee, D. J., and Agre, P. (1994) Human red cell *Aquaporin* CHIP, I. molecular characterization of ABH and Colton blood group antigens. *J. Clin. Invest.* **94**, 1043–1049.

18. Preston, G. M., Smith, B. L., Zeidel, M. L., Moulds, J. J., and Agre, P. (1994) Mutations in *aquaporin*-1 in phenotypically normal humans without functional CHIP water channels. *Science* **265**, 1585–1587.

19. Weyant, R. S., Edmonds, P., and Swaminathan, B. (1990) Effects of ionic and nonionic detergents on the *Taq* polymerase. *BioTechnology* **9**, 308,309.

20. Bookstein, R., Lai, C-C., To, H., and Lee, W-H. (1990) PCR-based detection of a polymorphic *Bam*HI site in intron 1 of the human retinoblastoma (RB) gene. *Nucleic Acids Res.* **18**, 1666.

21. Hung, T., Mak, K., and Fong, K. (1990) A specificity enhancer for polymerase chain reaction. *Nucleic Acids Res.* **18**, 4953.

22. Zhao, Z.-Y. and Joho, R. H. (1990) Isolation of distantly related members in a multigene family using the polymerase chain reaction technique. *Biochem. Biophys. Res. Comm.* **167**, 174–182.

23. Aelst, L. V., Hohmann, S., Zimmermann, F. K., Jans, A. W. H., and Thevelein, J. M. (1991) A yeast homologue of the bovine lens fiber MIP gene family complements the growth defect of a *Saccharomyces cerevisiae* mutant on fermentable sugars but not its defect in glucose-induced RAS-mediated cAMP signalling. *EMBO J.* **10**, 2095–2104.

24. Muramatsu, S. and Mizuno, T. (1989) Nucleotide sequence of the region encompassing the *glpKF* operon and its upstream region containing a bent DNA sequence of *Escherichia coli*. *Nucleic Acids Res.* **17**, 4378.

25. Höfte, H., Hubbard, L., Reizer, J., Ludevid, D., Herman, E. M., and Chrispeels, M. J. (1992) Vegetative and seed-specific forms of Tonoplast Intrinsic Protein in the vacuolar membrane of *Arabidopsis thaliana*. *Plant Physiol.* **99,** 561–570.

26. Fushimi, K., Uchida, S., Hara, Y., Hirata, Y., Marumo, F., and Sasaki, S. (1993) Cloning and expression of apical membrane water channel of rat kidney collecting tubule. *Nature* **361,** 549–552.

27. Gorin, M. B., Yancey, S. B., Cline, J., Revel, J.-R., and Horwitz, J. (1984) The major intrinsic protein (MIP) of the bovine lens fiber membrane: characterization and structure based on cDNA cloning. *Cell* **39,** 49–59.

28. Guerrero, F. D., Jones, J. T., and Mullet, J. E. (1990) Turgor-responsive gene transcription and RNA levels increase rapidly when pea shoots are wilted: sequence and expression of three induced genes. *Plant Mol. Biol.* **15,** 11–26.

29. Rao, Y., Jan, L. Y., and Jan, Y. N. (1990) Similarity of the product of the *Drosophila* neurogenic gene *big brain* to transmembrane channel proteins. *Nature* **345,** 163–167.

30. Fortin, M. G., Morrison, N. A., and Verma, D. P. S. (1987) Nodulin-26, a peribacteroid membrane nodulin is expressed independently of the development of the peribacteroid compartment. *Nucleic Acids Res.* **15,** 813–824.

31. Feng, D.-F. and Doolittle, R. F. (1990) Progressive alignment and phylogenetic tree construction of protein sequences, in *Methods in Enzymology, vol. 183: Molecular Evolution: Computer Analysis of Protein and Nucleic Acid Sequences* (Doolittle, R. F., ed.), Academic, New York, pp. 375–387.

# 42

## Amplification Using Degenerate Primers with Multiple Inosines to Isolate Genes with Minimal Sequence Similarity

**Simona Bartl**

### 1. Introduction

The use of degenerate primers in the polymerase chain reaction (PCR) is an effective method for identifying related genes that share limited sequence similarity. Other methods, using oligonucleotide or heterologous probes, for example, may fail to identify genes that are not highly conserved, have diverged over evolutionary time, or share only short regions of structural similarity. Degenerate primer pools of >$10^6$ have been successfully used in PCR to isolate novel gene family members or homologs from distantly related species (1–3). If inosines are used at ambiguous codon positions, the size of the primer pool decreases significantly (3–5). Smaller primer pools require less reagent for each PCR and usually titration of primer concentrations is not necessary to obtain optimal results. Since inosine (I) can pair with all four traditional nucleotides, these smaller pools of inosine-containing primers will anneal to the same large number of potential target sequences (PTS) in the template as larger pools of primers that do not contain inosines.

The method described here targets particular genes by using the least number of assumptions when designing primers. It is especially useful for the isolation of genes from distantly related species and from organisms for which no codon bias is known. It can also be used to target genes for proteins that share only short regions of sequence similarity within a common functional site (i.e., a ligand-binding site). The method assumes that protein regions that are important for function will be conserved between homologs and family members. No assumptions are made about conservation at the nucleotide level, which may be fortuitous.

From: *Methods in Molecular Biology, Vol. 67: PCR Cloning Protocols: From Molecular Cloning to Genetic Engineering* Edited by: B. A. White  Humana Press Inc., Totowa, NJ

Relatively long primers are designed to regions of 6–10 amino acids in length. These regions need not be highly conserved. If several functionally similar amino acids occur at a particular position, all possible codons for those amino acids are incorporated into the primer. If a particular position is not functionally important and can contain any of a number of amino acids, there is no need to guess which amino acid may be favored. Rather, an inosine triplet can be engineered at the corresponding site in the primer *(3)*.

The resultant primers can contain seven or more inosines *(6)*. Since inosine does not help or hinder the annealing of primer and template *(7)*, there has been some concern that long stretches of inosines would form regions of annealing instability. However, if a string of inosines, up to four in a row, is embedded within a longer primer, it can function well in PCR *(3)*. It is also believed that annealing at the 3' end of each primer is especially important for efficient PCR. However, inosines can be tolerated within three bases of the 3' ends of primers *(3)*. Therefore, it is not necessary to select for amino acids encoded by fewer codons at any particular sites within the primers.

The method is shown schematically in Fig. 1. It employs three degenerate, inosine-containing primers. PCR conditions are optimized using primers 1 and 2. The resultant product is used as template for primers 2 and 3 (primer 3 being internal to 1 and 2) *(8)*. This second amplification step helps to verify the correct product and can be omitted if no suitable third site can be targeted. Final analysis requires cloning and sequencing the products. If the second amplification step does not produce favorable results, new primers may need to be designed.

Alternative methods for the design of degenerate primers have been used successfully. These include primer pools that do not employ inosines and shorter primers that have lower PTS and require lower annealing temperatures. These methods work for well-conserved genes. The goal of the method described here is to design longer primers that tolerate high PCR annealing temperatures, have a low degeneracy (by using inosine) so small quantities are used, and allow for quite high PTS. This method allows for the amplification of genes that are less well conserved through evolutionary time or within gene families.

## 2. Materials

1. Control and experimental genomic DNA or cDNA templates.
2. Three oligonucleotide primers with restriction sites incorporated at the 5' ends.
3. Reagents for PCR.
4. Equilibrated phenol/chloroform:isoamyl alcohol (24:1), sodium acetate, ethanol.
5. Restriction enzymes, *Bam*HI and *Eco*RI.
6. Reagents for agarose gel electrophoresis.
7. A system for purifying DNA from agarose gels.
8. Plasmid vector suitable of the generation of ssDNA, cut with *Bam*HI and *Eco*RI.
9. Reagents for ssDNA preps and DNA sequencing.

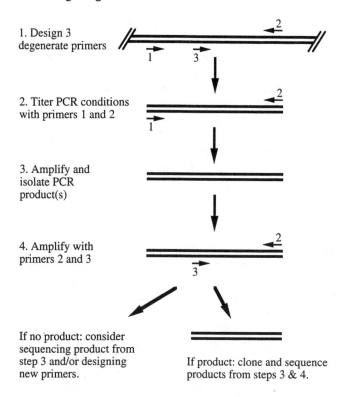

1. Design 3 degenerate primers

2. Titer PCR conditions with primers 1 and 2

3. Amplify and isolate PCR product(s)

4. Amplify with primers 2 and 3

If no product: consider sequencing product from step 3 and/or designing new primers.

If product: clone and sequence products from steps 3 & 4.

**Fig. 1.** A schematic diagram of the method. Primer 3 can be either in the sense or antisense orientation. The sense orientation is shown. If primer 3 were antisense, it would be used with primer 1 at the second amplification step.

## 3. Methods

### *3.1. Primer Design*

1. From alignments of protein sequences and information on protein function, choose three regions containing approx 6–10 amino acids that are reasonably well conserved. The best regions should contain amino acids that are encoded by the least number of codons (i.e., not serine, arginine, and lysine). The two outside regions should be separated by <700 amino acids, with the third region being internal. It is helpful if there are conserved amino acids within 60 amino acids of at least one of the primers that will easily identify the correct product on sequencing.
2. For any nucleotide positions that are fourfold degenerate, use inosines. For two- and threefold degenerate sites, incorporate those two or three nucleotides into the primer. If an amino acid position within a targeted region is not conserved, use an I triplet at this position. Sometimes one of two or more functionally equivalent amino acids are found at a particular site in the protein (i.e., aspartic acid or glutamic acid). Use all the codons necessary to all the possible amino acids.

3. Calculate the minimum $T_m$ for the primer pool. This is the $T_m$ for the primer that has an A or T at every degenerate position. Consider inosines to be neutral and, therefore, adding 0 degrees to the $T_m$ *(7)*. The example below shows the amino acid sequence to which a primer pool has been designed. This region was chosen because the cysteine residue forms an important disulfide bond that is required for the maintenance of the structure. One position is not well conserved (noted as a X), and at a second position three different amino acids can occur and all codons required to encode the three amino acids are used. In the primer sequence below an inosine triplet corresponds to the nonconserved amino acid. The minimum $T_m$ for the following oligonucleotide pool is 46°C. The number of degrees/nucleotide position is shown below the nucleotide sequence. Try to keep the minimum $T_m$ at >42°C.

| C | X | V | (TYS) | | G | F | Y | P |
|---|---|---|---|---|---|---|---|---|
| TG (TC) | III | GTI | (AT) (CAG) I | | GGI | TT (TC) | TA (TC) | CC |
| 24 2 | 000 | 420 | 2 | 2 0 | 440 | 22 2 | 22 2 | 44 |

4. Calculate the number of potential target sequences (PTS) to which the primer pool can anneal. This is the total number of primers in the pool multiplied by 4 for each I. In the example above the degeneracy is 48 ($2 \times 2 \times 3 \times 2 \times 2$) and the PTS are 196,608 ($48 \times 4 \times 4 \times 4 \times 4 \times 4 \times 4$). Primer pools with the PTS of approx 1,000,000 can be used successfully, especially if the second primer pool has a lower PTS.

5. Using this procedure, choose three primer pools that have the lowest PTS and the highest $T_m$ and that span a region >200 bp and <2 kb. Smaller products may be difficult to visualize on a gel and to purify. Larger products may be difficult to amplify. Design a primer pool internal to the first two that is either in the sense or antisense orientation that will yield a product of 100–2000 bp.

6. If only two primers can be designed that meet the above criteria, this method can still be applied. The third primer allows for verification of the PCR product in a second PCR reaction rather than solely at the sequencing step.

7. Add asymmetrical restriction enzyme sites and a two base spacer to the 5' ends of the primers. For example, the sequence CCGAATTC would be added to the primer shown above. The spacer allows PCR products that have incomplete ends to be cleaved by the restriction enzyme. Pick restriction enzymes that cleave well when they are positioned at the ends of DNA fragments; for example, *Eco*RI and *Bam*HI (a table is found in the back of the New England Biolabs catalog). Using different restriction sites on the 5' and 3' primers allows for unidirectional cloning, that will select for products containing primers from both primer pools and will aid in the orientation of resultant sequences.

8. When designing highly degenerate primers, it is difficult to abide by all the conventional rules of PCR primer design but the 5' and 3' ends of each primer pool should not be complementary and the 3' ends of the sense and antisense primer pools should not be complementary.

## *3.2. PCR*

1. Set up 50-µL reactions using standard PCR conditions (*see* Chapter 1). Use 1 µg of genomic DNA if the primers do not span an intron. Use 100 ng of cDNA if an

adequate guess can be made of the tissues that are expressing the gene of interest. It is recommended that a control template, containing one to the sequences to which the primers were designed, be used in conjunction with the experimental template throughout this protocol. The primers need to be assessed for their ability to amplify product from the genes to which they were designed. Use 100 pmol of each primer pool. If preferable, the reactions can be scaled down to 10 μL.

2. The most important condition to titer is the annealing temperature. *See* Note 1 for the effect of altering other parameters besides the annealing temperature. Start with an annealing temperature that is approx 5°C below the lowest minimum $T_m$. Perform five cycles at this annealing temperature and an additional 30–35 cycles at an annealing temperature that is 5°C higher. This allows for the increased $T_m$ resulting from the incorporation of the 5' tails containing the restriction sites into the template. Therefore, the PCR profile will be as follows: 7 min at 95°C (to denature the template, for genomic DNA only), add *Taq* polymerase after 95°C has been reached; 5 cycles of 1 min at minimum $T_m$–5°C, 2 min at 72°C, and 1 min at 94°C; 30–35 cycles of 1 min at minimum $T_m$, 2 min at 72°C, and 1 min at 94°C; 1 min at minimum $T_m$ and 10 min at 72°C (final extension step).

3. Continue increasing the annealing temperature by 5°C for each new PCR trial until a prominent product of the expected size is amplified from the control template. If the expected size of the PCR product from the experimental template is known, target this product. If the experimental template yields several bands or a prominent band of an unexpected size, do not give up hope; you may still have the correct product.

4. Monitor the temperature titration by running 5 μL of each reaction on an agarose gel of the appropriate concentration for the size of product that is expected (*see* Chapter 1). A quick way to load these gels is to dot approx 1 μL of 6X loading dye on a piece of parafilm for each sample. Reach through the oil layer in the PCR reaction and withdraw 5 μL of sample. Pipet up and down on the dot of dye to mix and withdraw 5 μL to load. The oil will form a halo around the dot of sample, and by pipeting up only five of the six total microliters the oil can be excluded from the inside of the tip. Oil on the outside of the tip will float harmlessly on the top of the gel buffer.

5. Once the best PCR conditions have been determined, run an aliquot of the reaction on a gel and isolate the promising band or bands. For a quick method, *see* Note 2. Use a small aliquot of the isolated DNA (1/1000 of the total volume) as template in a second PCR reaction with the appropriate initial primer and the third primer. Use the PCR conditions outlined above with an initial annealing temperature that is 5°C below the minimum $T_m$. This should yield a single strong band of the expected size. If multiple bands are seen, try using less template and/or a higher annealing temperature. A product of the correct size after the second amplification indicates that the original product probably contains the appropriate gene fragment.

6. Prepare both the original product and the product amplified with the internal primer for cloning. Clean up the remainder of each reaction (45 μL) by extracting

once with one vol of equilibrated phenol/chloroform/isoamyl alcohol (25:24:1) and one volume of chloroform/isoamyl alcohol (24:1). At the first extraction, the aqueous layer will be sandwiched between the organic layer and the oil layer. Any oil which is carried over will be removed at the second extraction. Precipitate DNA, resuspend in TE and digest with restriction enzymes for sites incorporated in the primers. Run digested PCR products on a preparative gel. Running the products on a gel after digestion rather than before, allows for the assessment of intact product (that does not have internal restriction enzyme sites). Excise the band of the expected size and/or the brightest intensity. Isolate DNA from the agarose using any standard method.

7. Estimate the amount of digested, purified DNA either as based on predicted yields as determined by manufacturer's instructions or by previous use. Alternatively, estimate the yield by comparing the staining intensity of a small fraction (1/10–1/50) of the sample on a minigel to that of several dilutions of a standard DNA solution (i.e., 1.0, 0.1, and 0.01 μg of a digested plasmid that yields fragments of known size) after resolving on a minigel.

8. Ligate the digested and isolated bands into predigested vector and sequence. Use 100 ng of predigested vector and a 1:3 to 1:10 molar ratio of vector to insert.

9. Characterize clones by DNA sequencing. Since many products may need to be sequenced to screen for the correct product, it is best to use a method that consistently gives reliable sequence data, such as the generation of ssDNA for sequencing template. One easy and reliable protocol (furnished by the manufacturer) for the generation of ssDNA uses pBLUESCRIPT phagemid (Stratagene, La Jolla, CA). Briefly, grow the inoculum from a single trans-formed colony in the presence of ampicillin and helper phage for 1–2 h. Then, add kanamycin to select for infected cells and grow culture to saturation (16–24 h). The next day, isolate phage particles (containing ssDNA) from the supernatant and purify ssDNA.

10. Experimental data indicates that even if the PCR yields several bands, one may contain the correct products. Sequencing 10 clones for each band should be suf-ficient to find the correct product.

## 4. Notes

1. Other PCR parameters that can be optimized include the $MgCl_2$, template, and primer concentrations. The length of time at each step in the PCR cycle can be decreased since the suggested times reflect maximum values for any sets of prim-ers. Experimental data using several sets of primers indicate that five initial cycles at the lower annealing temperature and a total of 35–40 cycles is optimal. The large number of cycles may be required to amplify specific products using slightly mismatched primers in the later cycles.

2. A quick method for the isolation of DNA fragments for preparative-quality agarose follows. A small hole is made in the bottom of a 0.5-mL microfuge tube. The hole is covered with a layer of siliconized glass beads or glass wool. The tube is placed into a 1.5-mL microfuge tube. The gel slice is placed in the 0.5-mL tube and the

setup is spun for 1–2 min at full speed in an microfuge. The liquid that has collected in the bottom tube can then be precipitated and used in a ligation reaction.

3. Sequence information that is complimentary to the inosines in the primer does not reflect the sequence in the template since *Taq* polymerase seems to preferentially pair Gs with Is.

## References

1. Gould, S. J., Subramani, S., and Scheffler, I. E. (1989) Use of the DNA polymerase chain reaction for homology probing: isolation of partial cDNA or genomic clones encoding the iron-sulfur protein of succinate dehydrogenase from several species. *Proc. Acad. Natl. Sci. USA* **86,** 1934–1938.
2. Bartl, S. and Weissman, I. L. (1994) Isolation and characterization of major histocompatibility complex class IIB genes from the nurse shark. *Proc. Acad. Natl. Sci. USA* **91,** 262–266.
3. Bartl, S. and Weissman, I. L. (1994) PCR primers containing a inosine triplet to complement a variable codon within a conserved protein-coding region. *BioTechniques* **16(2),** 246–250.
4. Huang, G. C., Page, M. J., Roberts, A. J., Malik, A. N., Spence, H., McGregor, A. M., and Banga, J. P. (1990) Molecular cloning of a human thyrotropin receptor cDNA fragment: use of highly degenerate, inosine containing primers derived from aligned amino acid sequences of a homologous family of glycoprotein hormone receptors. *FEBS Lett.* **264,** 193–197.
5. Knoth, K., Roberds, S., Poteet, C., and Tamkin, M. (1988) Highly degenerate, inosine-containing primers specially amplify rare cDNA using the polymerase chain reaction. *Nucleic Acids Res.* **16,** 10,932.
6. Patil, R. V. and Dekker, E. E. (1990) PCR-amplification of a *Eschericha coli* gene using mixed primers containing deoxyinosine at ambiguous positions in degenerate amino acid codons. *Nucleic Acids Res.* **18,** 3080.
7. Ohtsuka, E., Matsuki, S., Ikehara, M., Takahashi, Y., and Matsubara, K. (1985) An alternative approach to deoxyoligonucleotides as hybridization probes by insertion of deoxyinosine at ambiguous codon positions. *J. Biol. Chem.* **260,** 2605–2608.
8. Bartl, S. and Weissman, I. L. (1994) The isolation of putative major histocompatibility complex gene fragments from dogfish and nurse shark. *Ann. NY Acad. Sci.* **721,** 346–349.

# 43

## Designing PCR Primers to Amplify Specific Members or Subgroups of Multigene Families

### Robert M. Horton, Raghavanpillai Raju, and Bianca M. Conti-Fine

### 1. Introduction

Multigene families throw a particular twist into the problem of polymerase chain reaction (PCR) primer design. One frequently needs primers that will not only amplify the sequence of interest, but at the same time fail to amplify the set of very similar sequences that comprises the remaining members of the family. In a "normal" PCR experiment, one must worry about "random" priming on unrelated sequences. With multigene families, one must attempt to avoid nonrandom priming on related sequences by exploiting differences that distinguish the sequence of interest. Most "primer design" programs do not help with this problem. Here we discuss what is involved in designing subfamily-specific primers.

#### 1.1. Why This Chapter Focuses on Coding Regions

The simplest approach to amplifying a specific member of a multigene family usually involves focusing on noncoding regions, such as introns or flanking regions for genomic sequences, or 5' and 3' untranslated regions for cDNAs. Untranslated regions, not being subjected to evolutionary pressure to conserve amino acid sequences, are usually more divergent than coding regions, so it is easier to find sequences unique to individual family members. Such regions should certainly be considered for making primers to a single locus in one species, as might be used, for example, to amplify alleles or mutants (1–5). In this chapter, however, the emphasis will be on the more technically challenging use of protein coding regions, since there are instances in which some conservation of sequence among family members can be useful.

An example of how some degree of conservation among family members may be useful is given in Section 1.1. In this case, we did not want to make

From: Methods in Molecular Biology, Vol. 67: PCR Cloning Protocols: From Molecular Cloning to Genetic Engineering Edited by: B. A. White Humana Press Inc., Totowa, NJ

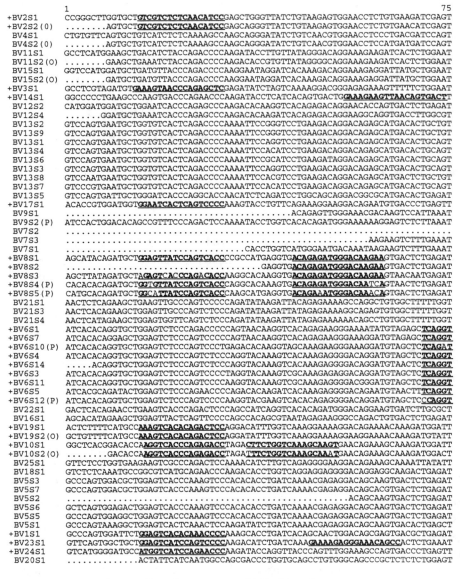

Fig. 1. Aligned sequences of TCR V-region genes.

primers for each V-region locus, because there are so many loci (*see* Fig. 1). Instead, we made primers for related groups or subfamilies of sequences. In this case again, untranslated regions with sequences unique to each locus are too divergent and too specific. Furthermore, it is commonly the case that not as many sequences are available in untranslated regions.

```
       76                                                                                    150
BV2S1        GCCGTTCCCTGGACTTTCAGGCCACAACTATGTTTTGGTATCGTCAGTTCCCGAAACAGAGTCTCATGCTGATGG
BV2S2(0)     GCCGTTCCCTGGACTTTCAGGCCACAACTATGTTTTGGTATCGTCAGCTCCGGAAACAGAGCCTCATGCTGATGG
+BV4S1       ...GTCAAGTCGATAGCCAAGTCACCATGATGTTCTGGTACCGTCAGCAACCTGGACAGAGCCTGACACTGATCG
+BV4S2(0)    ...GTCAAGTCGACAGCCAAGTCACCATGATGTTCTGGTACTGTCAGCAACCTGGACAGAGCCTGACACTGATTG
BV11S1       G...TTCTCAAACCATGGGCCATGACAAAATGTACTGGTATCAACAAGATCCAGGAATGGAACTACACCTCATC.
BV11S2(0)    G...TTCTCAAACCATGGGCCATGACAAAATGTACTGGT.....CAAGATCCAGGAATGGAATTACACCTCATC.
BV15S1       G...TTCTCAGACTAAGGGTCATGATAGAATGTACTGGTATCGACAAGACCCAGGACTGGGCCTACGGTTGATC.
BV15S2(0)    G...TTCTCAGACTAAGGGTCATGATAGAATGTACTGGTATCGACAAGACCCAGGACTGGGCCTACAGTTGATC.
BV3S1        G...TGTCCAGGATATGGACCATGAAAATATGTTCTGGTATCGACAAGACCCAGGTCTGGGGCTACGGCTGATC.
BV14S1       G...TTCTCAGAATATGAACCATGAGTATATGTCCTGGTATCGACAAGACCCAGGGCTGGGCTTAAGGCAGATC.
BV12S2       G...TCACCAGACTGAGAACCACCGCTATATGTACTGGTATCGACAAGACCCGGGGCATGGGCTGAGGCTGATC.
BV12S4       G...TCACCAGACTTGGAACCACAACAATATGTTCTGGTATCGACAAGACCTGGGACATGGGCTGAGGCTGATC.
BV13S2       G...TGCCCAGGATATGAACCATGAATACATGTACTGGTATCGACAAGACCCAGGCATGGGGCTGAGGCTGATT.
BV13S9       G...TGCCCAGGATATGAACCATGAATACATGTACTGGTATCGACAAGACCCAGGCATGGGGCTGAGGCTGATT.
BV13S1       G...TGCCCAGGATATGAACCATGAATACATGTCCTGGTATCGACAAGACCCAGGCATGGGGCTGAGGCTGATT.
BV13S4       G...TGCCCAGGATATGAACCATGGATACTTGTCCTGGTATCGACAAGACCCAGGCATGGGGCTGAGGCGCATT.
BV13S6       G...TGCCCAGGATATGAACCATAACTACATGTACTGGTATCGACAAGACCCAGGCATGGGGCTGAAGCTGATT.
BV13S3       G...TGCCCAGGATATGAACCATAACTCCATGTACTGGTATCGACAAGACCCAGGCATGGGACTGAGGCTGATT.
BV13S8       G...TGCCCAGGATATGAACCATGAATACATGTATCGGTATCGACAAGACCCAGGCAAGGGGCTGAGGCTGATT.
BV13S7       G...TGCCCAGGATATGAACCATGGATACATGTCCTGGTATCGACAAGACCCAGGCATGGGGCTGAGACTGATT.
BV13S5       G...TACCCAGGATATGAGACATAATGCCATGTACTGGTATAGACAAGATCTAGGACTGGGGCTAAGGCTCATC.
BV17S1       G...TGAACAGAATTTGAACCACGATGCCCATGTACTGGTACCGACAGGACCCAGGGCAAGGGCTGAGTGATCA.
+BV9S1       G...TGAACAAAATCTGGGCCATGATACTATGTATTGGTATAAACAGGACTCTAAGAAATTTCTGAAGATAATG.
+BV9S2(P)    G...AGAACAAAATCTGGGCCATAATGCTATGTATTGGTATAAACAGGACTCTAAGAAATTTCTGAAGACAATG.
BV7S2        G...........................................................................
BV7S3        G...TGAACAACATCTGGGGCATAACGCTATGTATTGGTACAAGCAAAGTGCTAAGAAGCCACTGGAGCTCATG.
BV7S1        G...TGAACAACATATGGGGCACAGGGCAATGTATTGGTACAAGCAGAAAGCTAAGAAGCCACCGGAGCTCATG.
BV8S1        G...TAAACCAATTTCAGGCCACAACTCCCTTTTCTGGTACAGACAGACCATGATGCGGGGACTGGAGTTGCTC.
BV8S2        G...TAAACCAATTTCAGGACACGACTACCTTTTCTGGTACAGACAGACCATGATGCGGGGACTGGAGTTGCTC.
BV8S3        G...TCAGCCAATTTTAGGCCACAATACTGTTTTCTGGTACAGACAGACCATGATGCTCAAGGACTGGAGTTGCTG.
BV8S4(P)     G...CGAACCAATTTCAGGCCACAATGATCTTCTCTGGTACAGACAGACCTTTGTGCAGGGACTGGAATTGCTG.
BV8S5(P)     G...TGAGCCAATTTTTGGCCACAATTTCCTTTTCTGGTACAGAGATACCTTCGTGCAGGGACTGGAATTGCTG.
BV21S1       G...TGATCCTATTTCTGGCCATGCTACCCTTTACTGGTACCGGCAGATCCTGGGACAGGGCCCGGAGCTTCTG.
BV21S3       G...CAATCCTATATCTGGCCATGCTACCCTTTACTGGTACCGGCAGATCCTGGGACAGGGCCCAAAGCTTCTG.
BV21S4       G...CAATCCTATTTCTGGCCACAATACCCTTTACTGGTACCTGCAGAACTTGGGACAGGGCCCGGAGCTTCTG.
+BV6S1       G...TGATCCAATTTCAGGTCATACTGCCCTTTACTGGTACCGACAAAGCCTGGGGCAGGGCCCAGAGTTTCTA.
+BV6S7       G...TGATCCAATTTCAGGTCATACTGCCCTTTACTGGTACCGACAGAGCCTGGGGCAGGGCCTGGAGTTTTTA.
+BV6S10(P)   A...TGATCCAATTTCAGGTCATAATGCCCTTTATTGGTACCGACAGAGCCTGGGGCAGGGCCCTGGAGTTTCCA.
+BV6S4       G...TGATCCAATCTCGGTCATGTATCCCTTTATTGGTACCGACAGGCCCTGGGGCAGGGCCCAGAGTTTCTG.
+BV6S14      G...TGATCCAATTTCGAGTCATGTAACCCTTTATTGGTATCAACAGGCCCTGGGGCAGGGCCCAGAGTTTCTG.
+BV6S3       G...TGATCCAATTTCGGGTCATGTATCCCTTTTTGGTACCAACAGGCCCTGGGGCAGGGCCCAGAGTTTCTG.
+BV6S11      G...TGATTCAATTTCGGGTCATGTAACCCTTTATTGGTACCGACAGACCCTGGGGCAGGGCTCAGAGGTTCTG.
+BV6S5       G...TGATCCAATTTCTGAACACAACCGCCTTTATTGGTACCGACAGACCCTGGGGCAGGGCCCAGAGTTTCTG.
+BV6S12(P)   G...TGATCCAATTTCGGGTCAGGTAACCCTTTATTGGTACCGACAGACCCT.GGGCAGGGCCAAGAGTTTCTG.
+BV22S1      G...TGTCCCATCTCTAATCACTTATACTTCTATTGGTACAGACAAATCTTGGGGCAGAAAGTCGAGTTTCTG.
BV16S1       G...TGACCCAATTTCTGGACATGATAATCTTTATTGGTATCGACGTGTTATGGGAAAAGAAATAAAATTTCTG.
+BV19S1      G...TACCCCCGAAAAGGACATACTTTTGTTTATTGGTATCAACAGAATCAGAATAAAGAGTTTATGCTTTTG.
+BV19S2(0)   G...TACCCCCAAAAACGGACATACTTTTTGTTTGTTGGTATCAGCAGAATCAGAATAAAGAGTTTATGTTTTTG.
BV10S1       G...TGTTCCTATAAAAGCACATAGTTATGTTTACTGGTATCGTAAGAAGCTGGAAGAAGAGCTCAAGTTTTTG.
BV10S2(0)    G...TGTTCCTATAAAAAGCACATAGTTATGTTTACTGGTATCATAAGACGCTGGAAGAAGAGCTCAAGTTTTTT.
BV25S1       G...TGCCCCAATAAAAGGACACAGTTATGTTTTTTGGTACTTCAACAGGTCCTGAAAAACGAGTTCAAGTTCTTG.
BV18S1       G...CAGCCCAATGAAAGGACACAGTCATGTTTACTGGTATCGGCAGCTCCCAGAGGAAGGTCTGAAATTCATG.
BV5S3        G...CTCTCCTATCTCTGAGCACAAGAGTGTGTCCTGGTACCAACAGGTCCTGGGTCAGGGGCCCCAGTTTATC.
BV5S7        G...CTCTCCTATCTCTGGGCACACCAGTGTGTCCTCGTACCAACAGGCCCTGGGTCAGGGGCCCCAGTTTATC.
BV5S2        G...CTCTCCTAAGTCTGGGCATGACACCTGTGTCCTGGTACCAACAGGCCCTGGGTCAGGGGCCCCAGTTTATC.
BV5S6        G...CTCTTCTCAGTCTGGGCACAACACTGTGTCCTGGTACCAACAGGCCCTGGGTCAGGGGCCCCAGTTTATC.
BV5S5        G...CTCTCCTATCTCTGGGCACAGCAGTGTGTCCTGGTACCAACAGGCCCCGGGTCAGGGGCCCCAGTTTATC.
BV5S1        G...CTCTCCTATCTCTGGGCATAGGAGTGTATCCTGGTACCAACAGACCCCAGGACAGGGCCTTCAGTTCCTC.
+BV1S1       G...CTCCCCTAGGTCTGGAGACCTCTCTGTGTACTGGTACCAACAGAGCCTGACCAGGGCCTCCAGTTCCTC.
BV23S1       G...CTATCCTATCCCTAGACACGACACTGTCTACTGGTACCAGCAGGGTCCAGGTCAGGACCCCCAGTTCCTC.
+BV24S1      G...TTCTCAGACTTTGAACCATAACGTCATGTACTGGTACCAGCAGAAGTCAAGTCAGGCCCCAAAGCTGCTG.
+BV20S1      GCACTGTGGAGGGAACATCAAACCCCAACCTATACTGGTACCGACAGGCTGCAGGCAGGGGCCTCCAGCTGCTC.
```

**Fig. 1.** *(continued)*

Also, you might not know the sequence of the gene you want in your species of interest. In this case, sticking to the coding regions gives you a better guess at what is conserved. An example would be the primers for the α3 acetylcholine receptor subunit (*see* Section 1.2.), which were designed from human and rat sequences, but work specifically on mouse α3 as well.

```
          151                                                                   225
+BV2S1       CAACTTCCAATGAGGGCTCCAAGGCCACATACGAGCAAGGCGTCGAGAAGGACAAGTTTCTCATCAACCATGCAA
+BV2S2(0)    CAACTTCCAATGAGGGCTCCGAGGTCACATACGAGCAAGGCGTCAAGAAGGACAAGTTTCCCATCAACCATCCAA
BV4S1        CAACTGCAAATCAGGGCTCTGAGGCCACATATGAGAGTGGATTTGTCATTGACAAGTTTCCCATCAGCCGCCCAA
BV4S2(0)     CAACTGCAAATCAGGGCTCTGAGGCCACATATGAGAGCAGATTTGTCATTGACAAGTTTCCCATCAGCCGCCCAA
+BV11S1      ..CACTATTCCTATGGAGTTAATTCCACAGAGAAGGGAGATCTTTC...CTCTGAGTCAACAGTCTCCAGAATAA
+BV11S2(0)   ..CACCATTCCTATGG.GTTAATTCCACAGAGAAGGGGAGATCTTTG...CTCTGAGTCAACAGTCTCCAGAATAA
+BV15S1      ..TATTACTCCTTTGATGTCAAAGATATAAACAAAGGAGAGATCTC...TGATGGATACAGTGTCTCTCGACAGG
+BV15S2(0)   ..TATTACTCCTTTGATGTCAAAGATATAAACAAAGGAGAGATCTC...TGATGGATACAGTGTCTCTCGACAGG
+BV3S1       ..TATTTCTCATATGATGTTAAAATGAAAGAAAAAGGAGATATTCC...TGAGGGGTACAGTGTCTCTAGAGAGA
BV14S1       ..TACTATTCAATGAATGTTGAGGTGACTGATAAAGGAGGATGTTCC...TGAAGGGTACAAAGTCTCTCGAAAAG
+BV12S2      ..CATTACTCATATGGTGTTAAAGATACTGACAAAGGAGAAGTCTC...AGATGGCTACAGTGTCTCTAGATCAA
+BV12S4      ..CATTACTCATATGGTGTTCAAGCACTAACAAAGGAGAAGTCTC...AGATGGCTACAGTGTCTCTAGATCAA
BV13S2       ..CATTACTCAGTTGGTGAGGGTACAACTGCCAAAGGAGAGGTCCC...TGATGGCTACAATGTCTCCAGATTAA
BV13S9       ..CATTACTCAGTTGGTGAGGGTACAACTGCCAAAGGAGAGGTCCC...TGATGGCTACAATGTCTCCAGATTAA
BV13S1       ..CATTACTCAGTTGGTGCTGGTATCACTGACCAAGGAGAAGTCCC...CAATGGCTACAATGTCTCCAGATCAA
BV13S4       ..CATTACTCAGTTGCTGCTGGTATCACTGACCAAGGAGAAGTCCC...CAATGGCTACAATGTCTCCAGATCAA
BV13S6       ..TATTATTCAGTTGGTGCTGGTATCACTGATAAAGGAGAAGTCCC...GAATGGCTACAACGTCTCCAGATCAA
BV13S3       ..TATTACTCAGCTTCTGAGGGGTACCACTGACAAAGGAGAAGTCCC...CAATGGCTACAATGTCTCCAGATTAA
BV13S8       ..TACTACTCAGTTGCTGCTGCTCTCACTGACAAAGGAGAAGTTCC...CAATGGCTACAATGTCTCCAGATCAA
BV13S7       ..TACTACTCAGCTGCTGCTGGTACTACTGACAA...AGAAGTCCC...CAATGGCTACAATGTCTCTAGATTAA
BV13S5       ..CATTATTCAAATACTGCAGGTACCACTGGCAAAGGAGAAGTCCC...TGATGGTTATAGTGTCTCCAGAATAA
+BV17S1      ..TACTACTCACAGATAGTAAATGACTTTCAGAAAGGAGATATAGC...TGAAGGGTACAGCGTCTCTCGGGAGA
BV9S1        ..TTTAGCTACAATAATAAGGAGCTCATTATAAATGAAACAGTTCCAA...ATCGCTTCTCACCTAAATCTCCAG
BV9S2(P)     ..TTTATCTACAGTAACAAGGAGCCAATTTTAAATGAAACAGTTCCAA...ATCGCTTCTCACCTGACTCTCCAG
+BV7S2       .......................................GAATTCGCTCCA...GTGGCTTCTCACCTGAATGCCCCA
+BV7S3       ..TTTGTCTACAACTTTAAAGAACAGACTGAAAACAACAGTGTGCCAA...GTCGCTTCTCACCTGAATGCCCCA
+BV7S1       ..TTTGTCTACAACTATGAGAAACTCTCTATAAATGAAAGTGTGCCAA...GTCGCTTCTCACCTGAATGCCCCA
BV8S1        ..ATTTACTTTAACAACAACGTTCCGATAGATGATTCAGGGATGCCCGAGGTCGATTCTCAGCTAAGATGCCTA
BV8S2        '..ATTTACTTTAACAACAACGTTCCGATAGATGATTCAGGGATGCCCGAGGATCGATTCTCAGCTAAGATGCCTA
BV8S3        ..GCTTACTTCCGCAACCGGGCTCCTCTAGATGATTCGGGGATGCCCGAAGGATCGATTCTCAGCAGAGATGCCTG
BV8S4(P)     ..AATTACTTCTGCAGCTGGACCCTCGTAGATGACTCAGGAGTGTCCAAGGATTGATTCTCAGCACAGATGCCTG
BV8S5(P)     ..AGTTACTTCCGGAGCTGATCTATTATAGATAATGCAGGTATGCCCACAGAGCGATTCTCAGCTGAGAGGCCTG
BV21S1       ..GTTCAATTTCAGGATGAGAGTGTAGTAGATGAATTCACAGTTGCCTAAGGATCGATTTTCTGCAGAGAGGCTCA
BV21S3       ..ATTCAGTTTCAGAATAACGGTGTAGTGGATGATTCACAGTTGCCTAAGGATCGATTTTCTGCAGAGAGGCTCA
BV21S4       ..ATTCAGATATGAGAATGAGGAAGCAGTAGACAGTTCACAGTTGCCTAAGGATCGATTTTCTGCAGAGAGGCTCA
BV6S1        ..ATTTACTTCCAAGGCACGGGTGCGGCAGATGACTCAGGGCTGCCCAACGATCGGTTCTTTGCAGTCAGGCCTG
BV6S7        ..ATTTACTTCCAAGGCAACAGTGCACCAGACAAATCAGGGCTGCCCAGTGATCGCTTCTCTGCAGAGGACTG
BV6S10(P)    ..ATTTACTTCCAAGGCAAGGATGCAGCACAAATCGGGGCTTCCCCGTGATCGGTTCTCTGCACAGAGGCTG
BV6S4        ..ACTTACTTCAATTATGAAGCTCAACAAGACAAATCAGGGCTGCCCAATGATCGGTTCTCTGCAGAGAGGCCTG
BV6S14       ..ACTTACTTCAATTATGAAGCTCAACCAGACAAATCAGGGCTGCCCAGTGATCGCTTCTCTGCAGAAAGGCCTG
BV6S3        ..ACTTATTTCCAAGGCAACAGTCAACTAGACAAATCGGGGCTGCCCAGTGATCGCTTCTTTGCAGAGAGGCCTG
BV6S11       ..ACTTACTCCCAGAGTGATGCTCAACGAGACAAATCAGGGCGGCCCAGTGGTCGGTTCTCTGCAGAGAGGCCTG
BV6S5        ..ACTTACTTCCAGAATGAAGCTCAACTAGAAAAATCAAGGCTGCTCAGTGATCGGTTCTCTGCAGAGAGGCCTA
BV6S12(P)    ..ACTTCCTTCCAGGATGAAACTCAACAAGATAAATCAGGGCTGCTCAGTGATCAATTCTCACAGAGAGGTCTG
BV22S1       ..GTTTCCTTTTATAATAATGAAATCTCAGAGAAGTCTGAAATATTCGATGATCAATTCTTAGCTGAAAGGCCTG
+BV16S1      ..TTACATTTTGTGAAAGAGTCTAAACAGGATGAGTGCCCACAATCGATTCTTAGCTGAAAGGACTG
BV19S1       ..ATTTCCTTTCAGAATGAACAAGTTCTTCAAGAAACGGAGATGCCAAGAAGCGATTCTCATCTCAATGCCCCA
BV19S2(0)    ..ATTTCCTTTCAGAATGAACAAGTTCTTCAAGAAATGGAGATGCAACGAAGCGATTCTCATCTCAATGCCCCA
BV10S1       ..GTTTACTTTCAGAATGAAGAACTTATTCAGAAAGCAGAAATAATCAATGAGCGATTTTTAGCCCAATGCTCCA
BV10S2(0)    ..ATTTACTTTCAGAATGAAGAAGTTATTCATTCAGAAAGCAGAAATAATCAATGAGCGATTTTCAGCCCAATGCCCCC
BV25S1       ..ATTTCCTTCCAGAATGAAAATGTCTTTGATGAAACAGGTATGCCCAAGGAAAGATTTTCAGCTAAGTGGCCTCC
+BV18S1      ..GTTTATCTCCAGAAAGAAAAATATCATAGATGAGTCAGGAATGCCCAAAGGAACGATTTTCTGCTGAATTTCCCA
+BV5S3       ..TTTTCAGTATTATGAGAAAGAAGAGAGAGGAAGAGGAAACTTCCC...TGATCGGATTCTCAGCTCGCCAGTTCC
+BV5S7       ..T  T CAGTATTATGAGAAAGAAGAGAGAGGAAGAGGAAACTTCCC...TGATCAATTCTCAGGTCACCAGTTCC
+BV5S2       ..T  T  CAGTATTATGAGAAAGAAGAGAGAGGAAGAGGCAACTTCCC...TGATCGGATTCTCAGGTCACCAGTTCC
+BV5S6       ..TTTTCAGTATTATAGGGAGGAAGAGAATGGCAGAGGAAACTTCCC...TCCTAGATTCTCAGGTCTCCAGTTCC
+BV5S5       ..TTTGAAATAGCTAATGAGTTAAGGAGATCAGAAGGAAACTTCCC...TAATCGATTCTCAGGGCGCCAGTTCC
+BV5S1       ..TTTGAAATACTTCAGTGAGACACAGAGAAACAAAGGAAACTTCCC...TGGTCGGATTCTCAGGGCGCCAGTTCT
BV1S1        ..ATTCAGTATTATAATGGAGAAGAGAGAGCAAAGGAAACATTCT...TGAACGATTCTCCGCACAACAGTTCC
BV23S1       ..ATTTCGTTTTATGAAAAGATGCAGACCGATAAAGGAAGCATCC...TGATCGATTCTCAGCTCAACAGTTCA
BV24S1       ..TTCCACTACTATGACAAAAGATTTTAACAATGAAGCAGACACCCC...TGATAACTTCCAATCCAGGAGGCCGA
BV20S1       ..TTCTACTCCATTGGTATTGACCAGATCAGCTCTGAGGTGCCCCAGAATC......TCTCAGCCTCCAGACCCC
```

Fig. 1. *(continued)*

## 1.2. "Subfamily" Broadly Defined

We will use the term "subfamily" to mean a group of related sequences that are more closely related to one another than they are to other members of the larger group. Such a situation could be reflected in isotypes of proteins, related subunits of hetero-oligomeric proteins, or specific loci in various species. The same approaches apply when attempting to amplify a single member of a family.

```
          226                                                          299
-BV2S1    GCCTGACCTTGTCCACTCTGACAGTGACCAGTGCCCATCCTGAAGACAGCAGCTTCTACATCTGCAGTGCTAGA
-BV2S2(O)  ACCTGACCTTCTCCGCTCTGACAGTGACCAGTGCCCATCCTGAAGACAGCAGCTTCTACATCTGCAGTGCTAGA
-BV4S1    ACCTAACATTCTCAACTCTGACTGTGAGCAACATGAGCCCTGAAGACAGCAGCATATATCTCTGCAGCTCCAGG
-BV4S2(O)  ACCTAACATTCTCAACTCTGACTGTGAGCAACAGGAGACCTGAAGACAGCAGCATATACCTCTGCAGCGTTGAA
BV11S1    GGACGGAGCATTTTCCCCTGACCCTGGAGTCTGCCAGGCCCTCACATACCTCTCAGTACCTCTGTGCCAGCATG
BV11S2(O)  GGATAGAGCGTTTTCCCCTGACCCTGGAGTCTGCCAGCCCCTCACATACCTCTCAGTACCTCTGTGCCAGCAGT
+BV15S1   CACAGGCTAAATTCTCCCTGTCCCTAGAGTCTGCCATCCCCAACCAGACAGCTCTTTACTTCTGTGCCACCAGT
+BV15S2(O) CACAGGCTAAATTCTCCCTGTCCCTAGAGTCTGCCATCCCCAACCAGACAGCTCTTTACTTCTGTGCCACCAGT
BV3S1     AGAAGGAGCGCTTCTCCCTGATTCTGGAGTCCGCCAGCACCAACCAGACATCTATGTACCTCTGTGCCACCAGT
BV14S1    AGAAGAGGAATTTCCCCCTGATCCTGGAGTCGCCCAGCCCCAACCAGACCTCTCTGTACTTCTGTGCCAGCAG.
BV12S2    AGACAGAGGATTTCCTCCTCACTCTGGAGTCCGCTACCAGCTCCCAGACATCTGTGTACTTCTGTGCCATCAGC
BV12S4    ACACAGAGGACCTCCCCCTCAC...............................................
+BV13S2   AAAAACAGAATTTCCTGCTGGGGTTGGAGTCGGCTGCTCCCTCCCAAACATCTGTGTACTTCTGTGCCAGCAGT
+BV13S9   AAAAACAGAATTTCCTGCTGGGGTTGGAGTCGGCTGCTCCCTCCCAAACA..................
+BV13S1   CCACAGAGGATTTCCCGCTCAGGCTGCTGTCGGCTGCTCCCTCCCAGACATCTGTGTACTTCTGTGCCAGCAGT
+BV13S4   ACACAGAGGATTTCCCGCTCAGGCTGGAGTCAGCTGCTCCCTCCCAGACATCTGTATACTTCTGTGCCAGCAGT
+BV13S6   CCACAGAGGATTTCCCGCTCAGGCTGGAGTTGGCTGCTCCCTCCCAGACATCTGTGTACTTCTGTGCCAGCAGT
+BV13S3   ACAAACGGGAGTTCCGCTCAGGCTGGAGTCGGCTGCTCCCTCCCAGACATCTGTGTACTTCTGTGCCAGCACC
+BV13S8   ACACAGAGGATTTCCCCCTCAAGCTGGAGTCAGCTGCTCCCTCTCAGACTT..................
+BV13S7   ACACAGAGGATTTCCACTCAGGCTGGTGTCGGC.............................
+BV13S5   ACACAGATGATTTCCCCCTCACGTTGGCGTCTGCTGTACCCTCTCAGACATCTGTGTACTTCTGTGCCAGCAGT
BV17S1    AGAAGGAATCCTTTCCTCTCACTGTGACATCGGCCCAAAAGAACCCGACAGCTTTCTATCTCTGTGCCAGTAGT
BV9S1     ACAAAGCTCACTTAAATCTTCACATCAATTCCCTGGAGCTTGGTGACTCTGCTGTGTATTTCTGTGCCAGCAGC
BV9S2(P)   ACAAAGTTCATTTAAATCTTCACATCAATTCCCTGGAGCTTGGTGACTCTGCTGTGTATTTCTGTGCCAGCAGC
+BV7S2    ACAGCTCTCACTTATTCCTTCACCTACACACCCTGCAGCCAGAAGACTCGGCCCTGTATCTCTGCGCCAGCAGC
+BV7S3    ACAGCTCTCACTTATTCCTTCACCTACACACCCTGCAGCCAGAAGACTCGGCCCTGTATCTCTGTGCCAGCAGC
+BV7S1    ACAGCTCTCTCTTAAACCTTCACCTACACGCCCTGCAGCCAGAAGACTCAGCCCTGTATCTCTGCGCCAGCAGC
BV8S1     ATGCATCATTCTCCACTCTGAAGATCCAGCCCTCAGAACCCAGGGACTCAGCTGTGTACTTCTGTGCCAGCAGT
BV8S2     ATGCATCATTCTCCACTCTGAAGATCCAGCCCTCAGAACCCAGGGACTCAGCTGTGTACTTCTGTGCCAGCAGT
-BV8S3    ATGCAACTTTAGCCACTCTGAAGATCCAGCCCTCAGAACCCAGGGACTCAGCTGTGTATTTTTGTGCTAGTGGT
BV8S4(P)   ATGTATCATTCTCCACTCTGAGGATCCAGCCCATGGAACCCAGGGACTTGGGCCTATATTTCTGTGCCAGCAGC
BV8S5(P)   ATGGATCATTCTCTACTCTGAAGATCCAGCCTGCAGAGCAGGGGGACTCGGCCGTGTATGTCTGTGCAAGTCGC
+BV21S1   AAGCAGTAGACTCCACTCTCAAGATCCAGCCTGCAGAGCTTGGGGACTCGGCCATGTATCTCTGTGCCAGCAGC
+BV21S3   AAGCAGTAGACTCCACTCTCAAGATCCAGCCTGCAAAGCTTGAGAACTCGGCCGTGTATCTCTGTGCCAGCAGT
+BV21S4   AAGCAGTAGACTCCACTCTCAAGATCCAGCCTGCAGAGCTTGGGGACTCGGCCGTGTATCTCTGTGCCAGCAGC
BV6S1     AGGGATCCGTCTCTACTCTGAAGATCCAGCGCACAGAGCGGGGGGACTCAGCCGTGTATCTCTGTGCCAGCAGC
BV6S7     GGGGATCCGTCTCCACTCTGACGATCCAGCGCACAGCAGAGGAGGACTCGGCCGTGTATCTCTGTGCCAGCAGC
BV6S10(P)  AGGGATCCATCTCCACTCTGAAGTTCCAGCGACACACAGCAGGGGGACTTGGCTGTGTATCTCTGTGCCAGCAGC
BV6S4     AGGGATCCATCTCCACTCTGACGATCCAGCGCACAGAGCAGCGGGACTCGGCCATGTATCGCTGTGCCAGCAGC
BV6S14    AGGGATCCATCTCCACTCTGACGATTCAGCGCACAGAGCAGCGGGACTCAGCCATGTATCGCTGTGCCAGCAGC
BV6S3     AGGGATCCGTCTCCACTCTGAAGATCCAGCGCACAGCAGCAGGAGGACTCCGCCGTGTATCTCTGTGCCAGCAGC
BV6S11    AGAGAT.CGTCTCCACTCCGAAGATCCAGCGCACAGAGCAGGGGGACTCA.....................
BV6S5     AGGGATCTTTCTCCACCTTGGAGATCCAGCGCACAGAGCAGGGGGACTCGGCCATGTATCTCTGTGCCAGCAGC
BV6S12(P)  AG.GATCTTTCTCCAC.CTGAAGATCCAGCGCACAGAGCAAGGGCGACTCG...................
BV22S1    ATGGATCAAATTTCACTCTGAAGATCCGGTCCACAAAGCTGGAGGACTCAGCCAGTGACTTCTGTGCCAGCAGA
BV16S1    GAGGGACGTATTCTACTCTGAAGGTGCAGCCTGCAGAACTGGAGGATTCTGGAGTTTATTTCTGTGCCAGCAGC
BV19S1    AGAACGCACCCTGCAGCCTGGCAATCCTGTCCTCAGAACCGGGAGACACGGCACTGTATCTCTGCGCCAGCAGT
BV19S2(O)  AGAACGCACCCTGCAGCCTGGCAATCCTGTCCTCAGAACCGGGAGACACGGCACTGTATCTCTGCGCCAGCAGT A.............................................................
BV10S1    AAAACTCATCCTGTACCTTGGAGATCCAGTCCACGGAGTCAGGGGACAGCAGCACTGTATTTCTGTGCCAGCAGC
BV10S2(O)  AAAACTCACCCTGTACCTTGGAGATCCAGTCCACGGAGTCAGGAGACACAGCACGGTATTTCTGTGCCAACAGC
BV25S1    CAAATTCACCCTGTACCTTGAGATCCAGGCTACGAAGCTTGAGGATTCAGCAGTGTATTTTTTGTGCCAGCAGT
BV18S1    AAGAGGGCCCCAGCATCCTGAGGATCCAGCAGGTAGTGCGAGGAGATTCGGCAGCTTATTTCTGTGCCAGCTCA
BV5S3     CTAACTATAGCTCTGAGCTGAATGTGAACAGCCTTGGTGCTGGGGGGACTCGGCCCTGTATCTCTGTGCCAGCAGC
BV5S7     CTAACTATAGCTCTGAGCTGAATGTGAACGCCTTGTTGCTAGGGGACTCGG....................
BV5S2     CTAACTATAGCTCTGAGCTGAATGTGAACGCCTTGTTGCTGGGGGACTCGGCCCTCTATCTCTGTGCCAGCAGC
BV5S6     CTAATTATAACTCTGAGCTGAATGTGAACGCCTTGGAGCTGGACGACTCGGCCCTGTATCTCTGTGCCAGCAGC
BV5S5     ATGACTGTTGCTCTGAGATGAATGTGAGTGCCTTGGAGCTGGGGGACTCGGCCCTGTATCTCTG.........
BV5S1     CTAACTCTCGCTCTGAGATGAATGTGAGCACCTTGGAGCTGGGGGACTCGGCCCTTTATCTTTGCGCCAGCAGC
BV1S1     CTGACTTGCACTCTGAACTAAACCTGAGCTCTCTGGAGCTGGGGGACTCAGCTTTGTATTTCTGTGCCAGCGGA
BV23S1    GTGACTATCATTCTGAACTGAACATGAGCTCCTTGGAGCTGGGGGACTCAGCCCGTGTACTTCTGTGCCAGCAGC
BV24S1    ACACTTCTTTCTGCTTTCTTGACATCCGCTCACCAGGCCTGGGGGACGCAGCCATGTACCTGTGTGCCACCAGC
BV20S1    AGGACCGGCAGTTCATTCTGAGTTCTAAGAAGCTCCTCCTCAGTGACTCTGGCTTCTATCTCTGCCTGGAGT
```

**Fig. 1.** *(continued)*

## 1.3. Examples

This chapter will focus on designing coding region primers that will amplify a set of sequences within one subfamily, but not members of related subfamilies. Our approach is to collect the known sequences of members of the multigene family and generate a sequence alignment, from which we choose appropriate primer sequences. We will illustrate this approach with examples from two multigene families.

## 1.3.1. T-Cell Receptor V-Regions

Our first primer-design examples are from human T-cell receptor β-chain variable regions. The T-cell receptor is similar to immunoglobulins in that the genes rearrange to combine V, D, and J regions with a C (or constant) region.

An alignment of the sequences of the known human T-cell receptor V-regions (6) is shown in Fig. 1. The GenBank accession numbers are given in Fig. 2. The positions of the subfamily-specific primers are underlined, and bases that match the primers are in boldface (some sequences do not match the primers exactly, so the mismatches are underlined but not bold). The primer sequences themselves are given in Fig. 3. The regions matching primers at the 5' end of an amplified region ("5' primers," whose names end in "e5" in Fig. 3) are indicated by a "+" in front of the sequence name on the line where they appear. Lines with 3' ("e3") primers have a "-" sign in front of the sequence name, and the sequence matching the primer is in italics. Two sets of primers are shown, a set of "cloning" primers and a set of "analytical" primers.

The cloning primers were designed to allow us to clone representative V-region sequences of all of the subfamilies. We started with "degenerate" primers, in the hope that they would allow us to clone most, if not all, of the sequences. We wanted these clones to form a "matched set" of similarly sized V-region clones, to be used in a hybridization-based dot-blot procedure for detecting V-region usage analysis, similar to that described in ref. 7. Unfortunately, these primers gave an overwhelming preponderance of one subfamily, and a few members of other families, but most subfamilies were not represented at all (see Note 1). We also encountered another general problem with amplifying many related sequences simultaneously: Random overlap extension at homologous regions in partially completed DNA strands can lead to inadvertent PCR-mediated recombination.

Thus, we designed additional primers with varying degrees of subunit specificity, to fish out the recalcitrant members. These are the cloning primers of Fig. 3. Figure 4 shows which of these primers were used to clone which V-regions. Note that only three specific e3 primers were used in the cloning set. The remaining V-regions were cloned using the degenerate primer hpVbe3.

The "analytical" set of primers was designed for PCR-based analysis of T-cell receptor V-region usage. In this approach, only one subfamily specific e5 primer is used, together with a constant region e3 primer (which is not located in the V-region, so it is not shown in Fig. 1).

Since all the specificity in this system comes from the V-region primer, this system can be quite informative in terms of primer design (see Note 1). For those readers fascinated by this sort of system for either biological or technical reasons, panels of T-cell receptor subfamily specific primers from other groups can be found in refs. 8–11.

```
gb_pr:M27381   /begin=50     /end=339    BV1S1
gb_pr:M12886   /begin=58     /end=356    BV2S1
gb_pr:L05149   /begin=1      /end=291    BV2S2(0)
gb_pr:M13843   /begin=16     /end=305    BV3S1
gb_pr:X04921   /begin=53     /end=348    BV4S1
gb_pr:L05150   /begin=1      /end=288    BV4S2(0)
gb_pr:X04927   /begin=50     /end=339    BV5S1
gb_pr:M13850   /begin=1      /end=235    BV5S2
gb_pr:X58801   /begin=121    /end=410    BV5S3
gb_pr:X61439   /begin=409    /end=688    BV5S5
gb_pr:X57615   /begin=57     /end=346    BV5S6
gb_pr:L26226   /begin=79     /end=345    BV5S7
gb_pr:X04934   /begin=50     /end=342    BV6S1
gb_pr:M11952   /begin=207    /end=499    BV6S3
gb_pr:X58806   /begin=59     /end=351    BV6S4
gb_pr:X58805   /begin=97     /end=389    BV6S5
gb_pr:M14262   /begin=50     /end=342    BV6S7
gb_pr:X61444   /begin=1324   /end=1616   BV6S10(P)
gb_pr:L13762   /begin=137    /end=404    BV6S11
gb_pr:M97503   /begin=149    /end=415    BV6S12(P)
gb_pr:X57607   /begin=1      /end=288    BV6S14
gb_pr:M13855   /begin=1      /end=255    BV7S1
gb_pr:M13856   /begin=1      /end=110    BV7S2
gb_pr:X57617   /begin=1      /end=231    BV7S3
gb_pr:K01571   /begin=87     /end=379    BV8S1
gb_pr:M13858   /begin=1      /end=255    BV8S2
gb_pr:X07223   /begin=384    /end=676    BV8S3
gb_pr:X07224   /begin=370    /end=662    BV8S4(P)
gb_pr:X06936   /begin=380    /end=672    BV8S5(P)
gb_pr:M13859   /begin=1      /end=246    BV9S1
gb_pr:X57608   /begin=127    /end=416    BV9S2(P)
gb_pr:M13860   /begin=51     /end=343    BV10S1
gb_pr:L05151   /begin=1      /end=285    BV10S2(O)
gb_pr:M13861   /begin=67     /end=356    BV11S1
gb_pr:L05152   /begin=1      /end=276    BV11S2(O)
gb_pr:M14268   /begin=35     /end=324    BV12S2
gb_pr:L26230   /begin=107    /end=337    BV12S4
gb_pr:X04932   /begin=50     /end=339    BV13S1
gb_pr:X61445   /begin=404    /end=693    BV13S2
gb_pr:X58809   /begin=80     /end=369    BV13S3
gb_pr:X61447   /begin=372    /end=661    BV13S4
gb_pr:X58810   /begin=67     /end=356    BV13S5
gb_pr:X58815   /begin=56     /end=345    BV13S6
gb_pr:L26228   /begin=86     /end=332    BV13S7
gb_pr:L26227   /begin=87     /end=353    BV13S8
gb_pr:L26229   /begin=106    /end=371    BV13S9
gb_pr:M13865   /begin=36     /end=324    BV14S1
gb_pr:M11951   /begin=181    /end=470    BV15S1
gb_pr:L05153   /begin=1      /end=282    BV15S2(O)
gb_pr:X57723   /begin=58     /end=350    BV16S1
gb_pr:M27388   /begin=50     /end=339    BV17S1
gb_pr:M27389   /begin=65     /end=357    BV18S1
gb_pr:M27390   /begin=68     /end=360    BV19S1
gb_pr:L26225   /begin=22     /end=241    BV19S2(O)
gb_pr:M13554   /begin=1      /end=276    BV20S1
gb_pr:M33233   /begin=739    /end=1031   BV21S1
gb_pr:M33235   /begin=163    /end=455    BV21S3
gb_pr:X56665   /begin=28     /end=320    BV21S4
gb_pr:X57727   /begin=69     /end=361    BV22S1
gb_pr:X58799   /begin=40     /end=329    BV23S1
gb_pr:X58800   /begin=77     /end=366    BV24S1
gb_pr:L26231   /begin=411    /end=703    BV25S1
```

**Fig. 2.** File of sequence names for generating the alignment in Fig. 1.

The primers in Fig. 3 have all been shown to amplify T-cell receptor cDNA fragments of the expected sizes from a heterogeneous population of T-cells (peripheral blood lymphocytes). For most of them, specificity has been demon-

| Analytical | primers | Cloning | primers |
|---|---|---|---|
| Vb1e5 | taggtctggagacctct | Vb1cpe5 | ggagtcacacaaacccc |
| Vb2e5 | gcaaggcgtcgagaag | Vb2cpe5 | gtcgtctctcaacatcc |
| Vb3e5 | ctcatatgatgttaaaatgaaa | Vb3cpe5 | gaaagtaacccagagctc |
| Vb4e5 | gatagccaagtcaccatg | Vb8-1cpe5 | ggagttatccagtcacc |
| Vb5e5 | acttccctggtcgattctcagg | Vb10cpe5 | aggtcacccagagacc |
| Vb6e5 | tcaggtgtgatccaatttc | Vb17cpe5 | ggaatcactcagtcccc |
| Vb7e5 | tgaatgccccaacagct | Vb18cpe5 | ggcgtcatgcagaaccc |
| Vb8e5 | acagagatgggacaagaa | Vb19cpe5 | aaagtcacacagactcca |
| Vb9e5 | aacaggactctaagaaattt | Vb20cpe5 | actattcatcaatggcc |
| Vb10e5 | cttctggtcaaagcaagt | Vb21acpe5 | ggagtggttcagtctcc |
| Vb11e5 | cctatggagttaattccac | Vb23cpe5 | ggagtcatccagtcccc |
| Vb12e5 | actcatatggtgttaaagat | Vb24cpe5 | atggtcatccagaaccc |
| Vb13e5 | gtcggctgctccctc | Vb2e3 | ctagcactgcagatgta |
| Vb14e5 | gaaagaagttaacagtgact | Vb4e3 | gctgcagagatatatgc |
| Vb15e5 | tgtctctcgacaggcac | Vb8-3e3 | ccactagcacaaaaatac |
| Vb16e5 | tttgtgaaagagtctaaaca | | |
| Vb17e5 | acagatagtaaatgactttc | hpVbe5*: | |
| Vb18e5 | atctccagaaagaaaatatc | gggaattcgtcgactggagtc(at)cccaga(ac)ncc | |
| Vb19e5 | gaatcagaataaagagtttat | | |
| Vb20e5 | actgtggagggaacatc | hpVbe3: | |
| Vbw21e5 | gagtagactccactctc | gggaattcgtcgactgctggcrcagarrta | |
| Vbw22e5 | tctctaatcacttatacttc | | |
| Vbw23e5 | gaaaagagggaaacagcc | | |
| Vbw24e5 | gactttgaaccataacgtc | | |
| Cbe3 | tgtgggagatctctgct | | |

Fig. 3. TCR V-region primer sequences. *from ref. 27, with an additional SalI site added. Notes: The "e5" and "e3" designations at the end of primer names indicate that they are at the 5' or 3' ends of the target sequence, respectively. That is, an "e5" primer contains the same sequence as the top strand of DNA, whereas an "e3' primer is the reverse complement of the sequence in the top strand. The "e3" primers are those shown in italics in Fig. 1. The Cbe3 constant region primer is not shown in Fig. 1, which only includes variable region sequences.

strated by using them to amplify cDNA sequences from T-cell clones or oligoclonal lines. All of the products of the cloning primer panel, and many of the products of the analytical panel, have been cloned and sequenced.

### 1.3.2. Acetylcholine Receptor Subunits

The nicotinic acetylcholine receptor is a ligand-gated ion channel that functions at neuronal and neuromuscular synapses. It is a transmembrane heteropentimer, consisting of several different subunits. Certain of these subunits are expressed in muscle, whereas others are expressed in nerve cells. We have used our primers to investigate expression of these molecules in unexpected places; for example, we have found "embryonic" subunits in adult extraocular muscles (12) and "ganglionic neuronal" subunits in skin

| Vb | 5' primer | 3' primer |
|------|-----------|-----------|
| 1 | hpVbe5 | hpVbe3 |
| 2* | Vb2cpe5 | Vb2cpe3 |
| 3* | Vb3cpe5 | hpVbe3 |
| 4* | hpVbe5 | Vb4cpe3 |
| 5 | hpVbe5 | hpVbe3 |
| 6 | hpVbe5 | hpVbe3 |
| 7 | hpVbe5 | hpVbe3 |
| 8.1* | Vb8.1cpe5 | hpVbe3 |
| 8.3* | hpVbe5 | Vb8.3cpe3 |
| 9*¶ | Vb9e5 | hpVbe3 |
| 10*¶ | Vbe5 | hpVbe3 |
| 11*¶ | Vbe5 | hpVbe3 |
| 12 | hpVbe5 | hpVbe3 |
| 13 | hpVbe5 | hpVbe3 |
| 14 | hpVbe5 | hpVbe3 |
| 15 | hpVbe5 | hpVbe3 |
| 16*¶ | Vbe5 | hpVbe3 |
| 17* | Vb17cpe5 | hpVbe3 |
| 18* | Vb18cpe5 | hpVbe3 |
| 19* | Vb19cpe5 | hpVbe3 |
| 20* | Vb20cpe5 | hpVbe3 |
| 21* | Vb21cpe5 | hpVbe3 |
| 22 | hpVbe5 | hpVbe3 |
| 23 | hpVbe5 | hpVbe3 |
| 24 | not done | |

Fig. 4. Specificity of V-region cloning primers. Sequences cloned using T-linkers (*see* Chapter 11). Only the degenerate "pan" primers were designed to include restriction sites, so T-linkers were used to add restriction sites to those sequences amplified with other primers at either end. Vβs 9, 10, 11, and 16 were amplified for cloning using the specific analytical primers.

keratinocytes *(13)*. Figure 5 gives the file of sequence names used to generate the alignment from which these primers were designed. Since the full alignment is too large to include here, we leave its generation as an exercise for the reader. The ha3cpe5 and ha3cpe3 primers shown in Fig. 6 have been used to clone full-length human α3 coding region sequence from epithelial cells (*see* Chapter XX).

We would prefer to have primers that would allow us to study several different species, since some tissues are more readily available from cows or mice than from people. Thus, in design of these primers, we have attempted to identify sequences that are characteristic of a given subunit, even in different species. The α3e5 and α3e3 primers allowed us to clone a segment of the mouse α3 subunit by virtue of its similarity to α3 in other species (Horton et al., unpublished).

## 2. Materials

1. A computer with access to the appropriate gene sequence databases, such as GenBank and EMBL.

```
gb_ro:musachrab    mouse a1    /begin=1      /end=1340
gb_ro:MUSACHRB     mouse b1    /begin=2      /end=1507
gb_ro:mmachrgr     mouse g     /begin=12     /end=1571
gb_ro:musachrd     mouse d     /begin=6      /end=1568
gb_ro:mmmnaes      mouse e     /begin=20     /end=1501
gb_ro:rnacra1      rat a1      /begin=62     /end=1435
gb_ro:rnacrb1      rat b1      /begin=71     /end=1576
gb_ro:rnacrg1      rat g       /begin=133    /end=1692
gb_ro:rnzcrd1      rat d       /begin=1      /end=1554
gb_ro:rnacre1      rat e       /begin=42     /end=1526
gb_ro:ratnnar      rat a2      /begin=305    /end=1840
gb_ro:rnachrar     rat a3      /begin=226    /end=1647
ratnarab.seq       rat a4
gb_ro:ratnachrr    rat a5      /begin=102    /end=1460
gb_ro:ratnara6s    rat a6      /begin=179    /end=1660
gb_ro:RATNARAD     rat a7      /begin=23     /end=1531
gb_ro:rru12336     rat a9      /begin=88     /end=1527
gb_ro:ratnarb      rat b2      /begin=180    /end=1691
gb_ro:ratnachr     rat b3      /begin=169    /end=1563
gb_ro:rnniacete    rat b4      /begin=41     /end=1594
gb_pr:Hsachra      human a1    /begin=49     /end=1422
gb_pr:Hsachrb      human b1    /begin=17     /end=1522
humang.seq         human g
gb_pr:Hsachrg      human d     /begin=5      /end=1558
gb_pr:Hsacetr      human e     /begin=12     /end=1493
gb_pr:Humnnar      human a3    /begin=70     /end=1581
gb_pr:huma5nicrc   human a5    /begin=149    /end=1555
gb_pr:hsara7a      human a7    /begin=104    /end=1612
gb_pr:hshachrb2    human b2    /begin=167    /end=1675
gb_pr:hsnachrb3    human b3    /begin=1      /end=1272
gb_pr:hsnicrb      human b4    /begin=1      /end=1357
```

**Fig. 5.** File of sequence names for generating acetylcholine receptor alignment. The human g chain cDNA sequence was inferred by joining together exon sequences from the following loci: HSACHG1, HSACHG2, HSACHG3, HSACHG4, HSACHG5, HSACHG6, HUMACHRG1, HUMACHRG2. A segment of the rat α4 sequence was left out because it interfered with the aligning of the other sequences (1051–1680 removed, cds = 1–1881).

2. Sequence analysis software, particularly for multiple sequence alignment. Examples given here will focus on the Wisconsin Package software from the Genetics Computer Group (GCG) *(14)*.
3. A computer printer.
4. Pencils, pens, and highlighters in various striking and invigorating colors.

## 3. Methods

Much of the work involved in designing primers for specific members of multigene families is done on a computer. We will give specific examples of how these steps are done using the GCG sequence analysis package; many other software packages are available for these sorts of things. GCG uses a "software tools" organization, where different parts of the job are done by a collection of separate programs.

| alpha1e5 | cgtctggtggcaaagct |
| alpha1e3 | ccgctctccatgaagtt |
| gamma-e5 | catcagcaagtacctgac |
| gamma-e3 | tgcttcaggctgccaca |
| epsilon-e5 | gaggacactgtcaccat |
| epsilon-e3b | cacgatgacgcaattcat |
| alpha3e5 | ccatgtctcagctggtg |
| alpha3e3 | gtccttgaggttcatgga |
| beta2e5b | cagctcatcagtgtgca |
| beta2e3b | gtgcggtcgtaggtcca |
| beta4e5 | ctgaaacaggaatggact |
| beta4e3 | ccatgtc(at)atctccgtgt |
| ha3cpe5 | ccagtggccagggcctcaga |
| ha3cpe3 | tatgcatcttccctggccatca |

**Fig. 6.** AChR subunit-specific primer sequences. **Notes:** The $\alpha 1$, $\gamma$, and $\varepsilon$ primers match bovine sequences as well *(12)*. The $\beta 2e3b$ primer is probably not terrible specific, because its 3' end is in a rather conserved site and it matches $\alpha 1$ and $\alpha 4$ quite nicely. However, it gives specific amplification together with $\beta 2e5b$, as judged by sequencing several clones. The $\alpha 3$, $\beta 2$, and $\beta 4$ primers are described in ref. *13*. Design of the ha3cp primers is further discussed in Chapter 11. The letter "b" in a primer name indicates that this was a second attempt at designing a specific primer.

## 3.1. Sequence Collection

The GenBank and EMBL gene sequence databases can be searched in several ways. The quickest way is to search the sequence annotations for key words. The GCG program that does this is called *stringsearch,* since it searches for a certain set of character strings. For example, searching the sequence annotations in the GenEMBL database for the sequences containing the phrase "nicotinic acetylcholine receptor" will return the file names of a number of nicotinic acetylcholine receptor subunits (other programs may give more sophisticated search options for "list refinement"). The other general approach to searching the database is to compare a given sequence to the sequences in the database, and finding those that are most similar. The GCG programs *fasta* and *blast* are for sequence comparisons of this sort. The *blast* program is interesting because it uses the resources of the Internet, and sends your query out to a computer at the NIH to do the actual search. This turns out to be very fast, and it does not load down your local system (be aware, however, that the *blast* algorithm is not as sensitive for nucleic acid sequences as *fasta*). Although searches for sequence similarity are much more computer-intensive than searches through annotations, they avoid the petty spelling and nomenclature discrepancies that can so easily defeat an annotation search. One might first search the annotations to find one member of the multigene family, and then use that sequence to search for similar sequences. Of course, if there is a recent review of the sequences in your multigene family, you may find the accession numbers of the sequences you need there. If you are really lucky, someone may have even published an alignment!

## 3.2. Sequence Alignment

Once you have identified the sequences to put into your alignment, you can use an automated program, such as GCG's *pileup,* to produce an alignment with maximal homology among the family members. Depending on your system, you may not need to collect the actual sequences in order to do the alignment; GCG, for example, can accept a file of just sequence names as input, and it will use them to look up the sequences to be aligned as it needs them. This can be quite useful because it helps keep your directories from getting overwhelmed by large numbers of sequence files, and it makes it easier to organize the data. A file of sequence names is a succinct way to summarize the list of sequences you want aligned. For example, Fig. 2 and 5 show the contents of the files of sequence names used to generate the V-region and acetylcholine receptor alignments, respectively. Note that there is some flexibility in the format of these files in GCG.

Since we were interested in the coding regions, we have added information to this file to indicate the regions of interest in each sequence. To find out where each coding region begins and ends, we used GCG's *fetch* to fetch the file, and looked in the annotations for the beginning and end of the coding regions. Setting beginning and ending limits on the regions of the sequences to focus on can help the program to produce a better alignment, because it reduces the amount of data the computer must consider. Since there are limits to the computer's resources, using limited sequence segments can allow you to align a larger number of sequences.

In some cases, it may be necessary to use a genomic sequence. The sequences in Fig. 1 are all from the same exon, so either cDNA or genomic sequence file could be used directly. The acetylcholine receptor sequences, however, are complete coding regions, spanning several exons. If a genomic sequence must be used in such a case, it must be edited to "splice" out the introns. This was done, for example, with the "humang.seq" file in Fig. 5. This file contains information from a variety of genomic GenBank sequence files, all spliced together into one file on the local directory.

You may also want to exclude regions of sequence that are not similar between the different family members. We have done this, for example, with the rat $\alpha 4$ sequence in Fig. 5. Here, the sequence file was copied into the local directory and edited to remove a portion of sequence that was unique to $\alpha 4$ and was long enough to throw off the alignment. Such a region might seem a perfect place to put an α4 specific primer. We chose not to do this for reasons similar to those that led us to focus on the coding regions in the first place; primarily, we were uncomfortable guessing that this unrelated insert would also be present in other species, such as mouse. Since we did not intend to use the α4 insert, leaving it out made it easier to get a good alignment, which boosts our confidence that the primers for the insertless subfamilies will not crossprime on the family with the insert. In general, however, if an insert has no similarity to regions of the other family members, and is not a repetitive element, by all means consider using it.

Quantitatively judging the accuracy of an alignment is beyond the scope of this chapter, but it is frequently obvious if an alignment is good. Things to look for are columns of aligned codons. Some codons are often conserved and easy to recognize, such as those encoding tryptophan (that has only one codon, TGG) and proline (CCN). Gaps that shift the reading frame and introduce nonsense mutations are probably bad.

### 3.3. Sequence Analysis

Several programs can help you visualize where the similarities and differences between your sequences lie.

### 3.3.1. Calculating Consensi and Displaying Deviations

A consensus sequence, consisting of the most common base at each sequence position, can be calculated with the GCG program *pretty* (using the "*-con*" option). This program is more sophisticated than some, in that it lets you set such parameters as the "threshold," or minimum number of times a base must appear at a position in order to be placed in the consensus. For example, with a threshold of 75%, a given position must have an A in three-fourths of the sequences in order for an A to be placed in the consensus. Otherwise, a null character is inserted. The program also gives you several options for displaying differences from the consensus on the individual sequences. For example, you can have any base that is the same as the consensus be represented by a dash, and only the different bases be represented by letters. This sort of thing can make it a lot easier to recognize sequence motifs unique to the sequence or subfamily of interest.

### 3.3.2. Variability (or Similarity) Plots

A Wu-Kabat *(15)* or related type of plot (such as GCG's *plotsimilarity*) can identify regions that are highly conserved among all superfamily members. You may wish to avoid these regions for making subunit-specific primers. A variation on this approach is to make a separate consensus sequence for each subfamily (using a low threshold, there is likely to be a base at each position), and run a variability plot on the consensus sequences *(16)*. This can help identify regions of subfamily specificity (where the consensus sequences differ) that may deserve special consideration for primer placement.

### 3.3.3. Search for Intrasequence Similarity

The program *dotplot* will allow you to draw a two-dimensional diagram of a sequence compared to itself. This will reveal if a sequence contains internal repeats. Such repeats can lead to intramolecular crosspriming in PCR. It makes sense to check at least a representative member of your gene family for such internal similarities. Similarly, you may want to check your representative sequence against its own complementary sequence to reveal inverted repeats. You may want to avoid such regions as candidates for primer design, or at least double-check that a primer designed to match one internal repeat does not match other repeats well enough to prime.

## 3.4. Primer Design

If by now you have suffered through the toil and heartache of obtaining a beautiful sequence alignment for your multigene family (or if you are lucky enough to find one in a review somewhere), you will be relieved to know that it is all downhill from here. The actual design of primers is essentially a matter of choosing the sequences to copy as primers. This is where the invigoratingly colored highlighters come in.

### 3.4.1. General Considerations

The basic rules of primer design still apply (*see*, for example, ref. *17*). If you are new to the fascinating world of primer design, or if you have a computer program that normally does it for you, it might be wise to review the rules. The most important rule is that you must have the proper orientations of the primers on the two DNA strands. Be assured that if you orient the primers backward, or put the bases in the 3'-5', rather than 5'-3' orientation, you stand little chance of obtaining the desired experimental result. It is very easy, and very common, to make this sort of mistake in designing primers for the first time. If you are at all uncertain about primer orientation, study the relationship between the primer sequences given in Fig. 3 and the DNA sequences in Fig. 1. Notice how the e3 primers, which define the 3' end of the target sequence, are the reverse complement of the bold italicized segments indicated on the DNA sequence alignment tables.

The second most important rule is that the two primers must not overlap each other to form intermolecular "primer-dimers." Careful examination of each primer pair will usually reveal likely 3' overlaps, and we often design primers on this "eyeball" assessment alone (but *see* Note 2). A more quantitative way to estimate 3' overlap stability is discussed in ref. *17;* several computer programs are available to check this.

Hairpin formation (intramolecular priming) is less of a problem, since, unlike primer-dimers, hairpinned molecules do not catalyze formation of more of themselves. Hairpins at the 3' end of a primer are of greater concern than those at the 5' end, since any molecules that are extended by polymerase in such a configuration will be ruined for priming on their specific template.

The predicted melting temperature of your primer should be appropriate for the application at hand: for amplifying moderately high abundance messages from cDNAs, portions of plasmids, or other templates of relatively low complexity, we typically use an estimated melting temperature of 52–54°C. For amplifications from genomic DNA, we aim for a higher temperature (near 65–70°C), to decrease the background from nonspecific priming. Both primers in a pair should be designed with similar annealing temperatures to increase the likelihood that they will each have maximal specificity at the reaction's annealing temperature.

Although more sophisticated approaches should theoretically be more accurate; refs. *18* and *19*) discusses a computer program that implements this method *[9]*, we have had good results by misapplying the formula:

$$T_m = \#(A + T) \times 2 + \#(G + C) \times 4 \text{ [in °C]} \tag{1}$$

That is, you count two degrees for each A or T and four degrees for each G or C *(20)*. This is a misapplication of the formula because it was really made to predict melting temperatures for hybridizations in a defined salt buffer, such as 5X SSPE. Not only is a typical PCR much less salty than 5X SSPE, but priming is not the same as hybridization; a primer presumably does not have to bind stably to a template in order to prime for polymerase, but can just "touch and go." Still, primers with $T_m$s around 52°C seem to work nicely at annealing temperatures up to about 55°C, and not much higher.

If a primer has a lot of Gs and Cs, it is frequently recommended that they be located as much as possible toward the 5' end *(17)*. This is supposed to make the 3' end of the primer less stable, and thus less tolerant to mismatches. Short di- or tri- nucleotide repeats, or long stretches of a single base, should be avoided, especially at a primer's 3' end.

Please note that obtaining specific and efficient amplifications is much easier for relatively short products (on the order of 500 bp or smaller). For longer products it is generally advisable to use slightly longer primers with higher annealing temperatures, and to more rigorously check them for adherence to the basic criteria of primer design, usually by using computer programs *(17)*. Also remember Note 2.

## 3.4.2. Subfamily Values

"Designing" the primers is really a matter of deciding where to position them, that is, which regions to make them match. The basic rule is that mismatches near the 3' end are much more likely to prevent an oligonucleotide from priming than are mismatches near the 5' end *(21)*.

Several groups have studied the effects of mismatches on primer specificity. Much of this work has been targeted at making "allele-specific" primers (so this is a good keyword to use when searching for references), but the concepts are valid in any circumstance in which minor variations must be used to specifically amplify one sequence over another.

Although a single base mismatch at the 3' end of a primer *(22,23)* or at the "penultimate 3'" base *(22)* can be used to preferentially amplify one allele over another, a second mismatch within the last four bases can add measurably to the specificity *(24)*. The specific base-pair of the mismatch has been reported to influence specificity; for example, primers ending with T may be more likely to crossprime *(24)*, whereas primer:template mismatches, such as A:A, A:G, C:C, or G:A, are particularly effective at inhibiting priming on the mismatched template *(24,25)*.

With this in mind, look for a sequence motif, which can be as small as two or three bases, that is unique to the individual sequence or subfamily of interest. Such motifs are candidates for the 3' end of your primer. The more bases unique to the subfamily of interest, the better. If you must use a region with a small number of differences between families, avoid Ts, and try to use those base pair mismatches that are most effective, if possible.

Be aware that the specificity of primers can be influenced by reaction conditions, such as magnesium chloride concentration *(22)* and annealing temperature. Extremely rapid cycling (speeds on the order of 35 s/cycle can be achieved using a special hot-air cycler) has been shown to increase the specificity of allele-specific primers *(26)*. The specificities of the primers described in this chapter have been studied using a Perkin-Elmer model 9600 thermal cycler, with cycle times of 2–3 min (which is fast for a block-type machine). Careful primer design can make it possible to achieve specificity under a wider range of reaction conditions.

## 3.4.3. Avoid Common Sequence Elements

Some primer design programs will check potential primer sequences against a database of common repetitive elements, to be sure that you have not placed your primer in the middle of an Alu-like repetitive element that occurs thousands of times in the genome. You can do a similar (and probably more thorough) check with prospective primer sequences by running *fasta* searches of each of them against the GenEMBL database. These searches should pull out good matches against the subfamily you want them to amplify; they may pull out poorer matches (you should have designed the primers so these will be heavy in 3' mismatches) with other supergene family members, and they should pull out little else. If you find exact matches with thousands of unrelated sequences, you have a problem.

### 3.4.4. Intended Cloning Strategy

Depending on how you intend to clone the amplified products, you may need to add to or modify your primer sequences. For example, you may wish to add restriction sites to the 5' ends of the primers for cloning purposes. If you intend to use the directional cloning potential of the T-linker approach, you should plan ahead for this (*see* Chapter 11). Directional cloning with T-linkers does not require extra bases to be added, but it does require that the 5'-most bases of the 5' and 3' primers be different. This usually can be accomplished through positioning of the primers rather than by adding bases. With other approaches, such as the use of a T-vector or nondirectional cloning with T-linkers, the primer sequence does not matter. If the product is to be cloned into an expression vector, the reading frame must be taken into account in deciding where to place the 5' ends of the primers.

## 4. Notes

1. The human pan V β 5' primer on which our hpVbe5 was based was reported to amplify a wide variety of V-regions *(27)*. However, when we used it in conjunction with our hpVbe3 primer, we found very biased amplification of a few families (particularly Vbeta6). This is despite the fact that hpVbe3 can amplify many other family members when used in conjunction with specific 5' primers (*see* Fig. 4). This illustrates that even slight biases on the part of each primer can lead to strong biases in amplification when both are used together. In other words, the specificity of each primer contributes synergistically to the overall specificity of the reaction.
2. Most of the authors' experience is in amplifying sequences from cDNAs. Amplifying from mammalian genomic DNA, for example, has different problems. In a sense, amplifying from cDNA is like cheating to get specificity in a certain member, since you can often choose a tissue or cell type that preferentially expresses that member of the family to start with. This is great if you can get away with it. Also, as mentioned above, it is probably a good idea to use longer primers (higher $T_m$) for genomic DNA.
3. If you are amplifying from cDNA, please note that a 52°C $T_m$ (primers around 18-bp long) really is enough in most cases, if you can find a region where the primer can be made to match exactly. Many investigators routinely make much longer primers. The shorter primers are not only cheaper to make, but are easier to fit into short regions of subfamily specificity.

4. It is usually easier to find a good alignment from protein sequences. Aligning proteins is generally more robust, for several reasons. First, the protein sequences are often more closely related than the corresponding gene sequences, because of silent nucleotide differences among the genes. Second, similarity among amino acids can be taken into account. Third, there is only one third as much information, so the computer can keep it all straight more easily. In sum, an alignment of protein sequences is less likely to be thrown off by such things as an insertion of unrelated sequence in one family member than is a gene sequence alignment. A protein alignment can be a useful first step before the DNA sequence alignment, since it will point out regions of nonhomology.

5. You may find several different opinions as to the real sequence of a given gene in the database. Such differences can represent actual polymorphisms (allelism), or they may be mistakes. Although, with enough library work, it is sometimes possible to determine which, if any, of the sequences are best, our approach is simply to avoid controversial bases. One way to do this is to include all the variants of each sequence in your alignment. If your gene family is too large to do this, you can make a separate alignment of all the variants in the subfamily of interest, and double-check your primer sequences against that.

6. A similar approach is used with a different experimental philosophy in ref. 8, where, a panel of primers was made for immunoglobulin sequences. These primers were not terribly specific, in that there is significant crosspriming between families. The goal was to use a panel of primers to amplify a broad range of sequences, without using degeneracy. In this type of application, a panel of primers to a variety of subfamilies can allow amplification of a wider range of family members, even if specificity is not a major goal. In this sense it would be possible, for example, to make a panel of degenerate primers, each patterned after a different subfamily (degeneracy and specificity are not mutually exclusive).

7. Hot Start, in which the reaction is prevented from starting until the tube is heated, can make a dramatic improvement in specificity (28–30). Since specificity is so important in amplifying specific members of a multigene family, you may want to try it. For best results when using a wax or grease meltable barrier, preheat the sample block to denaturing temperature before you put the samples in the machine. This makes the temperature at which your barrier melts less important, since the whole tube will be heated to denaturing temperature almost immediately.

8. If you are really uncertain about the primers you have designed, try asking someone to look them over, particularly to check for simple things like proper orientation. If you do not have a neighbor who does this sort of thing, I would like to volunteer the services of our virtual colleagues on the Internet, particularly the participants in the UseNet group "bionet.molbio.methds-reagnts." The eager PCR fans there will most likely offer to lend a hand. Do not forget to acknowledge them when you finally publish your work.

9. Although planning and forethought are all well and good, there comes a point at which you just need to try the primers and see if they work.

## Acknowledgments

The example primers shown in this chapter were made for projects supported by research grants from the Muscular Dystrophy Association of America and from the Council for Tobacco Research, by the NIH grant NS2319, and the NIDA Program Project grant DA05695. R. M. H. was the recipient of the Robert G. Sampson Neuromuscular Disease Research Fellowship from the Muscular Dystrophy Association. The authors' royalties from this chapter have been donated to the MDA.

## References

1. Horton, R. M., Hildebrand, W. H., Martinko, J. M., and Pease, L. R. (1990) Structural analysis of H-2K$^f$ and H-2K$^{fm1}$ by using H-2K locus-specific sequences. *J. Immunol.* **145,** 1782–1787.
2. Horton, R. M., Loveland, B. E., Parwani, A., Pease, L. R., and Fisher Lindahl, K. (1991) Characterization of the spontaneous mutant H-2K$^{bm29}$ indicates that gene conversion in H-2 occurs at a higher frequency than detected by skin grafting. *J. Immunol.* **147,** 3180–3184.
3. Cai, Z., Pullen, J. K., Horton, R. M., and Pease, L. R. (1992) Specific amplification of cDNA from targeted members of multigene families. *Methods Enzymol.* **216,** 100–108.
4. Pullen, J. K., Horton, R. M., Cai, Z., and Pease, L. R. (1992) Structural diversity of the classical H-2 genes: K, D, and L. (1992) *J. Immunol.* **148,** 953–967.
5. Hildebrand, W. H., Horton, R. M., Pease, L. R., and Martinko, J. M. (1992) Nucleotide sequence analysis of H-2Df and the spontaneous in vivo H-2D$^{fm2}$ mutation. *Mol. Immunol.* **29(1),** 61–69.
6. Wei, S., Charmley, P., Robinson, M. A., and Concannon, P. (1994) The extent of the human germline T-cell receptor V beta gene segment repertoire. *Immunogenetics* **40,** 27–36.
7. Champagne, E., Huchenq, A., Sevin, J., Casteran, N., and Rubin, B. (1993) An alternative method for T-cell receptor repertoire analysis: clustering of human V-beta subfamilies selected in responses to staphylococcal enterotoxins B and E. *Mol. Immunol.* **30,** 877–886.
8. Marks, J. D., Tristem, M., Karpas, A., and Winter, G. (1991) Oligonucleotide primers for polymerase chain reaction amplification of human immunoglobulin variable genes and design of family-specific oligonucleotide probes. *Eur. J. Immunol.* **21,** 985–991.
9. Genevee, C., Diu, A., Nierat, J., Caignard, A., Dietrich, P.-Y., Ferradini, L., Roman-Roman, S., Triebel, F., and Hercend, T. (1992) An experimentally validated panel of subfamily-specific oligonucleotide primers (Valpha1–w29/Vbeta1–w24) for the study of human T-cell receptor variable V gene segment usage by polymerase chain reaction. *Eur. J. Immunol.* **22,** 1261–1269.
10. Hall, B. L. and Finn, O. J. (1992) PCR-based analysis of the T-cell receptor V-beta multigene family: experimental parameters affecting its validity. *Biotechniques* **13,** 248–257.

11. Panzara, M. A., Gussoni, E., Steinman, L., and Oksenberg, J. R. (1992) Analysis of the T-cell repertoire using the PCR and specific oligonucleotide primers. *Biotechniques* **12,** 728–734.

12. Horton, R. M., Manfredi, A. A., and Conti-Tronconi, B. M. (1993) The "embryonic" gamma subunit of the nicotinic acetylcholine receptor is expressed in adult extraocular muscle. *Neurology* **43(5),** 983–986.

13. Grando, S. A., Horton, R. M., Pereira, E. F. R., Diethelm-Okita, B. M., George, P. M., Albuquerque, E. X., and Conti-Fine, B. M. (1995) A nicotinic acetylcholine receptor regulating cell adhesion and motility is expressed in human epidermal keratinocytes. *J. Invest. Dermatol.* **105(6),** 774–781.

14. Program Manual for the Wisconsin Package, Version 8, September 1994, Genetics Computer Group, 575 Science Drive, Madison, WI, USA 53711.

15. Wu, T. T. and Kabat, E. A. (1970) An analysis of the sequences of the variable regions of Bence Jones proteins and myeloma light chains and their implications for antibody complementarity. *J. Exp. Med.* **132,** 211–250.

16. Pease, L. R., Horton, R. M., Pullen, J. K., and Cai, Z. (1991) Structure and diversity of class I antigen presenting molecules in the mouse. *CRC Crit. Rev. Immunol.* **11(1),** 1–32.

17. Rychlik, W. (1993) Selection of primers for polymerase chain reaction, in *Methods in Molecular Biology,* vol. 15: *PCR Protocols: Current Methods and Applications* (White, B. A., ed.), Humana, Totowa, NJ, pp. 31–40.

18. Breslauer, K. J., Frank, R., Blocker, H., and Marky, L. A. (1986) Predicting DNA duplex stability from the base sequence. *Proc. Natl. Acad. Sci. USA* **83,** 3746–3750.

19. Rychlik, W. and Rhoads, R. E. (1989) A computer program for choosing optimal oligonucleotides for filter hybridization, sequencing and in vitro amplification of DNA. *Nucleic Acids Res.* **17,** 8543–8551.

20. Suggs, S. V., Hirose, T., Miyake, T., Kawashima, E. H., Johnson, M. J., Itakura, K., and Wallace R. B. (1983) Use of synthetic oligodeoxyribonucleotides for the isolation of specific cloned DNA sequences, in *Developmental Biology Using Purified Genes* (Brown, D. D. and Fox, G. F., eds.), Academic, New York, pp. 683–693.

21. Bottema, C. D. and Sommer, S. S. (1993) PCR-amplification of specific alleles: rapid detection of known mutations and polymorphisms. *Mutation Res.* **288,** 93–102.

22. Sarkar, G., Cassady, J., Bottema, C. D., and Sommer, S. S. (1993) Characterization of polymerase chain reaction amplification of specific alleles. *Anal. Biochem.* **186,** 64–68.

23. Wu, D. Y., Ugozzoli, L., Pal, B. K., and Wallace, R. B. (1989) Allele-specific enzymatic amplification of β-globin genomic DNA for diagnosis of sickle cell anemia. *Proc. Natl. Acad. Sci. USA* **86,** 2757–2760.

24. Kwok, S., Kellogg, D. E., McKinney, N., Spasic, D., Goda, L., Levenson, C., and Sninsky, J. J. (1990) Effects of primer-template mismatches on the polymerase chain reaction: human immunodeficiency virus type 1 model studies. *Nucleic Acids Res.* **18,** 999–1005.

25. Huang, M. M., Arnheim, N., and Goodman, M. F. (1992) Extension of base mispairs by *Taq* DNA polymerase: implications for single nucleotide discrimination in PCR. *Nucleic Acids Res.* **20,** 4567–4573.
26. Wittwer, C. T., Marshall, B. C., Reed, G. H., and Cherry, J. L. (1993) Rapid cycle allele-specific amplification: studies with the cystic fibrosis delta F508 locus. *Clin. Chem.* **39,** 804–809.
27. Broeren, C. P., Verjans, G. M., Van Eden, W., Kusters, J. G., Lenstra, J. A., Logtenberg, T. (1991) Conserved nucleotide sequences at the 5' end of T-cell receptor variable genes facilitate polymerase chain reaction amplification. *Eur. J. Immunol.* **21,** 569–575.
28. Mullis, K. B. (1991) The polymerase chain reaction in an anemic mode: how to avoid cold oligodeoxyribonuclear fusion. *PCR Methods Appl.* **1,** 1–4.
29. D'Aquila, R. T., Bechtel, L. J., Videler, J. A., Eron, J. J., Gorczyca, P., and Kaplan, J. C. (1991) Maximizing sensitivity and specificity of PCR by preamplification heating. *Nucleic Acids Res.* **19,** 37–49.
30. Horton, R. M., Hoppe, B. L., and Conti-Tronconi, B. M. (1994) AmpliGrease: Hot-start PCR using petroleum jelly. *Biotechniques* **16(1),** 42,43.

# 44

## Screening Gene Family-Enriched cDNA Sublibraries with an Unamplified cDNA Probe

*Focusing on Moderately to Abundantly Expressed Clones*

**Meimei Hu and Bruce A. White**

## 1. Introduction

The examination of gene expression and the search for novel and/or tissue-specific genes has been facilitated by the presence of conserved domains among the members of gene families (e.g., ref. *1*). As discussed in the previous chapters of this section, such a conserved domain can be exploited by designing degenerate primer pairs that allow for the polymerase chain reaction (PCR)-based amplification of this domain. Although the PCR product may appear as a single band after typical agarose gel electrophoresis, it is likely to be composed of multiple species of cDNA fragments. Cloning of these fragments (i.e., construction of a sublibrary of cDNA fragments) is necessary for the analysis of individual cDNA fragments. Initial screens (in addition to blue-white selection) can be employed to obviate numerous individual analyses of many clones and allow the investigator to focus on a fraction of clones that display certain characteristics.

One characteristic that might be sought after is that of a particular clone representing an mRNA that is moderately to relatively abundant in the cell or tissue type of interest. In this regard, one of the drawbacks of PCR is that its power of amplification can result in the cloning of DNA fragments that do not detect an mRNA by Northern blot hybridization or RNase protection assay. The problem of using PCR to clone a cDNA that may have no or very limited physiological relevance to the cell type of interest has been described previously. For example, the procedure of hybridization capture was developed to ascertain that a single band, as detected by the PCR-based differential display,

From: *Methods in Molecular Biology, Vol. 67: PCR Cloning Protocols: From Molecular Cloning to Genetic Engineering* Edited by: B. A. White  Humana Press Inc., Totowa, NJ

represents a single mRNA species that can be detected at the sensitivity of Northern blot hybridization *(2)*. We have faced this problem several times in our own research and have experienced the frustration of cloning and sequencing a cDNA fragment only to find that it is not expressed at a detectable level, even when assayed by RNase protection, in the cell types being studied. Although detection of an mRNA species after amplification by 20–30 cycles of RT-PCR may indicate the presence of a biologically important transcript (which may encode a stable, more abundant protein), it may be desirable in some cases to focus on more highly expressed genes.

We are currently examining protein tyrosine kinase (PTK) gene expression in rat pituitary $GH_3$ cells. Although approx 85 PTKs have been identified in the human, it has recently been estimated that about 1000 PTK genes exist *(3)*. We are interested in searching for tissue-specific PTKs as well as PTKs whose expression is regulated by fibroblast growth factor (FGF). To this end, we generated a cDNA sublibrary by PCR using degenerate primers to amplify the conserved kinase domain essentially as described *(4)*. In this chapter, we describe the steps for screening the sublibrary with a $^{32}$P-labeled, unamplified cDNA as a simple procedure for focusing on the more abundantly expressed gene products. Although we only present the steps for the simple determination of relative abundance, this procedure can be extended to differential screens for the detection of cell-specific transcripts, hormonally regulated genes, and so forth.

## 2. Materials

### 2.1. Sublibrary Construction

1. PCR product ligated into a suitable vector (e.g., T/A cloning; *see* Chapters 9 and 10).
2. Max Efficiency DH5α™-competent cells (Life Technologies, Gaithersburg, MD).
3. SOC media *(5)*: A small amount is supplied with the Max Efficiency cells. More can be prepared according to directions supplied with the cells.
4. LB agar plates containing 50 μg/mL of ampicillin.
5. X-gal: Prepare stock solution at a concentration of 20 mg/mL in *N,N*-dimethyl-formamide. Store at –20°C in the dark.
6. Nitrocellulose filters.
7. Sterile toothpicks.

### 2.2. Synthesis of Unamplified cDNA Probe

1. Poly(A)$^+$ RNA sample (*see* Note 1).
2. M-MLV reverse transcriptase, 5X first strand buffer, and 100 m*M* dithiothreitol (DTT) as supplied by the manufacturer (Life Technologies).
3. dA/G/T mix: dATP, dGTP, and dTTP, each at 20 m*M*. Prepare by diluting 100 m*M* solutions (Pharmacia Biotech, Piscataway, NJ) in sterile $H_2O$. Store at –20°C.
4. dCTP (120 μ*M*). Prepare as stated above.

5. $^{32}$P-dCTP (3000 Ci/mmol; Amersham, Arlington Heights, IL).
6. Oligo d(T)$_{16}$ (5 nmol/100 μL; Perkin Elmer, Foster City, CA).
7. 200 m$M$ EDTA, pH 8.0.
8. 10% SDS.
9. 10$N$ NaOH.
10. 1$M$ Tris-HCl, pH 7.5.
11. 2$N$ HCl.
12. Tris-buffered phenol.
13. Chloroform.
14. 95% ethanol.
15. TE buffer: 10 m$M$ Tris-HCl, pH 7.4, 0.1 m$M$ EDTA.
16. Microspin S-200 HR™ columns (Pharmacia).

## 2.3. Screening Sublibrary

1. 20X SSPE buffer: 200 m$M$ NaH$_2$PO$_4$, pH 7.4, 20 m$M$ EDTA, 3.6$M$ NaCl.
2. 10% SDS.
3. Deionized formamide.
4. 50X Denhardt's solution (1% BSA, 1% Ficoll 400, 1% polyvinylpyrrolidone).
5. Sonicated, sheared salmon sperm DNA (10 mg/mL; Life Technologies).
6. 2X SSC, 0.5% SDS.
7. 0.1X SSC, 0.5% SDS.
8. Hybridization oven, bottles, and nylon spacers.
9. Detection method (X-ray film or phosphoimaging device).

# 3. Methods

## 3.1. Sublibrary Construction

1. Warm LB-Amp plates and spread 80 μL of X-gal solution.
2. Transform 100 μL of DH5α bacteria with 1 μL of ligation mixture (*see* Note 2) according to supplier's directions in a 1.5-mL microfuge tube.
3. Add 900 μL of SOC medium to each ligation, mix gently, tape the tube to a platform of the bacterial incubator, and incubate for 1 h at 37°C shaking at 225 rpm.
4. Spread 100 μL of the cells onto 10 LB-Amp/X-gal plates and incubate at 37°C overnight (*see* Note 3).
5. Place a nitrocellulose filter, covered on both sides by its blue paper spacers, onto a hard surface. Imprint a grid onto the filter by drawing a grid on the top spacer paper with pencil and ruler. Keep the lines 5–7 mm apart.
6. Using forceps, transfer the filter onto a fresh LB-Amp plate.
7. Repeat steps 5 and 6, using the same blue spacer paper as a guide to draw the grid.
8. Transfer white colonies to the center of the corresponding grid squares on both plates using a sterile toothpick (*see* Note 4).
9. Grow colonies to 1–3 mm in diameter (*see* Note 3).
10. Process one of the duplicate filters for screening with a cDNA probe. Lyse bacteria and denature plasmid DNA by placing on two sheets on Whatman 3MM paper soaked with AL buffer for 5 min.

11. Blot filters on dry 3MM paper for about 5 min.
12. Transfer filters to two sheets of 3MM paper soaked in N buffer and incubate for 2 min.
13. Blot filters on dry 3MM for 2 min.
14. Soak filters on 3MM paper soaked in NS buffer for 10–15 min.
15. Blot filters on dry 3MM paper for 2 min.
16. Crosslink DNA to nitrocellulose with the Stratalinker using autocrosslink setting.
17. Wrap filters in plastic wrap and store at –20°C until ready for probing.
18. Wrap the unprocessed duplicate filter (on agar) with parafilm and store for up to 1 mo at 4°C as a backup.

## 3.2. Synthesis of Unamplified cDNA Probe

1. Adjust volume of 1 μg of poly(A)-enriched RNA plus 1 μL of oligo d(T) (*see* Note 5) to 5 μL with RNase-free $H_2O$.
2. Heat at 70°C for 5 min.
3. Chill in ice water.
4. Add the following: 1 μL 100 m*M* DDT, 1 μL RNasin, 5 μL 5X first strand buffer, 1 μL dA/G/T mix, 1 μL 120 μ*M* dCTP, 1 μL m-MLV reverse transcriptase, and 10 μL $^{32}$P-dCTP.
5. Incubate at 37°C for 1 h.
6. Add 2.5 μL 200 m*M* EDTA and 1 μL 10% SDS; mix gently.
7. Add 1 μL of NaOH and incubate at 68°C for 30 min.
8. Cool to room temperature and add 10 μL of 1*M* Tris-HCl, pH 7.5, 3 μL of 2*N* HCl, and 7.5 μL of sterile $H_2O$.
9. Extract with 25 μL phenol and 25 μL $CHCl_3$.
10. Spin in a microcentrifuge in a cold room for 10–15 min.
11. Purify labeled cDNA from the upper phase using a Microspin S-200 HR column as directed by the supplier (*see* Note 6).

## 3.3. Screening Sublibrary

1. Prepare 10 mL of hybridization mix by adding the following: 2.5 mL 20X SSPE (0.3*M* NaCitrate, pH 7.0, 3*M* NaCl), 100 μL 10% SDS, 5 mL deionized formamide, 1 mL 50X Denhardt's solution, 100 μL 10 mg/mL salmon sperm DNA (Life Technologies), and 1 mL sterile $H_2O$.
2. Assemble up to five filters with a nylon spacer in between filters and slide into the hybridization bottle.
3. Prehybridize for 6 h.
4. Add cDNA probe and hybridize overnight (i.e., 16 h) at 42°C.
5. Wash filters twice in 2X SSC, 0.5% SDS for 30 min at 50°C.
6. Wash filters twice in 0.1X SSC, 0.5% SDS for 30 min at 50°C.
7. Expose to X-ray film or phosphoimager. In our experience, colonies are either black or light gray. Pick black colonies from the duplicate filter and process further as desired (e.g., analysis by miniprep for correct sized insert, differential screening, and so forth).

## 4. Notes

1. For our own work, total RNA was isolated from 10 dishes (100 mm) of confluent GH$_3$ cells. Cells from each dish were scraped up, transferred into a 15-mL Falcon tube, and pelleted by centrifugation. The cell pellets were resuspended in 1 mL of cold PBS, transferred to a sterile microfuge tube, and centrifuged for 15 s in a microfuge. The supernatant washes were aspirated and the pellets resuspended in 180 μL of TE buffer containing 1 μL of RNasin. Cells were lysed by addition of 20 μL of 5% NP40 in TE buffer, incubated on ice for 1 min, and centrifuged for 1–2 min. The cytoplasmic lysates were quickly transferred to fresh microfuge tubes containing 200 μL of 2X PK buffer (200 m$M$ Tris-HCl, pH 7.5, 0.44$M$ NaCl, 2% SDS, 25 m$M$ EDTA) and 3 μL of 20 mg/mL proteinase K. After incubation at 37°C for 20–30 min, lysates were pooled, extracted with phenol/CHCl$_3$ and twice with CHCl$_3$, and then ethanol precipitated. Polyadenylated RNA was then isolated using the PolyATtract mRNA Isolation System IV (Promega, Madison, WI).

   Since RT-PCR is routinely performed with total RNA, it may not be necessary to isolate the poly(A)$^+$ fraction. On the other hand, the kits available for poly(A)$^+$ isolation are efficient and fast.
2. The amount of ligation required for transformation may need to be adjusted in order to yield numerous but separate colonies.
3. The time of incubation should be adjusted to yield small colonies (1–3 mm in diameter). This can be done by placing dishes at 37°C later in the day, or leaving them at room temperature overnight and shifting them to 37°C in the morning.
4. Since two filters are being generated in this screen, the classical procedure of replica plating can be used (e.g., ref. *6*). However, we have found that transferring colonies onto duplicate filters by simply touching both filters (at the corresponding grid positions) with the same toothpick creates more evenly sized colonies and does not require that much more time than making one master filter.
5. The use of oligo d(T), as opposed to the downstream gene family-specific degenerate primer, ensures an accurate representation of all cDNAs.
6. An ethanol precipitation can be included prior to this step *(6)* but it is not necessary.

## References

1. Preston, G. M. (1993) Use of degenerate oligonucleotide primers and the polymerase chain reaction to clone gene family members, in *PCR Protocols: Current Methods and Applications* (White, B. A., ed.), Humana, Totowa, NJ, pp. 317–338.
2. Li, F., Barnathan, E. S., and Kariko, K. (1994) Rapid method for screening and cloning cDNAs generated in differential mRNA display: application of Northern blot for affinity capturing of cDNAs. *Nucleic Acids Res.* **22,** 1764,1765.
3. Hunter, T. (1996) Tyrosine phosphorylation: past, present and future. *Biochem. Soc. Trans.* **24,** 307–328.

4. Kim, S. J., Sasaki, H., Takahashi, A., Katoh, M., Kakizoe, T., Kim, Y. K., Sugimura, T., and Terada, M. (1995) Rapid isolation of cell-type-specific protein tyrosine kinase by degenerate polymerase chain reaction combined with differential hybridization technique. *Biochem. Biophys. Res. Commun.* **214,** 60–68.
5. Hanrahan, D. (1983) Studies on transformation of *E. coli* with plasmids. *J. Mol. Biol.* 557–580.
6. Sambrook, J., Fritsch, E. F., and Maniatis, T., eds. (1989) *Molecular Cloning. A Laboratory Manual.* Cold Spring Harbor Laboratory, Cold Spring Harbor, NY.

# Index